AF478386

THE ORDER OF MAN

No Tiger in the mountain

THE ORDER OF MAN

A Biomathematical Anatomy of the Primates

CHARLES E. OXNARD
University Professor
Professor of Anatomy and Cell Biology,
and Biological Sciences, University of Southern California

Fellow of The American Association
for The Advancement of Science

Fellow of the New York
Academy of Sciences

HONG KONG UNIVERSITY PRESS
1983

ISBN 962-209-104-0 (Casebound)
962-209-047-8 (Limp)

The front-cover picture and the frontispiece are reproductions of a Chinese scroll painting by Xiao-gang dated Geng-shen (1920 The Year of the Monkey) of a Golden Monkey (*Rhinopithecus roxellanae roxellanae*) from Mount Omei.

The caption reads 'No Tiger in the Mountain', which is in itself a part of the well-known Chinese proverb

'no fish in the river
the crab is King
no tiger in the mountain
the monkey is King'

Printed in Hong Kong by Hing Yip Printing Co. Ltd.
44, Wong Chuk Hang Road, Block C & D
6th floor, Hong Kong

Contents

Preface

This book is an attempt to look broadly at the biological Order of Man. It reviews more than two decades of study of present-day primates using data and methods not hitherto made available in one place nor to the general reader.

It is the third book in a series. The first, *Form and Pattern in Human Evolution: Some Mathematical, Physical and Engineering Approaches*, describes some modern methods available for the study of biological form and function with especial reference to the primates.

The second book is *Uniqueness and Diversity in Human Evolution: Morphometric Studies of Australopithecines*. It takes the most well-developed of these techniques, multivariate morphometrics, and applies it to the particular problems raised by functional assessments of certain assumed human ancestors, the australopithecine fossils of Olduvai and Southern Africa.

But in the years since those publications, the scope of my investigations has expanded so that they now encompass a considerably wider range of methods, based in mathematics, physics and engineering, for the study of biological form and pattern. And the totality of my researches now cover a far wider range of anatomical regions: most parts of the body (shoulders and hips, arms and thighs, forearms and legs, hands and feet, trunk and head). The studies have, furthermore, been extended so that they now apply very widely throughout the entire Order Primates (humans, apes, Old World monkeys, New World monkeys, lemurs, bush-babies, lorises and tarsiers). Thus, this third volume has arisen naturally as an attempt to see how morphometric methods assess the entire Order, to discuss how these new evaluations meld, or how they do not, with the broad picture of what we already know, and to investigate how the results may influence future directions of thinking in this area of evolutionary morphology.

The new assessments can be viewed through study of localized anatomical regions; information about function mainly results. But they can also be viewed through investigation of entire anatomies; and this, it turns out, seems to speak more to our understanding of the overall relationships of the various primates.

None of this is to say, however, that the results of the present studies stand by themselves; on the contrary, perhaps the main interest in the present researches is in the pattern of concordances and discordances that they provide in relation to the prior body of knowledge of the Order. For though many different types of information are now available for aiding our understanding of the relationships of the primates, and even though the picture of the primates that they portray is similar overall, there are nevertheless a number of rather fundamental differences.

The detailed arrangements of the primates produced, for instance, by study of classical morphology and physiology of the whole organism differ in certain ways from those that appear when subcellular, genetic, biochemical and molecular data are examined. It is therefore of especial interest to see how the morphometric method fits, or does not fit, with each of these macro and micro assessments. A final consequence of these examinations, it turns out, may be that our understanding of human evolution is taking a new turn, with implications that go far beyond our current thinking about humans.

The reader is not to expect that this book follows conventional paths. I have felt constrained to seek new ways of obtaining data from old bones; I have felt obliged to follow where my data, duly checked, have lead; I have not been afraid to suggest ideas that challenge the conventional wisdom of this anthropological age. This is a book that some will be unable to accept. *See **nota bene** pp 334, i–xi.

In preparing this book, I have tried to present the new methods, the complex results and the broad speculations that stem from them, in general non-technical ways, so that they may be understood not only by immediate research colleagues, but also by beginning graduate students, by undergraduates, and indeed, by anyone interested in human evolution. There are, of course, many books that attempt to talk generally about human evolution. Without exception these books eschew the new methods and the new results that I am presenting here as being too difficult for the non-technical reader. I do not believe that this is the case.

It is true that the non-technical, general reader may have no interest in wading through the complex algebra of multivariate statistical methods, or the complicated scientific names of every living primate species, certainly not the massive compendium of terms used to describe mammalian anatomies.

But the general reader is certainly very capable of understanding the principles by which morphometric methods work. And I have tried to present many visual, geometrical examples. The general reader is well able to understand and enjoy the broad sweep of animal diversity that is represented in that fascinating group of animals to which we belong biologically: the Order Primates. And in this, too, readers may be aided by line drawings of individual primates in action, which, though not as direct as photographs taken in the field, actually show much more of the animals without blurring movement or obscuring foliage. Finally, more today than ever before, every reader is aware of the generalities of the anatomical structure of humans (and therefore, also, of the anatomies of apes, monkeys and lower primates). Even here, line diagrams of anatomical structures and the use of common anatomical names as much as possible may be extra helpful.

The scope of this book has required yet other short cuts. Thus, I occasionally reify anatomies and primates, arguing that 'primates arrange anatomies' and that 'anatomies arrange primates'. Of course, I know, and the reader knows, that anatomies and primates do not 'arrange' anything, save those anatomies and primates in the persons of scientists. But this short cut especially emphasizes these separate aspects of organisms that can be confused or misunderstood.

Such ploys may help the general reader to understand more easily how modern methods for the study of form and pattern operate when applied to us and our nearest living relatives. For a thorough understanding of the entire Order to which we belong is a prerequisite for any who desire to know of our own evolution, the pathway from which we came, the steps in which our feet may now be directed. Certainly this book attempts to take such a readership into areas that have usually been avoided on the presupposition, in my opinion wrong, that the new work is so difficult that it is for the specialist alone.

And it is especially my desire to present these results to students who already have interests in animal form and pattern but who do not wish to be limited to the older methods of observation and dissection. Thus may they see new possibilities unavailable to the unaided eye; and thus may they be introduced to the powers of new methodology in testing the many speculations that so readily arise when our own position is under discussion.

As I have explained, this book naturally follows from the earlier ones. But it also depends upon a series of invitations. The first was from Professor A.H. Schultz to contribute to the Karger Press Handbook of Primatology Series, with Jack T. Stern, Jr, a monograph entitled *Primate Locomotion: Some Links with Evolution and Morphology*. It rests also upon an invitation from John Van Doren, to contribute to The Great Ideas Today (pp. 92–153, Encyclopaedia Britannica Press, 1977) an article for general readership, *Human Fossils: The New Revolution*. It is based yet further on an invitation to give the keynote science lecture to the annual meeting of the Association of American Biology Teachers in Chicago in 1978; the thought required for that presentation, later published in the journal of the Association (The American Biology Teacher, 41: 264–276) as *Human Fossils: New Views of Old Bones* has flavoured my presentation here. And it depends, perhaps most of all, upon an invitation to participate in a symposium at Burg Wartenstein in 1975 under the auspices of the Wenner-Gren Foundation for Anthropological Research. That symposium, organized by Mary Ellen Morbeck, Holger Preuschoft and Neil Gomberg, and resulting in the volume, *Environment, Behavior and Morphology: Dynamic Interactions in Primates* and my own contributions to it, provides some of the basic discussion for this book.

A number of other invitations to present my work have been important in helping develop these ideas. A series of lectures (including, especially, the Lo Yuk Tong Foundation Lecture and the Shu Tzu Huang-Chan Memorial Lecture) in the past ten years in various departments at the University of Hong Kong through the invitations of Professor F.P. Lisowski have been especially valuable in this regard. And because I am not myself a statistician, I must especially acknowledge the help and collaboration that I have received from a number of individuals over the years who have especial expertise in this area (Professors Michael Healy, Roger Flinn, Peter Neeley, Paul Meier, David Wallace and William Kruskal). In addition to such personal discussion and help in the statistical area, I have also felt it most important to 'put my head in the lion's den' by accepting every invitation to present these studies in departments of statistics. In this regard, the Department of Statistics at the University of Chicago has been especially important through its invitations to me over the years. In more recent times, the Department of Statistics and its head, Professor John Aitchison, at the University of Hong Kong, and the Department of Statistics and its head, Professor Michael Healy, at the London School of Hygiene and Tropical Medicine, have likewise provided critical comment.

Another, different type of lecture participation has also been especially helpful. Those undergraduate, graduate and medical students and faculty at the University of Chicago who have taken my courses on the *Analysis of Biological Form and Pattern*, on the *Order of the Primates*, and on *Animal Mechanics* have contributed to this book in a manner that continually emphasizes to me the very close relationship and interaction that there is between teaching and research.

The University of Southern California has particularly made it possible for me to continue my academic work as University Professor and Professor of Anatomy and Biological Sciences, while yet allowing me to contribute to its academic development as Dean of the Graduate School. This interlinking of administration and research, started during my tenure of the Deanship of the College at the University of Chicago concurrently with my appointment as Professor of Anatomy, Anthropology and Evolutionary Biology, and now continued by the University of Southern California, has been a vital part of my efforts over many years.

This triple interplay between research, teaching and administration has, in recent years, been central in my enjoyment of academic life and has facilitated, rather than the reverse, the production of this book.

Perhaps most of all, however, this book, along with the most recent developments of my studies, has been stimulated by special colleagues in the new science of biological form. Among these, mention must be made of Professor Jack T. Stern, Jr. of the Department of Anatomy, State University of New York at Stony Brook, Professors James Hopson, R. Eric Lombard, Leonard Radinsky and Ronald Singer of the Department of Anatomy, The University of Chicago, Mr. Thomas F. Spence and Dr. Roger M. Flinn of the Department of Anatomy, University of Birmingham, England, Professor F. Peter Lisowski, Department of Anatomy, University of Hong Kong, and Dr. Francoise-K. Jouffroy and the late Dr. Jacques Lessertisseur of the National Museum of Natural History, Paris.

I am especially indebted to one person with whom I have collaborated at intervals throughout my entire career, and whose work stamps my work, Professor Eric H. Ashton, Department of Anatomy, University of Birmingham, U.K.

Especial thanks go to a group of individuals with whom I have studied very closely over the years, especially while they were graduate students at the University of Chicago, but also since. Professor Gene H. Albrecht, Department of Anatomy, University of Southern California must be first mentioned. He has contributed much to my recent work, not only during his period as a graduate student at the University of Chicago, but also in his academic appointments since. Likewise, thanks go to Dr. Betty J. Manaster, now in the Department of Radiology, University of Utah, to Professor John E. McArdle, now in the Department of Biology, Illinois Wesleyan University, and to Dr. Harry Yang now an academic surgeon and investigator in the Department of Surgery, the University of Chicago. A collaboration that is an especial delight to mention is with Miss Rebecca German, currently a graduate student at Harvard University, but for a number years involved with me in undergraduate research in the College at the University of Chicago and a persistent collaborator and colleague ever since.

It is ever appropriate to recognize the initial stimulus, and the continuing interest and collaboration over many years now, of my own Professor, Lord Zuckerman, OM., KCB., MD., DSc., FRS., previously Sands Cox Professor of Anatomy, University of Birmingham, England, and now Honorary Professor, University of East Anglia, and President, Zoological Society of London. Lord Zuckerman first saw the possibilities, then laid the foundation for the realities, and even now continues to support developments that have lead me into these new studies of biological form and pattern.

Many hands have contributed to this work through technical assistance in the laboratory, artistic and technical drawing, computational analysis and secretarial help. To be thanked in this regard are Mrs. Marsha Greaves, Misses Shirley Aumiller, La Vern Shatteen, Claire Vanderslice, Jacqueline Toy, Mesdames Eleanor Craycraft and Eleanor Oxnard, and Messrs Hugh and David Oxnard.

It is an especial pleasure to single out the work and expertise of my previous personal research assistant, Miss Joan Hives, who, though she has not seen the final stages of this book, took part in all of my studies during an especially seminal decade, and contributed so much to the basic work upon which this book rests.

My work has been much aided by the collegiality contributed through honorary appointments that I now hold: Honorary Professor of Anatomy, University of Hong Kong, Research Associate in Vertebrate Anatomy, Field Museum of Natural History, Chicago, and Overseas Associate in Anatomy, University of Birmingham.

A large number of institutions have kindly given me access to materials in their care: British Museum of Natural History, London, the Powell Cotton Museum, Birchington, the Departments of Anatomy and Zoology and the Duckworth Laboratory of Physical Anthropology of the University of Cambridge, the department of Anatomy, University of Leeds, the department of Anatomy of the University of Birmingham, the Musée Royale de l'Afrique Central, Tervuren, the Museum National d'Histoire Naturelle, Paris, the Institut Leon Fredericq, Universite de Liege, the Anatomischer Institut der Universität Gottingen, the Rijksmuseum van Natuurlijke Histoire, the American Museum of Natural History, the Duke University Primate Facility, Department of Anatomy, West Virginia University, Cleveland Museum of Natural History, Museum of Comparative Zoology, Harvard University, Lincoln Park Zoological Gardens, Chicago, National Museum of Natural History, Field Museum of Natural History, Chicago, Department of Anatomy, University of Chicago and Department of Anatomy, University of Southern California. The original studies and publications upon which a considerable portion of this book is based have been supported by funds from the United States Public Health Service, the National Institutes of Health, the National Science Foundation, the Wenner Gren Foundation, and research funds from the Universities of Birmingham, Chicago and Southern California.

CHAPTER 1

Human Fossils – The New Revolution

The excitement of new fossils – The excitement of new tools
Dating – Environment – Molecules – Behaviour – Structure
New views of old bones – Old doubts – New estimates.

These are most exciting times for all who have interests in the evolution of humans and our nearest living relatives, apes, monkeys and prosimians. This has been signalled, in a way that can have been missed by no one, by the spate of fossils that have been found in Africa and elsewhere in the past few years. A topic that once warranted little more than a column in the national presses during the silly season now occupies prime space and time in every national and international publicity medium.

The excitement of new fossils

Following many years during which all that was available was a tooth here, a fragment of skull there, datable with little accuracy, we are now confronted, it seems almost day by day, by extensive discoveries of fossil conglomerates. They often include remnants referable to several individuals, sometimes even many fragments of the same individual, and they come from a variety of geographic sites and with a wide range of determinable dates.

Such discoveries, in Olduvai and Southern Africa, at East Turkana, in the Omo, at Laetoli, in the Afar Valley, in Pakistan, in Burma, in Eastern Europe, and in several parts of China, for instance, have not come about by accident but by the assiduous work of teams of individuals led by such investigators as Philip Tobias, the late Louis Leakey, Richard Leakey, Clark Howell, Mary Leakey, Donald Johanson, Maurice Taieb, David Pilbeam, Elwyn Simons, Donald Savage and Wu Rukang. These finds increase, by orders of magnitude, over many thousands of new miles, and through millions of extra years, our knowledge of fossil primates presumed related to man.

Although these new fossils have by no means been fully studied, their mere existence suggests that the conventional notion of human evolution must now be heavily modified or even rejected, and that new concepts must be explored. No longer can the idea be held that there exists a single lineage from *Homo sapiens* at the present day and in prehistoric times, back through *Homo erectus* at up to half a million years, then through, at one to two million years, *Australopithecus* of Olduvai and Southern Africa, to, eventually, *Ramapithecus*, at ten to fifteen million years ago (Fig. 1.1). (The term '*Australopithecus*' includes here not only *A. africanus*, believed to be the direct human ancestor, but also *H. habilis* from which it has not been clearly differentiated, *H. africanus*, a recently invented synonym, and *A. robustus*, a closely related but probably parallel species.)

Now, whatever the details may turn out to be, we must be willing to envisage a number of different lines (an example is provided in Fig. 1.2), undoubtedly of different degrees of relatedness, and with the genus *Homo* itself going back several, indeed perhaps as many as five million years, perhaps even longer. It is highly likely that species of the genus *Homo* were entirely contemporaneous with two, possibly several, species of *Australopithecus*; some new finds that are older than the well-studied australopithecine specimens from Olduvai and Southern Africa may actually turn out to be more like *H. erectus* than like these australopithecines.

We must be willing to view these australopithecines as a series of genera at best only parallel to a series of human lineages; and we must be prepared to see even a radiation within this group, for others of the new finds may well be australopithecine but neither robust or gracile. This new, exciting uncertainty is indicated by multiple questions marks in Fig. 1.2. See also **nota bene** pp 334, i–xi.

We may even have to be willing to reassess our

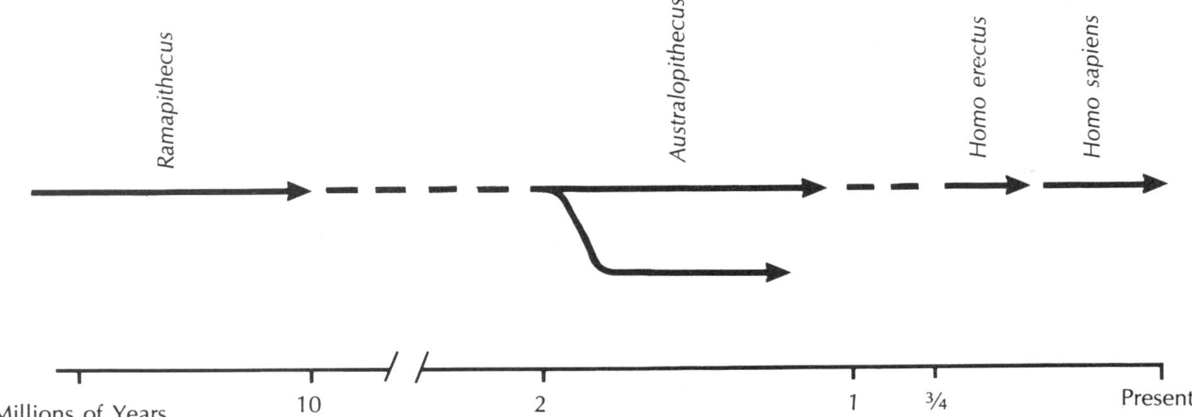

Fig. 1.1. The traditional idea of the human lineage. Solid lines denote known finds. Dotted lines define the presumed ancestral relationships. Diagrams not unlike this are found in many anthropological texts. Whatever the precise pattern that they display, the implication is (a) that a single lineage leading to humans is the main feature with side branches being few and minor, and (b) that the special fossil at issue is the stem form leading to humans.

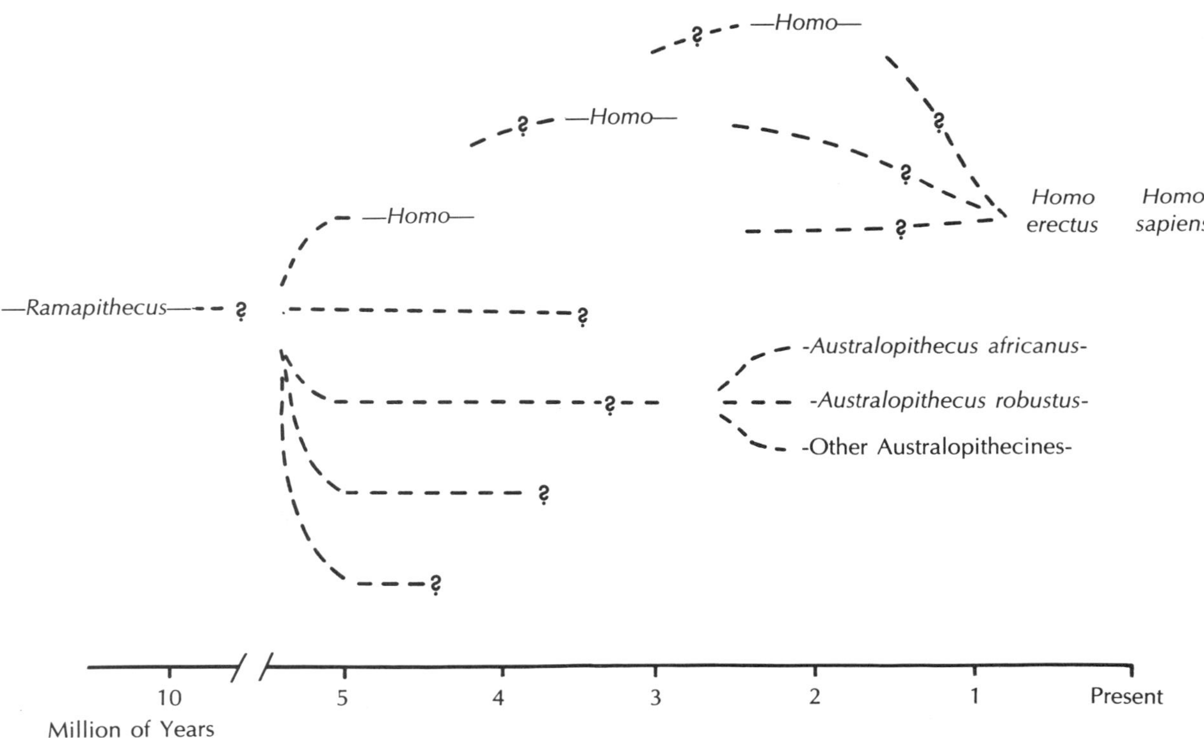

Fig. 1.2. One possible example of an evolutionary radiation as suggested by the new evidence. The names are aligned approximately with the dates of known finds. Dotted lines indicate possible relationships. Question marks demonstrate just how problematical all this is. Diagrams like this are rarely found in standard texts. It is not common to suggest the possibility that the various australopithecines may be only parallel species. Even *Ramapithecus*, though represented here by a single name, is likely to be a complex of species. And it is very uncommon to suggest that no individual fossil so far found bears any strict linear relationship to humans.

ideas about *Ramapithecus*. The finds of this fossil are so restricted anatomically (mainly teeth and jaw fragments, but see new finds in China, Wu, 1982) that we really have little basis upon which to say just where, in all of these lineages, ramapithecines may fit. But much other information concerning paleo-environmental and paleo-ecological data suggests that it is rather likely that a radiation of these forms also existed. See also NB pp 334, i–xv.

When, at this point, we remember that the various dryopithecine apes of twenty million years ago and more have long been known to form a radiation, then we may have a far better notion of the complexities of human evolution.

We may possibly have to take into account a newly announced discovery of remains, as yet unstudied, from Burma. These are dated tentatively at perhaps even more than forty million years. They are jaw fragments, classified in a preliminary manner as belonging to the genera *Pondaungia* and *Amphipithecus*, and are said to have a structure suggesting they are hominoids about the size of gibbons. It would appear that even these few remains represent more than a single species, perhaps a radiation.

And finally we must take into account a wide variety of fossils that have been found in China in the last two decades. There are now many different sites in China spread over many provinces: Yunnan, Hubei, Shaanxi, Anhui and Guangdong, and of course the famous Zhoukoudian. There are equally many different finds at each of the morphological levels: *Dryopithecus*, *Sivapithecus*, *Ramapithecus*, and *Gigantopithecus*, *Australopithecus*, *Homo erectus* and both early and late *Homo sapiens*. Though some of these specimens were found as long ago as 1957 and 1958, others are the result of searches as recent as 1980 and 1981. None have been studied outside China; and no doubt the designations of some may change as they are further studied; but of the multiplicity of finds there can be no uncertainty. Whatever these new fossils ultimately turn out to be, they further enhance the idea of a radiation, indeed of radiation*s*, of forms.

The task is thus changing from one in which it seemed necessary to fit each succeeding fossil into *a* single gradually changing lineage, to one in which radiations of many lineages, perhaps with sharp changes among them, must be assumed; from one in which each new find as it was announced was believed to be *the* crucial stem species, to one in which we generally expect new finds to fit on paths that do not lead directly on to humans; from an *idée fixe* on discovering *the* missing link, to the realization that the overwhelming majority of the creatures preserved by the rare accident of fossilisation will not, on statistical grounds alone, stand in any direct relationship to us. This concept of multiple radiations, apparent discontinuities and spotty representation fits, far better than that of a single gradual lineage, with biological ideas about evolution. It is thereby the more acceptable to biologists.

The new view of human evolution has a number of additional implications. The picture is turning inevitably away from one in which certain attributes were thought to be basically human and were believed, therefore, to have evolved only once (for example, bipedalism, tool-using and making, communication, high intelligence, perhaps even some social and cultural developments). It is turning towards one in which it is conceivable that many of these features have evolved more than once, albeit presumably in related species. Fossil evidence of such possibilities has now to be sought; they must be strongly denied before we can ever return to the older view. Certainly, new studies of the capabilities of the living great apes (toolmaking, tool usage, communication, even, perhaps, a limited degree of conceptual thought) suggest that this is not a wild idea.

This new picture also means that we have to return to a state of ignorance about the site of the origin of the genus *Homo*. Now that new finds presumed to be *Homo* are at least contemporaneous with, and indeed predate many australopithecine fossils, we have no certain progenitor for the genus *Homo* from Africa near this particular time range. And though human remains of this degree of antiquity are still best known from Africa, it must be acknowledged that Africa is the region that has been searched with the greatest degree of intensity. There is thus no especial reason to pinpoint Africa as the place where it all happened.

In fact, these various forms, *Homo sapiens*, *H. erectus*, *Australopithecus sp.*, *Ramapithecus sp.*, *Dryopithecus sp.*, are now all known (or in the case of *Australopithecus* believed by some) to exist not only in Africa but also each in Europe and Asia. New finds are being made in China; and though these are yet not fully studied, we must start to take them into account in our thinking. Thus, *Homo sapiens*, both late and early, has been found in many places (e.g. Guangxi and Hubei provinces). *Homo erectus* has been found at approximately one million years at Yuan Mou in Hunan, and in many other places (e.g. Anhui, Hubei and Shaanxi provinces). Teeth resembling those of *Australopithecus* have been found in Hubei and Guangxi provinces at even earlier dates. Preliminary reports suggest that eight million year old *Ramapithecus* and *Sivapithecus* have been located at several sites in the Southern Chinese province of Yunnan. *Gigantopithecus* has also been found in Yunnan. It may well be that yet other forms also exist somewhere in China. The possibility clearly opens up again that human origins may have been as lief in Asia as in Africa. Only further discoveries will provide evidence on these matters.

Finally, the new picture forces us to look again at the nature of our relationships with the African great apes, those creatures to which, among living forms, we are assuredly most closely related. The nature of that connection must be far more complex than previously thought, and in all likelihood must have been far more ancient than has been supposed by many in recent years. This inevitably follows if there is any truth in the idea of a series of radiations of pre-human and near human forms rather than a single lineage. One wonders whether those radiations might not have also involved the African great apes, or if those radiations were subsequent to links with African apes. One even wonders if the African apes themselves underwent a radiation of sorts, of which gorillas, chimpanzees, and bonobos (pygmy chimpanzees) are the only surviving forms. Perhaps some of the fossil remnants that seem at present to pose problems for the human radiation actually belong to extinct ape lineages.

There is thus much reason for excitement in the world of primate evolution, and much reason to anticipate the next few years with open minds.

The excitement of new tools

However, the discovery of new fossils is not the only source of the new visions that are opening up. Every bit as important, but less well-known to the wider public because of their more esoteric and difficult nature, are a series of developments in the many other disciplines that act as handmaidens in all evolutionary investigations.

Dating. One of these relates to our ability to make better assessments of time. Without a series of new tools for the study of absolute and relative time, much of the of the story would yet remain hidden (e.g. Jolly, 1978). The fossils themselves are sometimes dated; certainly, the geological layers related to them are dated, and by a series of different methods if possible. There are two basically different approaches to such dating.

One approach depends upon relative time scales in which recognizable sequences from successive layers are determined and correlated from one site to another. Such sequences may be merely those of the differential thickness of yearly deposits of mud on lake beds consequent upon climactic changes from year to year, century to century, or even over longer periods of climactic change, even as long as glaciations. Other sequences may involve recognizing changing patterns of associated animal and plant fossils, both in terms of longitudinal evolution within individual groups, and in terms of patterns of association among groups. In this regard the remains of small vertebrates such as rodents, invertebrates such as insects and plant fragments such as pollens and seeds may, because of increased samples, be far more valuable than those of visually more exciting animals such as large mammals. Some of these relative dating methods are chemical. They include nitrogen dating, uranium dating and the well-known fluorine dating that proved the downfall of Piltdown man. One of the most recent of the relative dating methods stems from discoveries in geomagnetic changes in the rocks. The earth's magnetism has reversed itself time and time again over the years. The patterns of these reversals have become 'fossilized' in some rocks, and their examination today offers a new basis for dating.

The other approach to dating is absolute and derives from changes over time in certain physical and chemical parameters of geological materials; one example includes dating based upon the decay of radio-isotopes, the potassium argon, argon-argon and carbon 14 methods; another example includes fission track studies which can date the time of origin of last heating of volcanic materials; a third results from a special change 'racemization' of what little protein material may be left in the fossil, a process that starts at death, and though temperature dependent, depends also upon absolute time; a fourth utilizes 'electron spin resonance' which can measure the 'damage' that the flux of time produces in crystal lattices such as hydroxyapatite, a normal crystalline component of bone.

As a result of these advances more precise dating is being achieved now than formerly. Many dates long accepted in the literature of human evolution are now known to be incorrect; and at the same time the new methods impress upon us the concept of 'error' that is inherent in the different dating methods. The impact has thus been to change many dates, to remind us that all dates are tentative, and, perhaps most important of all, to make us aware of the need for increased rigour in studies of primate evolution.

Environment. A second set of tools enhances our ability to assess fossils. New investigations pertinent to evaluating the biotic and physical environment of fossil creatures are freeing our ideas from the constraints of the strait-jacket resulting from convention (e.g. Coppens, Howell, Isaac and Leakey, 1976). The old concepts have led us to envision a relatively arid, certainly treeless or almost so, savannah-like environment for the australopithecine fossils. Such conceptions have compelled us to look only for evidence of savannah-like behaviours in the fossils; an analogue today includes the quadrupedalism of baboons together with the kinds of social structures that go with it. As a result, studies of baboon behaviour have been much utilized in order to make assessment of possible behaviours of our prehuman ancestors.

These traditional ideas prevent us from seeing, for instance, those possibilities that exist within heavily wooded or forested environments: hiding and stalking on the forest floor, leaping, climbing and acrobatic activities among the branches, and the different social, feeding and reproductive strategies that go with such environments.

Information about the types of animal and plant communities that may have existed in prior times may be preserved in geological samples and recognized through the associations of seeds and pollens, and animal and plant fossils. This is, of course, a different usage of associated remains from those mentioned in the last section on dating. As a result, such habitats as the lake shore, the river delta, the flood plain and other environments may be recognized.

The notion that forest environments cannot be sampled is probably a myth. A very small geographic area may well encompass perhaps a lake shore, the entrance of a river into it, a heavily forested area around such a riverbank, a small yet deep forest on each side of the immediately forested riverbank, leading to less dense woodland, to isolated forest stands and individual trees, to finally a rather savannah-like environment immediately adjacent. All these micro-environments may be present in a very small localized geographic area previously thought to be only a treeless savannah.

And another idea that is developing at the present time is some notion of what it is impossible to discover by these methods. It may be impossible to say that this or that fossil belongs to this, that or the other micro-environment of the overall habitat. We may well discover that we are confined to suggesting only a certain width for possible environments; we may have to agree, in the end, that environmental possibilities are very wide indeed.

Such advances in the rigour and clarity of paleoenvironmental and paleo-ecological investigations now mean that we have to change our ideas rather completely from the conventional picture; in particular, they suggest behavioural possibilities for many of the fossils that could stem from perhaps as wide a range of environments as exist anywhere in Africa today.

Molecules. Yet other methods have become important in the studies of human evolution. Information about changes over time of the various molecular and chemical materials that are the very stuff of evolution cannot be obtained (except in those rare cases where residues of actual materials are entrapped within the fossils). But knowledge of how molecular and chemical elements differ in living species tells us a considerable amount about what the changes may have been.

Such studies are really quite old (for instance, the first were carried out as long ago as the end of the last century by Friedenthal in 1900, and an early summary by Zuckerman in 1933 looked towards a time when we would have such information in abundance). Yet it has required the molecular revolution in biology in the 1950s, together with the extensive comparative biomolecular work of very recent years, to provide a sufficiently large data bank for these methods to contribute more fully to the solution of such evolutionary problems (e.g. Barnicott, 1969; Goodman and Tashian, 1976).

We are thus in a time when systematic discussion of an extant group routinely involves the biomolecular framework. Departures from the biomolecular framework on the part of studies involving structure or function at the organ-organismal level must supply their own glosses upon our knowledge of evolution.

These various methods involve a whole series of techiques which include, among others, amino acid sequencing of proteins, electrophoresis, nucleic acid hybridization, the techniques of immunology, studies of chromosomes and karyotypes and so on;

and concordances among these various molecular techniques are an important part of the entire story.

As with the new dating methods and the new paleo-ecological tools, the melding of information from the new biomolecular techniques is an important part of the corroboration and testing of evolutionary hypotheses. And as also with those dating and paleo-ecological investigations, the increased rigour that such experimental methods naturally bring to anthropology are an important element to be added to evolutionary studies of humans and other primates.

Behaviour. Even when we come to look at the evolution of behaviour there are major new data to examine (e.g. Chance and Jolly, 1970). Some of these stem from studies of morphology and relate to the very simple, perhaps we could call them primary, kinds of behaviours which have their effect upon morphology, mainly through their biomechanical associations with the bones, muscles and joints of the animals. Other parts result from investigations of many secondary, even tertiary, and yet more complex aspects of behaviour (such as social, sexual, communicative) now being studied in ways that were not well worked out years ago, although, again, earlier studies (Zuckerman, 1932; Yerkes, 1925; Carpenter, 1934) recognized many of them.

One behavioural possibility devolves from the tradition of discovering what is actually happening in the field and what may thus be said directly about the evolution of behaviour. A second possibility is the idea that such studies of behaviour and of social organization may provide indirect clues to the origin of human behaviour. A third is the notion that the experimental study of primate behaviour and psychology may help provide insights into the evolution of the complexities of the primate mind.

As in the other scientific disciplines which I have noted, knowledge is increasing; gaps in the data are being filled in year by year. But this particular area, though it started early, is perhaps, of all, the one in which it is most difficult to progress; it is just not at all easy to carry out these kinds of studies; and it is especially difficult to do them within the vast comparative framework that is necessary for the evolutionary overview. Yet it will certainly in the future contribute more and more to the new picture of primate evolution.

Structure. Finally, without new methods and ideas which have been developed for the study of the structure of the fossil skeletal specimens themselves, we would still remain fixed at the level of scientific deduction that is imposed by the conventional or classical way of looking at bones. These new methods include techniques for defining actual bony structures not only by measurements but also using various holistic ways of capturing the essence of forms and patterns. They involve not only statistical manipulations but also the use of a wide variety of other mathematical, physical and engineering approaches. They depend not only upon the technology of computers but also upon such inventions as electron microscopes, lasers, image analysers, indeed almost any part of the panoply of modern science that can be applied to problems of form and pattern (Oxnard, 1978a).

On the one hand, such new methods free us so that we can 'see' information in the skeleton that is not available to visual observation (Oxnard, 1973a). On the other hand, modes of interpretation are such that we no longer look immediately to genetic relationship as the primary explanation of morphological similarity in bones; rather do we now look towards similarities in the functions of bones within overall behaviour (Oxnard, 1975a). It is true that this results in part from the inheritance of plasticity of bone, rather than inheritance of any special bone shape. But it results in part also from the direct adaptation of bone to impressed mechanical forces that stem from activities other than those that might be genetically determined. This provides the possibility of the structural information melding with ecological, environmental, behavioural and other data obtained from the other modes of study.

New views of old bones

One clear evidence of these changes in techniques relates to changes in the way we view the structural relationships between the fossils and the living primates. Current views imply, for instance, that the australopithecine fossils are human-like rather than ape-like (or, as some would have it, human rather than ape). At the present time minds are much less open to a new view: the possibility of the existence of creatures falling into neither category.

Present views seize on human-like aspects of the fossil structures as meaning human-like functions (bipedalism and tool-making, for instance) for australopithecines. The newer recognition of other aspects of these structures means that we may come to envisage a range of functions for the australopithecines quite different from those seen in any present-day form, whether human, ape or even monkey. And this then allows us to include in our investigations all those pieces of information which were ignored in the prior attempts to make some of those fossils fit the conventional picture.

Our views about the ramapithecines from an earlier time period are undergoing similar transformations. The initial studies suggested to many investigators the notion that these species, too, were on the line leading to human evolution. Much was speculated about these creatures being early hominids possessing tool use and upright posture. New information suggests not only that such detailed speculation about the ramapithecines is not possible from fragments of jaws and teeth (very little post-cranial material is known), but also, to the degree that speculation is possible, to wit from the paleo-ecological and paleo-environmental evidence, that these fossils may represent many diverse creatures. At this time it may be most difficult to pinpoint any one of them as direct human ancestors. They may have lived in a wide variety of wooded and partly wooded environments that suggest non-human and probably arboreal habitats as their most likely living contexts.

And finally we will shortly have to come more scientifically to grips with the even older fragments that represent not only human but presumably many higher primate ancestors, in the persons of such forms as the twenty million year old dryopithecines and the forty million year old Burmese creatures from the Pondaung hills.

It is fascinating to see that in each scientific area-dating, paleo-ecology, molecular evolution, behaviour, fossil morphology, at first of course totally independent of one another, but now interdependent upon one another to greater and greater degrees various new results are appearing. Each area now provides tools that can be genuinely used in their own right and not merely presented as examples of what the future will bring. Each can now be used in conjunction with others to obtain new information greater than that stemming from each alone. Thus, the future does indeed shine brightly with the hope of what will be discovered as these investigational batteries are gradually brought to bear upon more materials, upon wider questions and with greater abilities for extending our inferences and recognizing our limitations in understanding the evolution of the primates. One big cloud on the horizon of such hopes stems from the increasing difficulties that new international situations pose for such investigations.

Old doubts. It is also of interest to realize that developments within each of the different academic areas contributing to our understanding of human evolution have already provided information predating the discoveries of the new fossils. Although it has taken the excitement of the new fossil finds of the last three or four years to bring to most minds a realization that human evolution must be far more complicated and the human lineage of far greater ancestry than previously thought, in fact the germs of these ideas were already detectable in prior studies.

Thus, examinations of the fragments of fossils that have been available for many years have continually suggested to some investigators the likelihood that the australopithecines from Olduvai and Sterkfontein were not all that they were made out to be by the conventional wisdom. Zuckerman and his colleagues (e.g. as reviewed in Zuckerman, 1970) have shown, for instance, through studies extending over many years now, that there must be real doubts about the conventional assessment of these particular australopithecines. Whenever these workers have attempted to compare such australopithecine structures with those of man and the living apes, they have always been able to confirm some points in which the fossils resemble man. But they have also always found other features in which the fossils resemble apes, and these features have usually been either totally ignored or given very little weight in the conventional assessment.

New estimates. Furthermore, within the last ten years, even newer assessments of these fragments have transformed Zuckerman's and colleagues' doubts into positive suggestions as to what, indeed, the australopithecines of Olduvai and Sterkfontein may have been, given that they are not direct human ancestors. Thus, the present author has been able to suggest that, although it is likely that the australopithecines may have been capable bipeds (but probably in a biomechanical mode quite different from that employed by humans), they may also have been capable quadrupedal animals, perhaps especially within a climbing mode in trees. Such a combination of capabilities is not found in any currently known creature and may well conform to the view that these forms, given their very late existence on the evolutionary time scale, are not direct human ancestors at all (e.g. Oxnard 1975b, 1977).

But such investigations do not have the excitement and publicity of the new fossil finds, and they presently depend upon a technology that is complex and appears difficult to understand. We have thus had to await the new fossils, such as, for example, the footprints from Laetoli which do seem to betoken bipedality and which are much older than the australopithecines of Olduvai and Sterkfontein, to confirm these suggestions.

Now, however, that confirmation is indeed at hand, it is worthwhile looking more closely at some of these more complex studies of primate morphology in order to understand more fully how they work and what new information they can supply about the problems of primate evolution.

CHAPTER 2

Distinguishing Primates

Primates – Convention and controversy – Evolutionary relationships
The basic groundwork – The organismic approach – The biomolecular picture
Points of controversy – Interpolating fossils – The piecemeal approach
Interpolating fossils – Studies of the whole – The new methods.

Although the new methods and the new logic introduced in Chapter 1 are most important in understanding primate evolution, we must never forget that current advances depend upon the very solid groundwork of many decades, indeed many centuries, of study of the Order Primates. It would be quite incorrect to think that this basic work should all be swept away. On the contrary, it is the meld of the broad picture that we all know and accept, with the new information and hypotheses that are coming forward, that presents the challenge of studying primate evolution in the coming decades. This broad picture is already very wellknown, but a summary is useful at this point to focus our minds upon its pattern.

Primates: convention and controversy

The primates are inevitably, and in some ways unfortunately, the most interesting of animals to that sapient species that also belongs to the Order. No other group of mammals has been studied by so many investigators and no other single living type is as wellknown as the human species. In consequence we know more about the relationships of humans to other primates, of the human place among the primates, than about most other vertebrates. The evidence upon which these relationships are grounded is extraordinarily widely based.

In spite of all of this, there is not total agreement about the classification of the primates. Although some believe that the amount of agreement there is is surprisingly high, others lament what they perceive as confusion.

That these two viewpoints can be held simultaneously is partly a tribute to the fact that investigators can be fairly well agreed about principles and concepts yet continue to use, often for pragmatic reasons, systems that differ widely. Yet it is also partly due to the fact that the importance of distinctions among the primates has, at different times and by a variety of researchers, been enormously exaggerated. Almost every difference in coat colour, every scrap of fossil tooth or bone, every aberrant individual, has been given a separate nomen; most living primates have had many alternate names; many distinct species have been called genera; many genera have been elevated to familial status (Simpson, 1945).

Even nowadays, although the general plan of primate relationships has been settled for quite a long period, individual points are challenged from time to time. Most work attempts to produce a marginally better picture; but, though in some minds marginal suggestions are deemed valid, others agree that to make such changes in nomenclature produces more confusion than help in our present state of knowledge.

One example of present controversy is perhaps to be seen in the basic division of the primates into the two groups, Prosimii and Anthropoidea. There is reality in the suborder Anthropoidea: monkeys, apes and humans. But some workers nowadays agree that the term Prosimii is a miscellany category. Among living forms it includes aye-ayes, bush-babies, lemurs, lorises, tarsiers and, even for some investigators, tree-shrews. Yet, for other workers, to do away with this grade appears less than useful, and so it is retained.

Another example of controversy in primate relationships is at the opposite end of the Order in the difficulties that exist over the generic names for gorillas and chimpanzees (*Gorilla* and *Pan* respectively). Here, also, most workers agree that gorillas and chimpanzees are far closer to one another than is represented by the use of separate generic designations; and a number of researchers have, on reasonable evidence, suggested that they be grouped as a single genus, *Pan* (e.g. Tuttle, 1975). The usage is starting to catch on. But the consensus about the basic information is probably good enough that most workers believe that there is no real need to make the nomenclatorial change; most investigators will undoubtedly try to reduce confusion in the literature by retaining the older terms while yet accepting the newer relationship.

In fact, there is probably no group of primates about which there is no argument. Individual species, even individual genera have been given many vernacular and scientific names over the years. Several particular genera have been made the sole representatives of subfamilial, even on occasion familial, groupings depending upon the investigator. The apparently unified groups of New and Old World monkeys contain numerous points of controversy. The major subdivision into Prosimii and Anthropoidea is clouded by discussion as to whether the spectral tarsier belongs with the one or the other. It is not even completely agreed what constitutes the entire Order; are tree-shrews primates?

When, therefore, we view the relationships of the living primates, we are looking at a series of pragmatic compromises as well as attempts to carry such studies as accurately and as far as possible.

Evolutionary relationships: the basic groundwork

Much of the data base upon which primate relationships are grounded is classical in nature; it depends upon assessment of the information contained within a series of morphological characters at the organ and organismic level. These include such features as external appearances of the face, hands, feet and genitalia, particularly the complex structures of the skull and teeth, together with some information from superficial markers such as coat colour and other pelage characteristics.

Organismal morphology is not, of course, static though at any given time a structure may present a 'frozen' appearance. In fact the earlier in development that we view the stages of vertebrate anatomy, the greater the similarities that we find. This is inevitable because evolution consists not only of the evolution of adult forms but also of the evolution of the developmental systems producing adult forms. A long literature spells out some of these ideas, including, in the last half century, investigations and discussions by de Beer, 1940, Medawar, 1945, Keith, 1949, Simpson, 1953, Zuckerman, 1954, Bonner, 1965, and Gould, 1977.

The 'developmental transformation' that occurs during the life cycle of an individual organism results in an increasing gradient of structural difference (Fig. 2.1, heavy arrow in the diagram) between the original fertilized egg and the adult individual.

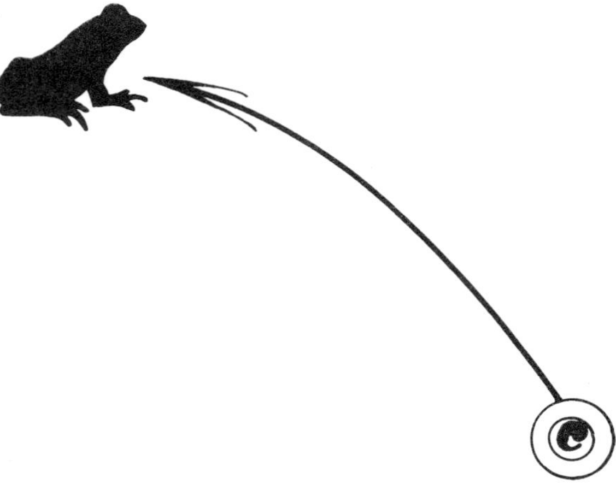

Fig. 2.1. The 'developmental transformation' that occurs from zygote to adult during the life cycle of an individual organism. The change in structure that occurs from early to late developmental stages is indicated by the heavy arrow. This is, of course, a real change due to growth and development in a given organism.

In comparing the 'developmental transformation' for an ancestor with that for a descendant, we notice a new gradient in the difference between the ancestral and descendant developmental transformations. This 'evolutionary gradient' (thin, approximately horizontal arrows in Fig. 2.2) often results in lesser differences (shorter arrows) between earlier developmental stages of ancestor and descendant, and greater differences (longer arrows) between later, adult forms of ancestor and descendant.

When we view two 'sibling' descendants derived from an intermediate common ancestor (Fig. 2.3) resulting from a split in the evolutionary lineage, we can see four developmental transformations, one each for the two linearly related ancestors and the two sibling descendants (four approximately vertical arrows in the diagram). This follows as an extension of the ideas in Fig. 2.1.

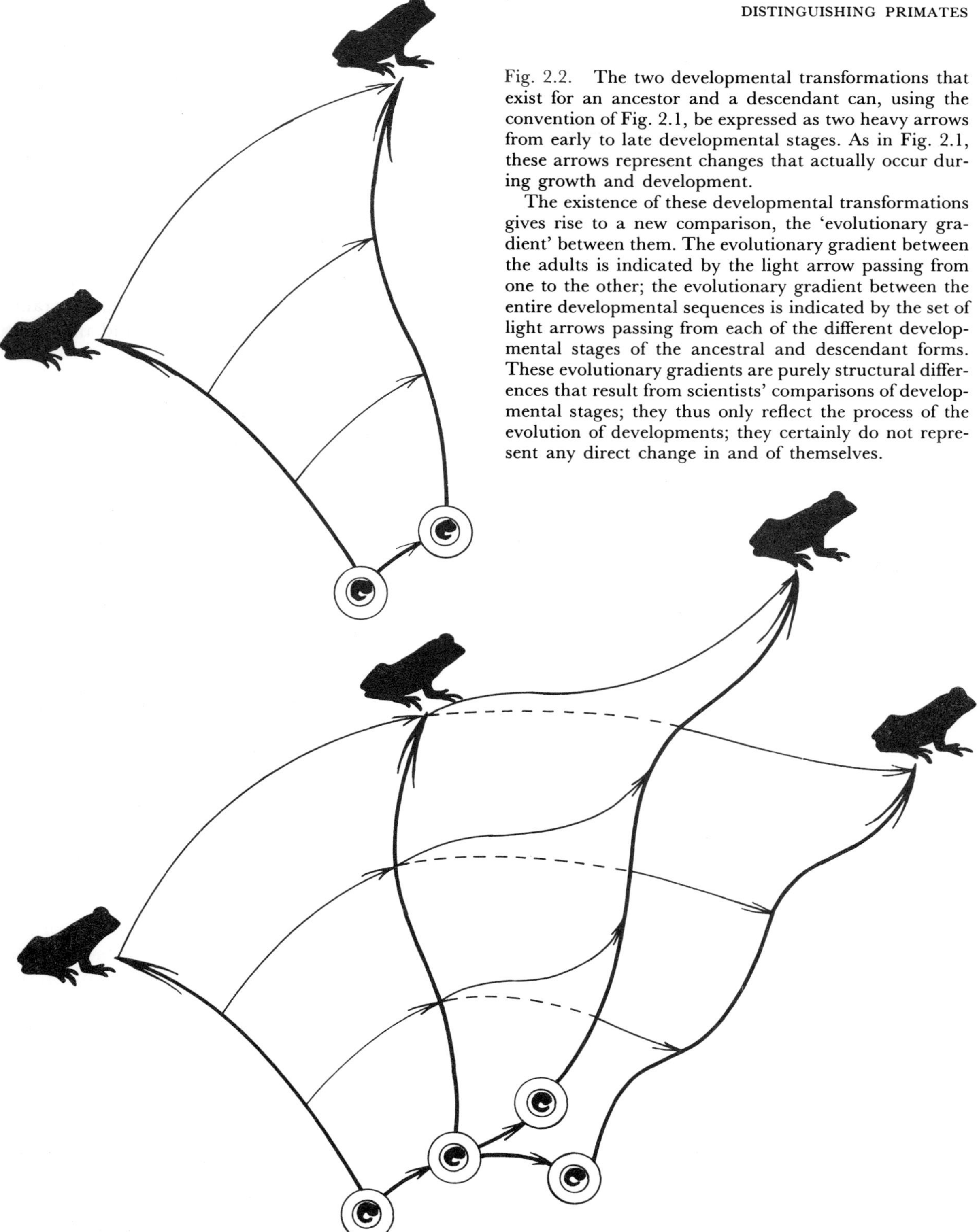

Fig. 2.2. The two developmental transformations that exist for an ancestor and a descendant can, using the convention of Fig. 2.1, be expressed as two heavy arrows from early to late developmental stages. As in Fig. 2.1, these arrows represent changes that actually occur during growth and development.

The existence of these developmental transformations gives rise to a new comparison, the 'evolutionary gradient' between them. The evolutionary gradient between the adults is indicated by the light arrow passing from one to the other; the evolutionary gradient between the entire developmental sequences is indicated by the set of light arrows passing from each of the different developmental stages of the ancestral and descendant forms. These evolutionary gradients are purely structural differences that result from scientists' comparisons of developmental stages; they thus only reflect the process of the evolution of developments; they certainly do not represent any direct change in and of themselves.

Fig. 2.3. Two linearly related ancestors and two further parallel descendants can be described through four developmental transformations (approximately vertical heavy arrows from cells to adults). Using the conventions of Fig. 2.2, these transformations give rise to three evolutionary gradients (horizontal light arrows from each developmental sequence to the next from ancestors to descendants). Again, the developmental transformations are transformations that have really occurred during the lives of the particular generations. The evolutionary gradients are descriptive devices for comparing differences between the developmental transformations.

In addition, however, we can see three evolutionary gradients between ancestors and descendants (three sets of approximately horizontal arrows in the diagram). These evolutionary gradients comprise the differences stemming from the starting developmental transformation of the original ancestor, through the intermediate developmental transformation of the intermediate ancestor from which splitting occurs, to the two new developmental transformations for the two sibling descendants. These new evolutionary gradients also naturally exhibit lesser differences between the embryonic stages of the ancestors and descendants (shorter horizontal arrows between early stages) and greater differences between the adult stages of the ancestors and descendants (longer horizontal arrows between adult stages).

The problem of evolutionary biology stems from the fact, of course, that we can rarely see the four developmental transformations or the three evolutionary gradients (hence in Fig. 2.4 the arrows representing these parts are drawn with thin lines

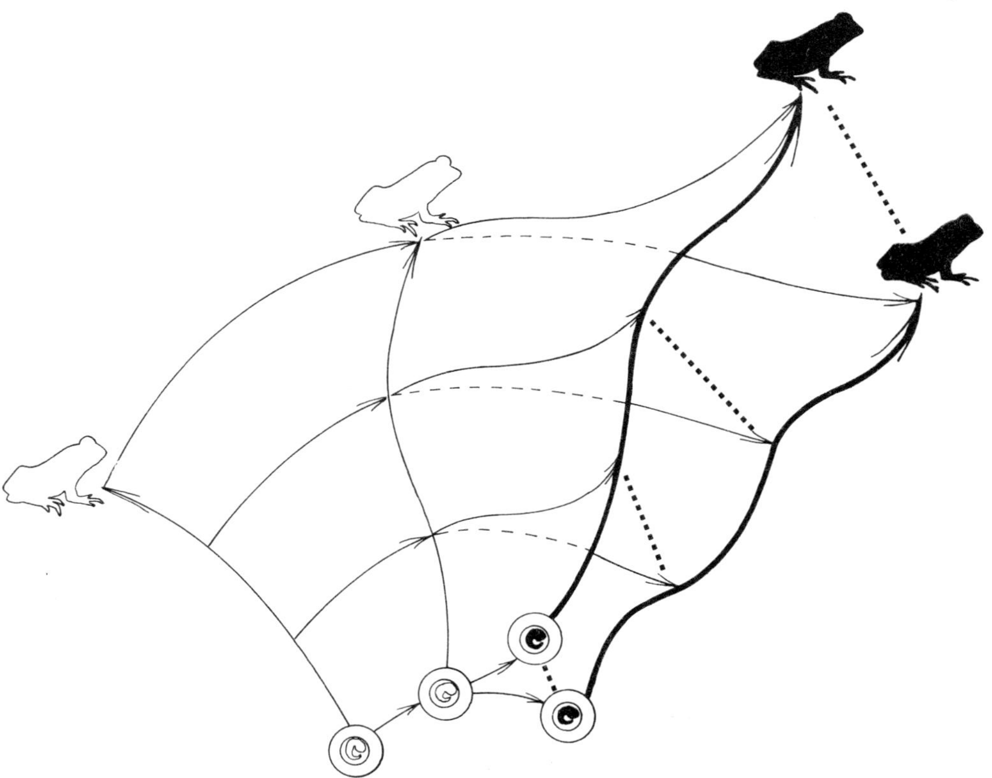

Fig. 2.4. The developmental and evolutionary processes that produce the descendants in Fig. 2.3. also allow a further descriptive comparison: that between the daughter groups themselves. This comparison differs from the evolutionary gradient because the two organisms are not linearly related to one another; it can be called a comparative trend. The comparative trend is thus, the comparison of the parallel descendant developmental sequences (comparison of the two heavy arrows). Our diagram is, therefore, essentially the same as Fig. 2.3, but the new comparison is emphasized by heavier lines and shading and by the new dotted lines that represent the comparison of the comparative trend.

Again, the comparative trend does not express any mechanism or process that actually occurs. It demonstrates a comparison that can be made. Such a comparison can always be made between any unrelated pair of animals. But in this case the comparison has special meaning because it is a comparison between developmental sequences that are specially related by the existence of the prior ancestral developmental transformations and the resulting descriptive evolutionary gradients. Its importance is that it may well be the only evidence that we have about the previously existing 'transformations' and the resulting 'gradients'.

Indeed, we may have even less information because it is so often the case that we do not have developmental information even for the two parallel descendants; we may have only the comparative trend between the adult forms (i.e. between the black figures). Estimating all of the rest of the information contained in this complex of changes (matrix of arrows) from the mere single difference between the two descendant adult forms is difficult indeed. Thus it is that evolutionary problems are so hard to study and give rise to so much untestable speculation.

and without shading). But we can often compare most easily and completely the descendant groups themselves because they may both be alive at the present time. This results in a third kind of comparison, a trend that is the product of the previous two: a 'comparative trend' between organisms of common ancestry. This new trend may be the rather simpler one resulting from the comparison of the adult descendant forms alone (the heavy dotted line joining the adult forms on Fig. 2.4). A more complex rendering of the 'comparative trend' between the two descendant forms derives from the comparison of the two complete developmental transformations of each (as shown by the full suite of heavy arrows, dark shading and dotted lines in Fig. 2.4).

Usually, of course, real evolutionary situations are far more complicated than this example and many different patterns can be perceived among differently related animals. But given these complications, it is possible to understand, in principal, what may have occurred and to look, therefore, for the kinds of information that may indicate, in practice, what has actually happened.

Of course, many other mechanisms have impact upon these transformations, gradients and trends, but the general picture remains viable; and it results in early forms of different animals showing somewhat greater similarities than later forms of those same animals. That this is the case for many vertebrates can be seen from Fig. 2.5 in which young embryos of turtles and turkeys, mice and men are less different than are older stages of those same species (Hildebrand, 1974).

Within the primates, differences in structures are likewise based upon the same notion. For although the upper limbs of different primates are well known to be very different from one another as adults, Fig. 2.6 shows that there is a much smaller difference to be observed in their structures at birth (Schultz, 1969).

In fact, most primates are remarkably uniform when compared with other living mammals which contain far more extreme and diverse specializations within individual orders. Thus, such extreme morphological adaptations as relate to the biomechanical demands of, say, swimming, gliding, climbing and burrowing are all found within single groups of mammals such as rodents and marsupials. And other drastic specializations of somewhat different types such as in energy consumption, speed of movement and type of diet are

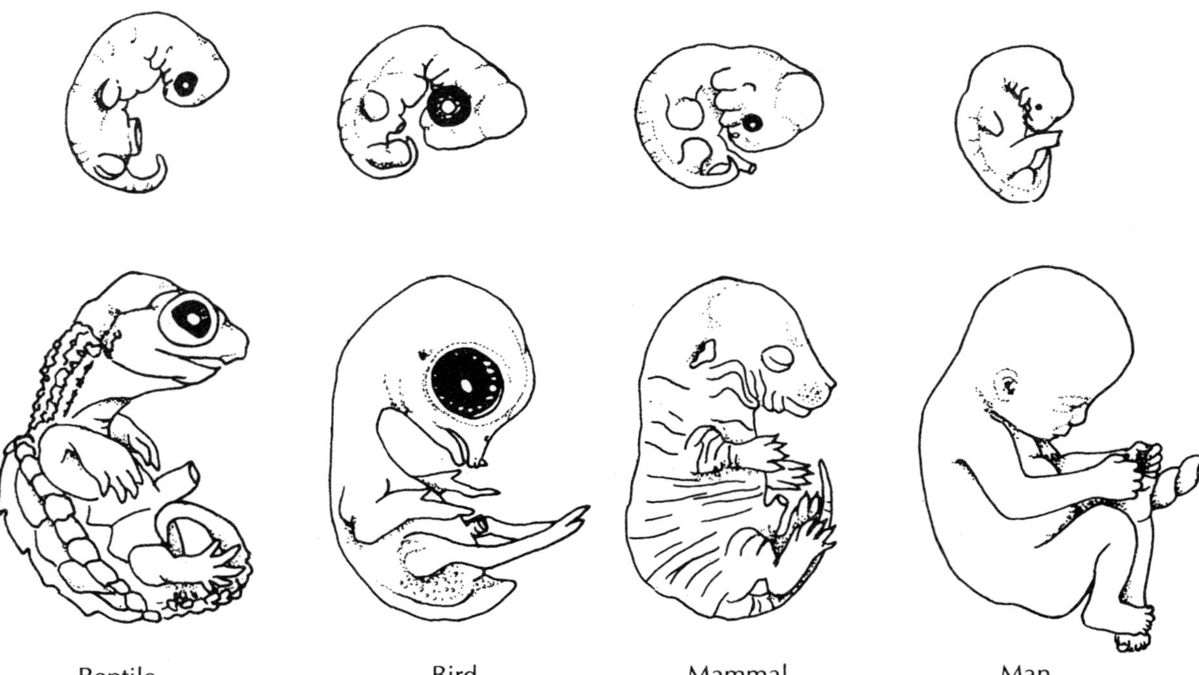

Fig. 2.5. Differences are less between younger embryos and greater between older ones (after Hildebrand).

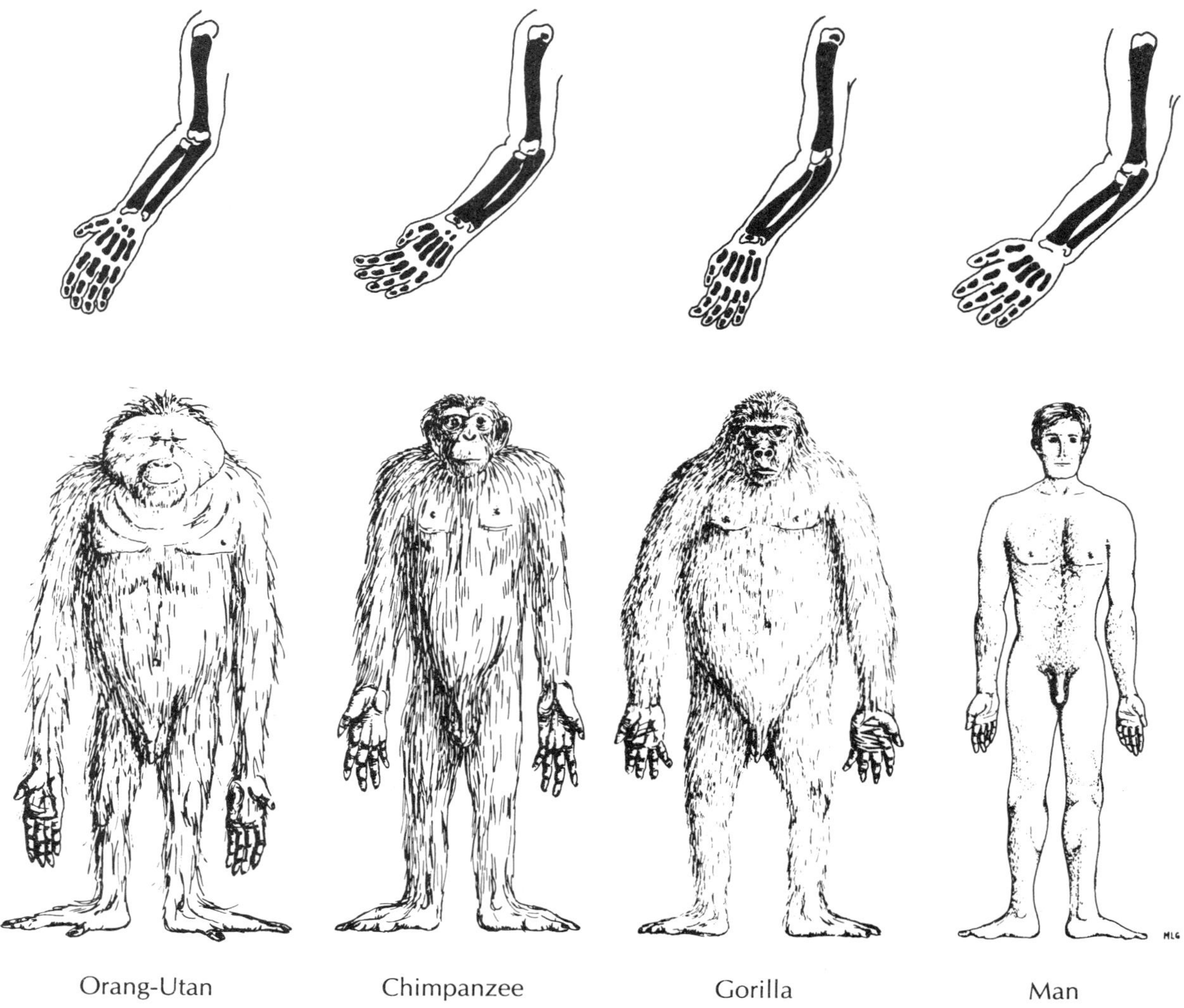

Fig. 2.6. Upper limbs of adult primates are far more different from each other than at birth (as seen from diagrams of radiographs, modified from Schultz).

well known among individual mammalian groups; each specialization results in metabolic and physiological adaptations with profound structural implications.

In contrast, the defining specializations of primates are generally rather few and comprise adaptations to a degree of arboreal life that nevertheless rarely prevents the species from moving on the ground when necessary. Indeed the very generality of the primates makes the provision of a morphological definition of the group rather difficult; the definition consists mainly of the absence of highly specialized traits that characterize other mammalian orders.

Primates contain no species, for instance, with feet highly adapted for cursorial locomotion through marked reduction in the numbers and sizes of digits, nor with dentitions highly modified by heavy grinding teeth for dealing with herbivorous diets, such as are found in the ungulates. In contradistinction to such a highly specialized group, most primates maintain a decidedly generalized limb construction preserving what is believed to be the primitive five-digit condition; and almost all have a relatively unspecialized dentition capable of dealing with many different kinds of food.

Notwithstanding this generality, however, there have been some specializations among the individual members of the Order. Five free digits have been reduced to four in the hands of a few members, such as spider monkeys and colobus monkeys, and this seems related to special arboreal functions of the hands of these creatures. One dentition, that of the aye-aye, has been so markedly modified that it includes extremely large ever-growing teeth somewhat like those of rodents and related to peculiarities of foraging and dietary habits whereby the animal is able to seek grubs hidden within tree trunks.

And a few typical primate specializations are recognizable; for example, reductions in apparatus for olfaction and increases in visual systems especially in relation to binocular and colour vision. And probably of greatest significance of all is the expansion and elaboration of the cerebral cortex and brain in general, evident from a very early stage in the evolution of the Order. Modifications of brains have gone very far in one member of the Order, the genus *Homo*.

The organismic approach

The general body of organismic data about the primates presents a pattern of associations that, with appropriate gaps and groupings, sees the living primates as a series of linearly related grades or levels of structure. Many prosimian forms are least like humans, the monkeys are intermediate, and, of course, the apes are most like us. In such studies it is always clearly presented that, structurally, humans are indeed situated at one extreme of the primates. This has been summarized in a diagram prepared after one by Professor Le Gros Clark (1959). It demonstrates an assessment about primate phylogeny based upon organismic information primarily from study of the teeth, jaws and crania (Fig. 2.7).

But there is a great deal of evidence stemming from many different anatomical systems that confirms this linear relationship among the various primates. For instance, the structures of primate hands and feet, when viewed in their entirety, suggest a similar linear grading of the primates. This is especially well seen from the studies of Professor Schultz and it is to his work we may turn for visual examples of the general picture. No one could, I think, disagree that this is the assessment presented by primate hands in Fig. 2.8. In a similar vein, such evidence stems from examination of the skeleton of the trunk, and even from such a restricted area as the form of the palate. Figs. 2.9, and 2.10 confirm these arrangements even although not all primates are represented in each of these examples.

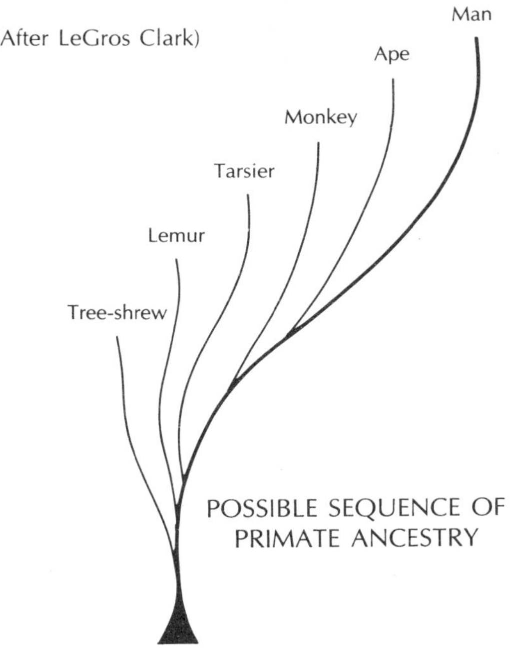

Fig. 2.7. Organismic studies suggest a linear arrangement of grades of primates from prosimians, through various monkeys, to apes and humans (modified after Clark).

This result is even evident when we look at primate brains, whether through examination of their relative size, a very crude measure, or whether we examine the more complicated kind of picture presented through studying their structure. Two examples are shown in Figs. 2.11 and 2.12 and again, although not all primates are represented, we can see that the overall generalization holds true.

In addition to the view of primate relationships that obtains from studying anatomical structure, a large enough number of organismic investigations have now been carried out on physiological processes of one kind and another that the view they represent is also evident.

For instance, studies of development have been undertaken over many years and the results have been summarized in recent reports by Luckett (1975). Whether we look at such an important physiological phenomenon as blastocyst implantation (Fig. 2.13) or whether we study a wide range of different reproductive features (Fig. 2.14) the general picture is one of a linear array of the primates. This array mirrors, largely, that stemming from the morphological data, with prosimians at one end, apes and humans at the other and monkeys lying between.

Fig. 2.8. Examination of the hands of primates demonstrates their linear arrangement from prosimians, through various monkeys, to apes and humans (modified after Schultz).

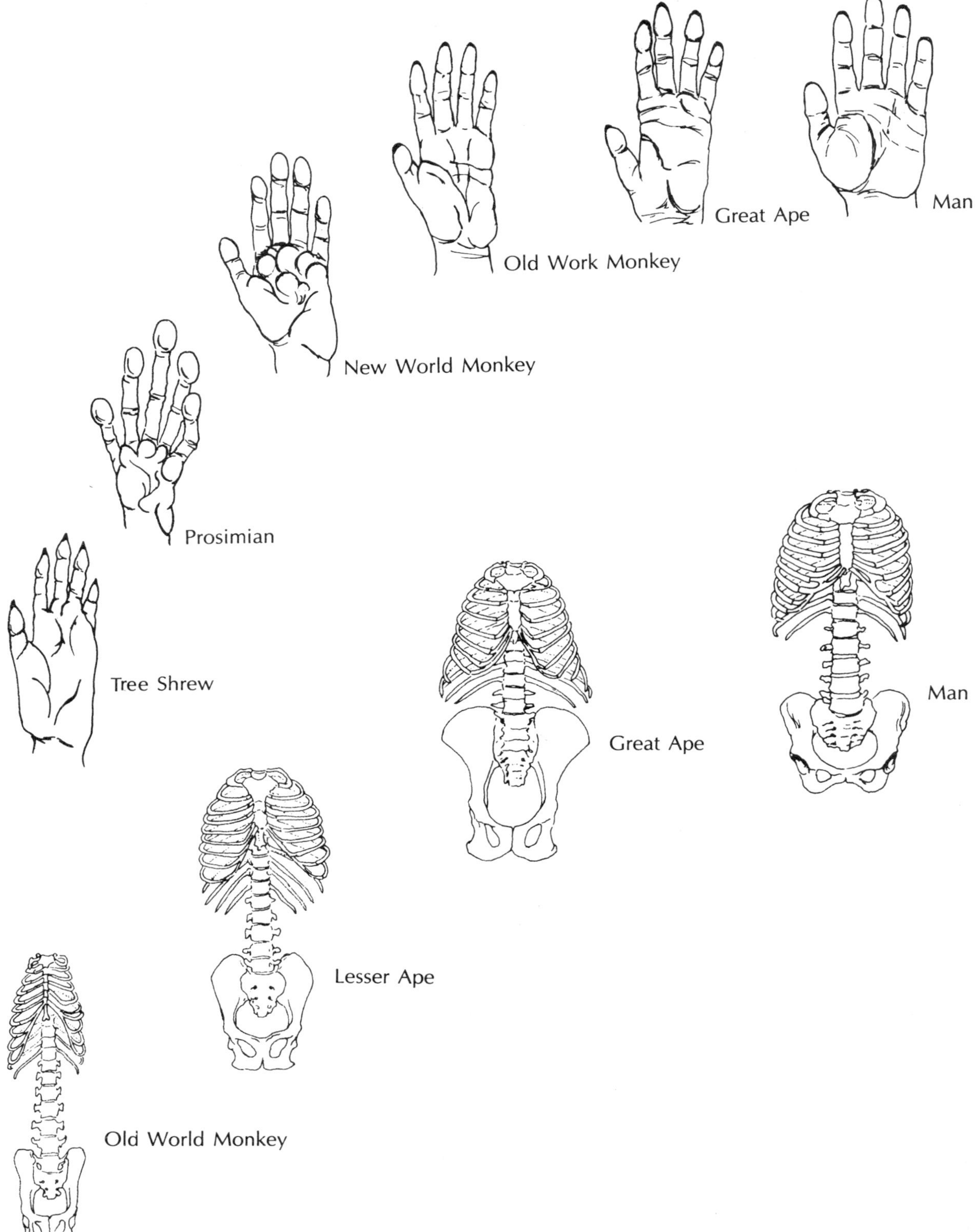

Fig. 2.9. Linear relationships in the skeleton of the trunk of primates (modified after Schultz).

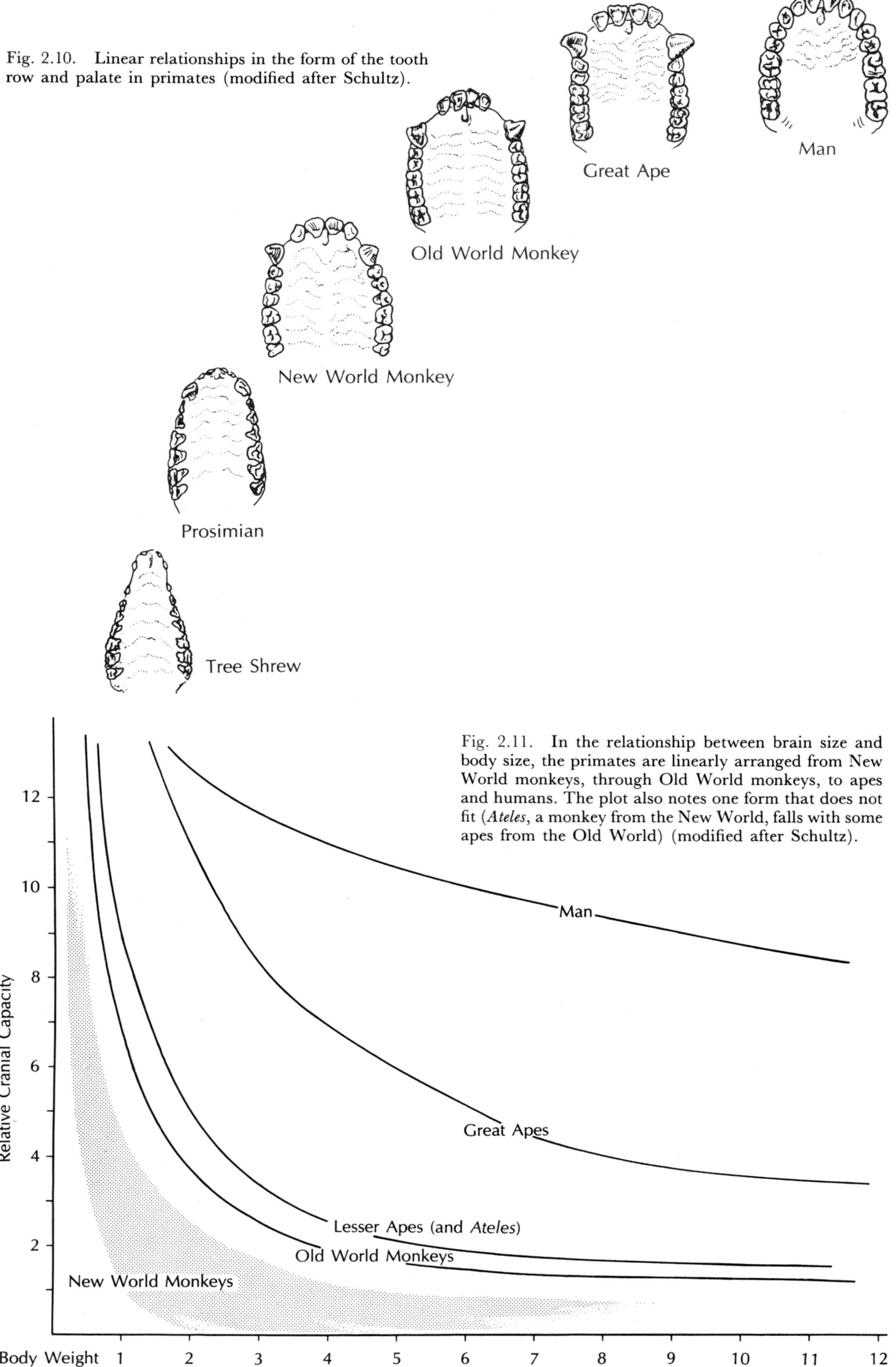

Fig. 2.10. Linear relationships in the form of the tooth row and palate in primates (modified after Schultz).

Fig. 2.11. In the relationship between brain size and body size, the primates are linearly arranged from New World monkeys, through Old World monkeys, to apes and humans. The plot also notes one form that does not fit (*Ateles*, a monkey from the New World, falls with some apes from the Old World) (modified after Schultz).

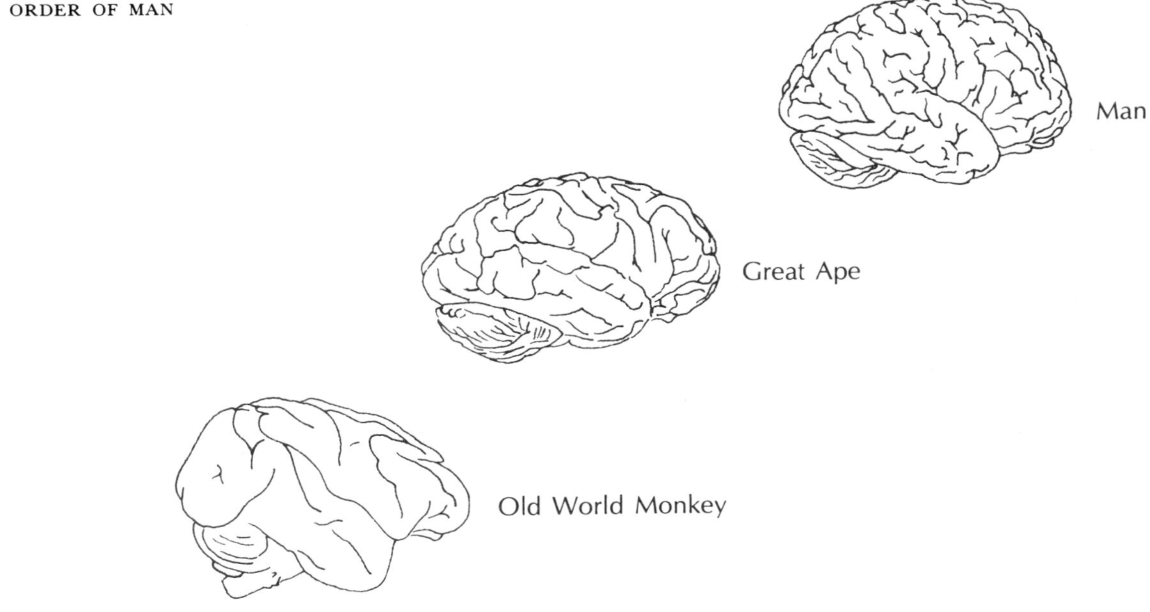

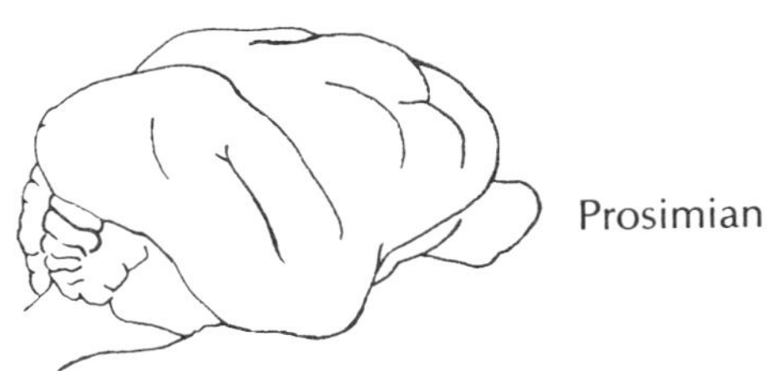

Fig. 2.12. The external surface of the brain shows increasing complexity in the linear series from prosimians, through monkeys, to apes and humans. Again, however, this feature is not so simple as it appears, brain complexity being related to absolute brain size (modified after Schultz).

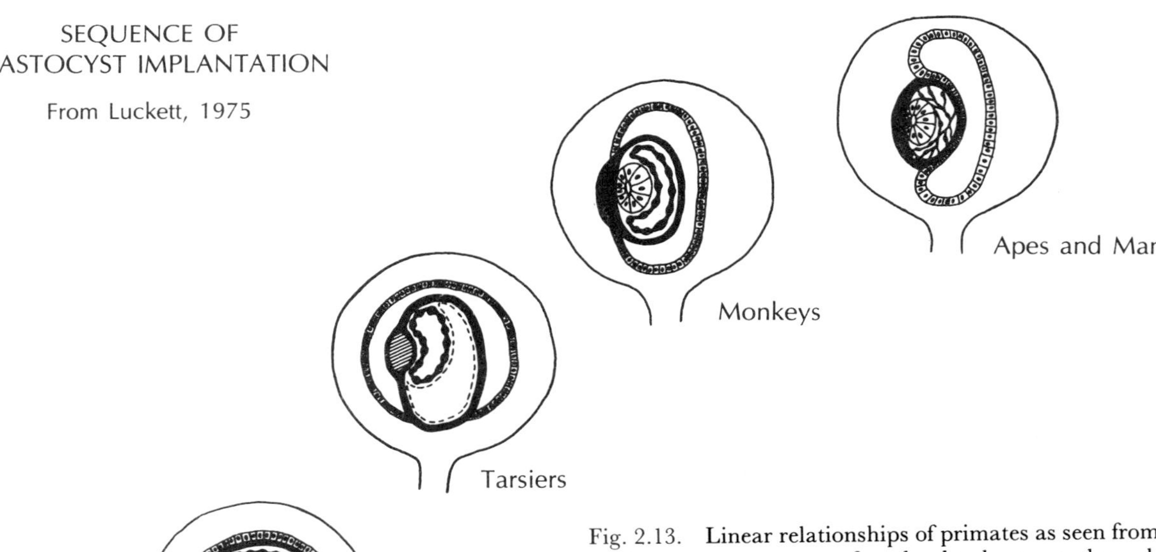

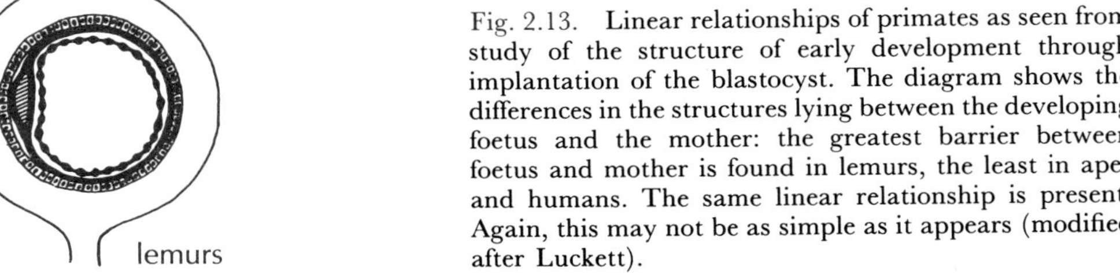

Fig. 2.13. Linear relationships of primates as seen from study of the structure of early development through implantation of the blastocyst. The diagram shows the differences in the structures lying between the developing foetus and the mother: the greatest barrier between foetus and mother is found in lemurs, the least in apes and humans. The same linear relationship is present. Again, this may not be as simple as it appears (modified after Luckett).

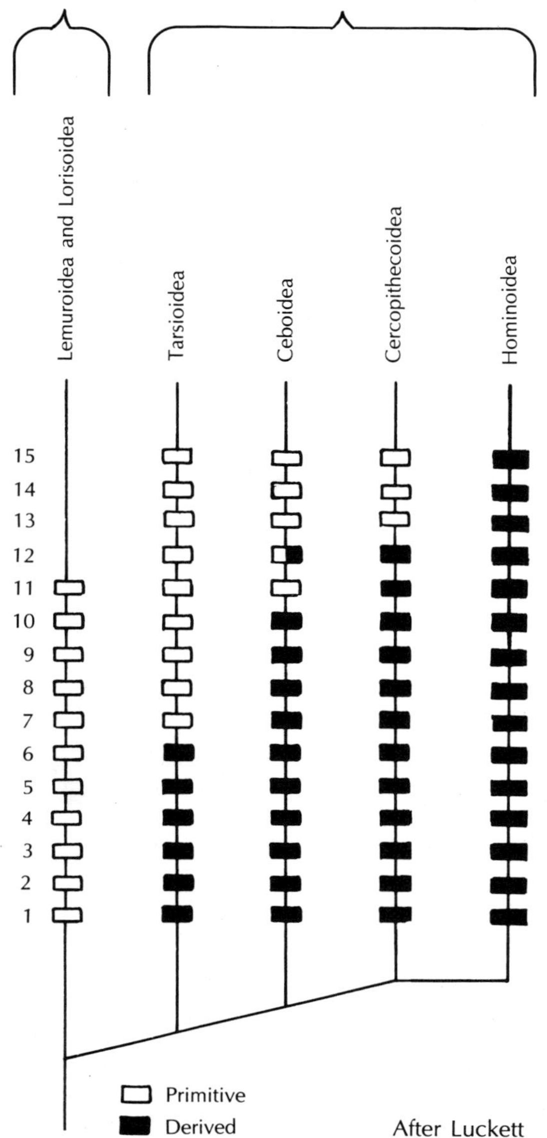

Fig. 2.14. Linear relationships among primates when an analysis is made of a wide variety of reproductive features in terms of the degree to which they are thought to be 'primitive' or 'derived' characteristics.

The features (numbered 1 to 15) include various structural states of the foetal membranes, and the full list is given in the original publication. Open squares represent primitive states of the foetal membrane characters, solid squares represent derived states of the same characters. There is an increasingly linear arrangement of shared derived states from lemurs through tarsiers, through New and Old World monkeys, to apes and humans (modified after Luckett).

This information from reproduction and prenatal development is fully confirmed when we come to look at postnatal development and growth. Here we can return to the studies of Professor A.H. Schultz (1969) for a view of the results and it too confirms the essentially linear array of the primates (Fig. 2.15).

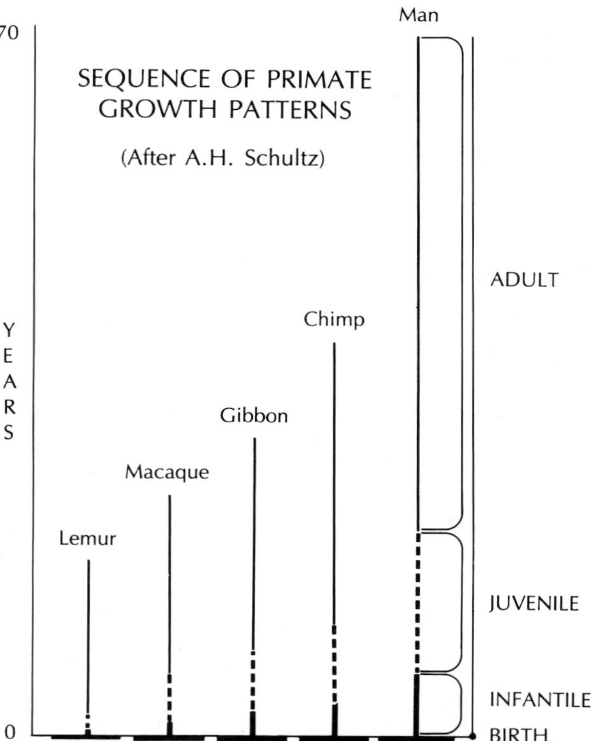

Fig. 2.15. Linear relationships, from prosimians, through monkeys, apes and humans as evidenced by primate growth patterns (modified after Schultz).

The biomolecular picture

However, the relationships of primates do not rest upon organismic (morphological and physiological) data alone. Whereas some organismic information has been known for centuries, data of a different type, relating to biomolecular entities, have become available in the last several decades (although it is only in the last two that the bulk of these data has loomed large enough to be used in a practical manner for helping delineate primate relationships, e.g. Goodman and Tashian, 1976).

There is little doubt that this information, in general, supports the classical picture to a quite remarkable degree. The evidence here consists of a wide range of types of information: biochemistry and serology, blood group distributions, chromosome analyses, protein sequencings, nucleotide changes, immunological tolerances and so on.

The general picture derived from such studies

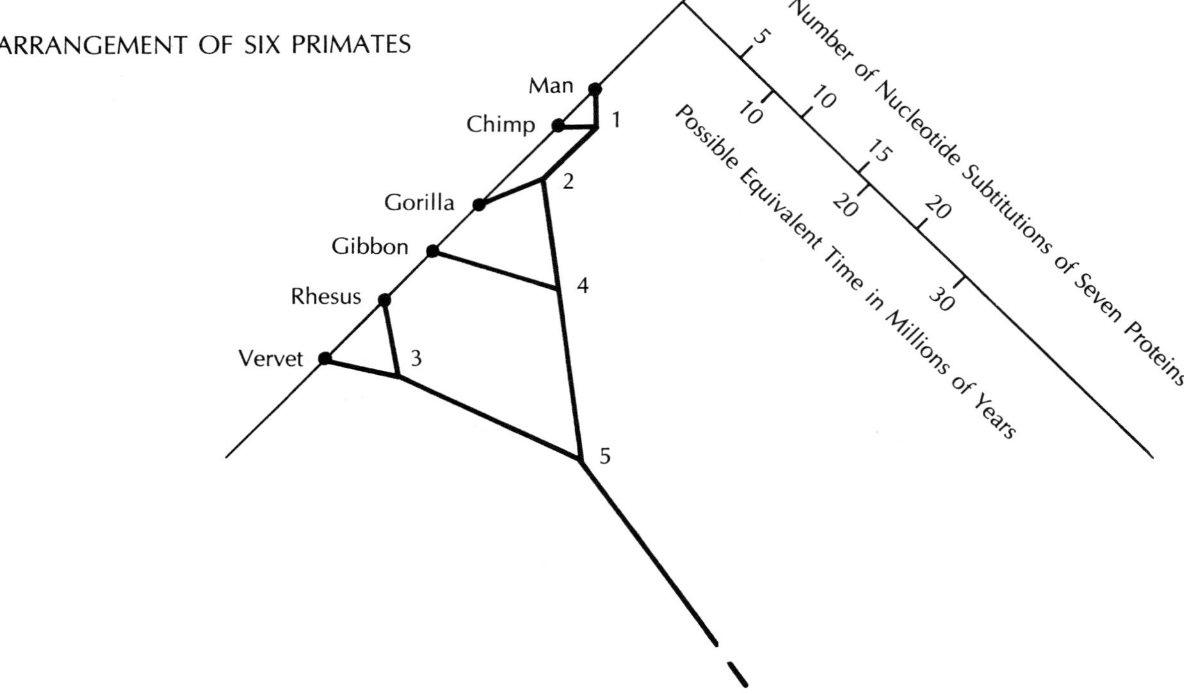

Fig. 2.16. Arrangement of six primates by study of nucleotide substitutions of seven proteins results in the same linear relationship from monkeys, through lesser apes, through great apes, to humans. This plot, normally presented as a tree diagram within x and y axes is here rotated in order to emphasize the linear relationships that are the same as in the previous figures. No other changes have been made (modified after Fitch and Langley).

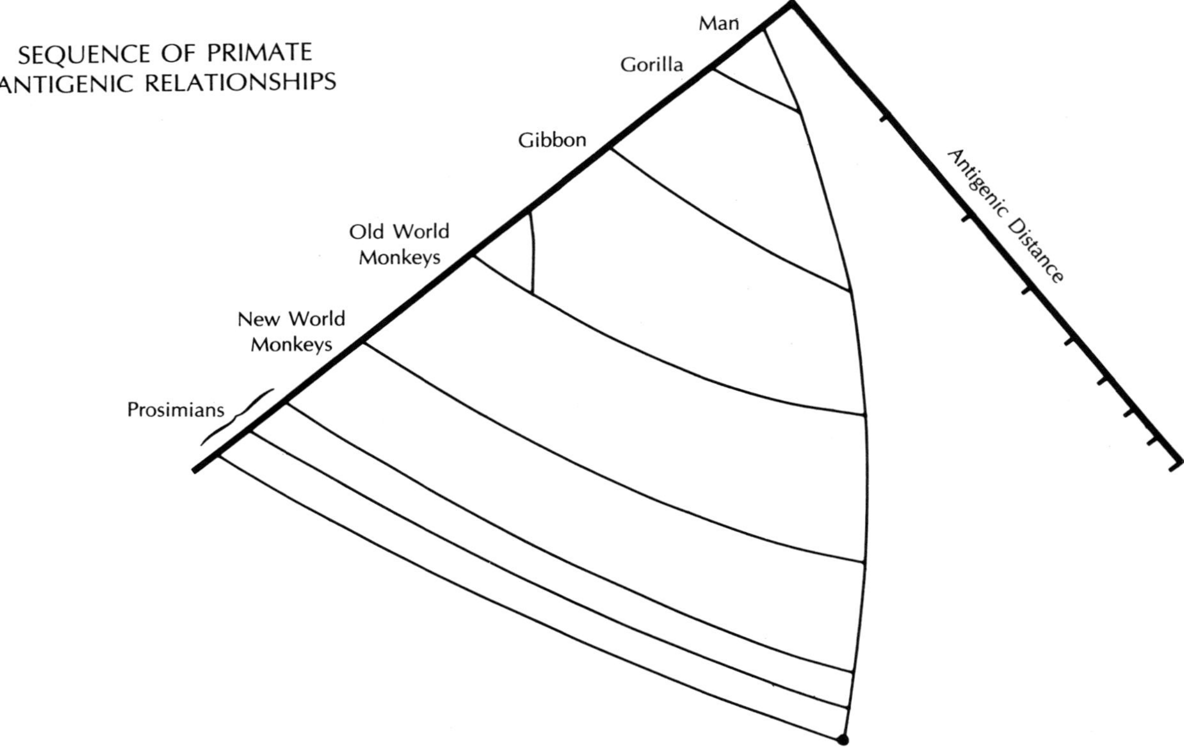

Fig. 2.17. A similar arrangement of the primates is revealed by some immunogenetic sequences reported by Morris Goodman. Again, the plot is rotated to emphasize the similarity with the previous figures (modified after Goodman).

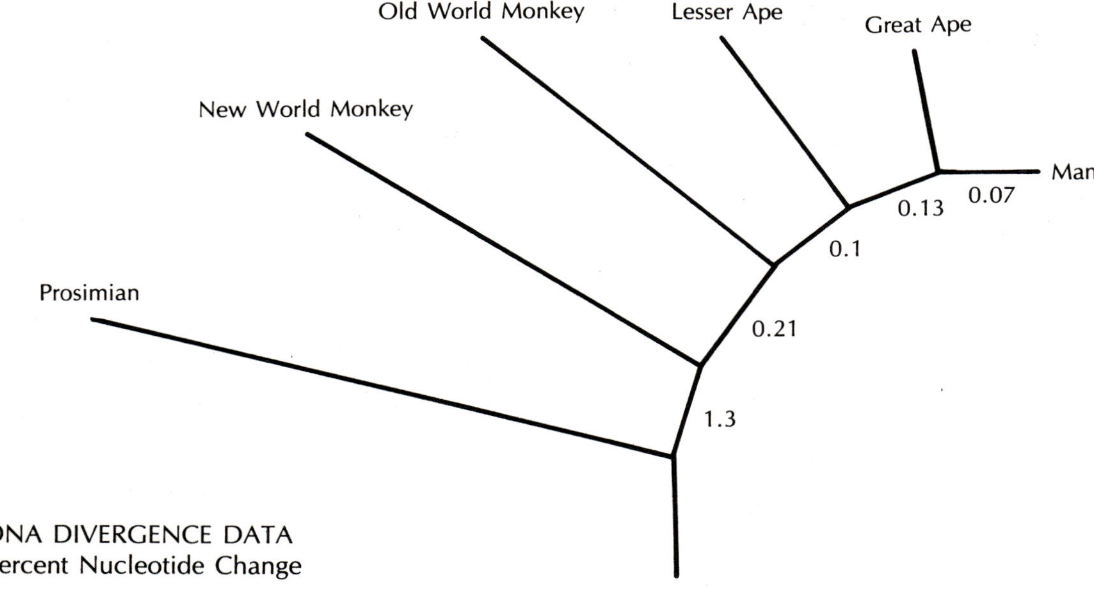

Fig. 2.18. DNA divergence data of Kohne further confirm the linear relationship. Again, the diagram is modified from that actually published by Kohne so that the linear arrangement is visually emphasized. But no other changes have been made in the information contained within Kohne's figure.

can be summarized in Figs. 2.16 and 2.17 which show relationships of primates as defined through the numbers of nucleotide substitutions of seven proteins (Fitch and Langley, 1976; Fig. 2.16) and through some antigenic distances discovered by Professor Morris Goodman (1976; Fig. 2.17). The picture is further confirmed through study of the amino acids of gamma chains of foetal haemoglobin (Fitch, 1977) and by results of percentage nucleotide changes from DNA divergence data (Kohne, 1975; Fig. 2.18). (In each diagram the results of these investigators have been rotated so that comparison with the oblique linear arrangements of the prior diagrams is more obvious).

It is remarkable how closely these different biomolecular relationships parallel those summarized in the organismic series. Species are arranged in a linear sequence, with some gaps to be sure; in particular, humans lie at one end of this sequence, an end shared, as before, with the apes. Prosimians are at the other extreme. New World monkeys lie closer to prosimians, Old World monkeys closer to apes. Many other examples of investigations of such biomolecular data provide generally similar arrangements, although it is true that, as with morphology and physiology, there are some detailed points of controversy.

Points of controversy

Of course, these two pictures, the organismic and the biomolecular, are not identical. What are the points of difference? There are indeed some and they are concentrated in particular places.

For instance, the biomolecular techniques are more likely to place humans together with chimpanzees and gorillas as being more closely related to each other as compared with the Asiatic apes. This contrasts with most morphological studies which place the Asiatic great apes with the African apes as pongids, humans separate from them at one extreme as hominids, and gibbons and siamangs separate at the other as hylobatids.

Similarly, the biomolecular techniques usually produce different sub-groupings of Old and New World monkeys as compared with those traditionally arrived at from the standpoint of most organismic studies that recognize the families, subfamilies, genera and many species of monkeys of the present day.

Even among the Prosimii problems arise. Perhaps the most prominent stem from the consideration of tarsiers and tree-shrews. Most organismic assessments place these two genera among the prosimian primates. But biomolecular studies usually place the spectral tarsier as a primate either in a group of its own of equal weight to the traditional Prosimii and Anthropoidea, or even as a member of the Anthropoidea itself. And many modern investigations deny the tree-shrew a place within the Primates at all.

Yet it should not be thought that there is a total dichotomy between the older organismic and the newer biomolecular pictures.

There are at least some studies in which, in particular controversies, biomolecular evidence supports the traditional assessment derived from whole organisms. One example is the classical arrangement of the orang-utan with the African great apes; this is supported by particular electrophoretic studies of 23 genetically controlled proteins in the Hominoidea (Bruce and Ayala, 1978).

And there are other studies where the opposite is the case, where the organismic data have predicted or supported new assessments derived from biomolecular studies. An example of this is the case of *Tarsius*; the classical picture assesses *Tarsius* as a prosimian; the new view stems from a variety of biomolecular evidences and is that *Tarsius* can best be thought of as a member of the Anthropoidea (e.g. Dene, Goodman, Prychodko and Moore, 1976). New organismic evidence from a number of sources (e.g. the structure of the nasal fossa, Cave, 1973, the anatomical relationships of the base of the skull, Szalay, 1975a, the structure of the placenta and related structures, Luckett, 1977, and morphometric studies of overall bodily proportions of primates, Oxnard, 1978b) are all organismic investigations that support the unconventional, biomolecular view.

Interpolating fossils: the piecemeal approach

Notwithstanding these points of controversy, there is, as I have said, enormous agreement about the relationships of the living primates. But only a smaller measure of agreement is evident in the consideration of fossil primates. This is legitimate because, for the majority of fossil primates, there is a relative paucity of information as compared with vast bodies of data for fossil representatives of many other vertebrate groups.

When it comes to judging fossil data, it is not possible to include them in most of the above ways of arranging the primates. Of the various items useful in making judgements on the basis of organismic methods, mostly only small parts – teeth, jaws and crania – have been available in the past. From the viewpoint of physiological or biomolecular investigations, little or no information whatsoever is known for most fossil materials. On the other hand, fossil materials do provide some data about time and to this extent new information is available which is of value in discussing primate relationships.

But the new information that is now becoming available is the result of the discovery of many more fossil fragments of the post-cranial skeleton. This means that we must now clearly understand the individual post-cranial parts in living primates before we can assess the fossils. In the first instance, therefore, the approach is piecemeal. The fact that so many classical studies of living species indicate a linear arrangement of the primates (Figs. 2.7 through 2.18) suggests that new studies of individual post-cranial anatomical regions should, likewise, present a linear arrangement. But this cannot be assumed to be the case; as later chapters will show, we need to keep open minds. For it is possible that analyses of anatomical parts may not always mirror the linear spread of whole primates.

We can easily see this by consideration of even those classical characters that we have already examined. For although, in general, there are indeed features of the comparisons of hands and trunks, of palates even of brains and placentae, that support the idea of a linear arrangement of the primates, we already know enough of the details of these various anatomical regions to see that the picture is actually more complicated. For example, biased selection of particular hands or particular trunks could render totally different verdicts. That we reject these verdicts stems from our overall knowledge of the forms. Yet the existence of examples like these demonstrates that in each case the situation is complicated. So complicated is it that, though our eyes may see the linear arrangement as the great generality, they may, in fact, be blind to more complex patterns of individual anatomies.

Interpolating fossils: studies of the whole

Once examinations of individual regions have been achieved, it is then possible for us to add the data from different anatomical regions together; but of course this must be done in such a way that information from one worker about one region is not used to help to make judgements about a second region when data from that second region has already been used to help make the judgement about the first. Mutually spurious reinforcement is always possible. The circularity in reasoning is not always obvious.

Moreover, when we add together osteological data from many different anatomical regions in order to make judgements about fossils, it is important that we should be aware of the consequences of what we do. Such a test of the consequences can be made by attempting such additions for living species alone where the results can be compared with the vast body of other data that exists. This allows us to test the result without, at the same time, compromising the fossil answer.

The fact that studies of small anatomical parts of extant species appear to confirm the linear arrangement of the primates (Figs. 2.7 through 2.18) suggests that the addition of such regions to one another should also present a similar, perhaps much clearer linear arrangement. But this also cannot be assumed to be the case. For each anatomical part, the additions bring in new information, the interactions between the parts being added. This is an expression of the notion that the whole is greater than the sum of its parts. It is at least theoretically possible that these additions could produce views at apparent variance with the information from each separate part. As we shall see later, this is the practical finding.

The new methods

In order, therefore, to investigate these aspects of form and pattern among the primates, it is necessary to use methods that are capable of finer discrimination and different insights. The methods depend upon advances of two kinds.

One is the ability to study biological structures so that more of the information contained within them can be revealed to allow better comparisons than previously. This requires quantitative and holistic handling of biological shapes and patterns so that shades of difference between shapes may be more readily perceived and so that variation and covariation among shapes may be taken into account.

The other, as we shall see, is the realization that, for many structures information is as much to be inferred about certain aspects of their function as about their immediate hereditary associations.

The application of ideas such as these is, thus, a most important part of the study of the relationships of the human place among the living primates. It is this relationship that forms the broad pattern against which the details of particular anatomical parts in each fossil must be compared. It is the information from this process that is providing the new ideas now being corroborated by the discovery of the new fossils. And it is to a study of these methods and the results that flow from their use that the rest of this book is devoted.

CHAPTER 3

Mathematical 'Dissection' of Anatomies

Abstract. In this chapter we review a variety of methods for describing structural differences between organisms. Assessment by the human eye and analysis by the mental acuity of the observer are briefly considered, as are some of the deficits of this approach. Structural analysis is considered first theoretically and various problems noted. The question of the very existence of such problems in practice in biology and anthropology is confirmed by examples stemming from studies of the Primates.

We then move to methods that involve simplifying visual assessment by reducing it to measurement. Such methods often involve, however, more complex analyses using computational methods of one kind or another; one of the most well-developed of such analytical approaches is the multivariate statistic. Some of the deficiencies of measurement and analysis are reviewed.

This discussion leads us to consider methods (as the observational) that depend upon visual data, but which apply methods of analysis able to obtain information from a picture over and above that available to the eye and mind. A few of these methods have been used to investigate real biological problems; many have so far only been used as exemplars of what may be done in the future.

Finally, we return to the best used of the newer techniques, multivariate statistical analysis, and discuss in some further detail problems in its usage, including the importance of the testing of all methods, whether complex or simple.

Introduction

Understanding anatomical fragments in the evolutionary context depends upon *first* obtaining information about the structural differences that truly exist and *second* attempting to make judgements about the biological meaning of the discovered differences. Classically, when assessment by the human eye and judgement by the human mind are the main tools involved, these two phases may not appear to be clearly separated from one another; the entire procedure may be done in one intuitive leap, as it were, a method that is nevertheless rather powerful. But as other techniques for describing and discovering structure are added to the powers of human observation, and as a series of different arguments at many different levels becomes part of the mental process of the biological judgement, so it becomes more and more important that the logic of the two phases be separate, or where not separate at least clearly identifiable within the overall process. It is the first of these two, describing structures, that forms the subject of this chapter.

Discovering structural differences: simple observation

The traditional method of assessing anatomical fragments has, on the whole, been visual, sometimes aided, it is true, by measurement to assess overall sizes of specimens and for use in taxonomic 'keys'. Such findings as have been made thereby have provided major information over the years, over the decades, indeed over the centuries; but such studies result nowadays in little more than the closure of relatively minor gaps in already fairly well-known patterns, rather than in the creation of new vistas. (We must draw the caveat that comparative studies of joints and comparative histological investigations are almost untouched fields). However, apart from such exceptions, there is little doubt that using dissection and observation prevents researchers from trying new methods. Both of these techniques are time-consuming and laborious, yet neither can be relegated to technical help; both require the complete attention, the experience and the expertise of the investigator alone and leave little time for other experimentation.

Observation is also often preferred because some of the newer analytical methods are thought to be excessively laborious; indeed, the earlier analytical work was. It was not so very long ago that an investigator might spend months doing 't' tests and regressions on mechanical calculating machines. Much of this older analysis is rejected by some workers as mere number-grinding; indeed some anatomists and anthropologists still look upon quantitative morphological studies as extravagant expenditure of time and energy, often for results that seem to them to add little to our knowledge.

There can be no doubt that criticisms like this are superficial; without the earlier laborious but pioneering studies, often rendered more difficult by the lack of techniques and equipment nowadays regarded as indispensable, it would not be possible now to go beyond the confines provided by interpretation based upon personal observation and dissection. For it is only the use of available formulae to the extreme of their capabilities that confers upon investigators the competence to help propose features and criteria for the development of yet better approaches.

But one of the chief stumbling blocks in the adoption of newer techniques is the belief that they are difficult to understand. Yet if physicists can explain to an educated public about quarks and black holes, then anthropologists ought to be able to explain the application of mathematical, physical and engineering tools in the study of human structure.

The earlier and simpler method of observation of anatomies has enormous strengths and these rest in the fine powers of recognition and discrimination that are shown by the human eye and mind. But it also has weaknesses and these reside especially in the following matters.

The human eye has difficulty when faced with voluminous data (difficulties that result from the many specimens and many observations taken upon each specimen, specimens arranged in many groups, and the complicated interactions among observations, specimens and groups). The visual method is not good at assessing variation among specimens and groups; that is better done by quantification and statistical evaluation. Simple observation is unable to provide assessments of the more complicated interrelationships of data such as association and regression between features, auto-correlation and cross-correlation among observations and so on. And finally, although visual assessment is often fairly good at recognizing differences between discrete groups, it is much less well able to recognize situations where data are arranged in more or less continuous fashion, where there are overlaps between groups, where groups may have complex shapes, or where other even more complicated data interrelationships may exist.

The concept of a group: of data or of organisms

We are all accustomed to the concept of a group in biology, whether we are discussing a set of organisms, or a suite of data. The human eye can readily pick out groups in two-dimensional views of animals, or in two-dimensional plots of data. We also easily recognize some of the defining measures of a group, for instance, the mean and the standard deviation. Of course, the distributions of cases within groups may not be statistically *normal* but we usually tacitly take little notice of this. Fig. 3.1 reminds us that perhaps we should not take such perturbations from *normality* for granted (Duda and Hart, 1973). The figure shows four different samples of two-dimensional data: one sample is a single spherical group; one consists of two elliptical sub-groups; one comprises two rectangular sub-groups; and one is a single doughnut-shaped group. These totally different groups have a special similarity. Each has been constructed so that the calculated mean and variance is identical. We would never mistake any one of these for any other, given that we actually inspected the two-dimensional picture. Yet if we suppose that these similarities existed in a multi-dimensional data space, composed perhaps

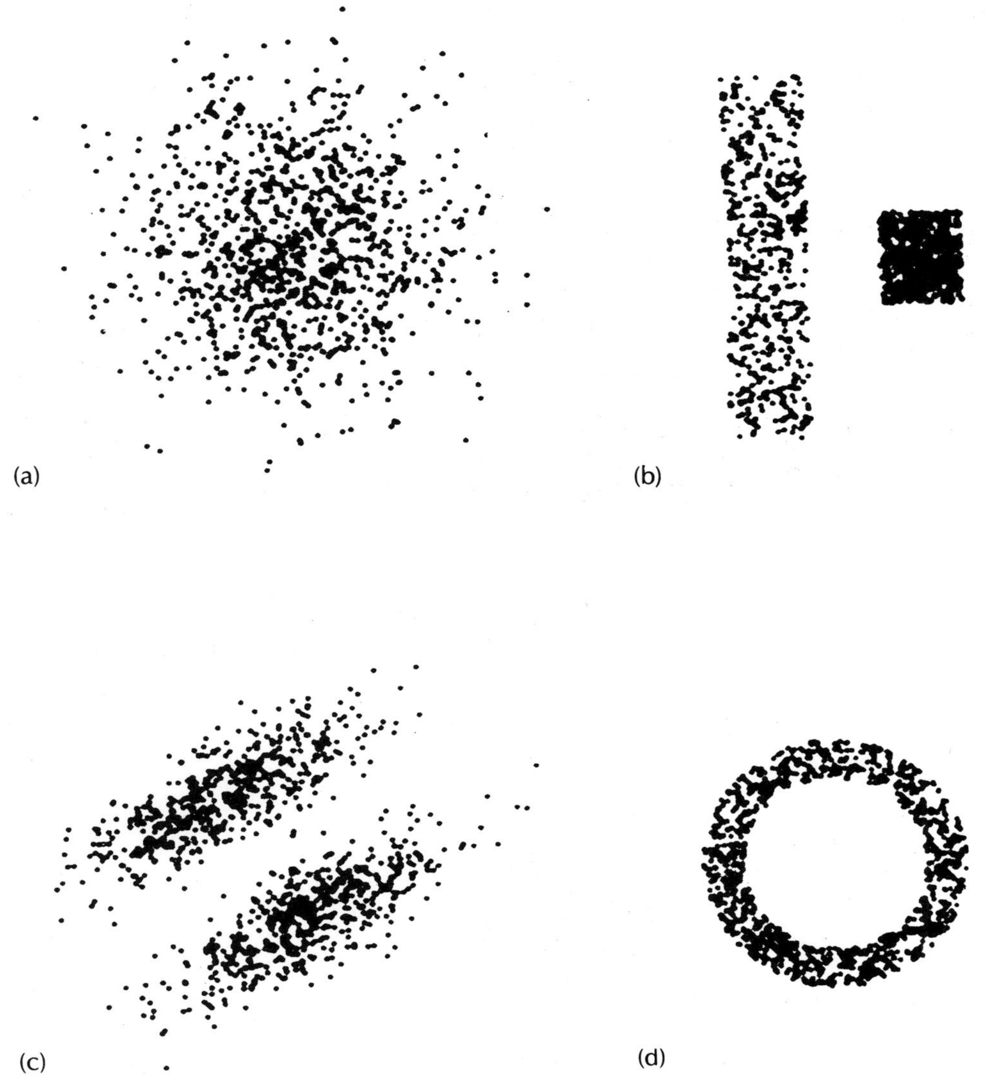

Fig. 3.1. Four sets of data, each represented by two measures and plotted within the two dimensions of the x and y directions of the figures. Each has been specially generated so as to have the same mean value and the same standard deviation (after Duda and Hart, 1973). It is the fact that the dimensionality is only two that allows us to 'see' that these groups are different even when the statistical parameters are the same. If the data had been multi-dimensional (x, y, z and yet other variables) it would not have been so easy for us to check the statistical result by 'eye'.

of the many anatomical attributes of organisms, I wonder if it would be as easy for us to 'see' the differences.

We are also accustomed in biology to the notion of fitting a curve to a group of points in a sample. Again, we are all aware that such fitting is merely one of an infinite number of other fits that may be made. But we often tacitly assume that the simple fit is the correct fit. Again, Fig. 3.2 reminds us that this may not always be the case (Duda and Hart, 1973). The figure shows a straight line, a parabola and a tenth-degree polynomial, each represented by only five data points. The parabola has been used to generate the five points so that, theoretically, it is the right 'fit' to the group. The tenth degree polynomial has been chosen to pass through each point perfectly. Yet it is easy for us to 'see' only the 'fitted' straight line (which is the least perfect) and

Fig. 3.2. A single group of five points again plotted as x and y values (such as weight against height, or sepal width against petal width) within the frame of the diagram. Some curves that may be thought to represent them have also been drawn (after Duda and Hart, 1973). It is the fact that the dimensionality is only two that allows us to 'see' that the polynomial fit is the best one even though the statistical best fit is the simpler curve. And, likewise, had the data been in more than three dimensions, we could not have as easily checked the finding by 'eye'.

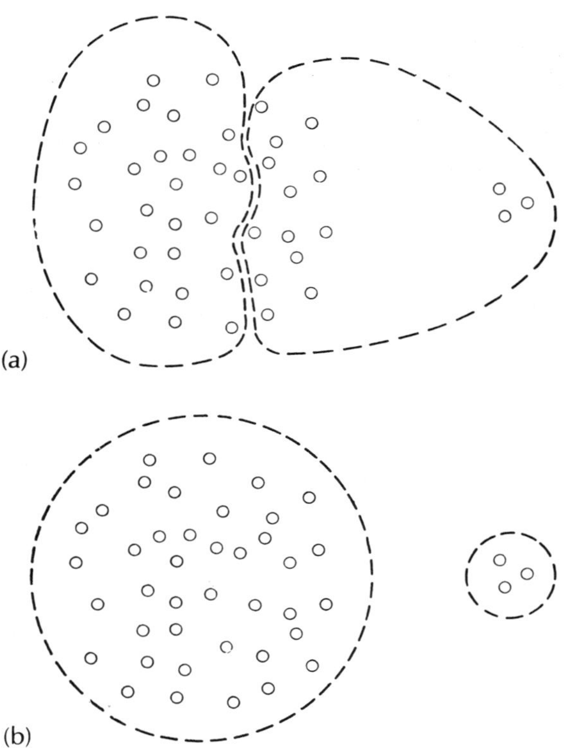

Fig. 3.3. Two clusters of data, one small and one very large, plotted in the x and y directions of the frame of the diagram. These might represent the plots of jaw length against jaw width for two organisms of very different variability. The diagram shows two ways in which they may be separated as described in the text (after Duda and Hart, 1973). It is the fact that the dimensionality is only two that allows us to 'see' that the first separation is not good. Had the dimensionality been higher we could not have checked the result in this way.

to be blind to the others.

The human eye and mind have a very limited view of a group unless forewarned.

The concept of several groups: of data or of organisms

Many of the major questions underlying the analysis of form lie in the definition of multiple groups and in the discovery of the relationships among them. In making such studies, we usually assume, tacitly, that our *a priori* defined groups are roughly circular or elliptical, and of approximately equal size and orientation, because this is what we most easily imagine and recognize. Fig. 3.3 reminds us of the trap that we may run into. Here are two circular groups of markedly different size. The separation shown above is the one that is statistically good (in the sense of having the least squared error). But visual inspection, in this case easy because the data are two-dimensional, indicates that the lower solution is the more correct. It would not be so simple for the human mind to recognize this situation in the multi-dimensional case.

This problem is further enhanced when groups become markedly non-circular or non-elliptical. In Fig. 3.4 the upper frame shows two separate circular clusters, the intermediate frame a single cluster that is dumb-bell shaped, the lower frame two parallel linear clusters. These are the judgements arrived at by visual inspection. But Fig. 3.5 demonstrates that some methods of analysis (such as furthest neighbour classification) may see the clusters differently. Thus, when the clusters are approximately circular, standard analyses discover the correct groups (upper frame), but such methods may not find the dumb-bell shaped group (middle frame), finding instead two approximately circular

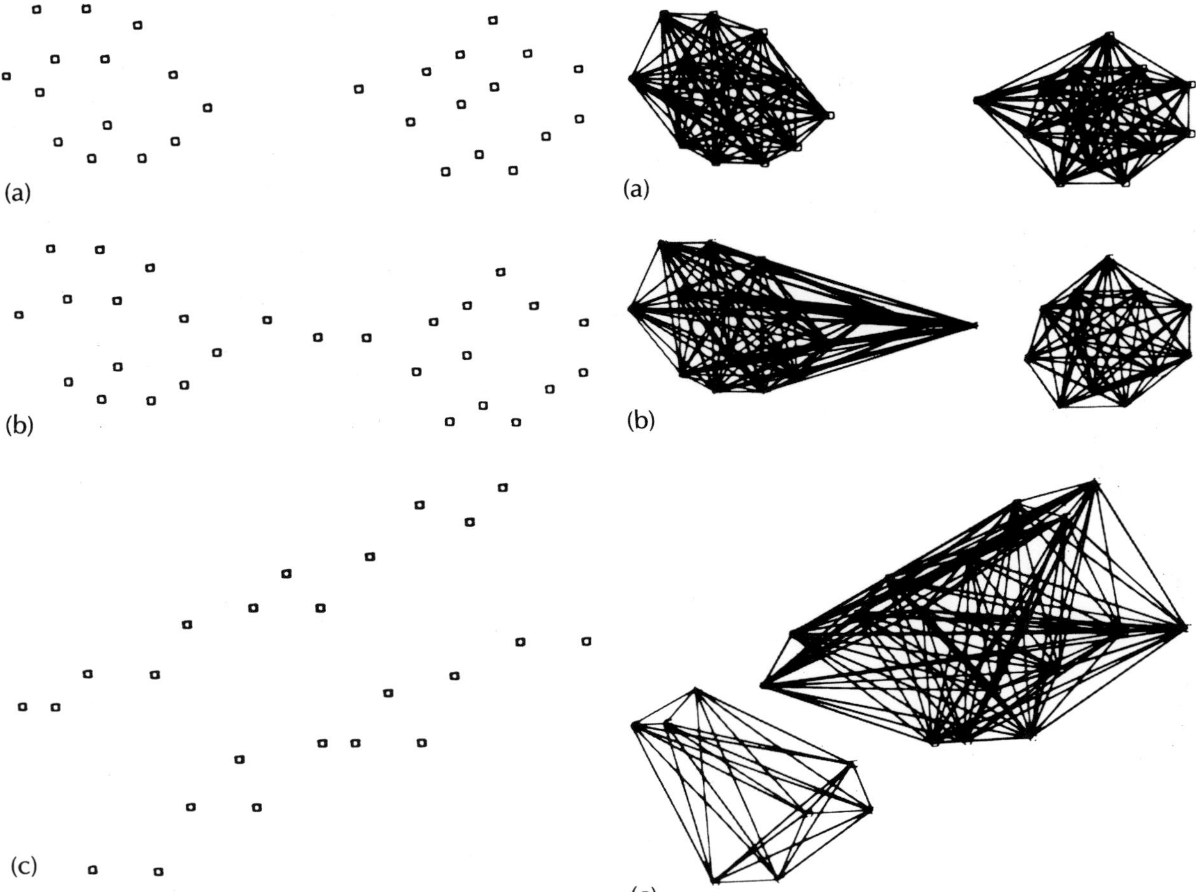

Fig. 3.4. Artificial data, plotted in the x and y directions of the frame of the diagrams, showing, to our eyes, two roughly circular groups (top), a single dumb-bell shaped group (middle) and two linear parallel groups (bottom; after Duda and Hart, 1973).

Fig. 3.5. The application to the data of Fig. 3.4 of a group-finding algorithm. The upper frame shows that two circular groups are readily found. The middle frame shows that a single dumb-bell shaped group is 'discovered' to be two approximately equal circular groups. The bottom frame demonstrates that the two linear parallel groups are 'found' to be two quite different unequal approximately circular groups. Other cluster finding methods perform differently. Because the data are two-dimensional, it is easy for us to 'see' the differences between the visual and computational results; had the data been in more than two dimensions, however, it would have been very difficult for us to have known of the first set of possibilities (after Duda and Hart, 1973).

groups; and they may perform especially badly with the two parallel linear groups (lower frame), finding instead two somewhat compacted, approximately circular groups. The tendency to delineate circular groups in preference to others is a distinct disadvantage of some methods. And when data are more complex than a simple two-dimensional plot, this is a disadvantage that applies especially to visual assessment. It is at least possible that groups may be joined to one another by bridges, that groups may be separated from one another by chasms, and that groups may be of quite unequal size and shape.

Yet another such problem is shown by Fig. 3.6. The upper frame shows a single group with a very dense centre and a very diffuse periphery. Or does it show two groups, one small and dense superimposed upon another large and diffuse? At least one clustering method produces yet a third possibility from those theoretical data for us: the star-shaped or 'crab-like' figure of the lower frame of Fig. 3.6.

The concept of interfaces between groups

There may, furthermore, be problems in the nature of interfaces between groups rather than in the nature of the groups themselves. Thus Fig. 3.7

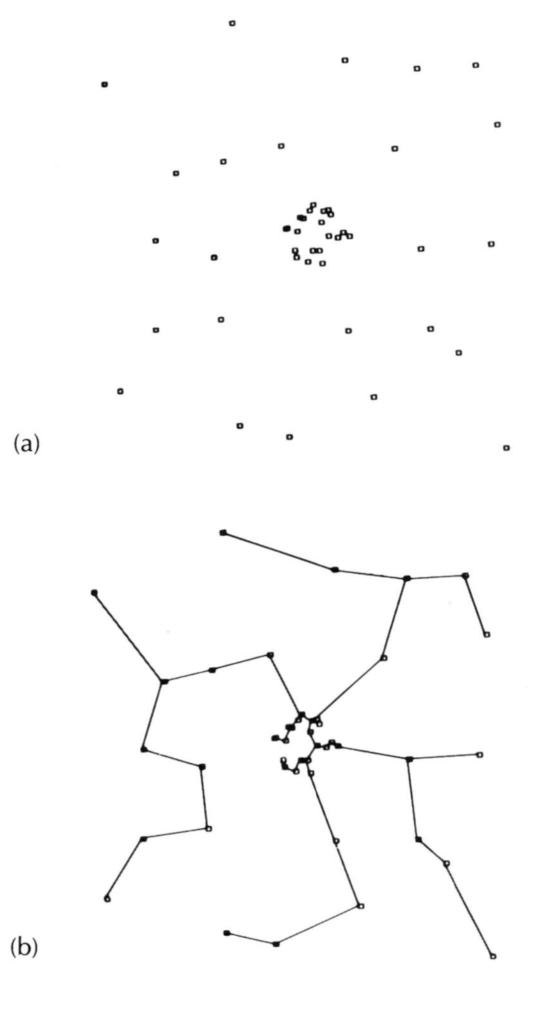

demonstrates a separation between groups that progresses from a relatively simple, linear interface, through a more complex curved one, to one in which there are two phases, as in the admixture of oil and water — two immiscible liquids. How are we to recognize when such situations exist?

Fig. 3.8 demonstrates another possibility in the nature of the boundaries between groups. A linear separation may lead from a simple boundary, through a contorted boundary as before, to a solution-like relationship, as between two miscible liquids such as water and alcohol.

Yet a third type of boundary between groups is shown in Fig. 3.9. Here the differences among the groups ('sixes', 'fives', 'threes' and 'eights') are not represented by the distances in the diagram but by the changes in the nature of the individuals of the

Fig. 3.6. The top frame shows a set of data plotted in the x and y axes of the frame of the diagram. We may well think *either* that this represents a single group with a tight centre and dispersed periphery *or* two groups, one very tight superimposed upon one very loose. At least one group-finding algorithm provides a picture suggesting yet a third, star-like arrangement (lower frame; after Duda and Hart, 1973). Once again, these possibilities are easily recognized visually because there are only two variables, x and y. Had the data been represented by several variables, with the plots therefore in a several variable space, we could not have 'seen' these differences nearly so easily.

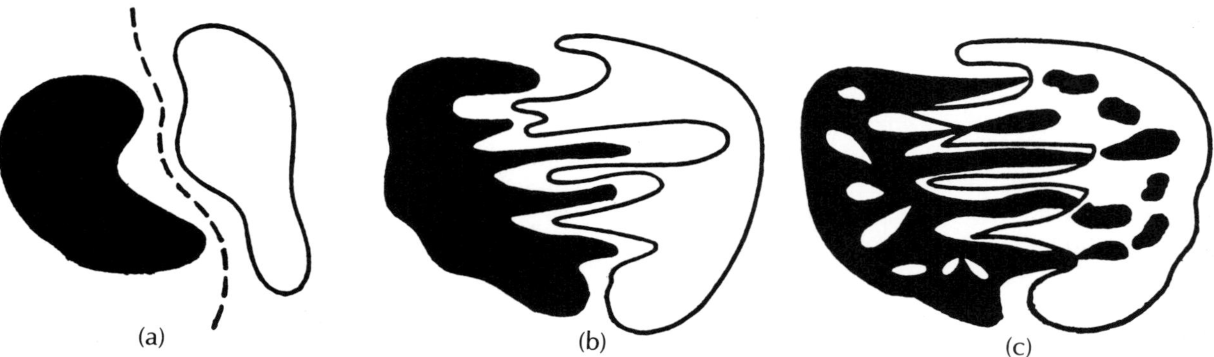

Fig. 3.7. Possible interfaces between two groups represented by two variables plotted along the directions of the axes of the diagram; left frame: a simple linear interface; middle frame: an interdigitated interface; right frame: complete intermixture but with recognition of the two phases as in the separations between two immiscible liquids such as oil and water (redrawn after J. Cowan, personal communication).

The first interface is of a type with which we are familiar in anthropology, e.g. the interface between, shall we say, two races separated by a geographic barrier. The second interface is not so familiar and we therefore look for it far less frequently: it might be found in a situation like the differences between gait patterns of different mammals (see Hildebrand, 1967). The third type of interface is even less familiar to us, and it is, consequently, even more difficult to provide an example, but some differences between human racial groups might be of this form. In many dimensions such possibilities would indeed be very hard to recognize.

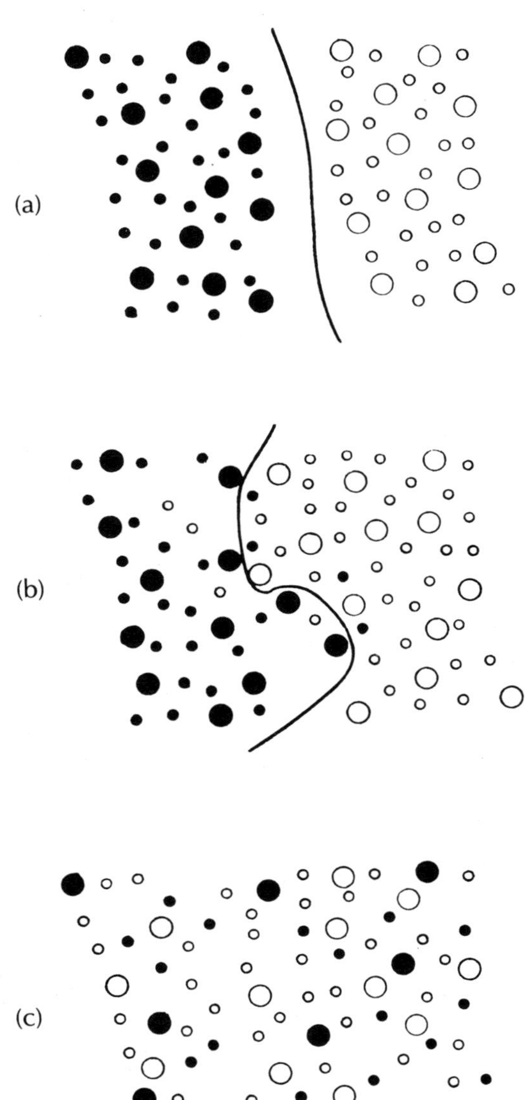

Fig. 3.8. Another set of possible interfaces between two groups plotted on the basis of two variables, x and y, along the edges of the picture; upper frame: a simple linear interface with intermediate forms; middle frame: a curvilinear interface with intermediate forms; lower frame: an intermixture something like that between two soluble liquids with intermediate forms (redrawn after J. Cowan, personal communication).

The interfaces in the first two frames are not unlike those in the previous figure, though there are some differences. The interface in the third frame is completely different. Again, we rarely think of such a complex possibility; once we have realized that it may exist, however, we might again look towards some of the complex relationships between human racial groups for real examples. In many dimensions such possibilities would indeed be hard to recognize using visual methods.

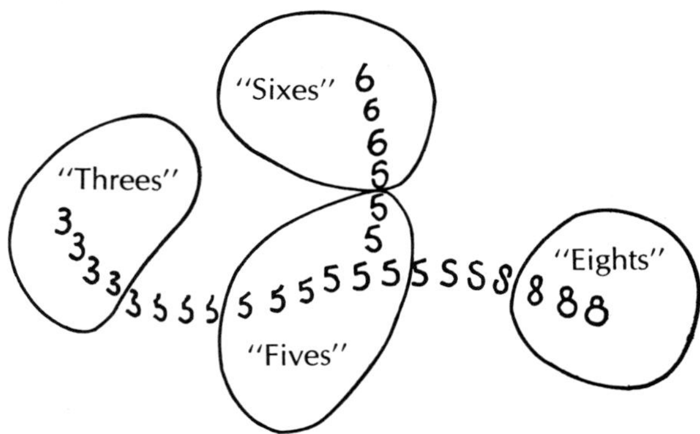

Fig. 3.9. Possible interfaces between groups of numbers: 'sixes', 'threes', 'fives' and 'eights'. This example is of continuous variation, which is rather common in biology. But when the intermediate forms have no names, they may be especially hard to identify. Certainly, they pose problems for the 'eye' or the 'computer' in 'reading' bad calligraphy. And again, though easy to visualize once explained in two dimensions, these figures in many dimensions would provide especial difficulties for the 'eye'.

groups. In biology we are very accustomed to continuous variation. Yet this question poses a problem that we may miss because we have names for the 'threes', 'fives', 'sixes' and 'eights'; we have no names for the various intermediate forms. We may therefore fail to recognize them.

Again, unless the data are actually visualizable in two dimensions, as in each of the examples above, the human eye, looking for interfaces between groups, readily remembers the simple situations and does not easily think of the more complex cases.

Examples from real biological data

We have looked briefly in a theoretical manner at a variety of problems to do with single groups, with many groups and with the interfaces between groups. And we have seen that they may present problems for the eye and for simple non-visual analytical methods. Before we examine such matters in more detail it would be worth knowing if we merely have a straw man here — are such matters actually of concern to 'real' biology? In other words, do problems of this sort actually exist in real biological situations? Let me present a small number of examples resulting from my own studies (Oxnard and Neely, 1969) that demonstrate unequivocally that they do and that we ignore them at our peril. The data are taken from one of the first investigations in which I was involved: the biometric characterization of the primate shoulder through measurements.

One suite of specimens is shown in the clustering diagram in Fig. 3.10. There seems little doubt that the groups 'A', 'B', 'C' and 'D' truly exist as shown. The reality of these super-groupings of specimens and species has been confirmed and displayed through a variety of methods: canonical variate analysis, neighbourhood limited classification, minimum spanning trees of generalized distances and high-dimensional analyses of canonical coefficients. However, the result of closer examination of the groups and their interfaces is shown in Fig. 3.11 in which multi-dimensional groups are drawn in their two-dimensionally equivalent shapes, and in which the number and lengths of linkages across interfaces between groups are approximately as shown (but the figure is diagrammatic for expository purposes — it is logically impossible to employ the actual lengths and relationships of a multi-dimensional situation in a two-dimensional display).

This examination demonstrates that the groups are not all circular (circular group 'A' versus sausage-shaped group 'C') and that they are not of equal density (group 'A' is small and tight, group 'D' is large and diffuse). The interfaces between them differ (the interface between groups 'A' and 'B', though narrow, includes only two links; that between group 'C' and 'D', though much more distant, comprises six links). Already, then, we see some of the conditions mentioned before as theoretical possibilities.

The biological reality of these four groups lies in the fact that they define groups of primates with similar functions of the shoulder: terrestrial primates, terrestrio-arboreal primates, regular arboreal primates and highly arboreal primates.

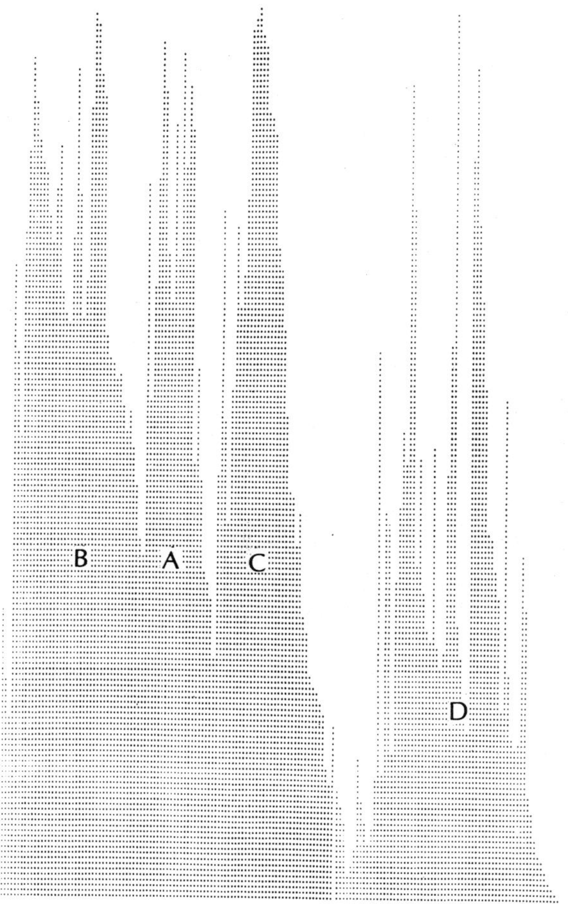

Fig. 3.10. A group-finding procedure (neighbourhood limited classification, Oxnard and Neely, 1969, and Oxnard, 1973) readily demonstrates four groups (A, B, C, D) in data taken from the analysis of the primate shoulder (Ashton, Healy, Oxnard and Spence, 1965).

In this case the data are truly many-dimensional (nine variables on each specimen). The eye just does not find these four groups.

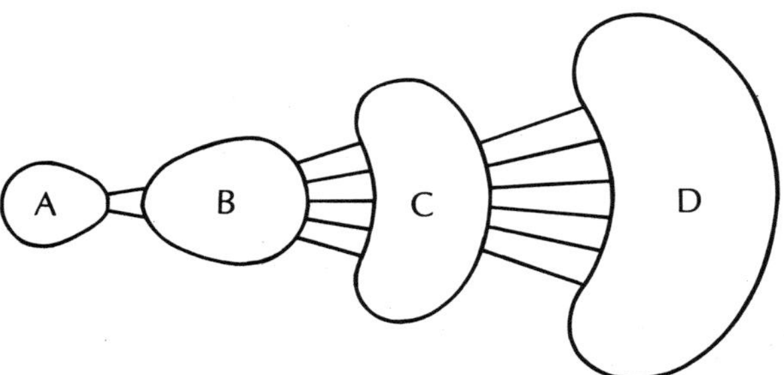

Fig. 3.11. A closer study of the numerical part of the computer output of the groups and interfaces in Fig. 3.10 demonstrates differences in the sizes and shapes of groups and in the numbers of links across interfaces (from Oxnard, 1973).

This representation can be shown in two dimensions, as here, for demonstration purposes, but because the data is actually nine-dimensional, the real 'shape' of the set of linked groups cannot be shown in two dimensions on a page.

Nevertheless, the two-dimensional representation gives correct information about relationships between neighbouring pairs of groups. It shows that A and B, though close together, are sharply separated; that C and D, though far apart, have many links in the interface between them and thus are less clearly separated. It demonstrates that groups other than spherical (hyperspherical) A exist, i.e. sausage-shaped (hyper-sausage-shaped) D.

A second example is contained in the grouping diagram shown in the upper frame of Fig. 3.12 in which two major groups seem to exist: 'A' and 'B', each of which contains two sub-groups, 'A_1' and 'A_2', and 'B_1' and B_2', respectively (for the purposes of this discussion we can ignore the small very distant group 'C'). Biologically, these two groups, 'A' and 'B', do not seem to make much sense; each consists of a mixture of prosimians and anthropoids.

The lower frame of Fig. 3.12 demonstrates that, notwithstanding the appearance of the prior frame, the relationships between 'A_1' and 'A_2', though short, are through a single link only. This frame also shows that the relationships between 'B_1' and 'B_2' are likewise minimal, only two links, even though the distances involved are close. In complete contrast, this frame also demonstrates that, although 'A' and 'B' are far distant from one another, there are ten links between them. And of the ten links that join these two super-groups, five are between new sub-groups 'A_1' and 'B_1', and five are between new sub-groups, 'A_2' and 'B_2'. This suggests most strongly that the real grouping in these data are the new group 'A_1-plus-B_1', and the new group 'A_2-plus-B_2'. There are really two major groups shaped like two dumb-bells lying next to one another.

The biological reality in this analysis is that the two new groups comprise, separately, the two major subdivisions of the primates: the Prosimii and the Anthropoidea.

Yet a third example can be found in the shoulder data in Figs. 3.13 and 3.14. Here we have five groups with a variety of non-circular, non-homogeneous arrangements. Neighbourhood limited classification has been able to separate out their differing densities and different forms of interfaces. The resulting picture makes good biological sense, in this case separating the primates into both functional and taxonomic groups.

A final example from those same data shows the five groups displayed in the top part of Fig. 3.15. The clustering study of the lower part of that figure shows unequivocally that, though four of the groups are real, the fifth group consists of a series of specimens (solid stars) peripherally arranged around one of the other groups (solid circles), a

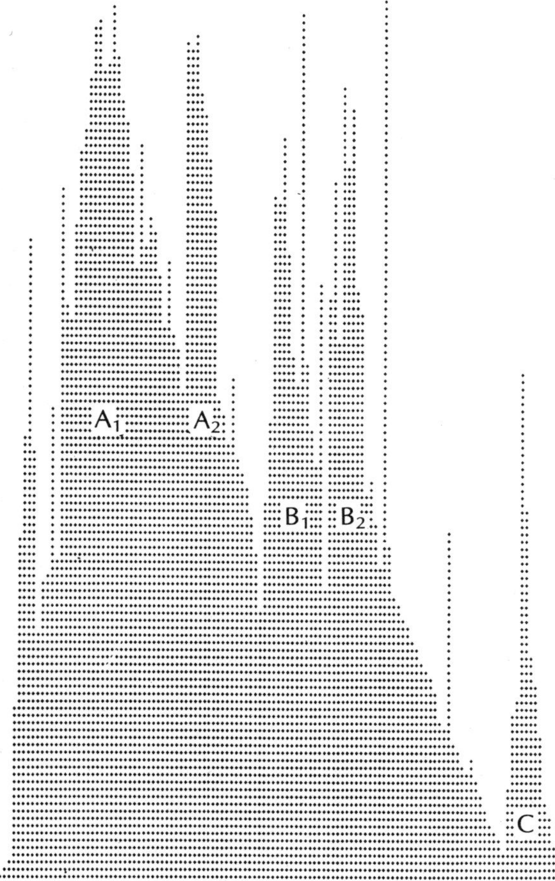

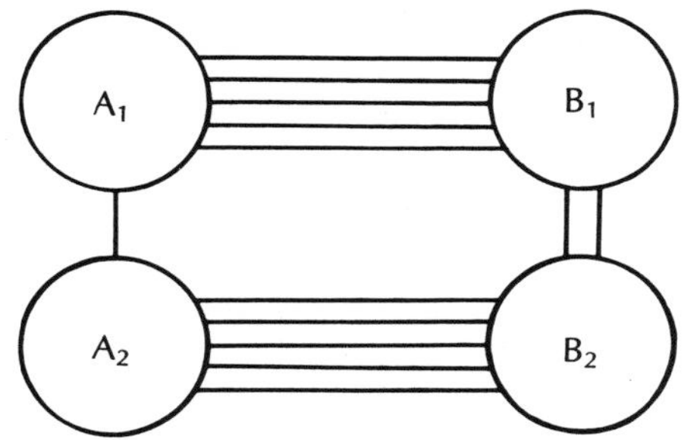

Fig. 3.12. A second example from the nine-dimensional data on the primate shoulder. Neighbourhood limited classification shows that two supergroups, 'A' and 'B', each contain two sub-groups: 'A_1' and 'A_2', and 'B_1' and 'B_2', respectively (upper frame). But study of the details of the interface links shows that in reality these data contain two dumb-bell shaped groups: one is 'A_1' together with 'B_1' and the other is 'A_2' together with 'B_2'.

This is an interesting result because the two major groups, 'A' and 'B', each contained curious mixtures of prosimians and anthropoids that did not make much biological sense. The two dumb-bells, on the other hand, are organized in such a way that one contains only prosimians and the other only anthropoids. This is independent information suggesting that the dumb-bell view is sound (from Oxnard, 1973).

The conventions of the diagrams are as in the previous two figures.

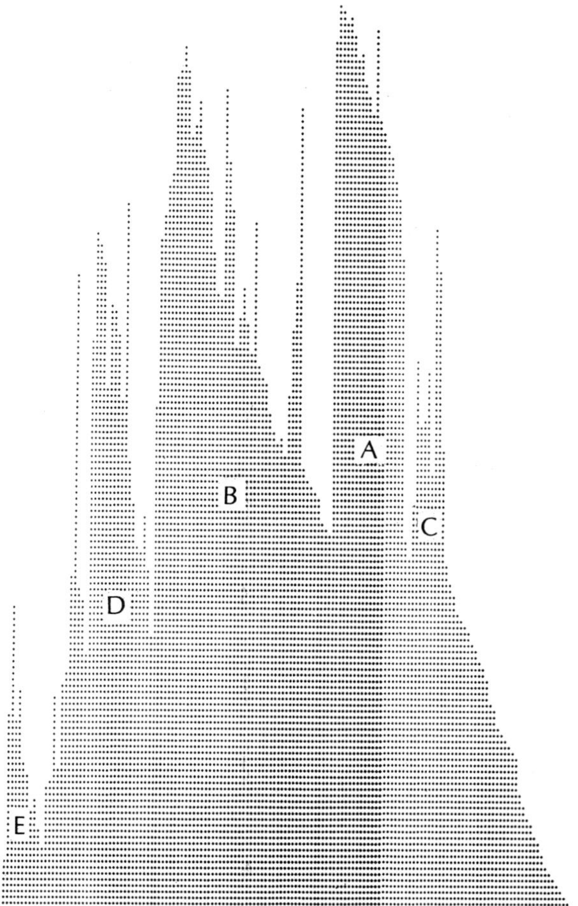

Fig. 3.13. A third example from the nine-dimensional data on the primate shoulder. In this case neighbourhood limited classification demonstrates five groups. Again, conventions are as in the previous figures.

situation not unlike the theoretical one posited in Fig. 3.6. Again, the results make good taxonomic and functional sense.

It seems apparent, then, at least in the data of one investigator, that complex arrangements are present. It behooves us to watch out for equivalent complexities in the groupings (of data or of animals) presented in other studies of form. Biological form in particular, and biological mechanisms in general, probably only rarely correspond with linearities, isotropicities and gradualisms. Far more likely, to my mind, are curvilinearities and irregularities, anisotropicities and non-homogeneities, quasi-discontinuities and true discontinuities. We should be prepared to take such features into account in our analyses even though it is easier to aim first at the simpler concepts (Oxnard, 1980a).

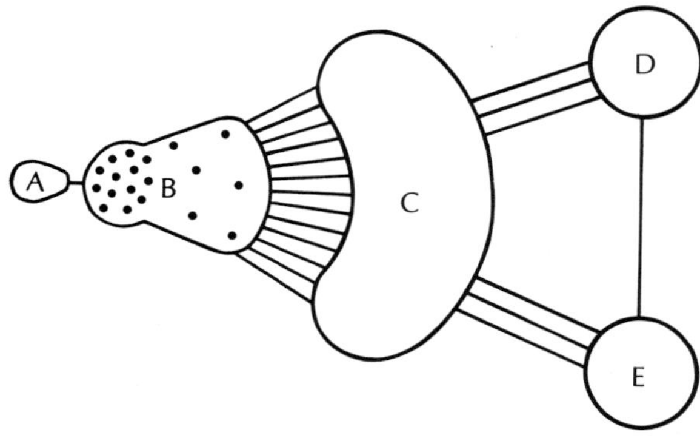

Fig. 3.14. Study of the details of the five groups in Fig. 3.13 shows the nature of the complexities of the sizes, shapes and interface relationships of the groups. The forms of these groups and their interfaces, not previously recognized in visual inspection of the raw data, seem to provide useful biological information as explained in the text.

Again, conventions are as in the previous figures.

NEIGHBORHOOD LIMITED CLASSIFICATION
HOMINOIDEA SHOULDER DATA

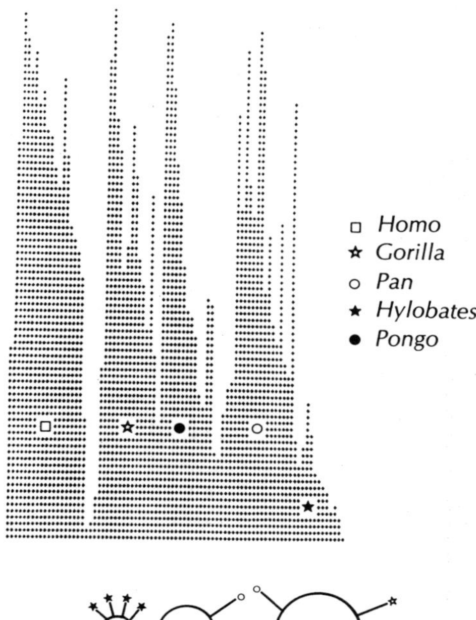

Fig. 3.15. A final example from the nine-dimensional data on the primate shoulder. Neighbourhood limited classification shows five groups. The linkages are such that two of these comprise a genuine group (solid dots) with a surrounding outlying periphery (solid stars). This recalls the theoretical star-like idea of Fig. 3.6.

Again, the conventions are as in the previous figures.

Discovering structural differences: new methods

Criticisms of the observational method have in fact been apparent for many years. And a few really great minds have been brought to bear upon the problems. D'Arcy Thompson's (1915, 1917) method of transformations reveals how a part of one creature may be described as a distortion of another, the 'comparative trend' of Chapter 2 (Fig. 3.16); Fisher's (1936) multiple discriminant functions show how quantitative data can distinguish between closely related plants and animals through statistical analysis of measurements (Fig. 3.17); Woodger (1945) proposes comparing living things by mapping, mathematically, from one to another (Fig. 3.18); such historical examples readily spring to mind.

But since those studies, the manner of investigating evolutionary change in shape has scarcely improved until the last few years.

In some ways it is easy to see the reason for the lack of progress. The earlier masters (especially D'Arcy Thompson) include within their writings clear indications of their difficulties; one of these is an inability to make large numbers of algebraic calculations with excessive reliance on the geometric approach. He had gone as far as was possible with the tools available to him. Just as microscopists awaited the electron microscope before being able to venture from the microscope to the ultrastructural level, so morphologists needed new tools before venturing far from assessment of gross shape by observation (occasionally backed by measurements and simple analysis) towards more complex evaluation of underlying factors of shape. However, the evolution of a number of modern tools (especially the electronic computer and computer software) provides new mechanisms for characterizing and comparing complex morphologies.

There are, in fact, many different methods by which we may carry out this characterization and comparison of structure, many different types of structure to be characterized and many different ways in which the results may be used within evolution. The methods range as widely as from the use of a ruler and univariate statistics to make and analyse individual measurements through to holographic techniques for characterizing entire structures. The forms to be characterized can vary from the real, external dimensions of some biological object, through the complex internal patterns that may be revealed by methods such as dissection

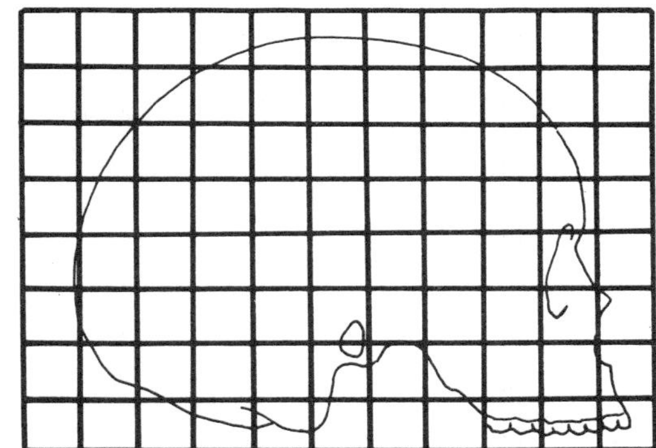

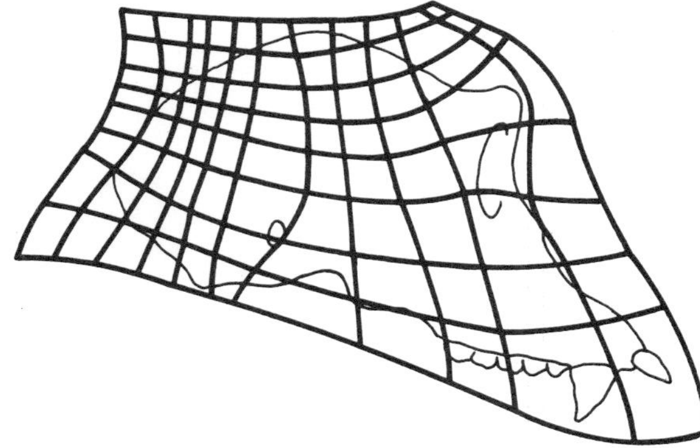

Fig. 3.16. The comparative trend (see Chapter 2, Fig. 2.4) between the skulls of a human and a chimpanzee when Cartesian coordinates for the human are deformed to fit the chimpanzee. The face of the chimpanzee is much expanded compared with that of the human, the braincase much compressed (after D'Arcy Thompson).

or radiography, to the complicated and multidimensional 'structure' of data referable to biological problems. The modes of attack on problems of morphology and evolution can range from attempts to obtain insights by straightforward descriptions of the actual objects themselves, through studies aimed at elucidating developmental, genetic, functional, environmental, populational, geographic, and yet other issues that relate to forms and patterns and at the same time have implications for evolution.

Though the general theme of morphometrics has existed for many years now, it is only in the last two decades or so that the familiarity of biologists with mathematical and computational tools has grown sufficiently that these methods promise to be of major practical use.

Many of the earlier investigations do little more than introduce us to the problems and difficulties that arise in their use. For instance, some of the

Frame 1

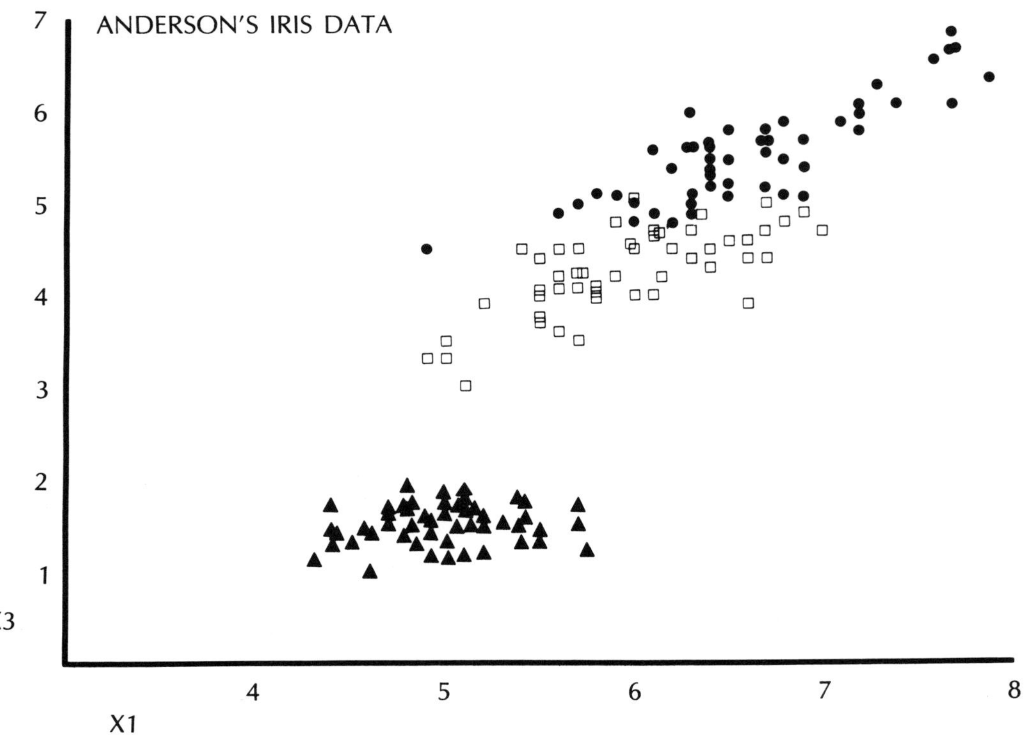

Frame 2

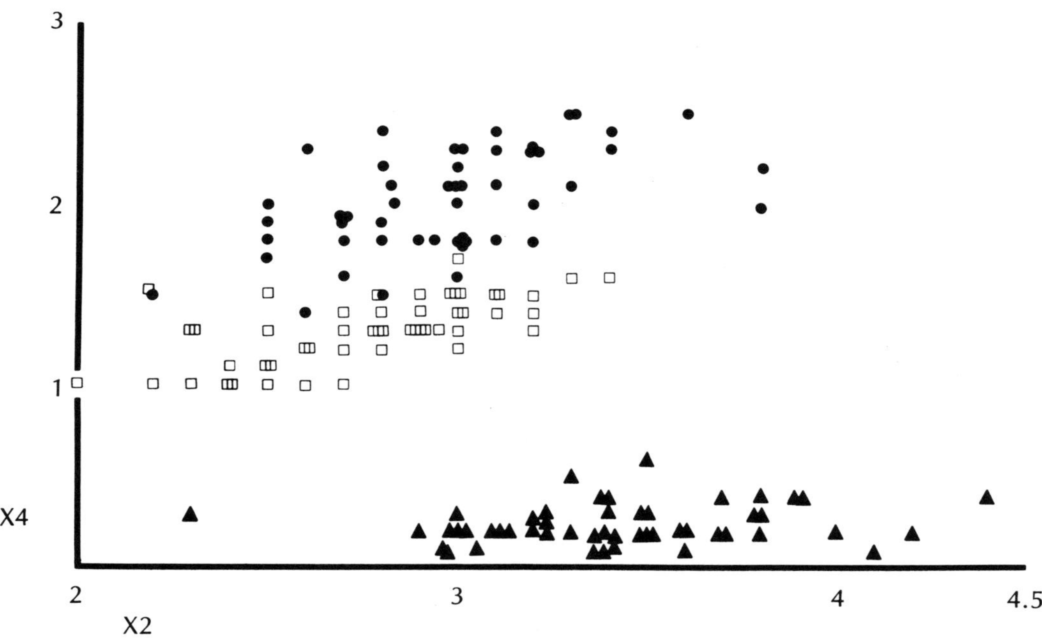

Fig. 3.17. The discrimination obtained by Fisher using a discriminant function (lower frame) is far clearer than that provided by the raw data, measures of sepals and petals (upper frames), of three groups of iris measured by Anderson.

Frame 3

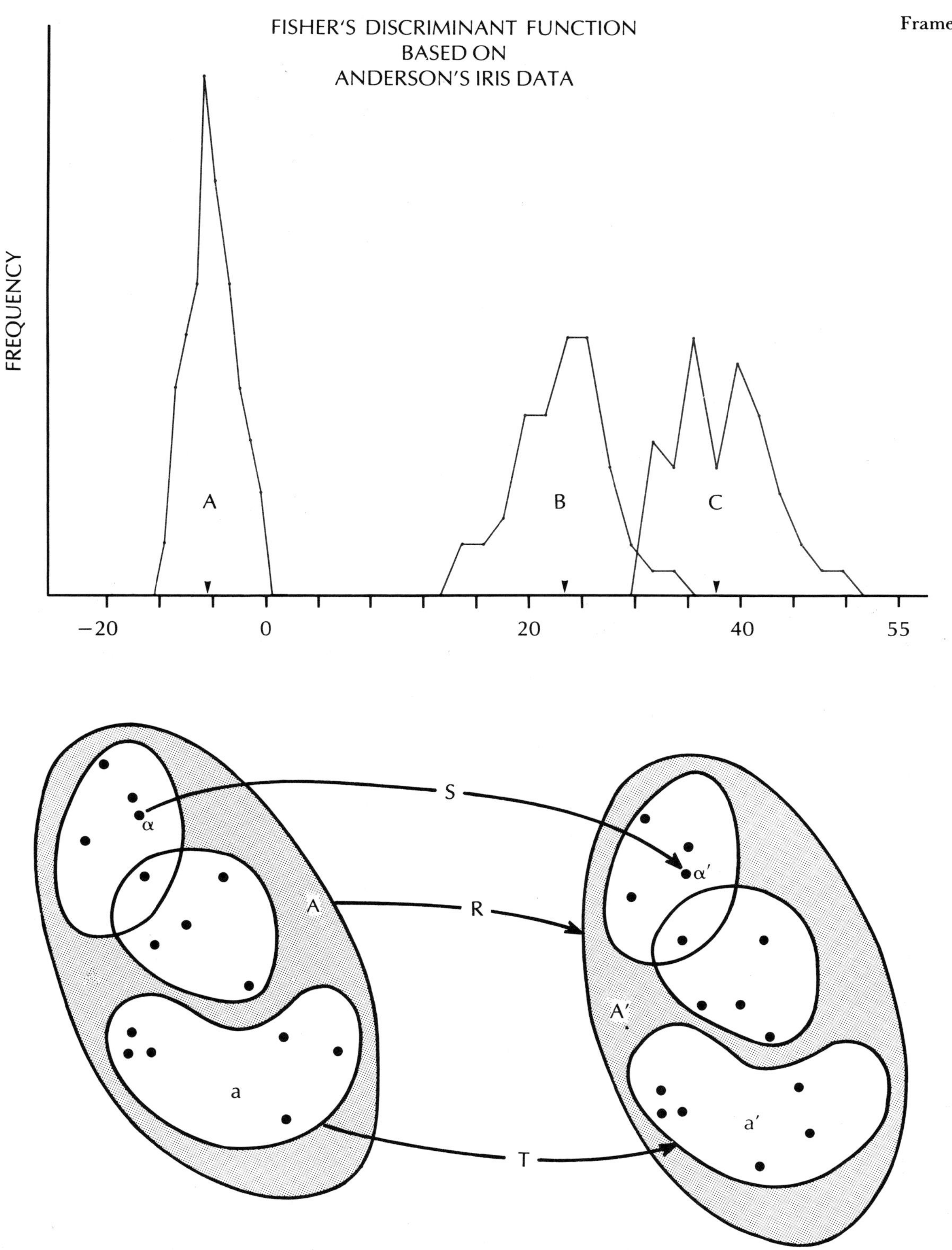

Fig. 3.18. A diagrammatic representation of the comparison of organisms by mapping from one to another, after the manner of Woodger. The mapping may include R, that between organisms A and A[1]; T, that between organs a and a[1]; and S, that between organelles alpha and alpha[1].

methods are applied to entirely artificial data sets (sometimes computer-generated) in order to allow us to discover and demonstrate possibilities. One creative example here is the study of theoretical organisms, "Caminalcules', invented by Camin and examined by Sokal (1966).

Other investigations are indeed applied to 'real' biological data, but to data rather specially chosen to help reveal the properties of the various methods. Anderson's measurements on the Iris, used first by Fisher (1936) in the development of discriminant functions, stand here as most useful data for displaying a variety of morphometric analytical procedures. Thus, they have been used for illustrating (Fig. 3.19) the applications of multivariate analyses (Oxnard, 1973a), of cluster-finding procedures (Rubin and Friedman, 1967), of 'fuzzy set' theory (Zadeh, 1965) and of neighbourhood limited classification (Oxnard and Neely, 1969).

Yet other investigations are aimed at 'real' biological problems but at ones for which an answer is already known (or presumed to be known). Here, the rationale is to supply confirmation that the morphometric method is not producing answers totally at variance with what can be learned by more classical methods. An excellent example of this is the demonstration (Albrecht, 1978) that cranio-facial multivariate statistical morphometrics define the same seven groups of Sulawesi macaques that were obtained by traditional observational methods (Fooden, 1969). (Albrecht then

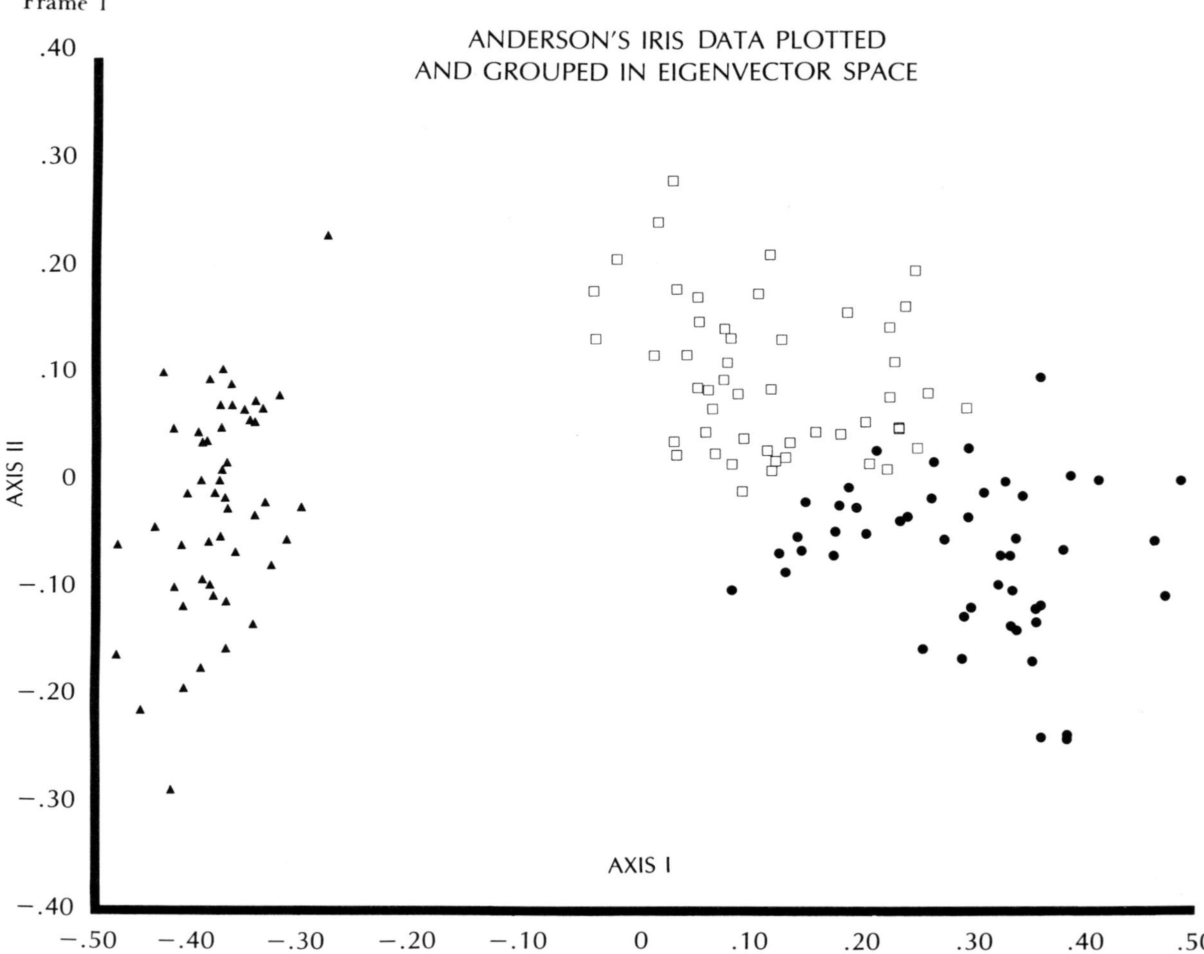

Fig. 3.19. Anderson's iris data, already shown in Fig. 3.17, has also been investigated by the multivariate method of canonical variate analysis (first frame), by fuzzy set theory (second frame), and by neighbourhood limited classification (third frame) page 40. In each case a far better impression of the three groups of iris, two overlapping, is obtained than is seen from the raw data. In particular, the number of equivocal specimens (specimens not clearly seen as belonging to a particular group) is progressively reduced in these more complex analyses.

goes on to examine new relationships, including those between the macaques of Sulawesi and neighbouring portions of Asia).

Finally, yet further investigations are pointed at true biological problems but so far at only a small number of them; they therefore act, at present, as little more than examples of what may be achieved; they are scarcely extensive enough to contribute in a major way to the broad spectrum of evolutionary biology.

Yet these essays are all essential steps in the development of new methods and from them it is becoming clear that a wide range of techniques is now available for characterizing and comparing complex shapes and structures. These methods stem from developments in many different fields so that we must be willing to look to workers in disciplines as disparate as statistics (Gnanadesikan, 1977), electronic engineering (Dunn, 1975), communications (Andrews and Pratt, 1969), optics (Almir and Shamir, 1976), applied mathematics (Cutrona, 1965, Andrews, 1970), stereology (Underwood, 1970), image analysis (Hildich, 1969), computer graphics (Welsch, 1976), pattern recognition (Cheng, 1969), and so on. In the same manner we must remember to keep our eyes open for the work of other investigators who may use these methods as we would, for instance geologists (Dobrin, Ingalls and Long, 1965; Dobrin, 1968; Pincus, 1969; Davis, 1970, 1973), microscopists (Prensky, 1971), metallurgists (e.g. see Underwood, 1970), meteorologists (e.g. see Rosenfeld,

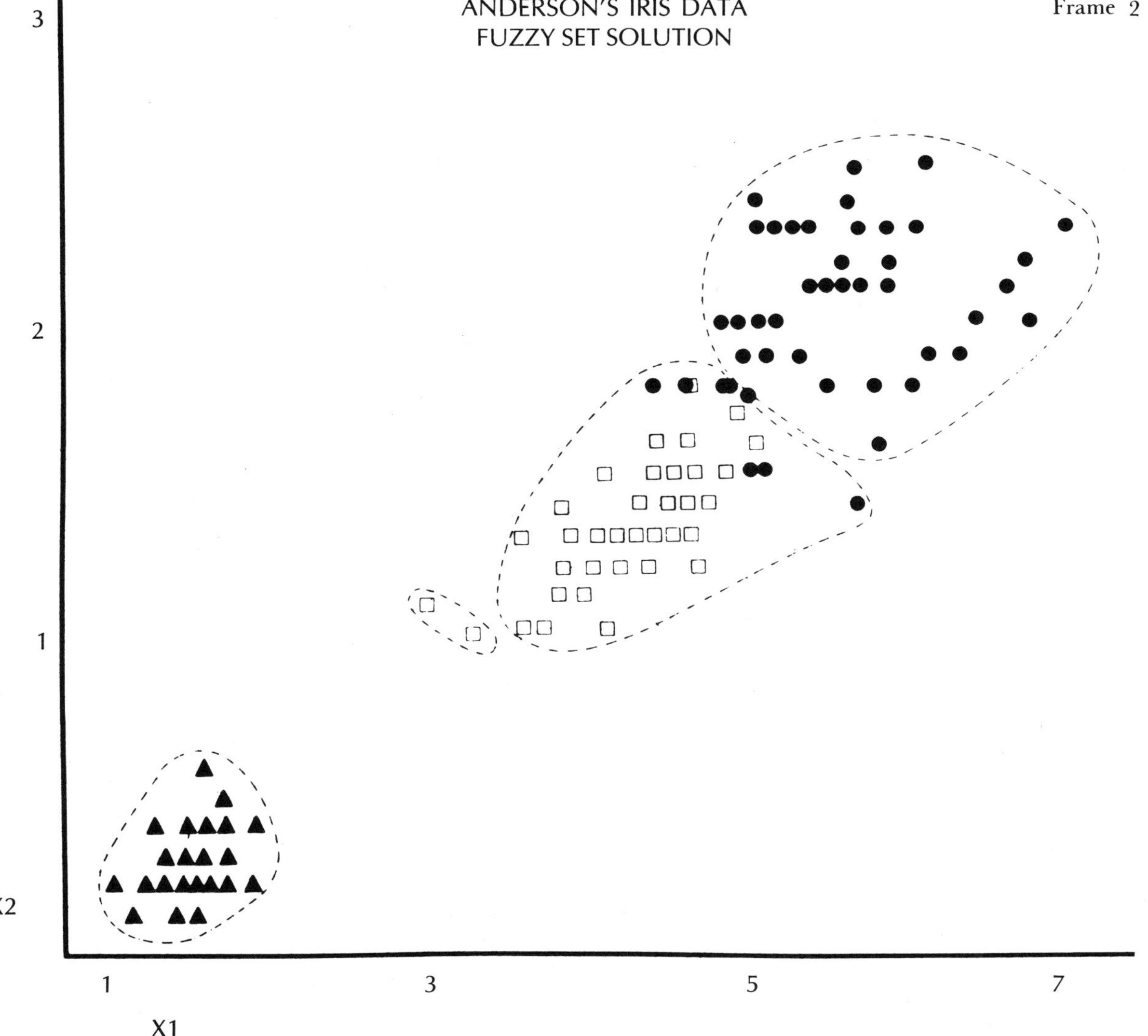

1969), astronomers (White, 1970), geographers (Pincus, Power and Woodzick, 1973), biomedical scientists (Chmielewski and Varner, 1969; Wechsler, 1976), and so on. It is likely that many of the questions facing these investigators may well have required the new technique before the problems of the anthropologist. Indeed, it may only be seeing how methods are applied in other disciplines that raises consciousness towards applications in our own.

It is possible to discuss these techniques in two broad groups: those that utilize characterizations of structure based upon measurements of individual parts of structure, and those that involve manipulation of the entire structure itself.

Frame 3

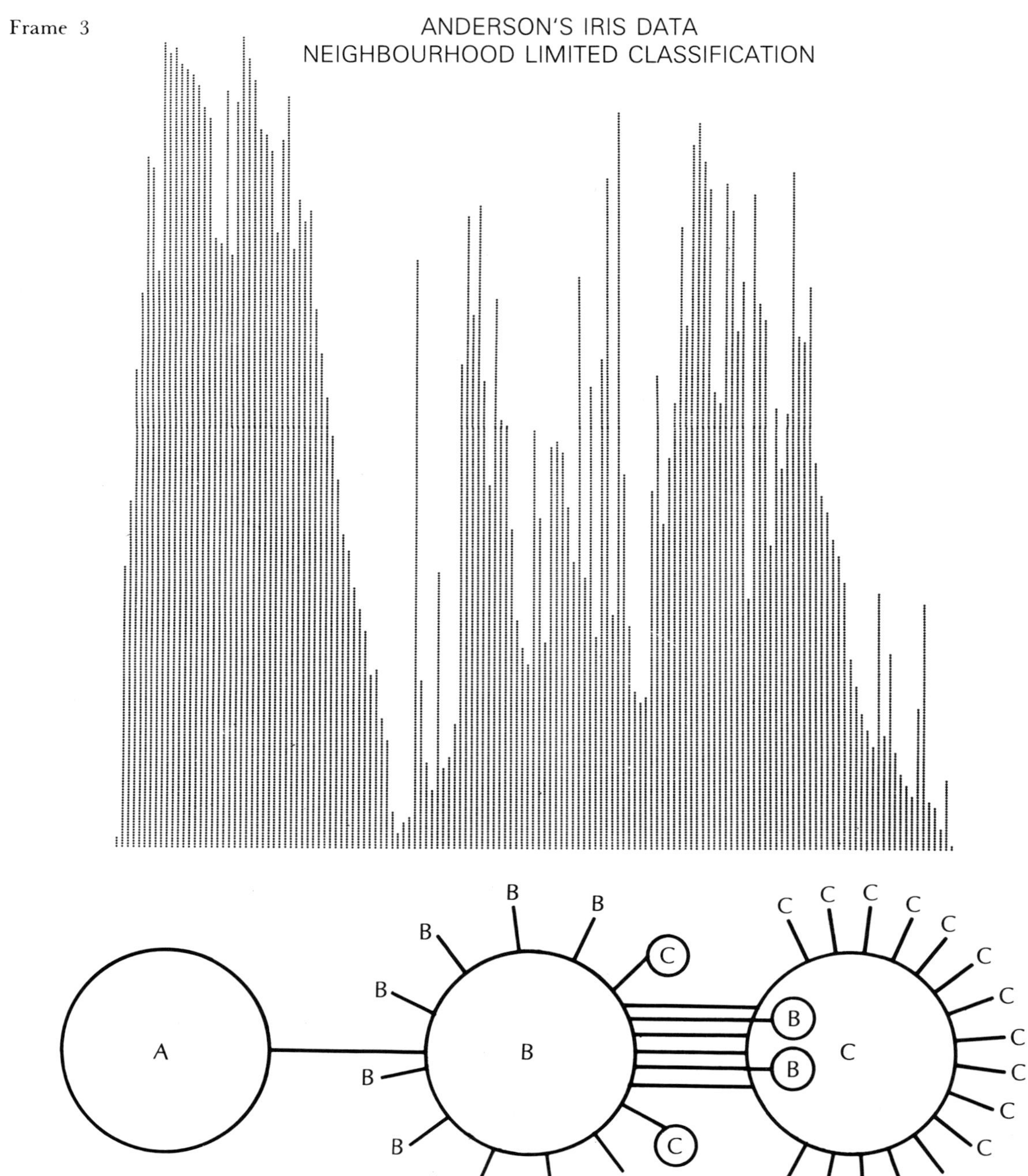

One method: the biometric approximation

Characterizing and comparing structures can be done with measurement. This may merely involve rather simple metrical descriptions of the bones of living and fossil species, often through a single measure or other descriptor, or a very small number of measurements or other parameters. It is common, for instance, to study growth through analysis of body length or height only (or of body weight); to describe the shape of a petal, or a shell, or a bone, or a tooth by means of greatest lengths, breadths or heights; to encapsulate the form of entire animals in the lengths of the head, the trunk, the limbs and the tail.

Although such overall measurements are indeed hallowed, they have, in fact, been only rarely useful in solving real biological problems. They are apparently not as good as the human eye for describing most biological forms and patterns. Their greatest value is often little more than as components in 'keys' that can be use for subsequent classifications once the real biological problems have been elucidated (but see the multivariate statistical analyses of simple overall bodily proportions of primates, Chapters 5 through 9).

However, the descriptive powers of measurement can be greatly increased (a) by taking more of them, (b) by selecting them carefully using underlying biological information and (c) by analysing them in ways that not only view each measurement separately but also look at the interactions among them.

The problems of taking more measurements are slowly being solved by the introduction of automation because, as the number of measurements increases, so the difficulty of taking them manually increases. Rulers and callipers give way to other kinds of apparatus to maintain accuracy and increase speed. One simple device that can be used to obtain information about a complexly shaped edge is the artist's 'copy cat' described for use in characterizing transects of faces as a 'contourometer' (Steegman, 1970). A somewhat similar 'profile measuring machine' was invented by Preston (1953) to delineate the form of bird's eggs. Other standard equipment such as osteometric boards, depth gauges, diagraphs, goniometers and so on have been used to produce particular kinds of measurements. A variety of novel instruments such as 'craniometers' (Ashton, 1949), 'pelvimeters' (Chopra, 1958), 'cranial radiometers' (Creel and Preuschoft, 1970) and 'magnetic goniometers' (Cartmill, 1970) have been devised to aid in the special measurement of complexly shaped parts. Perhaps the ultimate in such physical measuring devices is the 'stereometric craniostat' of Oyen and Walker (1975) and the 'head measuring device' of Claus, McManus and Durand (1974), both of which resemble gentle forms of an 'iron maiden'. Compared with non-physical methods mentioned below, most such physical inventions, although once useful, have now become overly cumbersome.

A next step, then, is the combination of physical with electronic measuring devices. These run the gamut from callipers that are electronically linked to recording instruments (Cable and Van Haagen, 1967) through to 'on-line automated osteometric data collectors' (Day and Pitcher-Wilmott, 1975), in which impulses from digital callipers are relayed through a number of recording and analytical devices to produce a direct computer print-out of a complex statistical analysis. (Such methods go too far whenever their automatic nature, or usage as a package, prevents examination of the original data and understanding of intermediate steps of the analyses).

Nowadays, a variety of approaches are available for measuring biological objects that avoid the use of any invasive instrument such as callipers however they may be augmented electronically. Some of these methods involve ultrasonic and optical instruments, and all are capable of producing numerous accurate measurements, in two or three dimensions if required.

Thus, in our own laboratory the 'Graf Pen' utilizes ultrasonic signals to produce two-dimensional (for photographs, radiographs, sections) or three-dimensional (for actual objects) coordinates of points or lines. Other methods use stereo-pairs of photographs of biological objects and a variety of tools are available for obtaining measurements from them, including the use of stereo-plotters for measuring heights and distances on maps by geographers. Creel (1976) is applying such methods and instruments to the analysis of photographs of primate skulls.

The problem of making better measurements now rests in using information from the newer biomechanical studies of form to 'invent' measures that more clearly reflect biomechanical factors.

The problems of making better sense of the resulting large data sets now devolve on analyses such as multivariate statistics and instruments such as computers and computational programs.

Thus, it is that the study of form and pattern through analysis of measurement is yielding gradually to new orders of investigation. The more refined descriptions that result from many new measurements, together with the succinct summaries and sometimes the deeper insights that stem from the new analyses, demonstrate unequivocally their enormous advantage over the use of a small number of measurements studied using univariate or bivariate investigative procedures.

Many examples abound and they stem from every stage of investigation. Not only may the investigation of correlation and covariance, the underlying point of most of the statistical methods, provide information about a data set that is not evident from simply viewing means and variances, but also the very creation of larger numbers of new measurements of biological objects, rather than mere acceptance of a few traditional descriptors, may help show new views of old objects.

For instance, although study of traditional morphological descriptors of various prosimian primates such as bush-babies, mouse lemurs and indris suggest that the structure of their hip and thigh is related to their ability to leap (Napier and Walker, 1967), study using new biometric descriptions and multivariate statistical methods shows rather clearly that each of these groups of animals has a different structure of the hip and thigh. This suggests that several different anatomical arrangements for leaping exist (Oxnard, German and McArdle, 1981). And this, in turn suggests the behavioural hypothesis that the different animals may in fact leap in rather different biomechanical ways; such information about the behaviour of these rare animals as is currently available suggests that this is indeed true (see Chapter 6).

Furthermore, although an increase in the numbers of dimensions required to describe shape may result in a more creative choice of measurements, it turns out that the powers of the analytical tools themselves are such that sometimes even rather non-descript measures, such as overall lengths, widths and heights or overall body proportions, may yield new insights when studied by such methods. Thus, multivariate statistical investigation of simple lengths and breadths of teeth have supplied quite unexpected information about the overall dimensions of some of the fossil teeth found in Africa (Ashton, Healy and Lipton, 1957, and see Chapter 10). And the new results of studying the detailed structure of the hip and thigh in leaping prosimians described above are confirmed by multivariate statistical analyses of simple measures of the overall proportions of the upper and lower limbs of these creatures (Oxnard, German, Jouffroy and Lessertisseur, 1981, and see Chapter 8).

Many excellent expository texts are available for supplying understanding, in all of its algebraic detail, of the mathematical and computational procedures that comprise these approaches to analysing morphology (e.g. Cooly and Lohnes, 1971; Gnanadesikan, 1977). But it is worth having an intuitive, pictorial, non-mathematical description of the methods and of some of the ways in which they have been used in studies of human evolution.

The core of these techniques, the multivariate statistical approach, is the following. If we suppose that a single object can be defined by two measurements, then that object can be represented as a point on a two-dimensional plot or graph, the axes of which represent the original two measurements. A group of similar objects will then appear as a cloud of points lying relatively close together on the graph. Because the two measurements are likely to be correlated, such clouds will generally be elliptical in shape. Three different groups of objects will appear as three elliptical clouds of points separated from one another on the graph (Figs. 3.20 and 3.21, upper frames after Albrecht, 1980).

If the original measurements defining the groups are uncorrelated with one another, then the original system of axes on the graph will best describe the arrangement of the clouds. If, however, the original measurements defining the groups are correlated with one another to some or other degree, then the true relations among the clouds may be best seen from the vantage point of other systems of axes obtained by rotating the system of clouds.

The rotation may be performed in different ways. One rotation emphasizes the dimensions of the overall universe of clouds: that is, the total dispersion of the points (in biological terms the total variation). This is shown in both Figs. 3.20 and 3.21 where elliptical clouds are rotated to place the long diameter of the universe of ellipses into the first (horizontal) axis. There is still a smaller amount of structure in the new second (vertical) axis.

A different rotation emphasizes the distances between the individual clouds: that is, the dispersion of the clouds from one another (in biological terms, the variation between the groups). This is also shown in Figs. 3.20 and 3.21 where elliptical

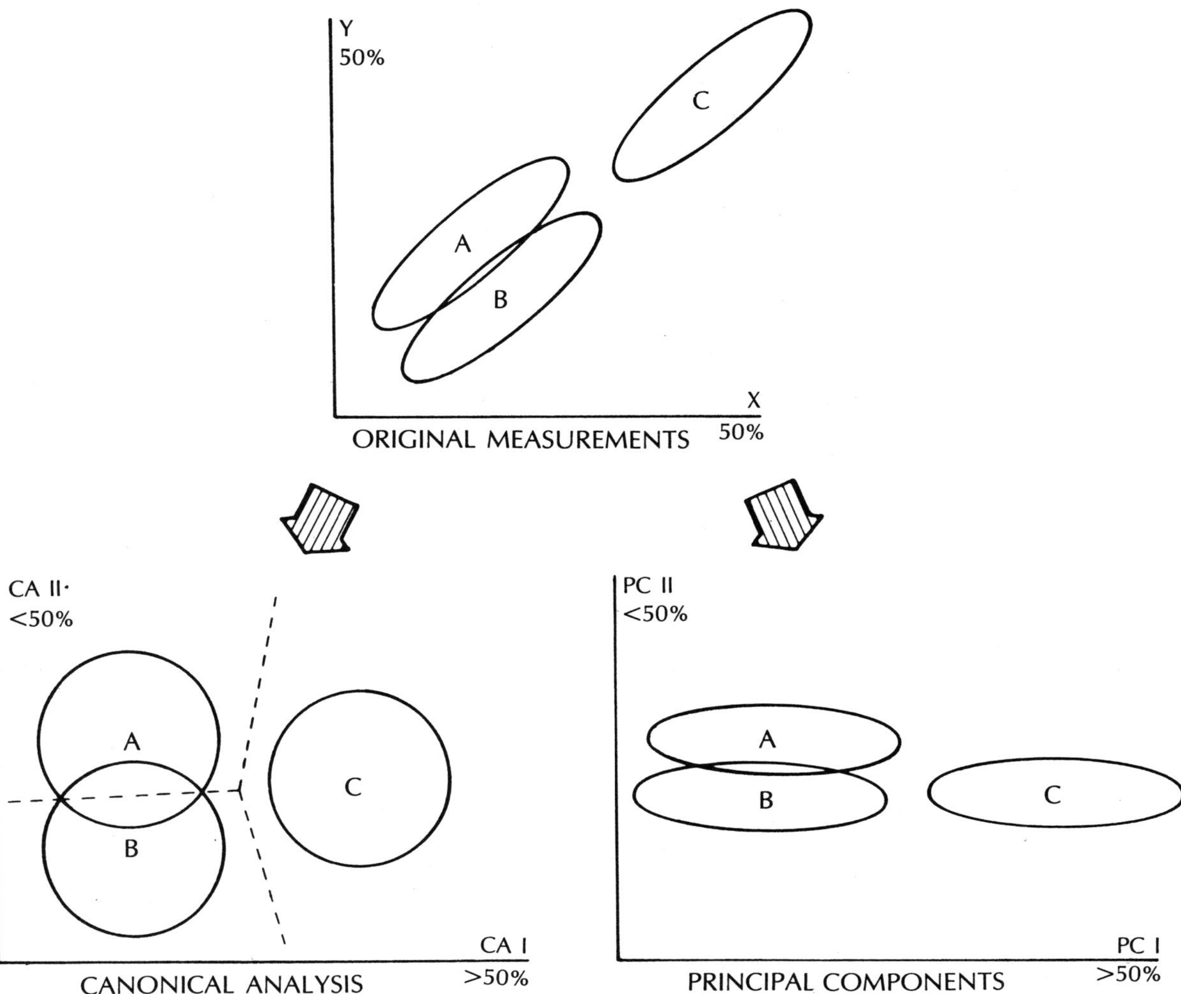

Fig. 3.20. Three groups of animals may be represented by each of two measurements. When these measurements are plotted so that the measures are the x and y axes, then the data may appear as in the top frame. The various points representing the animals may fall into three clouds of points (groups of animals). Very often, with biological data, the clouds are elliptical in shape and oriented at some angle to the original data axes. Rotation of the three clouds can be carried out so that the major distances between the clouds are aligned along a first horizontal (lower frames) axis.

This can be done in two ways. If we retain the original scaling of the variables, then the ellipses are rotated without changing their shape. This is shown in the lower right-hand frame and is the geometric description of principal components analysis.

However, the second way involves first rescaling the original variables so that they become circles (i.e. the spreads of the three clouds are now the same in all directions). Then rotating the clouds produces the changed relationships of circles shown in the lower left frame. This is the geometric description of canonical variate analysis.

If it we wish to understand or concentrate upon the original shape of the entire system of clouds (i.e. upon the original variation of the measurements), then principal components analysis is used, the clouds remain elliptical and the rotation is such that the new first axis is a measure of the greatest distance across the universe of clouds. The first axis is, therefore, parallel (or as closely parallel as possible) to the long axes of the ellipses irrespective of the actual differences between the ellipses.

If we desire to understand the relationships *between* the groups, then canonical variate analysis is used, the transformation renders the groups circular in form (i.e. to make the within-group variation the same in all directions) and the rotation is such that the new first axis is a measure of the greatest distance *between* the clouds. The first axis is, therefore, parallel to the greatest rescaled distance between the groups (circles).

In each of these processes, the defining characteristics of the groups may be reduced from two original axes containing approximately equal information from each variable to two new axes of which the first alone may contain most of the significant information. Such a new first axis may be a simpler summary of the original data.

Also, in this theoretical example, it appears that there has been no change in the groups (C is separated from overlapping A and B by the first axis in both studies, but see next figure).

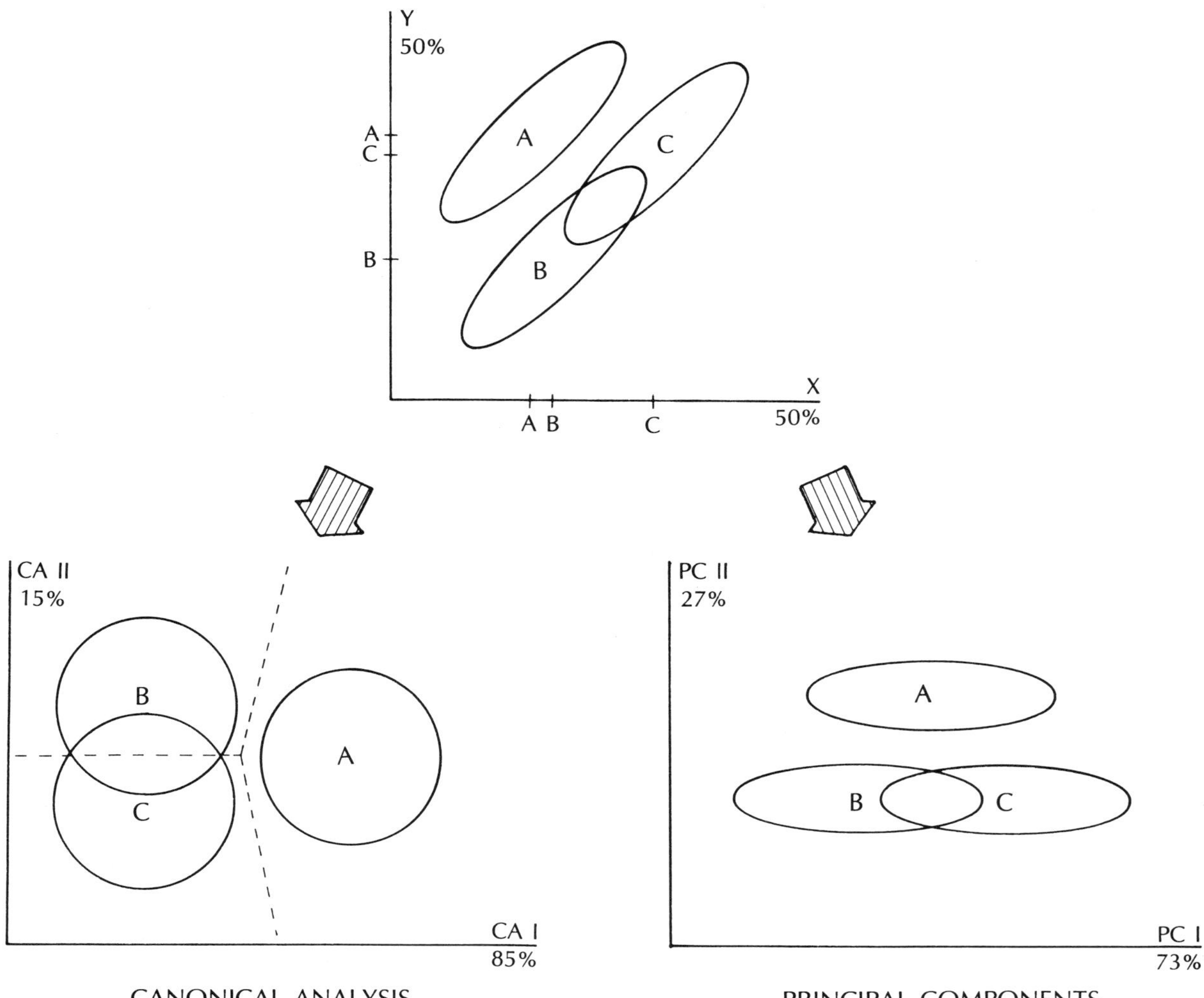

Fig. 3.21. In contrast to the previous figure, we can study two original variables defining three elliptical clouds (groups of organisms) in the new situation shown in the upper frame. Here the major length of the cloud system is still at 45 degrees to the main axes, but in this case the complete separation is between A and the other two.

Principal components analysis (lower right) makes the major length of the entire system of clouds in the first axis as before. But canonical analysis places the major distance *between* the clouds, i.e. between A and the others, into the first axis. Principal components analysis is still telling us about the total structure of the cloud's systems. Canonical variate analysis is telling us about the main distances between the clouds. And in this example these are not the same thing. Thus canonical analysis is what is important for our purposes in learning about the relationships between animals.

clouds are first rescaled to circular groups. This standardizes the variance within the groups to be the same in all directions. The standardized circular groups are then rotated to maximize the distances between them (maximize the between group variation) in the first (horizontal) axis. There is still a small amount of between group variation in the new second (vertical) axis. However, in this way most of the separation may be reduced from two axes to one, a situation easier to understand. This results, however, in different pictures in each figure because the inter-group relationships are truly different.

In terms of three measurements taken on each object, this procedure is the equivalent of constructing and viewing from one position the three-dimensional model of the circular clouds, and then rotating and viewing the model from a new position that best separates the clouds (Fig. 3.22). Again, the new view may reduce from three to two or even one the number of new axes necessary to describe the position of the clouds.

If we extrapolate such two- and three-dimensional descriptions to an example where we have taken many measurements on each object, then the problem is many-dimensional. We cannot draw or construct a geometric many-dimensional model for ourselves, but we can represent such a model algebraically in the computer. The computer can then rotate this model to give us the new view that best separates the multi-dimensional clouds; in so doing, it may discover that the many dimensions can be reduced to a smaller number that better allows us to understand the relationship.

Once such analyses have been performed however, these techniques may still produce results that are truly of dimensionality higher than three and that, accordingly, are rather difficult to display. Demonstrating the results inherent in these more complicated situations requires the use of a variety of group-finding procedures; the simplest of these is a technique that examines the shortest distances between the various groups (e.g. Gower, 1967; Gower and Ross, 1969). This is known as the 'minimum spanning tree' and it was evolved, in part, from the very practical problem of discovering the minimum length of wire required to link a group of towns into a telephone grid. In exactly the same way the minimum spanning tree can provide a picture of the links between data representing the structures of groups of animals even when those data are high-dimensional. The traditional dendrogram gives such a view of data but it does not contain as much information as a minimum spanning tree and may, if drawn injudiciously, actually allow incorrect inferences to be made as shown in Fig. 3.23.

Yet other methods of display may be required. One of these is to embed multi-dimensional information from a multivariate statistical analysis within the infinite-dimensional space of some mathematical function. This can be readily achieved using the well-known sine-cosine functions (Andrews, 1972, 1973) but others are also appropriate. If we suppose that it takes, say, seven multivariate dimensions to represent the mean position of a group of animals, then we can represent that mean as a single wavy line on a graph using the seven dimensions as the first seven coefficients of a sine-cosine plot. Several similar groups, even though characterized by a space of seven dimensions, are displayed by several similar wavy lines in such a plot. A group that is totally different from them is represented by a totally different wavy line. Another group that is in-

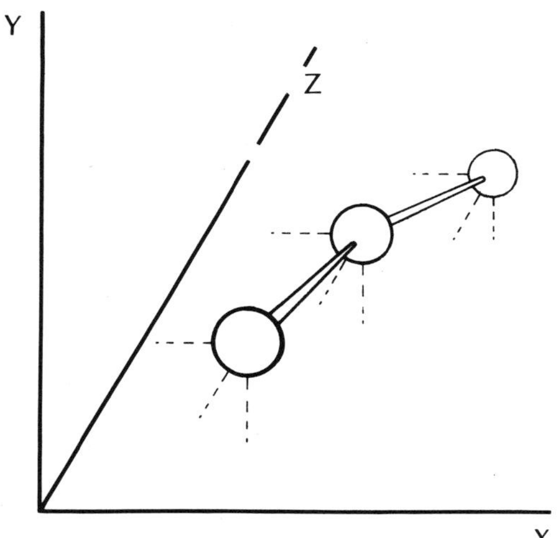

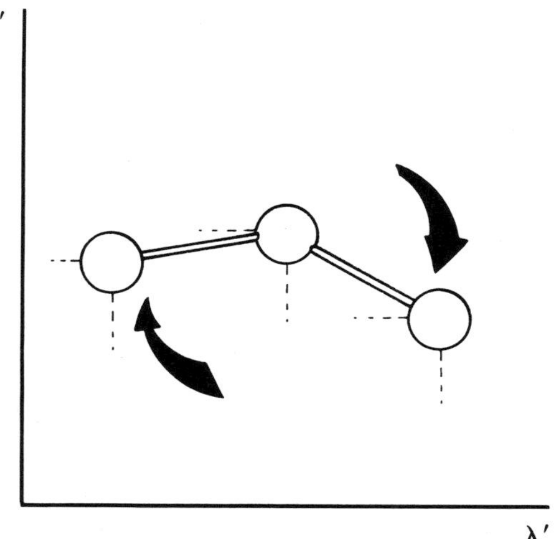

Fig. 3.22. Here we take the step into a third dimension – i.e. a third measure taken upon each of the animals – and look only at canonical variate analysis. Three clouds, already transformed into spheres, are separated by three variables, X, Y and Z, as shown in the first frame. Rotation of three clouds as indicated by the arrows onto two new axes (lambda[1] and lambda[11]) provides a better view of the separation of the three clouds. The defining characteristics in this case have been reduced from three axes to only two, and of these the first contains most of the information about the separations between the groups. A considerable summary of the data has thus been achieved.

Fig. 3.23. The analysis of measures of the primate shoulder. The first frame provides the dendrogram (tree diagram) of the relationships between the various primates. It shows that humans are more different from all the non-human primates than any non-human primate is from any other non-human primate.

This is confirmed by the second frame, which gives the minimum spanning tree of the relationships. Humans are located inside the large circle; the other squares, circles, diamonds and triangles are the non-human primates.

The information about distances in these pictures is the same, but the dendrogram may well suggest to the uninitiated that humans lie at one extreme of the various primates – the link for humans, being the longest, must logically be drawn at one or the other end of the tree. Indeed, the dendrogram suggests visually that humans are closest to apes.

The minimum spanning tree makes it clear that this is not so; though humans are indeed separated by the longest distance from other species, it is from particular species that lie fairly centrally placed within the primates.

Frame 1

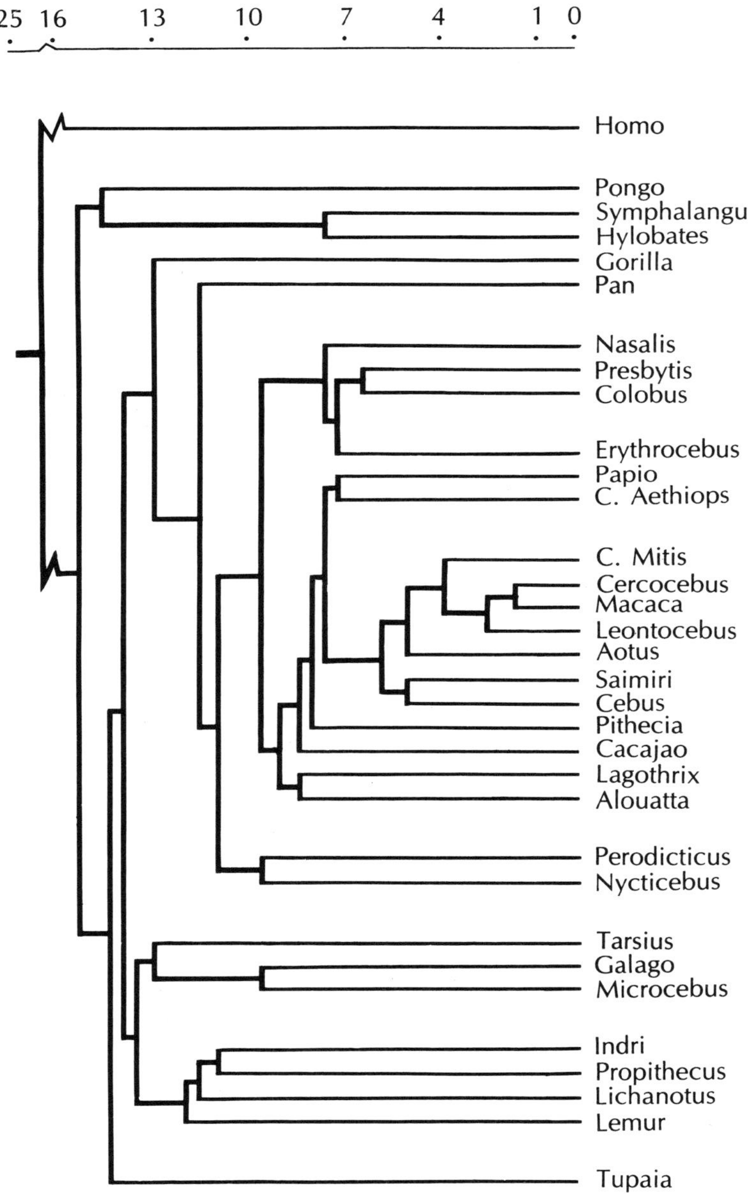

Frame 2

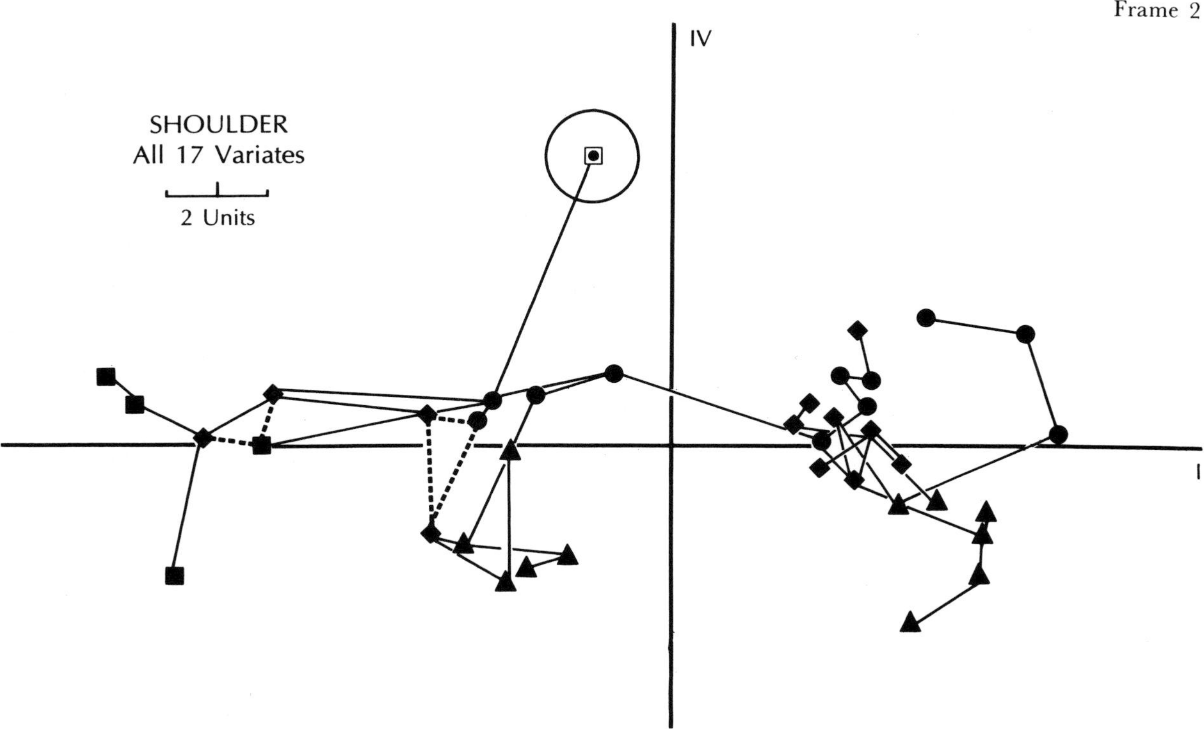

termediate is visualized by a line of intermediate waviness. We have avoided the impossible task of 'seeing' a seven-dimensional model (Fig. 3.24). We have substituted for that task the comparison of two-dimensional curves, something that the human eye readily does.

These methods are of value not only for the examination of a series of known groups, i.e. data from living animals, but also for investigating the positions, relative to such known groups, of unknown specimens, i.e. fossils. Clearly a fossil specimen can be interpolated so that the relationships pertinent to the biological investigation are preserved and can be examined. Of course, biological speculations about unknown groups are more easily made from these analyses when the unknown groups are reasonably close to one or another of the extant groups in the original analyses. If the new fossil group is far distant from all the groups then we can say little except that it is uniquely different. But the problem of knowing when a fossil group really is out of place among a series of extant forms is somewhat more easily solvable with these analyses and displays (Fig. 3.25).

However, even with these approaches, it is still not easy to view more than a small number of groups at one time unless the data are fairly strongly clustered. Again, therefore, we recognize a limitation in our abilities to visualize data in many dimensions. And the problems of data display are so urgent that a number of different methods have been invented in recent years. Thus seven-dimensional data conventionally seen as three clusters through a regular tree diagram may also be represented through displays as clusters of similarly shaped 'stars' or 'faces' (Fig. 3.26, Chernoff, 1973; Welsch, 1976). In complicated examples it may well be the case that such two-dimensional displays representing seven dimensional data are easier to see than the original seven dimensions themselves.

A second method: the pictorial approach

A second group of tools are also being developed to help reveal the information hidden within complex biological shapes and structures. These include some of the techniques of form, pattern, image and texture analysis. Here the rationale is to attempt to avoid the problems inherent in the measurement of structure, but to use the information held in the total 'pictorial' view that is lost in the use of measurement and is too complex to be seen from inspection alone.

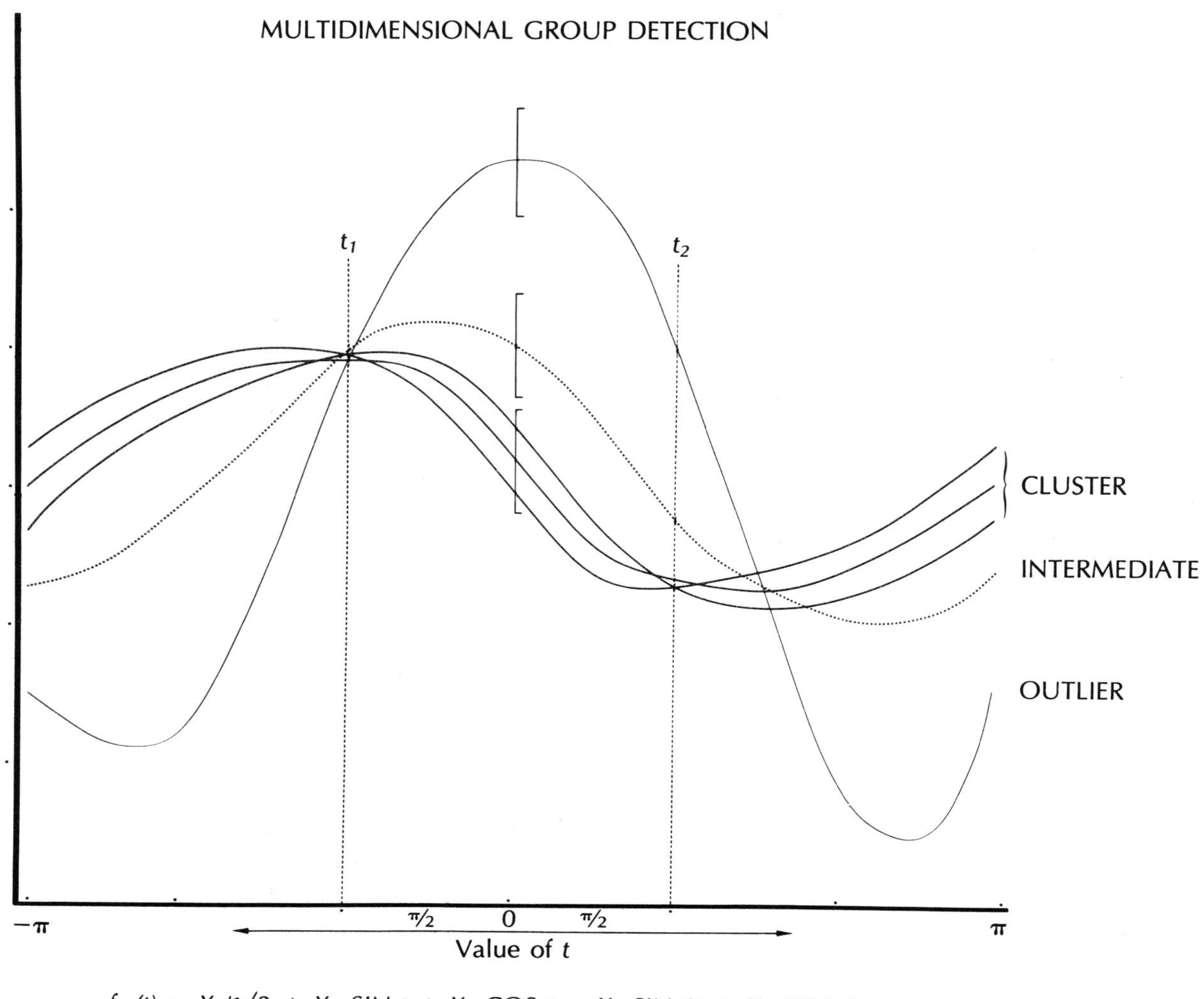

$$f_x(t) = X_1/\sqrt{2} + X_2 \, SIN \, t + X_3 \, COS \, t + X_4 \, SIN \, 2t + X_5 \, COS \, 2t + \ldots$$

Fig. 3.24. The display of seven-dimensional data using sine-cosine plots. Each of the seven dimensions is used as the x_1, x_2, x_3– ... x_7 coefficients to generate the sine-cosine wave. A single wave thus represents the position of a given form in seven-dimensional space as characterized by the seven coefficients. For technical reasons, these dimensions must be independent (such as the new independent canonical variate calculated by the multivariate statistic) and not correlated (e.g. not the original correlated variables taken upon each specimen).

A cluster of similar specimens is evident as a cluster of similar waves; an outlying specimen is clear as a markedly different wave; an intermediate specimen is represented by a wave lying in an intermediate position. This is one of the techniques for the display of high-dimensional data invented by Andrews.

This may be done by scanning the picture using photosensitive instruments in some way or other. In this way a complex shape may be characterized in a manner that allows its quantitative handling for purposes of biological comparison. One example is the recognition, by a computer, of the shape of a structure through the definition of which parts of a picture contain the shape and which parts do not when a series of random lines are thrown across the picture (Fig. 3.27 after J. Cowan, personal communication). Another is through the delineation of a perimeter function using lengths and angles of lines which are tangent to the shape of interest (Fig. 3.28, e.g. Attneave and Arnoult, 1966). Yet another characterizes a shape through a central axis function obtained by collapsing the shape into its midline (Fig. 3.29, e.g. Blum, 1962). Such methods have been developed mainly in fields outside human evolution (e.g. Udupa and Murphy, 1977, in image processing) but, as the exam-

Fig. 3.25. **The first frame supplies an assessment of the unknown form (solid ball) that places it among the various known forms (shaded balls) by a multivariate statistical study resembling the explanations of Figs. 3.20, 3.21 and 3.22.**

The second frame demonstrates that when full information is available for the unknown, its position is actually far distant from the known groups as indicated by a new axis in the model.

Use of the sine-cosine function prevents the mistake of the first frame. Thus, the third frame shows a high-dimensional plot of the information from the model in the second frame; and from this it is as equally clear that the unknown (dotted line) is very different from the known forms (solid lines).

The fourth frame (next page) shows the high-dimensional plot for the information from the model of the first frame; though the model did not show that the unknown was different, the high-dimensional analysis does (although the difference is represented in a different way – markedly increased curvature). This result was hidden in consideration of the first model alone.

The fifth frame (next page) takes the information in the third frame and places the first coefficient (the one responsible for the unique position of the unknown) into the position of the fifth coefficient. Note how the uniqueness of the unknown is now represented in the high-dimensional plot: it is remarkably similar to the fourth frame.

Frame 1

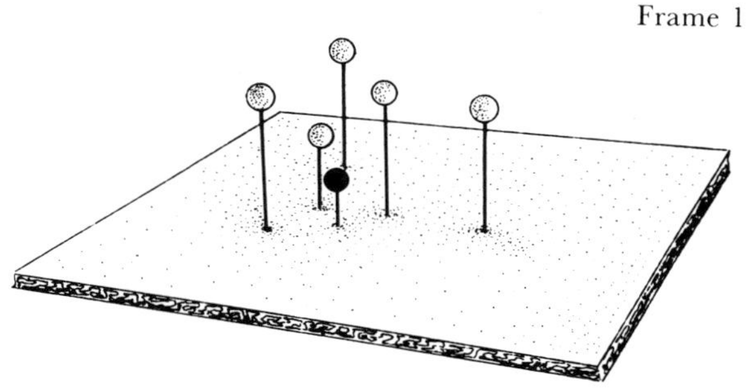

Frame 2

Frame 3

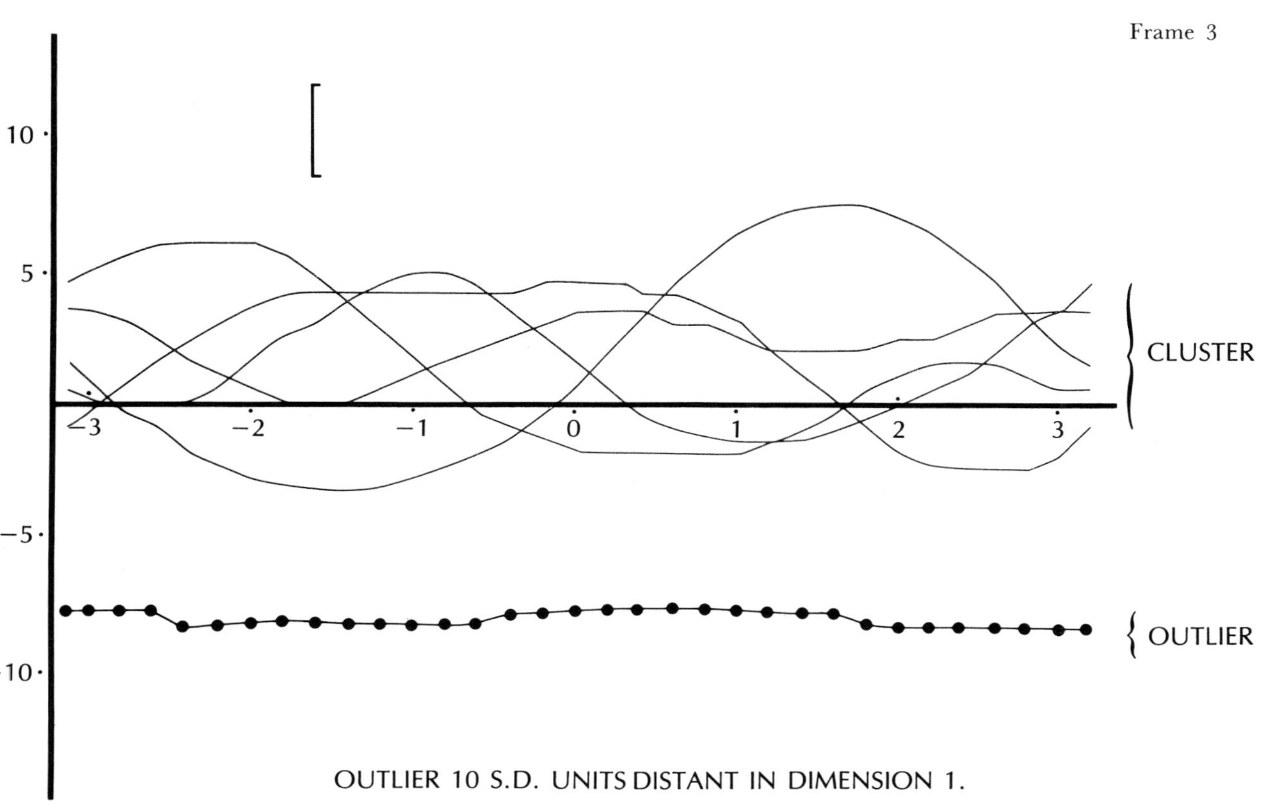

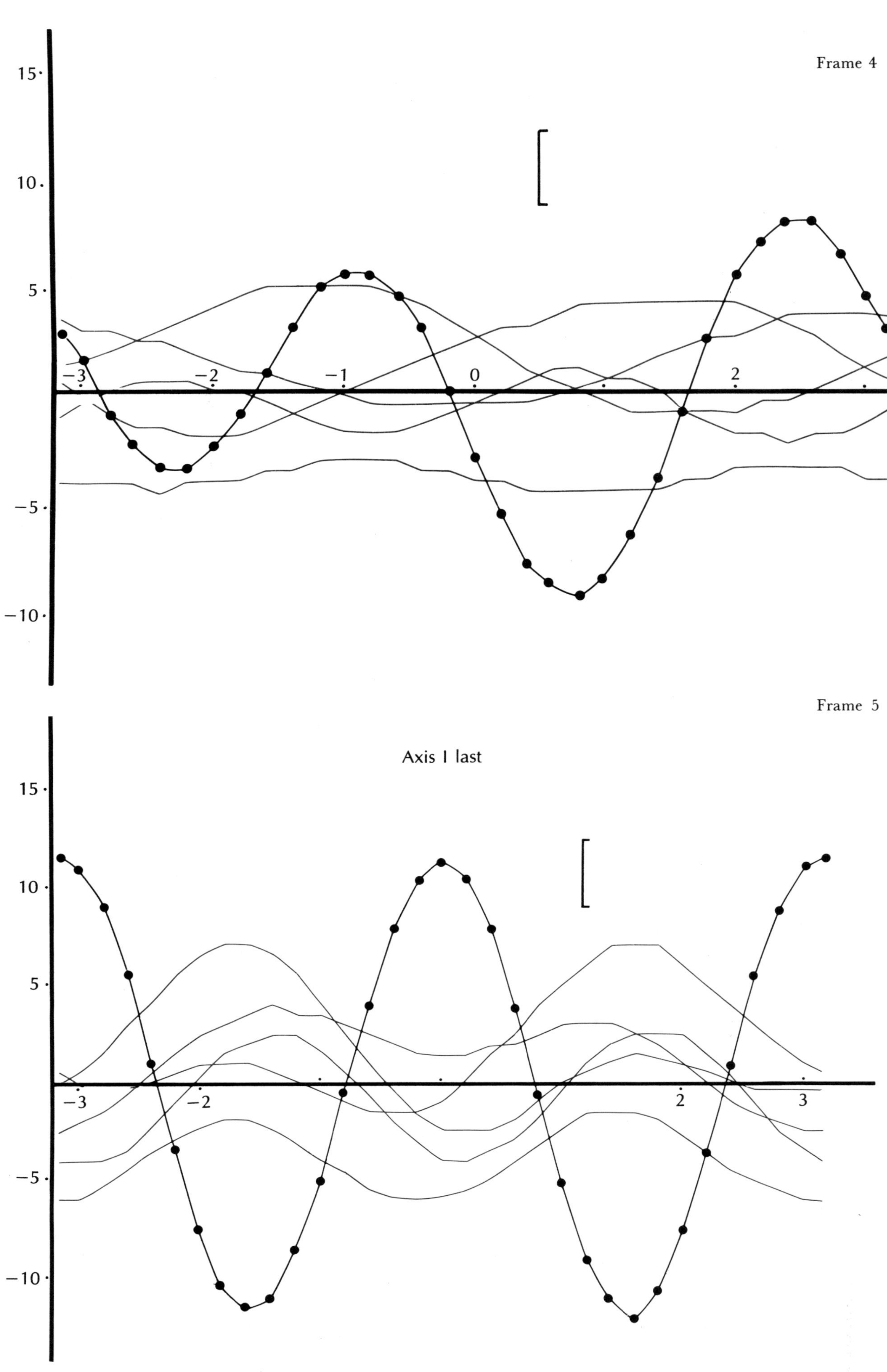

Frame 4
15
10
5
−5
−10
−3
−2
−1
0
2
Frame 5
Axis I last
15
10
5
−5
−10
−3
−2
2
3

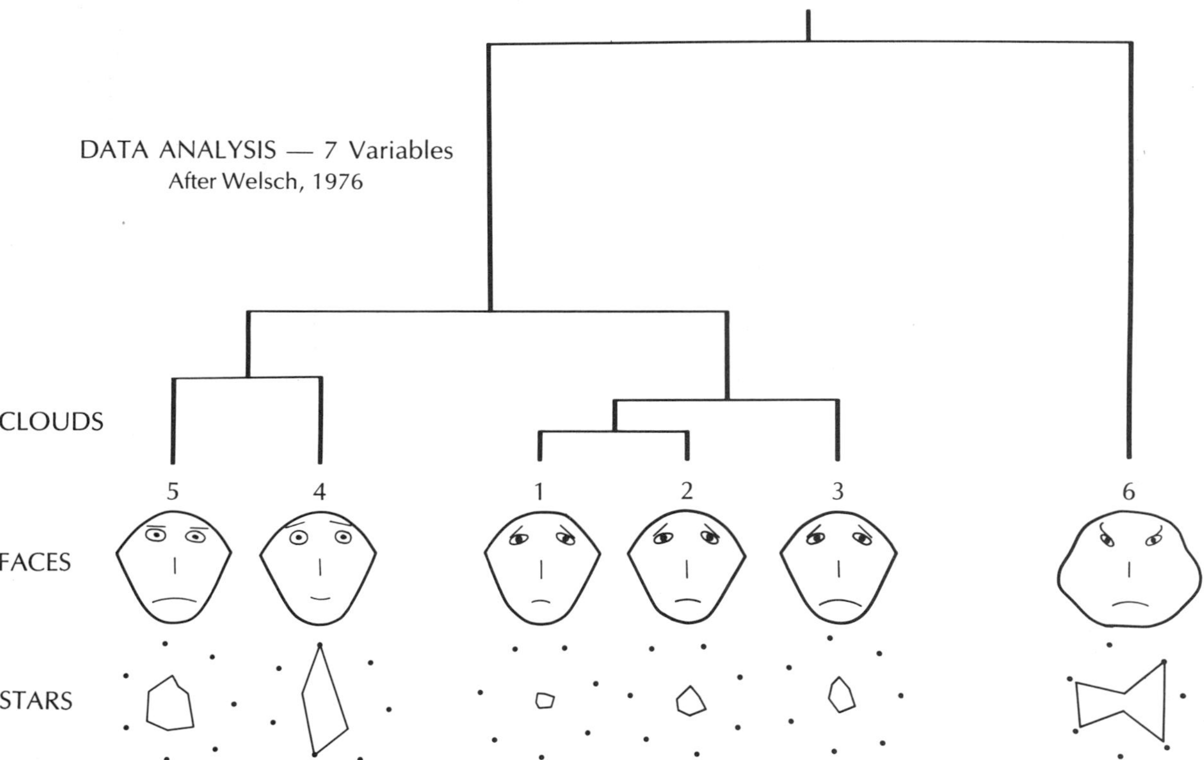

Fig. 3.26. A set of seven-dimensional data have been analysed by Welsch demonstrating their arrangement by a dendrogram into the three groups at the top of the diagram (the group comprising items 5 and 4, the group containing items 1 through 3, and the group including only the solitary item 6, respectively).

A computer program that plots the same data as a series of different faces allows the same three groups to be recognized.

So does another computer program that plots the data as a series of stars. Any of these methods may have especial value in rcognizing groups in different studies.

ples show, all are capable, with differing degrees of success, of characterizing aspects of the forms of bones.

The 'pictorial' problem may also be tackled by transforming the entire information present in a picture using optical techniques. One attempt is Moiré fringe analysis (stemming from discoveries in optics known for many years now) for contouring and comparing complex shapes (e.g. Takasaki, 1970; Weinberger and Almi, 1971). Certainly in theory at any rate, a complicated form like the pelvic girdle, or a complex surface such as that of a joint cavity or the convoluted surface of a tooth (e.g. Elliot and Morris, 1978) may be characterized by this method (Fig. 3.30). Often Moiré fringe analysis can be used together with stereophotogrammetry (e.g. Savara, 1965) or holography (Zelenka and Varner, 1968) mentioned earlier. Reductions such as these are achieved without defining special points upon the surfaces of the bone, although if it be necessary to incorporate information about such points, for example points of possible biological import such as the edge of an articular surface or the margin of a muscular scar, this may be done.

When, however, interest in defining form and pattern reaches into such complex forms as these, it perhaps ought then to include even more of the information presented by a bone. For a bone consists of very much more than the two-dimensional representation in a picture, or the three-dimensional envelope of its outer surface. One of the persistent problems that has vexed those interested in the functional significance of bone form over the years has been the description of the complex architecture *within* a bone. This architecture includes the network of bony spicules (trabeculae) that can be seen with the eye, or, at most, a hand lens or low-power dissecting microscope. It is a structure especially visible under low

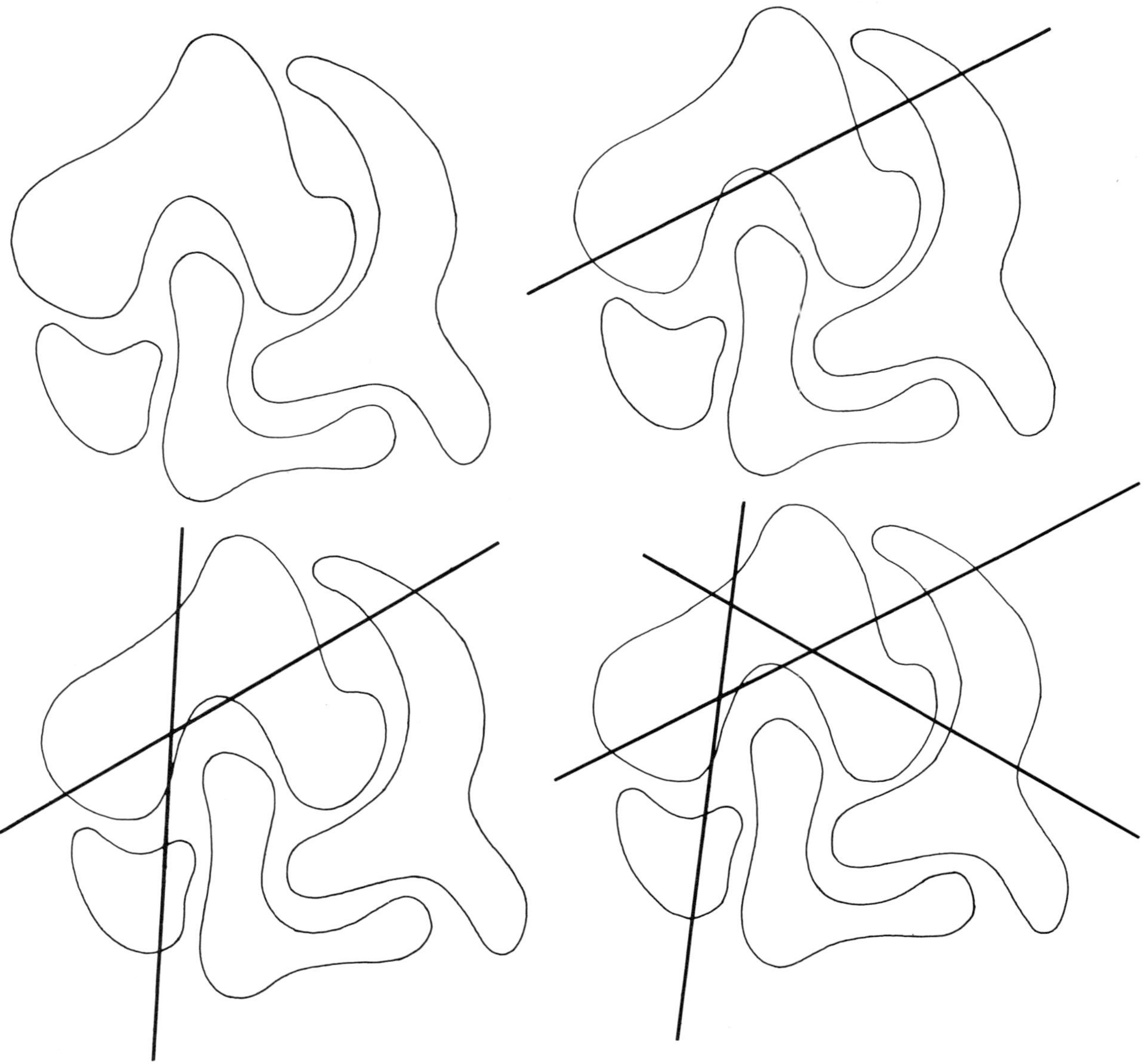

Fig. 3.27. A complex set of shapes is presented in the first frame. In the second frame some information is obtained about those shapes through consideration of how many of the shapes are cut by a first 'random' line. The amount of information about the shapes is increased when a second random line cuts the shapes (third frame), and yet again with the addition of another random cut (fourth frame). Repetition of this process eventually defines the original complex rather well. This type of group-finding procedure can be carried out using a computer.

magnifications of the scanning electron microscope. And it is evident from the information contained within sections and radiographs of bones and some fossils (Fig. 3.31). How can these more complex patterns be characterized? For they must be characterized before we can understand them.

Usually the delineation of such a pattern depends upon defining, visually or through densitometry, the major bundles of the bony spicules and the most prominent parts of the compact external shell of the bones. The primary problem, a more complex characterization of the internal trabecular network, has scarcely been tackled. And it is clear why this is so. For instance, one way to characterize such patterns is to measure the length, width and orientation of each trabecular element within a given section of bone. But it is most time-consuming to measure hundreds of trabeculae for even one section of bone; the comparison of many such sections within a single bone to obtain the

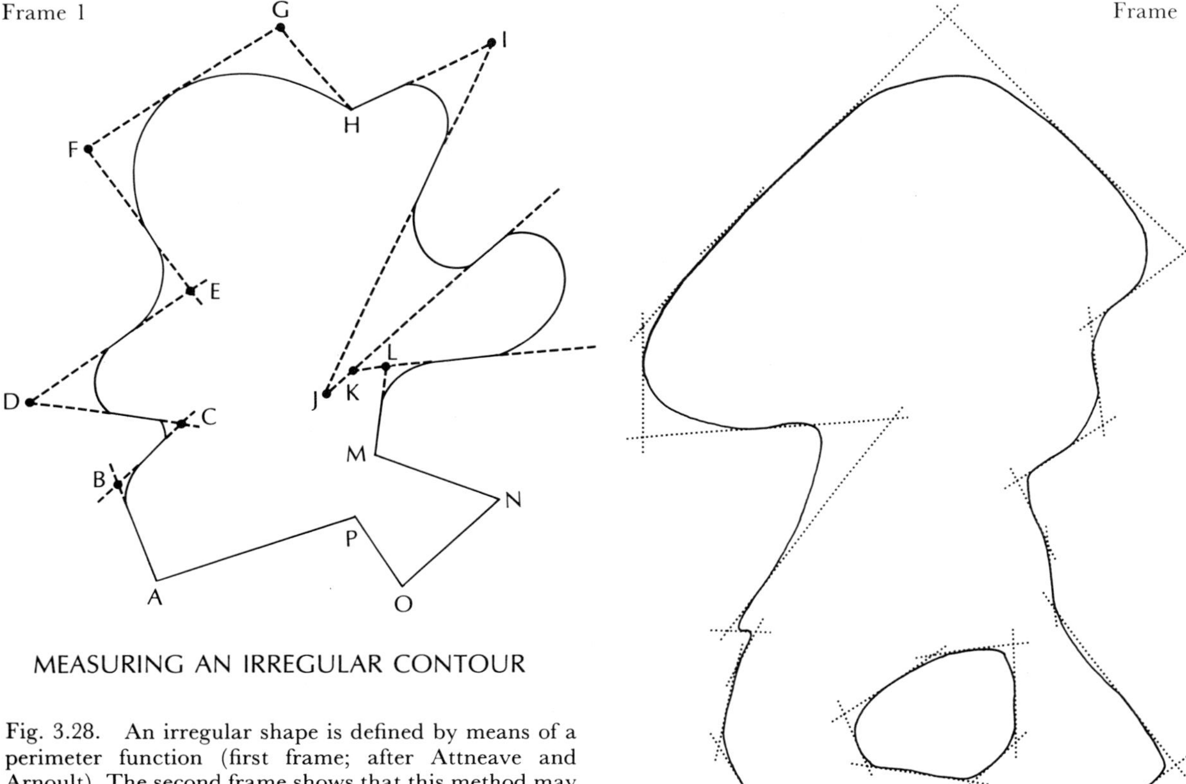

Fig. 3.28. An irregular shape is defined by means of a perimeter function (first frame; after Attneave and Arnoult). The second frame shows that this method may not work very well with a bone, the pelvis. In part this is because the complexity of the shape of the bone is rather greater than that of the nonsense shape in the first frame.

three-dimensional view is a proportionately greater task; the final comparison of many single bones of one group of animals with many other single bones of each of several other groups of animals becomes virtually impossible.

However, it happens that other scientists are also interested in avoiding all of this work. Computational and optical methods for studying images have evolved as fallout from modern technological advances related to such problems as the exploration of space and the development of instruments such as lasers and computers. The best known examples of this are found in the transmission and improvement of pictures taken by artificial satellites in space probes (Andrews, 1970), in pattern recognition studies for recognizing writing or fingerprints using powerful computers (Duda and Hart, 1973), and in holographic and photographic investigations using optical data processing (Holeman, 1968). A by-product of many of these methods is the realization that, in the procedure of the reconstruction of an improved picture, an intermediate stage exists in which the pictorial data are transformed into a non-pictorial form (Fig. 3.32). Both computational and optical processing, for instance, transform the original picture in some quantitative manner. This intermediate transformation has been used simply for reproducing a better final image of the pattern. But it has also been used as a key for identifying patterns. And it has been used for helping to understand patterns. All of these may be useful for clarifying patterns within bones.

In the case of pictures of X-rays of bones, this intermediate transformation may supply succinct yet comprehensive information about the details contained within complex trabecular lattices. And although this can be done using mathematical manipulations with a computer to create Fourier transforms of a picture, it can also be done optically using a laser (Goodman, 1968; Davis, 1970; Lipson, 1972). For one property of a lens system forming a real image is that it performs an optical transform on the input signal. Using such optical equipment, specified visual items in the original picture that are defined by size and orientation can be easily identified in the transform. And this identification can be quantified so that the contributions of the specified items relative to the whole picture may be obtained (Fig. 3.33).

Such procedures can be used in an exploratory manner: that is, the technique may be employed in an empirical way as a searching tool. The Fourier analysis may be used as a 'fingerprint' of the bony patterns for the purposes of recognizing individual patterns and of comparing one bony pattern with another.

The technique may also be used explicitly as an 'hypothesis testing' device: for instance, in the studies of bone a guiding theory might be that the trabecular network is the realization of a random process. From the Fourier analysis, unique and sufficient parameters may be obtained that characterize the network and afford a test of the actuality of the random nature of the bone fabric.

An alternative model that may be more useful to test relates to the idea that the bony plates lie principally at right angles to one another; for some

Fig. 3.29. The first frame presents a series of diagrams showing how a central axis function displays the essential points of a shape (after Blum). The second frame shows that this may work rather well with the shape of a bone, again the pelvis.

Frame 1

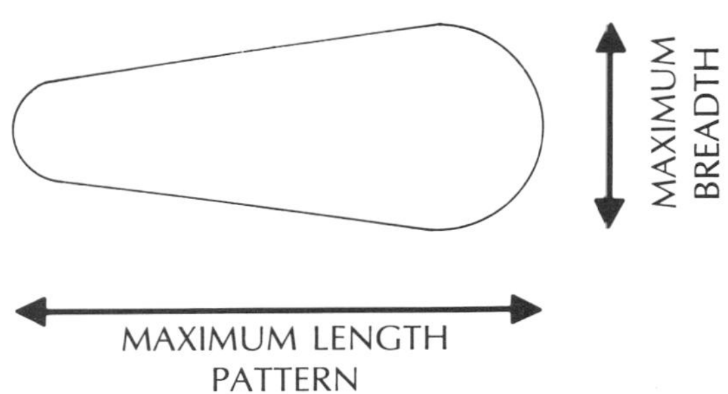

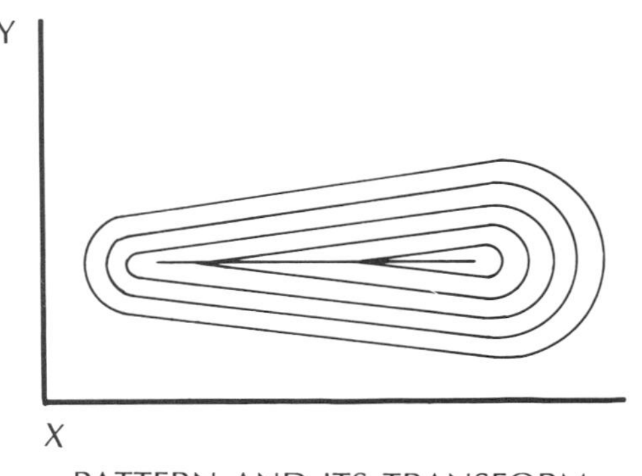

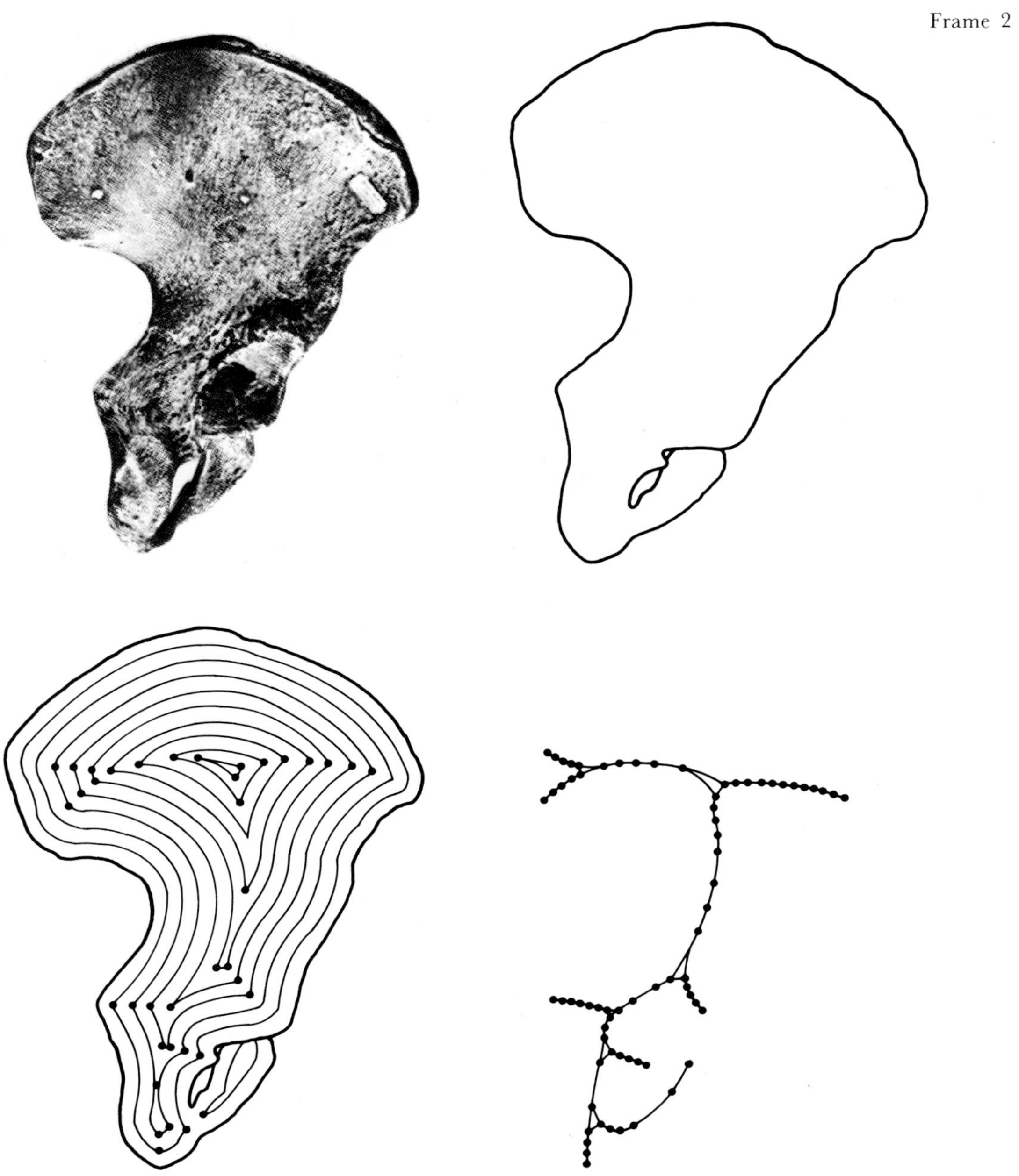

investigators believe that this pattern relates to the right-angle (orthogonal) network of principal stresses that can be computed inside a bone (or any object) during load-bearing functions (e.g. Murray, 1936). Such a model may be a more realistic idea to test than the notion that bone is the realisation of some random process.

Thus, Fig. 3.34 presents a comparison between sections of the fourth and second lumbar vertebrae. There are some differences in pattern to be sure, but the chief elements seem to be vertical and horizontal. Figs. 3.35 and 3.36 confirm this view for tomograms (the middle few millimetres) and radiographs (the full thickness of these same vertebrae) the actual picture is different in each case, but each contains similar vertical and horizontal information.

Optical Fourier transforms are presented in Fig. 3.37 for the sections of Fig. 3.34. The transform for the fourth lumbar vertebra is shaped like a regular American football with chief axes vertical and horizontal. But the transform for the second lum-

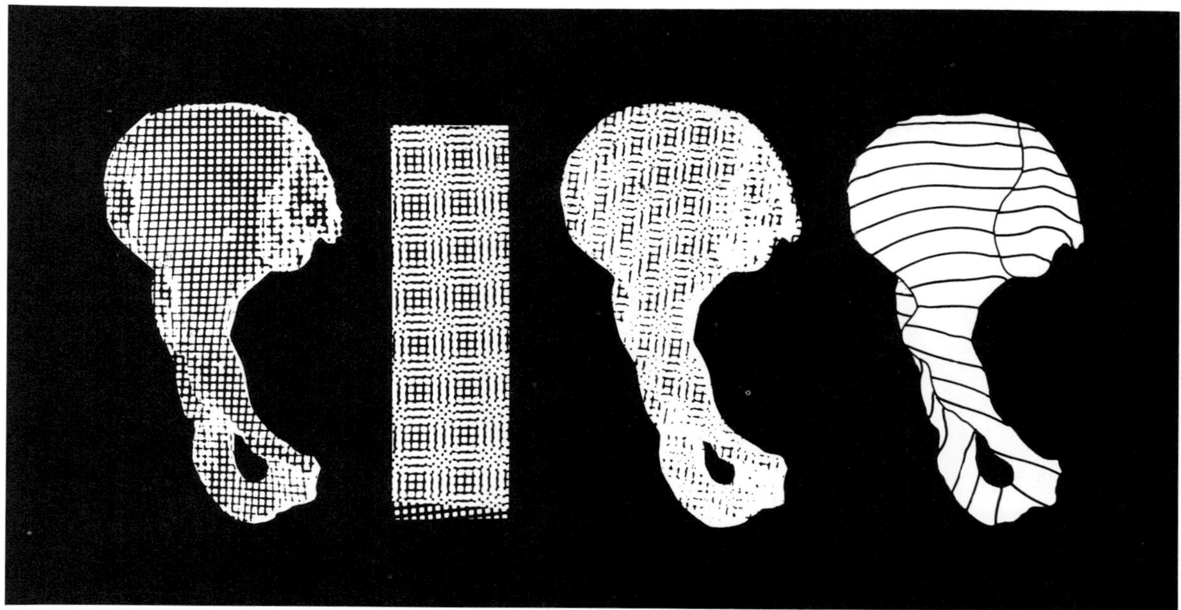

Fig. 3.30. A pelvis can also be examined by shadowing upon it a grid (first frame). The shadow of the grid is deformed because of the three-dimensional nature of the bone. However, just as series of Moiré fringes are formed when two grids are superimposed at a slight angle to one another (second frame), so, too, Moiré fringes are formed when the deformed grid on the pelvis is superimposed upon the plane undeformed shadow of the grid (third frame). The Moiré fringes are redrawn for clarity in the fourth frame. This example was a rather crude, early attempt by the author. Recent developments allow such Moiré patterns to define complex forms in a highly detailed and extremely clear manner (e.g. see Takasaki, 1970).

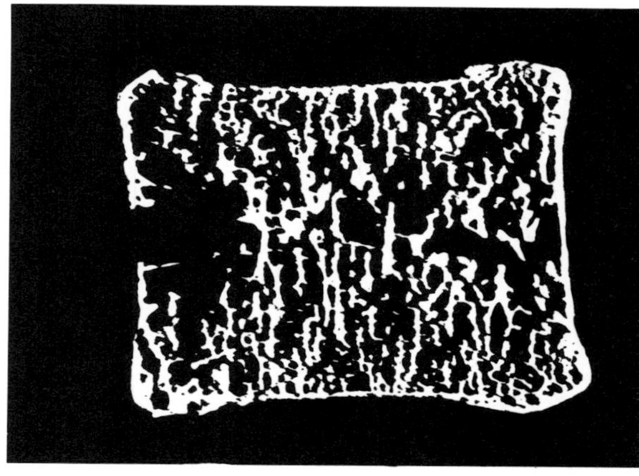

Fig. 3.31. A section of the body of a human vertebra demonstrates its complex internal trabecular structure.

bar vertebra has additional elements lying at other angles. This implies additional elements lying at angles other than vertical and horizontal in the original section. Fig. 3.38 gives the contours for the transforms making it all the clearer that some off-orthogonal elements are present in the second that are absent in the fourth. Attention is drawn specifically to this in Fig. 3.39, the contoured transform for the second lumbar vertebra.

A similar analysis is presented in Fig. 3.40 for the tomogram and the radiograph though the data are presented in the form of a plot. For both tomogram and radiograph of the second lumbar vertebra, the curve representing the orientation of elements shows a peak about 30 degrees out of phase from the 0 and 90 degree peaks that exist for both second and fourth vertebrae. The 30 degree elements are real and can be identified in a photograph of an actual vertebra (Fig. 3.41)

An even more startling example of the power of this method to reveal information not obvious upon inspection stems from the study of radiographic patterns in vertebrae of great apes (Oxnard and Yang, 1981). As with humans so with the great apes: a radiograph of a vertebral body seems to evidence mainly vertical and horizontal shadows representing vertically and horizontally aligned trabeculae, long thought to be related to the vertical and horizontal pattern of principal stresses existing in the vertebra during function. In the case of the chimpanzee (Fig. 3.42) the Fourier transform is cruciate, and this confirms that most of the ele-

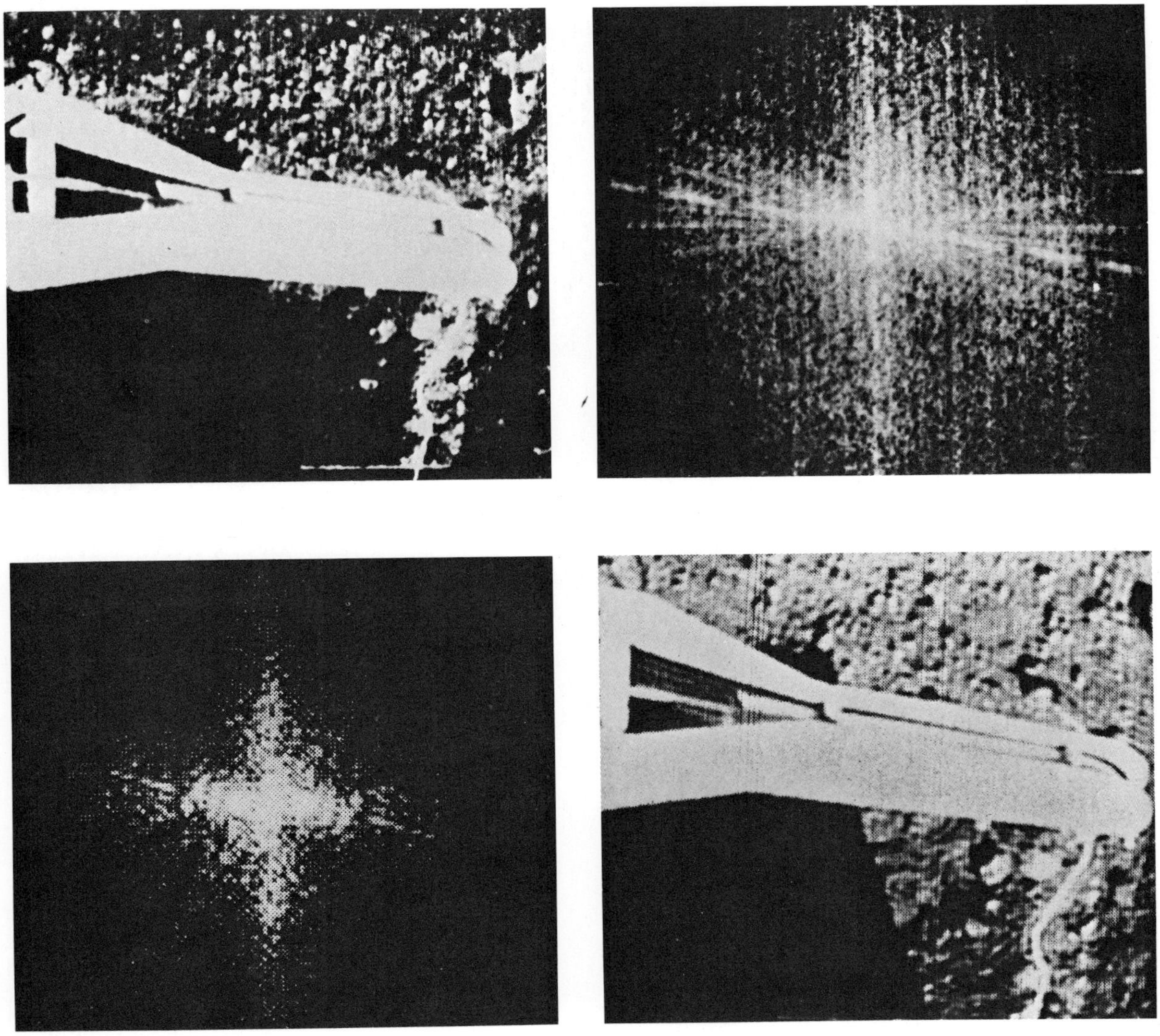

Fig. 3.32. The original photograph of the surveyor spacecraft, upper left. Upper right is its computational Fourier transform; lower left the computationally 'cleaned' transform. Lower right is the improved picture: note the clarity of detail that can be obtained by retransforming the 'cleaned' transform. (Courtesy of H.C. Andrews)

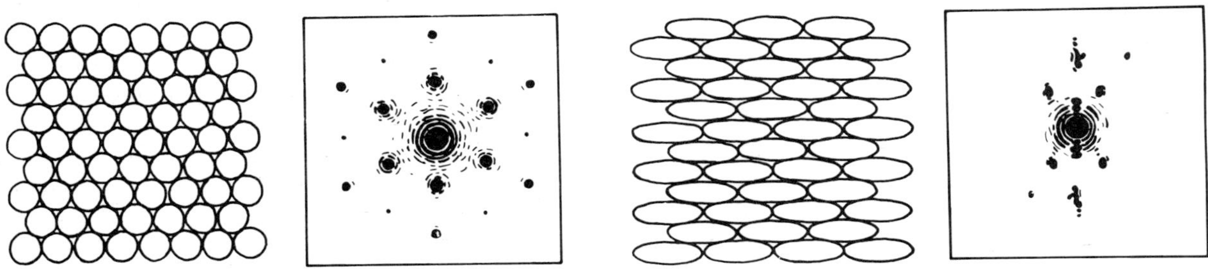

Fig. 3.33. Optical data analysis of a pattern of touching circles (left) and of a uniformly squashed pattern (right). The orientational changes evident in the patterns are as easily seen in the transforms (in each case on the right of the pattern) as in the original patterns (after Pincus).

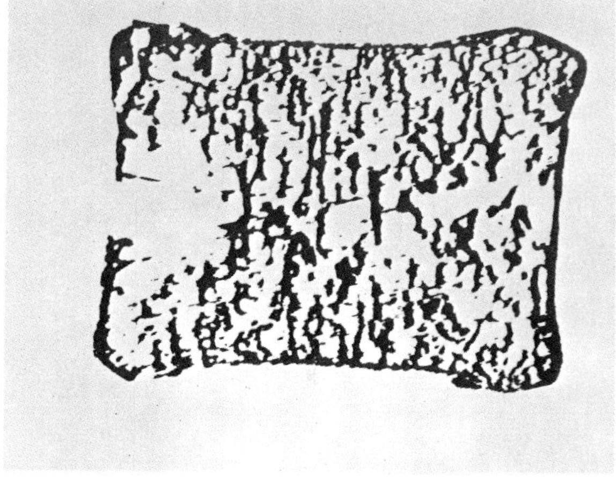

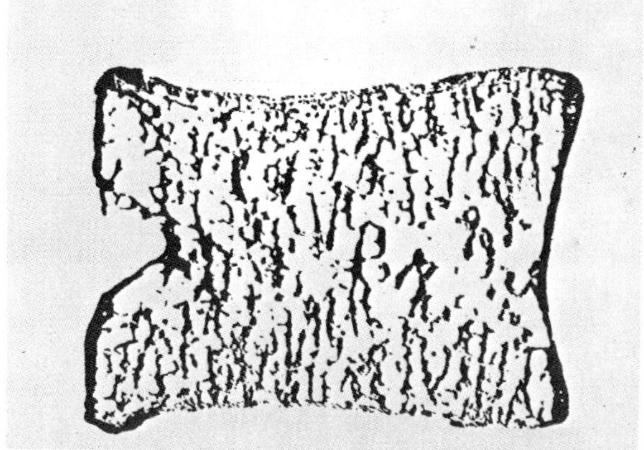

Fig. 3.34. A comparison between saggital sections of the second (above) and fourth (below) human lumbar vertebrae. Visual inspection suggests that the primary elements of the trabecular network inside both of these bones is vertical and horizontal.

Fig. 3.35. A comparison of tomograms (laminograms) of the same lumbar vertebrae as in Fig. 3.34. In this case the tomogram is focused to provide information about the trabecular radiographic shadows of the middle few millimetres of the vertebrae. Again visual inspection confirms the vertical and horizontal features of the internal pattern suggested by the previous sections.

ments in the original radiograph are indeed vertical and horizontal. A section of a chimpanzee vertebra (Fig. 3.43) confirms these basically vertical and horizontal patterns.

Comparison of radiographs of pygmy chimpanzees (bonobos), orang-utans, gorillas and humans (Fig. 3.44) suggests that the vertical and horizontal patterns of the chimpanzee are also the major part of the patterns in these other species (even though small numbers of off-orthogonal elements exist in some of them). This is amply confirmed for bonobos, gorillas and humans — their transforms are basically cruciate (Fig. 3.45).

But Fig. 3.45 especially demonstrates that whatever may be the appearance of the radiograph in the orang-utan, its Fourier transform — a large star-shaped figure — indicates a totally different arrangement (also Fig. 3.45). What pattern could possibly exist in the orang-utan to produce such a transform?

I asked this question of my freshman undergraduate class. They came up with a variety of different possibilities. The pattern might merely be a random mish-mash of trabecular shadows crossing in all directions, said one. This would deny the hypothesis about the association between orthogonnality of architecture and stress. The pattern might consist truly of orthogonal sets of shadows, but with the angles of the different orthogonal localities within the vertebrae changing from point to point, proposed another. This would confirm the theory in a more complexly functioning region. The pattern might consist of a series of trabeculae lying in

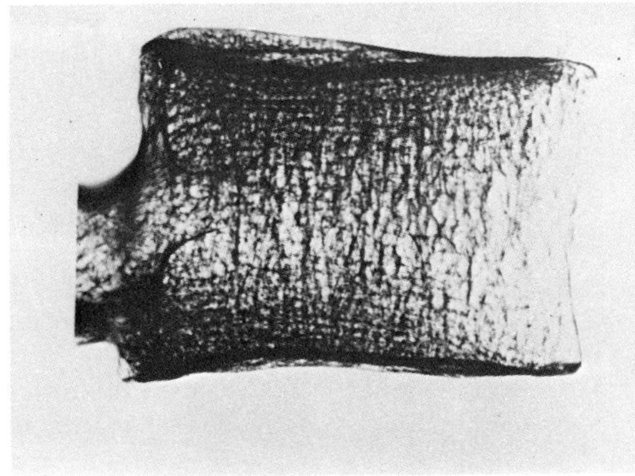

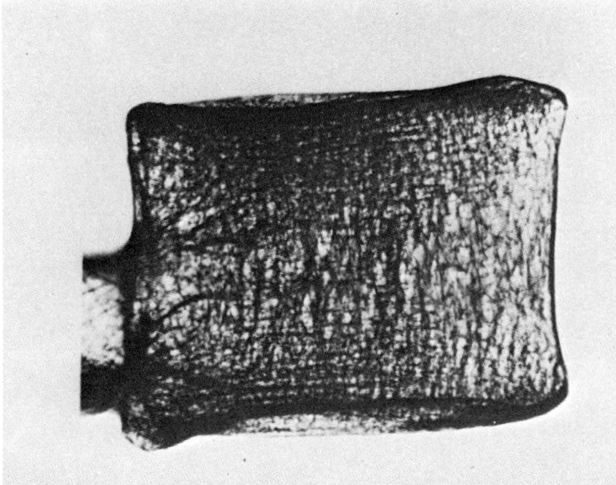

Fig. 3.36. A comparison of full thickness radiographs of the same two lumbar vertebrae. Once again, though the pattern is quite different from that in the section and the tomogram, the overall visual impression is of patterns with primarily vertical and horizontal features.

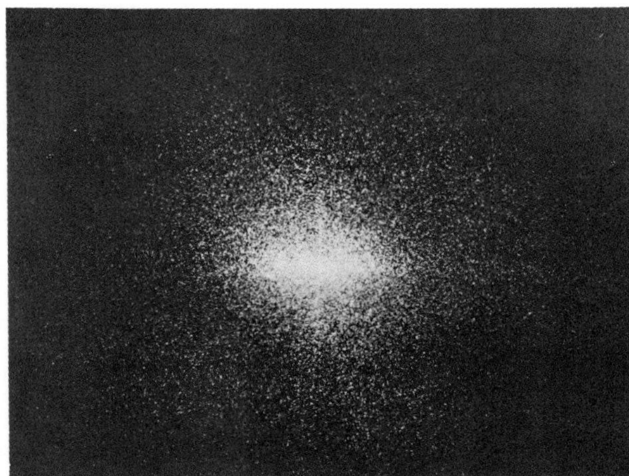

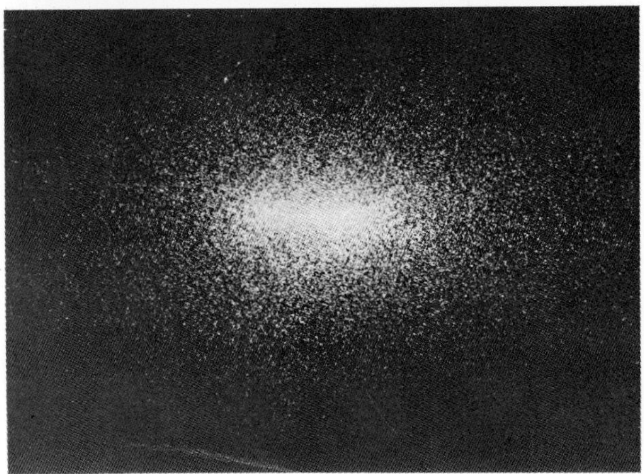

Fig. 3.37. The optical Fourier transform obtained from the vertebral sections of Fig. 3.34. Both have their long axes horizontal, thus confirming the primarily vertical arrangements of the trabeculae in the original sections. But the transform above seems to have some other elements within it that we shall examine in Figs. 3.38 and 3.39.

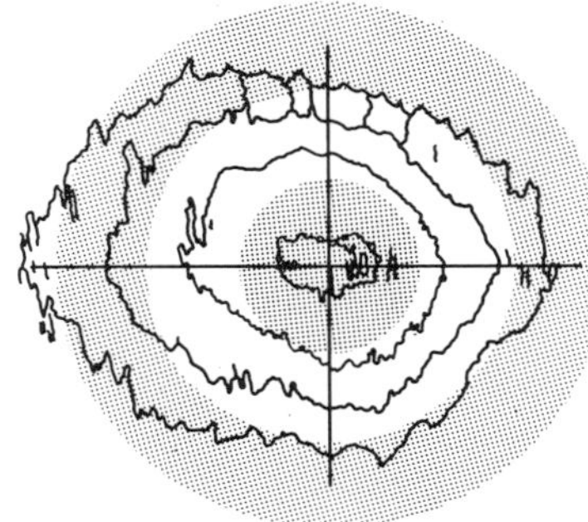

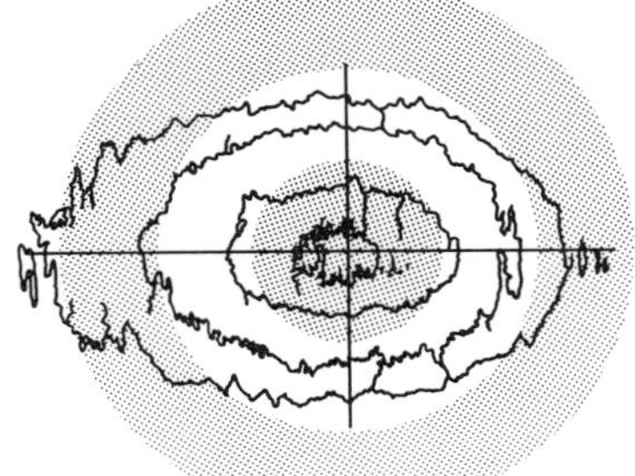

Fig. 3.38. The optical transforms can be contoured to provide quantitative information about them. When this process is carried out (above) we can see that though indeed, the transform of the fourth lumbar vertebra is mainly an oval, indicating that the primary elements in the radiograph are vertical and horizontal, the contours for the second lumbar vertebra are not uniform. This indicates that there are additional elements in the second lumbar vertebra that do not exist in the fourth. And consideration of Fig. 3.39 demonstrates where they lie.

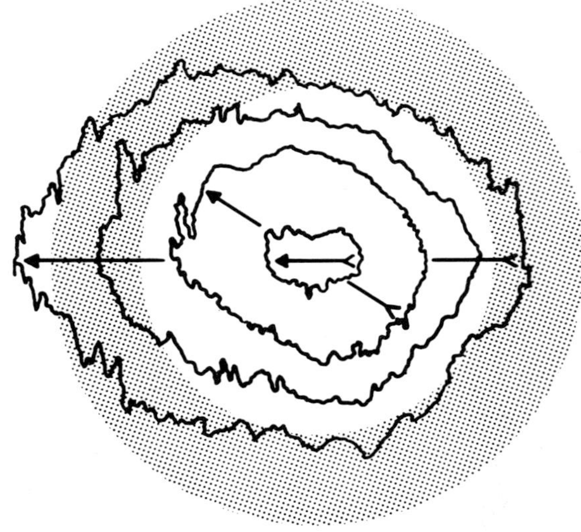

Fig. 3.39. The optical transform of the second lumbar vertebra with arrows drawn to indicate that though the smallest contours and the largest contours are indeed horizontally aligned, certain intermediate contours are aligned at about 30 degrees. This implies that appropriately angled trabeculae are present in considerable numbers in the original section.

many different directions but joined at their centres like small stars, said a third. Or the pattern might consist of a series of trabeculae also arranged so as to be lying at many different angles but joined at their ends, as in a honeycomb, suggested a fourth. Both of these latter arrangements would deny the beginning hypothesis.

The answer turned out to be the last. Fig. 3.46 shows a section of two orang-utan vertebrae in which there exist generally honeycomb networks of trabeculae.

This finding does not deny that there is a relationship between stress and architecture, but it does suggest that the precise form of that relationship must be more complex than a simple one-to-one relationship between orthogonal arrangements of trabeculae and linear principal stresses (Oxnard and Yang, 1981). This hypothesis testing mode of optical data analysis may be most useful.

These techniques can be taken further. It is possible to utilize filters to screen out some of the data in a transform, thereby allowing other information in the picture to reveal itself the more readily (e.g. Shulman, 1970). For instance, if it is obvious that a certain number of shadows in a radiograph are clearly oriented in a particular direction, a directional sector-shaped filter can be inserted which will screen out all items oriented in that way (Becker, Meyers and Nice, 1969). The resulting reconstruction may allow us to see far more clearly what remains (Fig. 3.47).

Again, it may be that the patterns of certain sizes of radiographic shadows are very obvious; once more an appropriate filter (this time, one that is band-shaped) will screen out all items of those sizes, permitting easier inspection of the pattern of other, less obvious, smaller shadows such as edges (Fig. 3.48, and Pfeiler, 1969).

Filtering techniques may be taken yet further. The Fourier analysis of one specimen can itself be used as a filter for the examination of a second. This process then results in a reconstruction of the image of the actual difference between the specimens. From this may be measured the parameters of the difference, and such comparisons consist of the true differences of all the many elements in the picture, not just those of the major trends that are evident to the naked eye (Fig. 3.49, and Scheibengraber, 1974).

Although the example just cited is confined to the comparison of two-dimensional photographs of primate skulls by subtraction, three-dimensional characteristics can be readily obtained by examining, not the Fourier transform of the picture of the object itself (two-dimensional) but the transform of the effect of its three-dimensionality upon some regular pattern shadowed onto the object. Thus Fig. 3.50 demonstrates the difference between a pattern of small ovals shadowed upon a plane, and that same pattern shadowed upon a cube. A more realistic example is obtained by examining the transform of the pattern of ovals when distorted through being shadowed upon a complex shape, a primate skull (Fig. 3.51, and Scheibengraber, 1977, 1979).

Finally, there is the very special application of filtering methods for the reconstruction of patterns in fossils. Radiographs of fossils often delineate quite clearly trabecular and other architectural patterns. But in addition, the picture of internal structure relating to the original bone of the fossil is often obscured by crystalline and other patterns associated with geological processes of fossilization. With the technique of optical filtering described above, it is possible to subtract the unwanted (for biological purposes) geological part of the pattern from that of the entire fossil. What is left then relates to those elements of the fossil that are associated with the original architecture of the bones (Oxnard, 1975a).

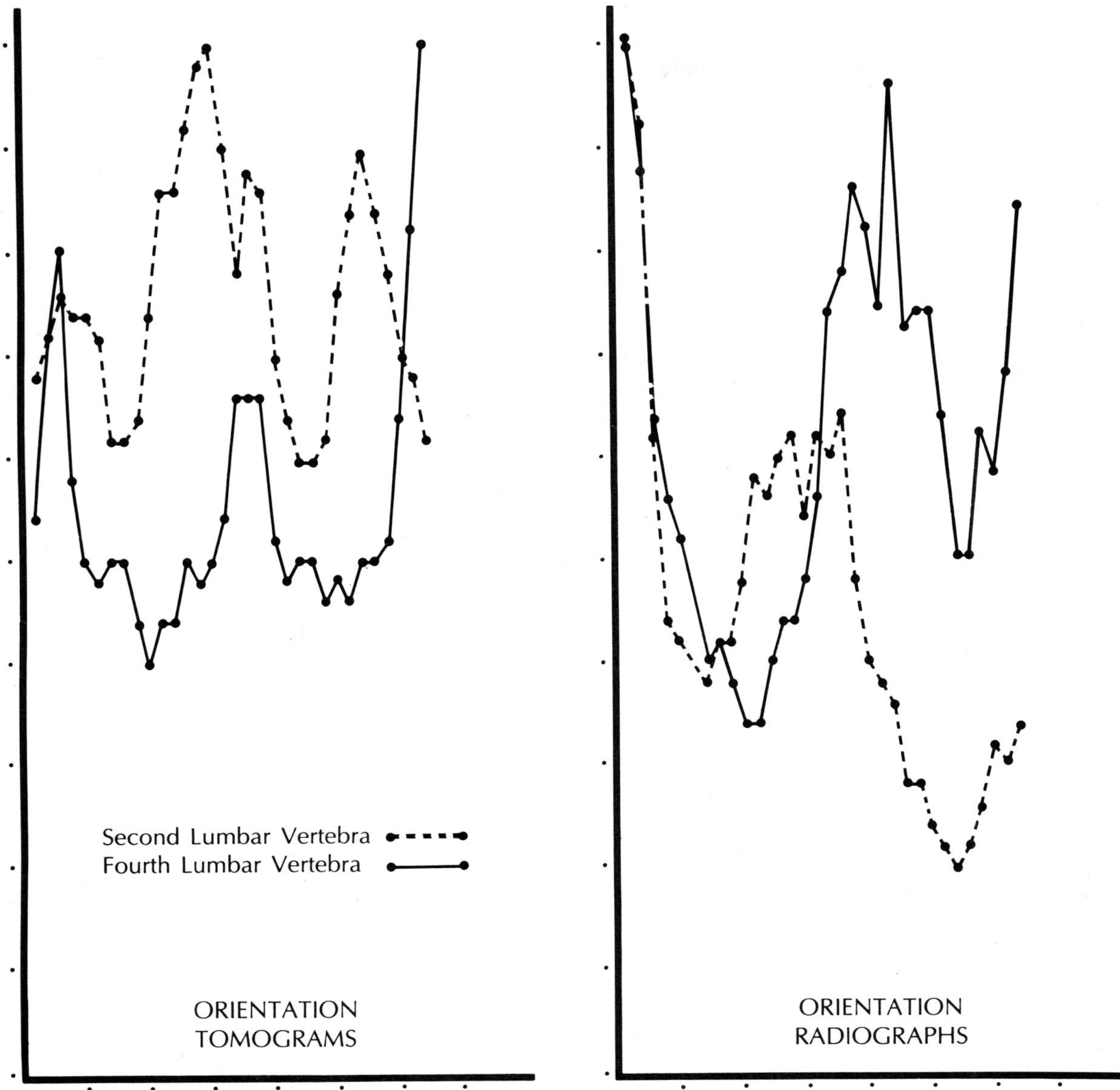

Fig. 3.40. A similar analysis is performed for the optical transform of the tomogram and the radiograph but is displayed in a slightly different manner. The vertical axis is the flux of light across the transform: it tells us, in a transformed manner, about the amount of material in the original picture. The horizontal axis is the angle of orientation of the structural elements on a scale from 0 to 180 degrees by 30 degree steps. The vertebrae have peaks at 0 and 90 degrees in both tomograms and radiographs. In addition, however, the graphs clearly indicate that the second lumbar vertebra has additional elements at about 30 degrees out of phase with those in the fourth vertebra, thus confirming that what is revealed by Fourier transforms in the section also exists in the tomogram and the radiograph.

A last question that may be worth our while considering is: why use a technique such as Fourier analysis at all? Once we become enmeshed in a field like Fourier optics, two things immediately become obvious: (a) Fourier transformations are only one of a whole series of mathematically related transformations and (b) all of these transformations can be implemented, in principle at any rate, by digital, electronic or optical techniques.

Thus, while Fourier transforms can be looked upon as transformations from spatial to frequency domains, so Walsh transformations and Good transformations may also be used (Andrews and Pratt, 1969). Walsh manipulations produce trans-

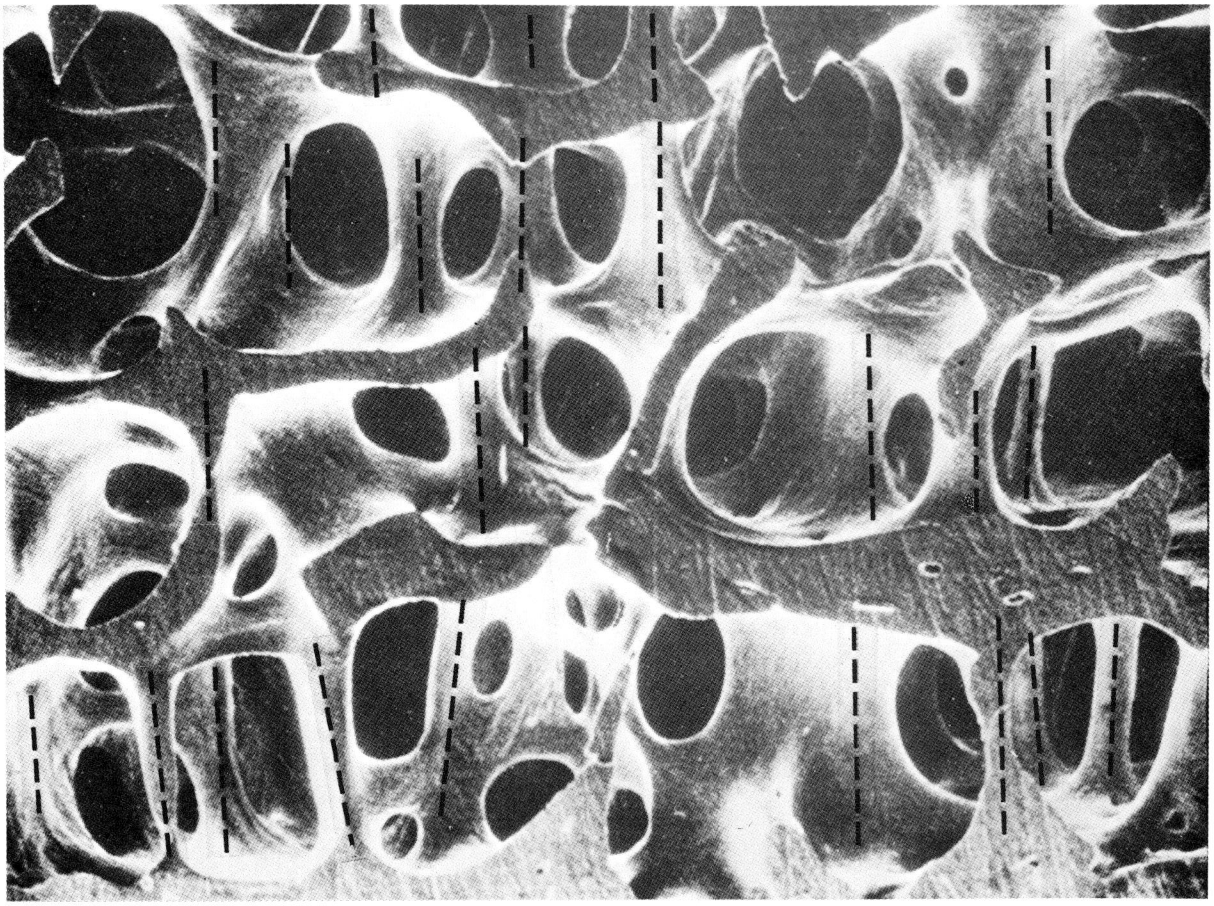

Fig. 3.41. This displays photographs from scanning electron microscopy of the actual second lumbar vertebra. Although the vertical and horizontal architectural features are most clear (upper frame), the additional approximately 30 degree trabeculae (lower frame) can be discerned.

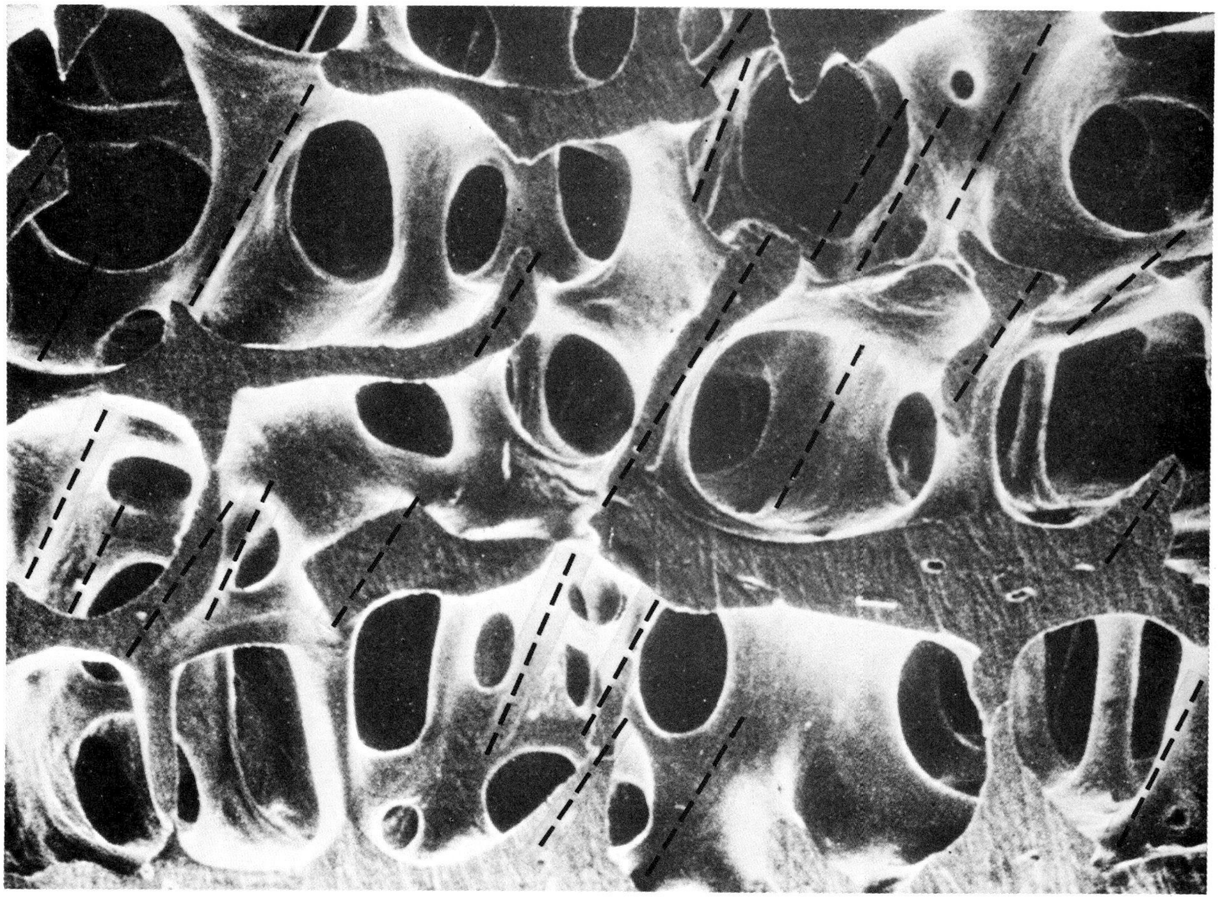

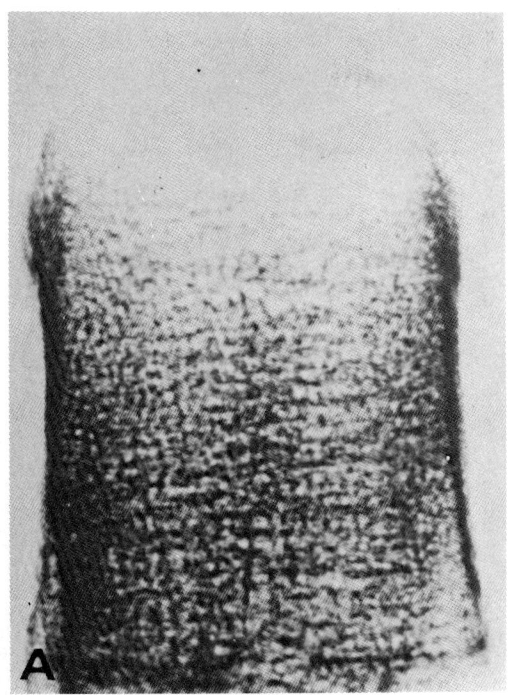

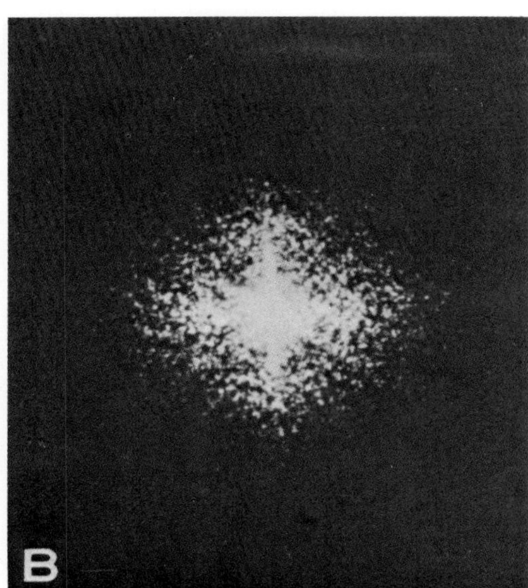

Fig. 3.42. The upper frame is a radiograph, in lateral view, of the body of a lumbar vertebra in a chimpanzee. The lower frame is its optical Fourier transform. The cruciate shape of the transform implies vertical and horizontal elements predominating in the radiograph.

formations from the spatial to the 'sequency' domain, where sequency is a property of 'square' waves analogous to the frequency of the 'curved' waves associated with Fourier transforms. In the Good analysis, the new domain relates to the comparison of points in pairs and hence differs from the preceding in that locality information is revealed (Fig. 3.52).

These general methods are very widely applicable throughout science. They are already well established in a variety of non-biological studies which include automatic processing of aerial photographs, distinction between ground and cloud patterns in meteorological studies, automatic tracking of particles in spark, bubble and cloud chambers and automatic signal analysis for monitoring explosions (e.g. Cheng, 1969; White, 1970). They are now starting to become powerful weapons for biologists (e.g. Meltzer, Searle and Brown, 1967, in botany; Seltzer, 1969, in radiology; Taylor and Ranniko, 1974, in electron microscopy; Kopp, Lisa, Mendelsohn, Pernick, Stone and Wohlers, 1976, in oncology) and especially for biological morphologists interested in the identification, characterization and comparison of complex biological patterns present in bones and fossils.

Yet other methods: the rest of the spectrum

By describing the quantitative handling of form and pattern achievable (a) through discrete measurement and statistical analysis and (b) through continuous methods that describe the envelope of a form in its external entirety, the weave of a pattern in all of its internal complexity, I do not mean to imply an absolute dichotomy of method. There can be little doubt that those two modes of description are merely the well-developed ends of a continuous spectrum. In between are many techniques that utilize a smaller amount of information about a shape or pattern, say through measurement, but that attempt to provide a field picture by interpolation between the points actually examined. Some of these are especially pleasing to biologists, presenting, as they do, attractive geometrical pictures of biological shape comparisons.

D'Arcy Thompson's method of transformations was one of the early attempts. Here, a shape from one animal, characterized by discrete points within itself, is compared with a shape from a different animal through the medium of the distortion of some set of coordinates characterizing the first shape (Fig. 3.53). The work involved in this procedure has been eased through the development of

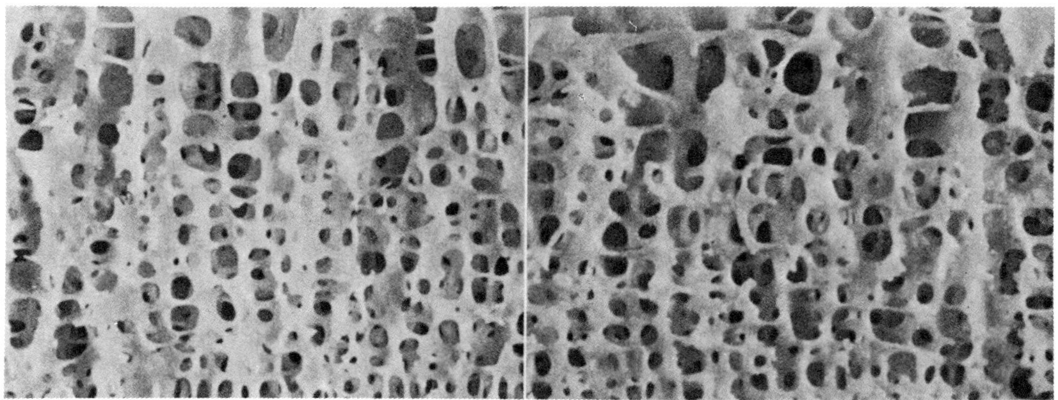

Fig. 3.43. Saggital sections of two chimpanzee lumbar vertebral bodies. The vertical and horizontal pattern of trabeculae predominate as predicted by the transforms.

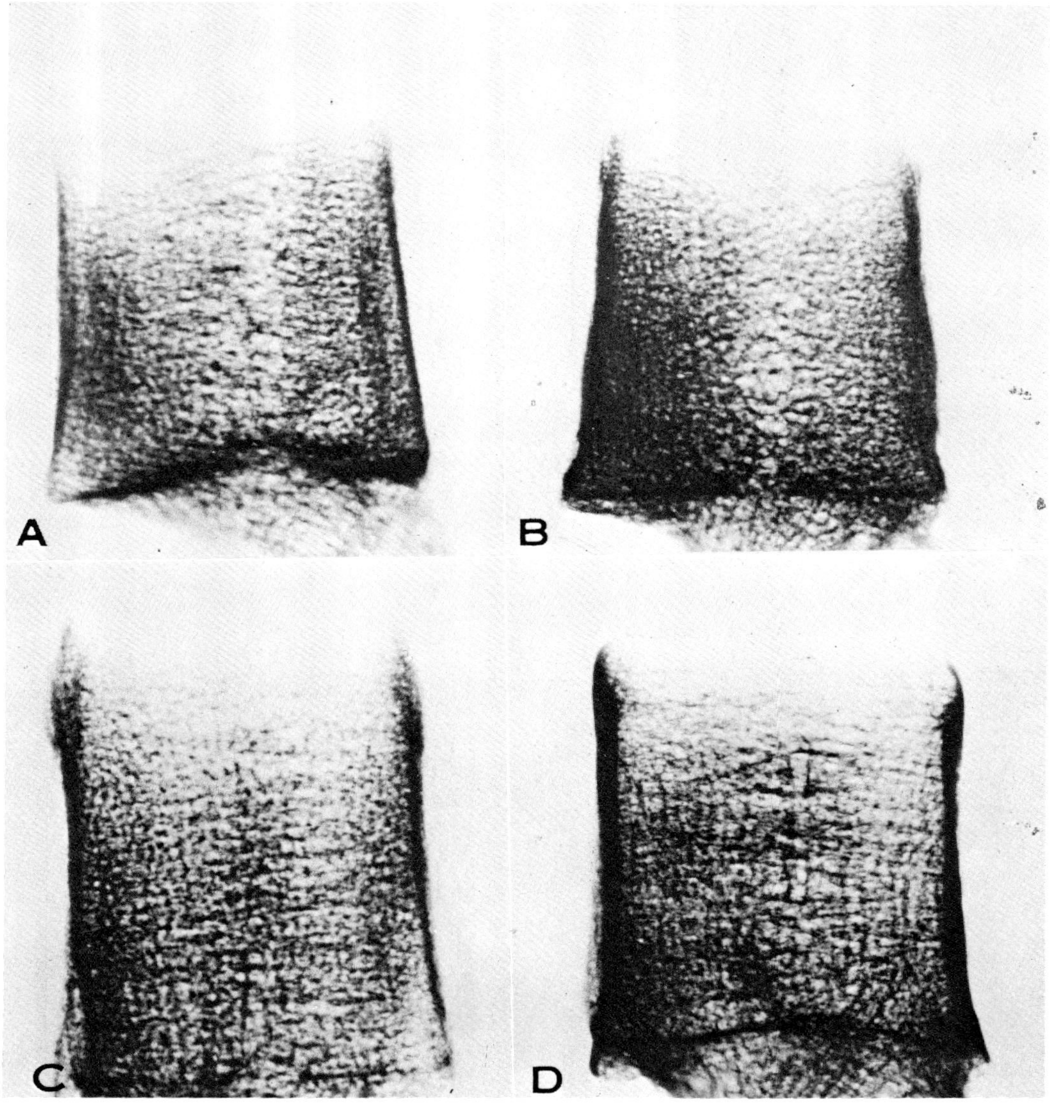

Fig. 3.44. Radiographs of the bodies of lumbar vertebrae in A, bonobo, B, orang-utan, C, gorilla and D, human. Vertical and horizontal patterns predominate in all four.

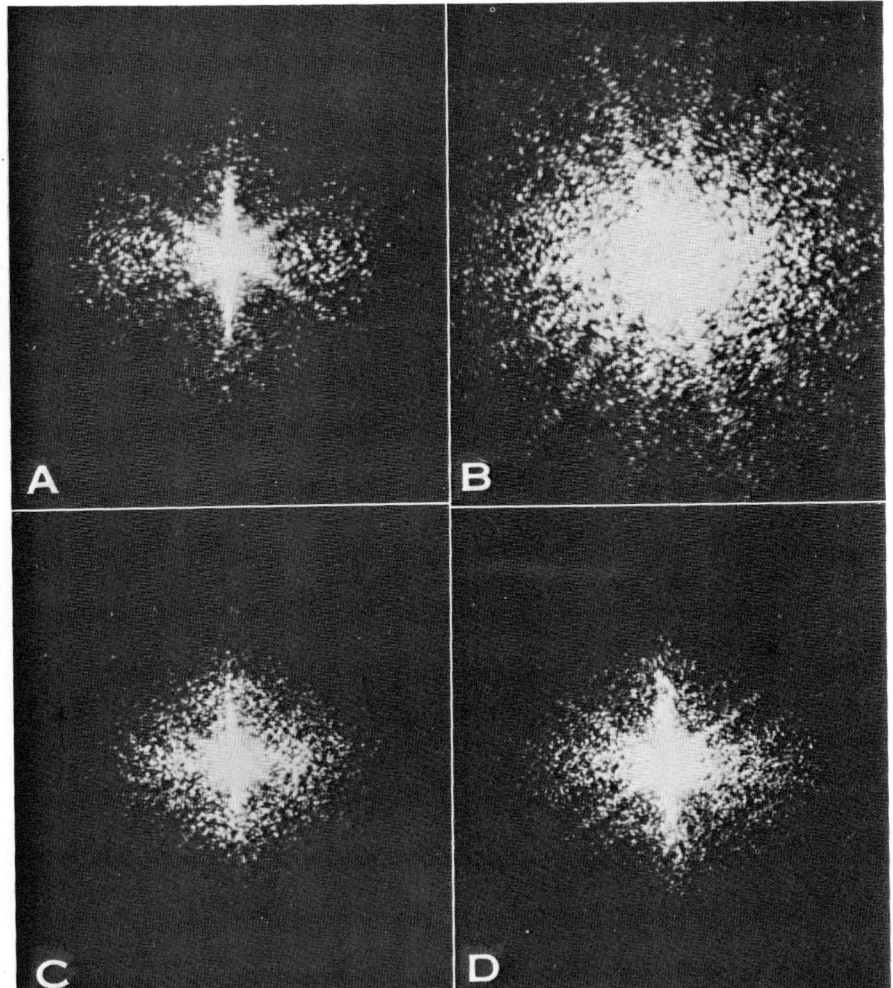

Fig. 3.45. Transforms for the vertebrae in Fig. 3.44. Three of the radiographs (A, C and D) have cruciate transforms, implying that our visual assessment that the radiographic patterns are chiefly vertical and horizontal is correct. But the fourth radiograph (B for the orang-utan) has a star-shaped transform, implying shadows oriented at many different directions within the radiograph.

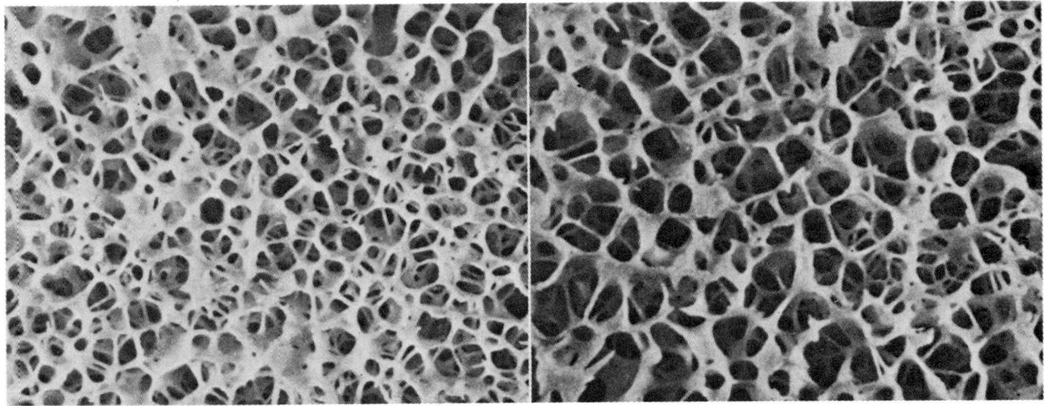

Fig. 3.46. Saggital sections of two orang-utan lumbar vertebral bodies. Elements lying at many angles, as predicted by the transform, truly exist as a honeycomb pattern. This is notwithstanding the contrary visual impression provided by the radiograph.

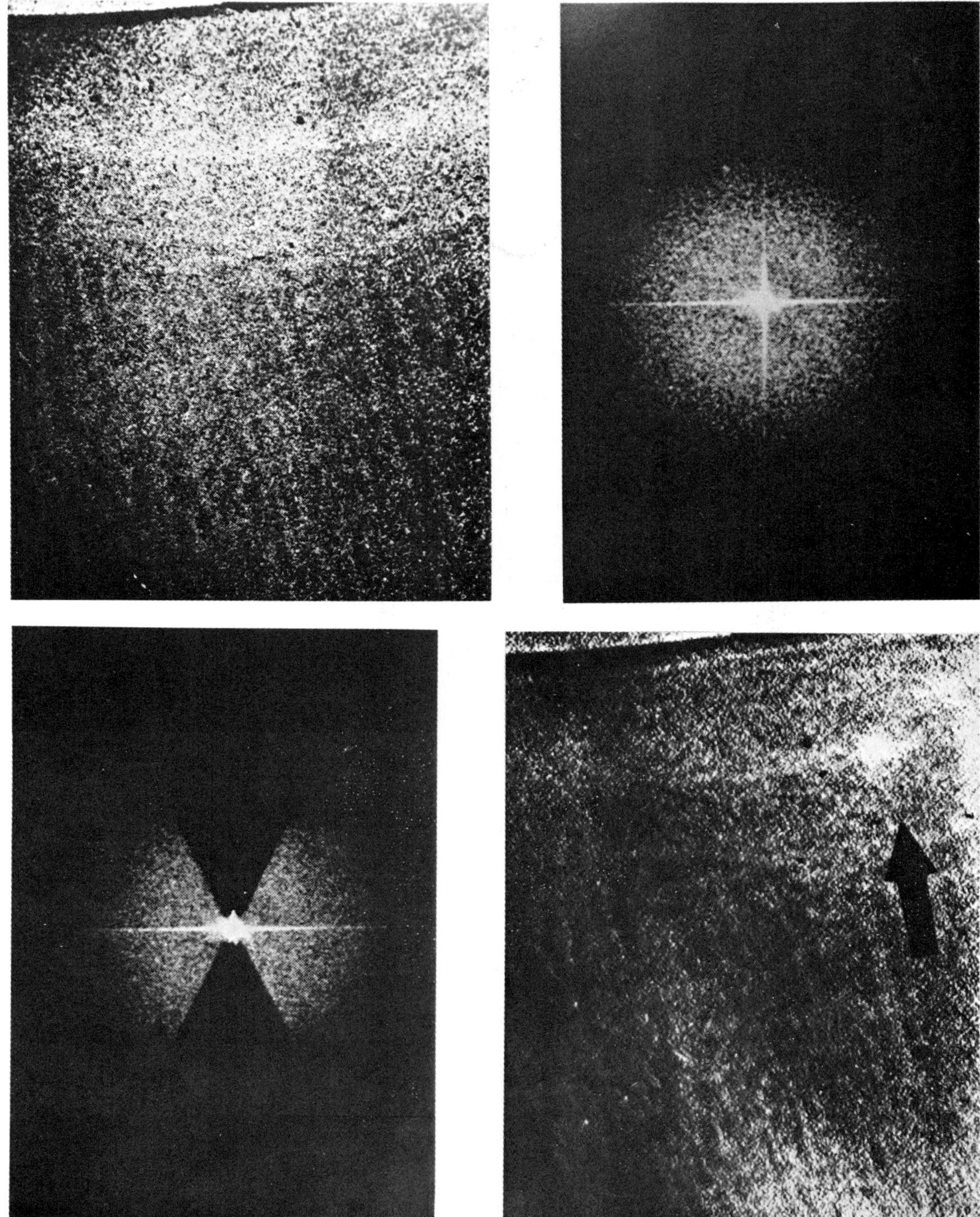

Fig. 3.47. Optical filtering of a radiograph of a small piece of tibia, much enlarged. The top left frame reproduces the trabecular shadows within the bone; their main directions are horizontal. The top right frame shows the optical transform. The bottom left frame shows the transform with a vertical filter in place to screen out horizontally oriented features in the original. The bottom right frame shows the filtered reconstruction. The horizontal shadows have been removed and there is thus revealed a small round shadow shown by subsequent histological examination to be a small (approximately 1 mm) ivory osteoma. This work was carried out by Becker, Meyers and Nice and is reproduced with permission.

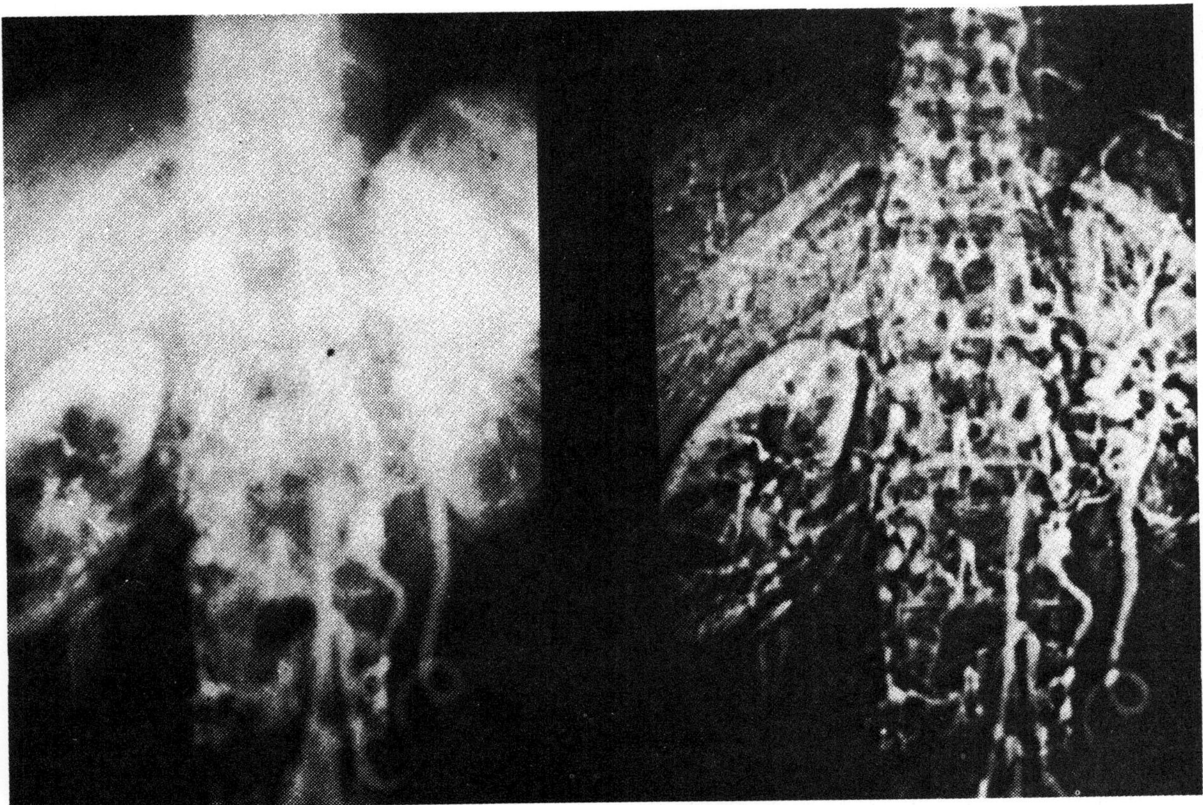

Fig. 3.48. Optical filtering of a radiograph of the abdomen in which one ureter is filled with a radio-opaque medium, thus making it visible.

The left frame is the usual view of such a radiograph, demonstrating that the ureter is still obscured by a variety of other soft tissue shadows (e.g. of the abdominal wall, muscles, other organs and so on).

The right frame shows a reconstruction where a filter has been used to remove the low-frequency shadows resulting from the soft tissues. The reconstruction emphasizes the high-frequency shadows, the edges of objects. In addition, therefore, to revealing much of the internal structure of the spinal column, it also especially outlines the edges of the ureter, making it more visible. It particularly demonstrates that in this individual the ureter is abnormally coiled. This work was carried out by Pfeiler and is reproduced with permission.

interactive computer graphics (Appleby and Jones, 1976; Gabrielson, 1977).

Using the notion of the distortion of a grid by applying the method of trend-surface analysis from geology, Sneath (1967) also compares these shapes with one another (Fig. 3.54). In this case what appears to be a deformed grid is really a deformed surface, the deformation of which represents the trend of difference between the specimens that is based upon measurements from discrete points within the shape. Simple visual examination of these curved surfaces demonstrates the overall similarity between, for instance, the australopithecine and chimpanzee specimens, and the complete difference of both from a modern human.

Yet another development of this representation of differences between specimens is provided by Bookstein (1977a, b, 1978) and Tobler (1977). In this case the differences as expressed in the grids do not depend, as do the prior methods, upon the particular coordinate system used (Cartesian, polar and so on were all used by Thompson as referrent grids). In this new method a coordinate system is used which corresponds to the actual geometry of the objects being compared. It is as though it were imagined that one shape had 'grown' into the other and the coordinate system of grid lines is the one that outlines those local growth changes that are linear. As another analogy, it is as though one of the objects were being deformed into the other object by some set of forces; the grid that is drawn is the grid of the linear principal strains of the distortion (Fig. 3.55). Again, simple visual observation of these grids displays the similarity

Frame 1

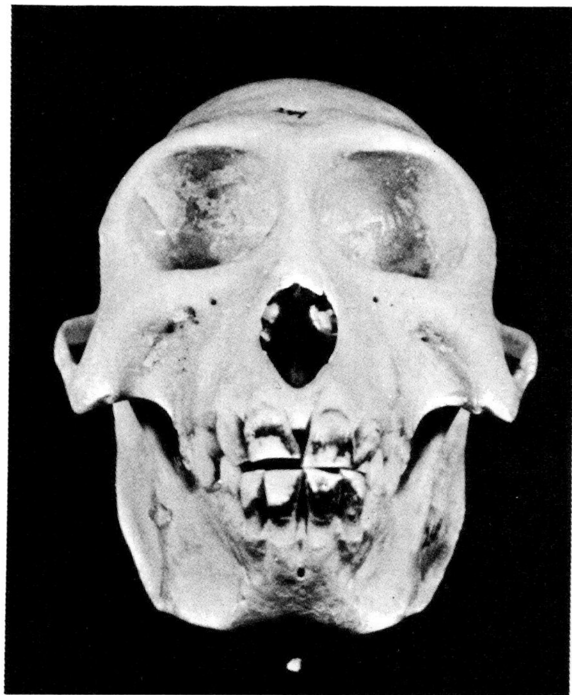

Cercocebus

Frame 2

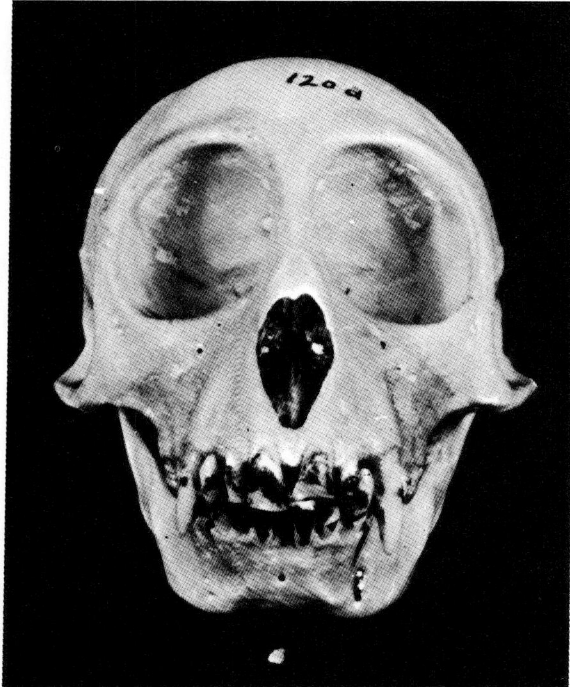

Cercopithecus

Frame 3

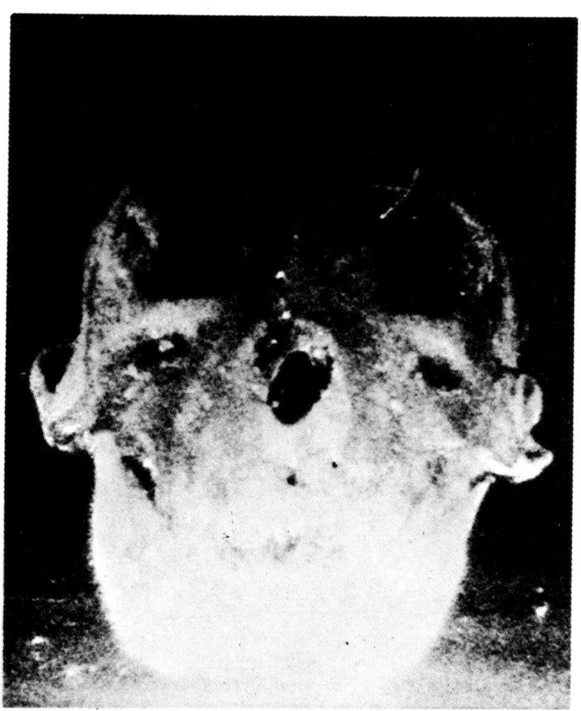

Cercocebus minus Cercopithecus

Fig. 3.49. The top frames are photographs of the skulls of a mangabey and a cercopitheque. The bottom frame is the picture produced by subtraction, using optical Fourier methods, of the cercopitheque from the mangabey. The overall shape of the subtracted picture is that of the mangabey, but the optical density of that picture in its different localities is a measure of the difference between the two skulls.

Thus, although the eye would normally see the photographs as differing greatly in the areas of the orbits and forehead, the fact that the subtraction is least dense there tells us that the difference between the skulls is actually least in that region. The fact that the optical density is greatest in the regions of the jaws and teeth tells us that the photographs are actually most different in this region. The eye is misleading in assessing the differences between the photographs because it more easily sees the large but few features in the orbits and forehead. It does not see nearly so easily the many differences between the jaws and teeth, perhaps because there are so many of them, and perhaps also because they are of a smaller scale. In totality, however, these photographs are far more different in the region of the jaws and teeth.

The reverse subtraction, of the mangabey from the cercopitheque, shows the same pattern of optical densities; but, of course, the framework against which they are presented is that of the cercopitheque. This work was carried out by Carl Scheibengraber and is reproduced with permission.

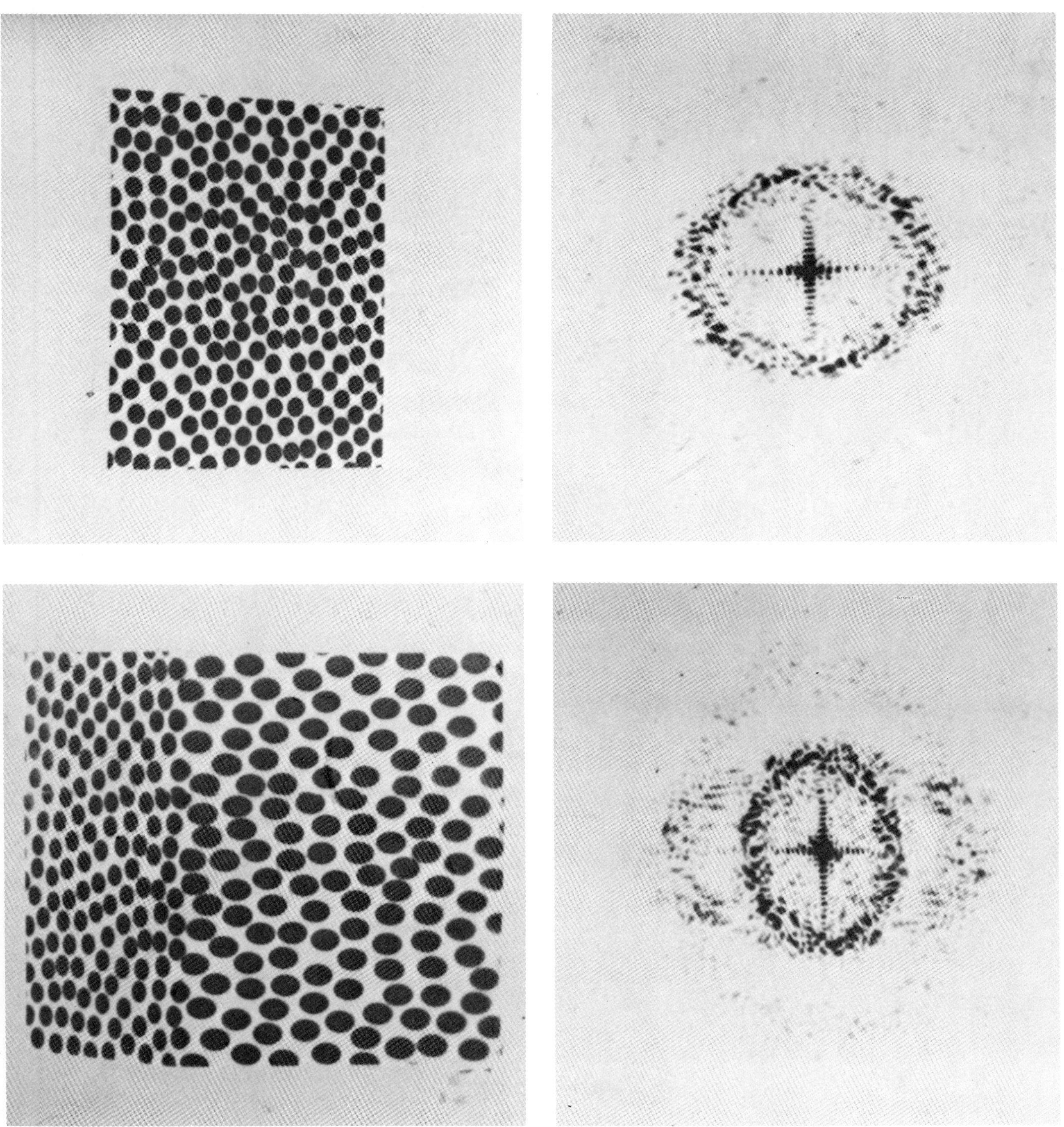

Fig. 3.50. The frames at the top show a pattern of ovals shadowed onto a square and its optical Fourier transform. The frames below show a similar pattern of ovals but this time shadowed onto a three-dimensional cube. The corresponding transform indicates clearly that additional information is present. This work was carried out by Carl Scheibengraber and is reproduced with permission.

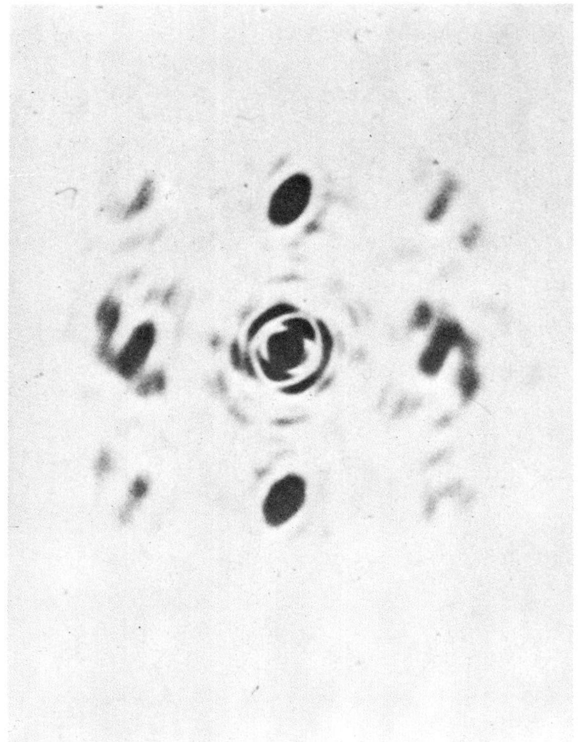

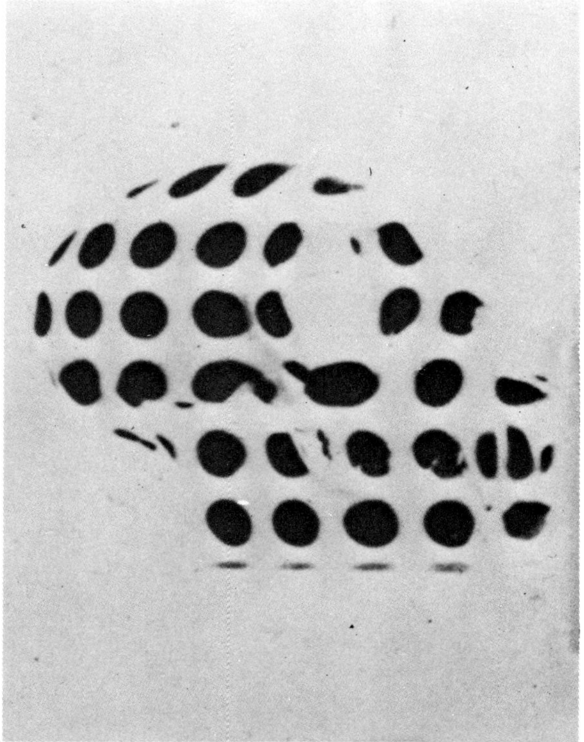

Fig. 3.51. As Fig. 3.50, a three-dimensional object, a primate skull, is examined through optical data analysis of the two-dimensional deformation of a pattern of ovals shadowed upon the skull. The transform (left) is a more ready source of information about the deformation than is examination of the pattern of ovals itself (right).

Of course, this example is not sensitive enough to be useful in the practical case, but it could easily be modified appropriately. The work was carried out by Carl Scheibengraber and is reproduced with permission.

between the skulls of australopithecines and chimpanzees. Because no one specimen starts with a regular grid (as is the case for humans in the kind of comparison done by Thompson) there is no special bias towards one form. Even more can be done with the method in the way of suggesting when one form lies 'between' two others (Bookstein, 1980).

Though dependent upon underlying discrete points, such field descriptors of shape may be most valuable for handling forms and patterns.

And a variety of methods have arisen where complex shapes and patterns are characterized through digital information. These are rather less dependent upon the biological details of the shape or pattern and rather more upon the sensitivity of the grating or grid system that is available for its digital characterization. Many of these tools use information obtained by 'showing' the shape or pattern through an image-reading device to a computer. One of the simplest of these mechanisms, a system something like the raster of a television set, scans the picture (or radiograph or section or other representation) of the object. In this way the pictorial data may be digitized and the digital information can then be operated on by computational methods.

Other ways of putting the shape or pattern into a form that can be handled computationally depend upon throwing a set of points, or lines or fields over it. The devices are capable of recognizing when the object does or does not touch these points, lines or fields. The shape or pattern can also be described using a flying spot scanner which recognizes the outline of a shape by hewing to it until the perimeter is covered before it passes on to study the next shape with which it comes into contact.

Examples of the use of such methods abound outside biology, e.g. in physics for understanding the nature of patterns in bubble chamber experiments (White, 1970); in geophysics for studying pore spacing and pattern in sedimentary rocks (Rink, 1976), in meteorology for the study of aerial

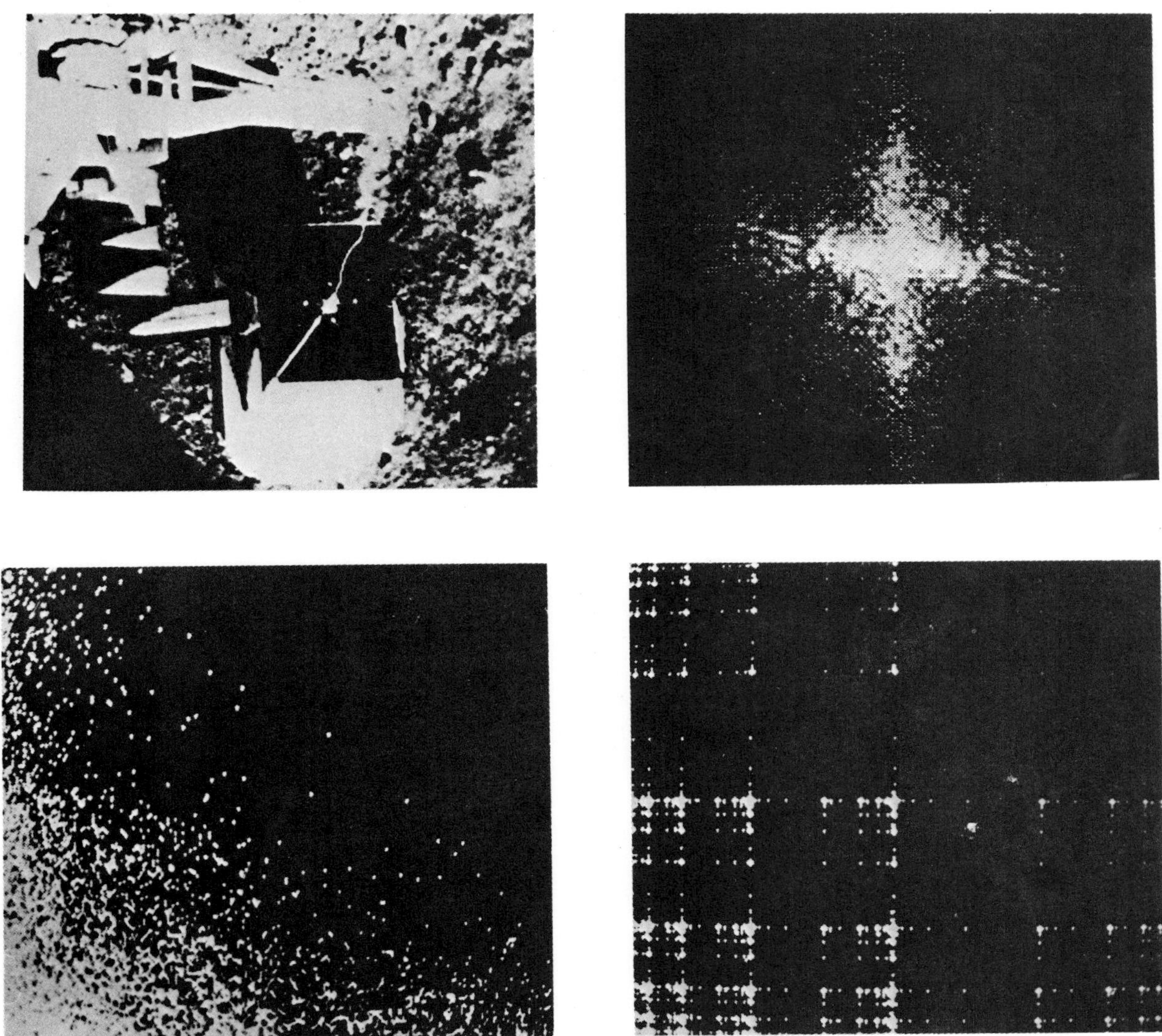

Fig. 3.52. A photograph of the surveyor spacecraft experimental box, upper left. A Fourier transform, upper right, a Hadamard transform, lower left, and a Good transform, lower right, are all appropriate ways of characterizing the original picture. Reproduced by permission from work carried out by Harry Andrews.

and cloud photographs (Cheng, 1969). However, many of these methods have also been applied in biology, e.g. for the characterization of chromosomes (Butler, Butler and Marczynska, 1969; Hildich, 1969), of fingerprints (Cheng, 1969; Shelman, 1972), of animal cell movements *in vitro* (Barski, Butler and Thomas, 1969), and for a variety of microscopic biological studies: autoradiography (Prensky, 1971), nerve fibres (Potts et al, 1972), bone structure (McQueen et al, 1973) and muscle patterns (Eccles, McQueen and Rosen, 1977). They have even been used to study living, growing specimens (Prothero, Tamarin and Pickering, 1974). But again, although capable of descriptions that might be most useful in evolutionary studies of man and other primates, they have scarcely, if at all, been so used.

Many of these studies have deduced a wide variety of features that can be extracted from pictures. Thus, rather than simple measures such as greatest widths and lengths, diameters and perimeters and so on, such studies have isolated features inherent in structure. Thus, analyses of edges, corners, nodes, vectors, tangents, inflection points, central axes, curve-fitted images and so on, may all provide important form and pattern de-

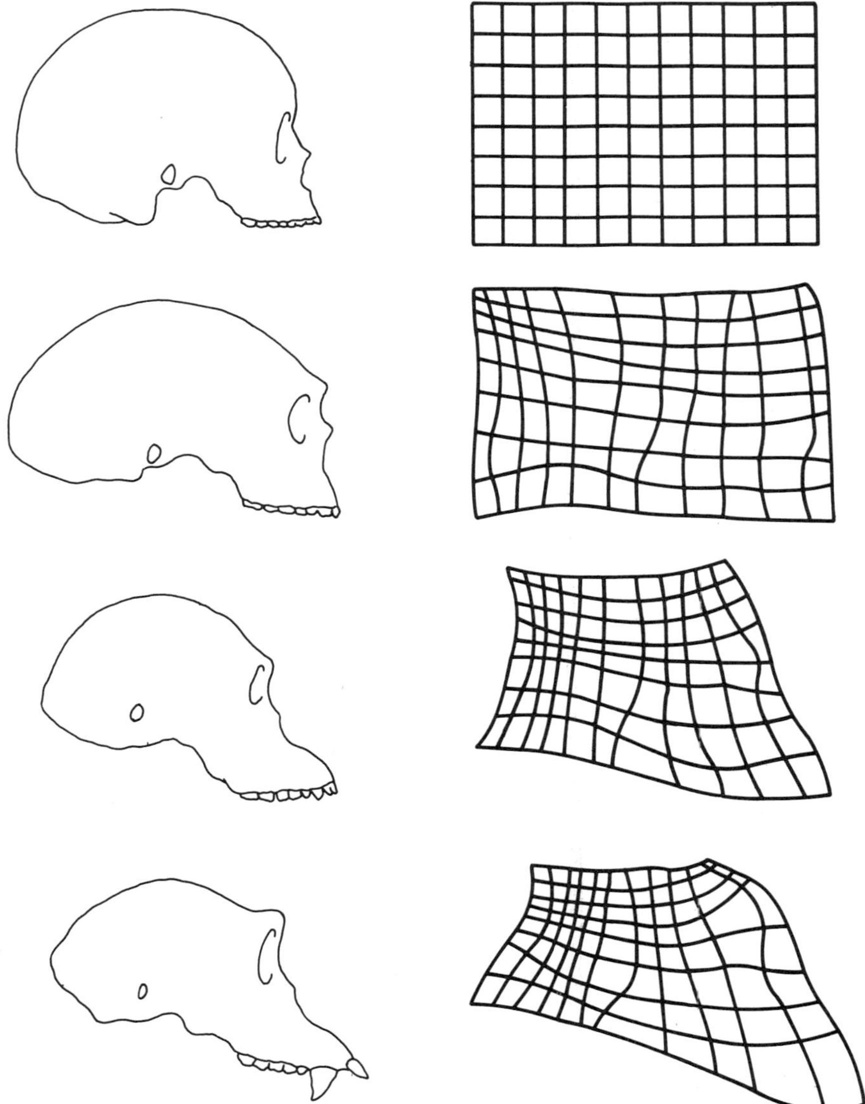

Fig. 3.53. The deformation of Cartesian coordinates inscribed within an adult human skull (top) when applied to a Neanderthal skull, second, an australopithecine skull, third, and a chimpanzee skull (bottom). The deformed coordinates have been separated from the skulls themselves to allow the reader to see the differences more easily.
The similarities are clearly between humans and Neanderthalers, and between australopithecines and chimpanzees. Redrawn after D'Arcy Thompson and P.H.A. Sneath.

scriptors for use in a variety of situations. On the whole, these have not been much applied in biological situations (e.g. Dunn, 1975, detects corners and edges in engineering problems). But the use of tangents and inflection points is described by Attneave and Arnoult (1966) and illustrated in a biological example by Oxnard (1973b; Fig. 3.28). Likewise the central axis function is described by Blum (1962, 1967) and Philbrick (1968) using as some of the examples simplified biological shapes; actual demonstrations of the method (Fig. 3.29) for describing real biological objects are noted by Oxnard (1973b); its further possibilities are emphasized by Waddington (1977); and a practical example on the mandible is available from Webber and Blum (1979); but the method yet remains to be applied to extended biological problems. Curve-fitted images have been used, if not on real biological objects, at least on a representation of one (a Barbie doll, Agin and Binford, 1976).

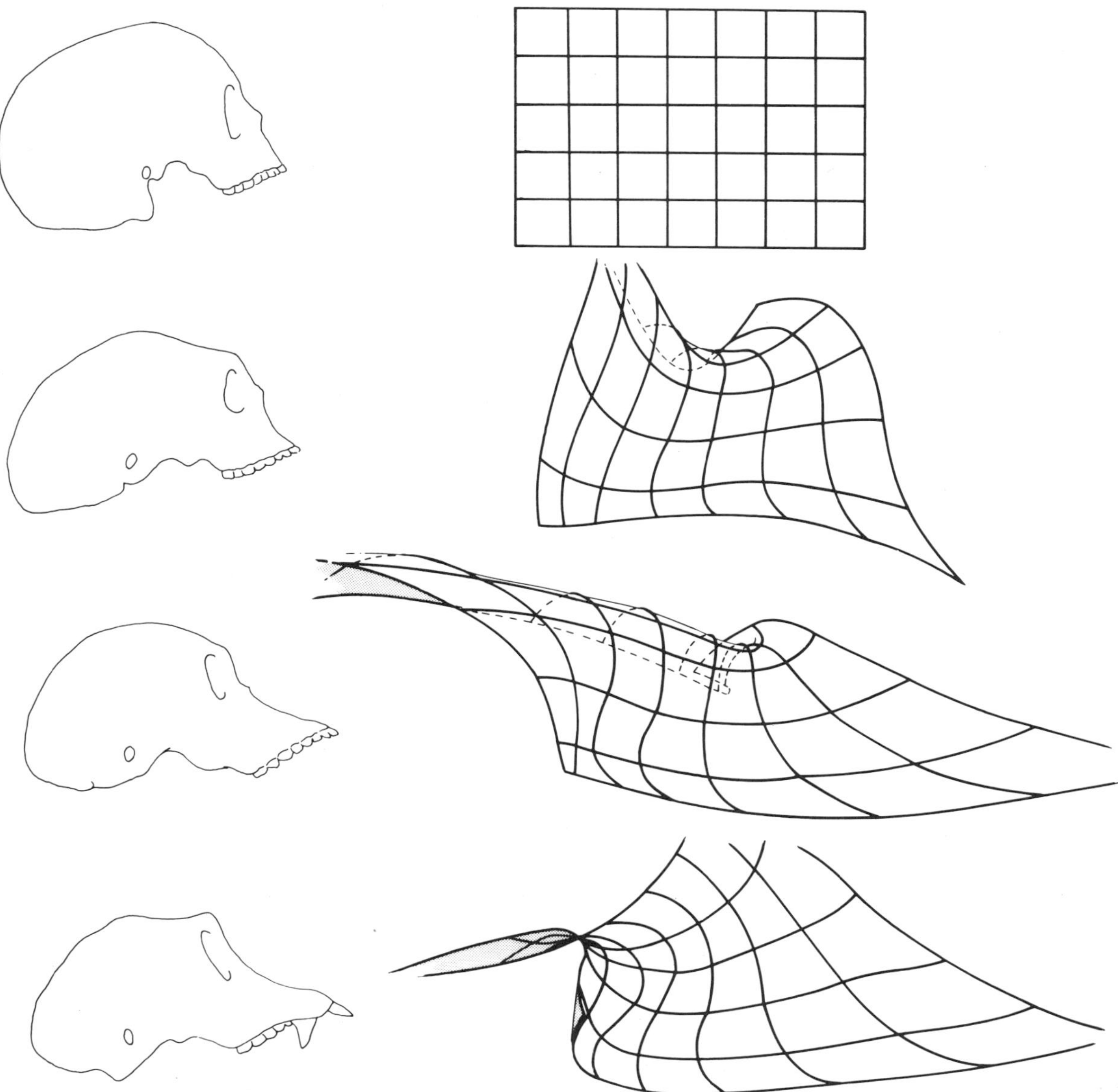

Fig. 3.54. The deformations when a coordinate system for humans is deformed in accordance with the method of trend surface analysis. The same skulls are examined as in Fig. 3.53 and the skull shapes have again been separated from the transformation diagrams so that the reader can more easily see the expressed differences.
Again, the similarities are between humans and Neanderthalers on the one hand, and between australopithecines and chimpanzees on the other, although the nature of the comparison is more complex than in Fig. 3.53. (Redrawn after P.H.A. Sneath.)

None of these approaches supersedes the more usual techniques: visual assessment and simple analysis of a few measurements. Rather do they add enormously to our armamentarium for studying shape and pattern. All are able to demonstrate that there is a great deal of information within bone form and pattern that cannot be obtained by observation and simple measurement alone.

Problems: few measurements versus many

Although our survey of new approaches, necessarily brief and incomplete, indicates that shape may be characterized by any number of measurements, in fact, much discussion stems from considerations about the number of measurements that

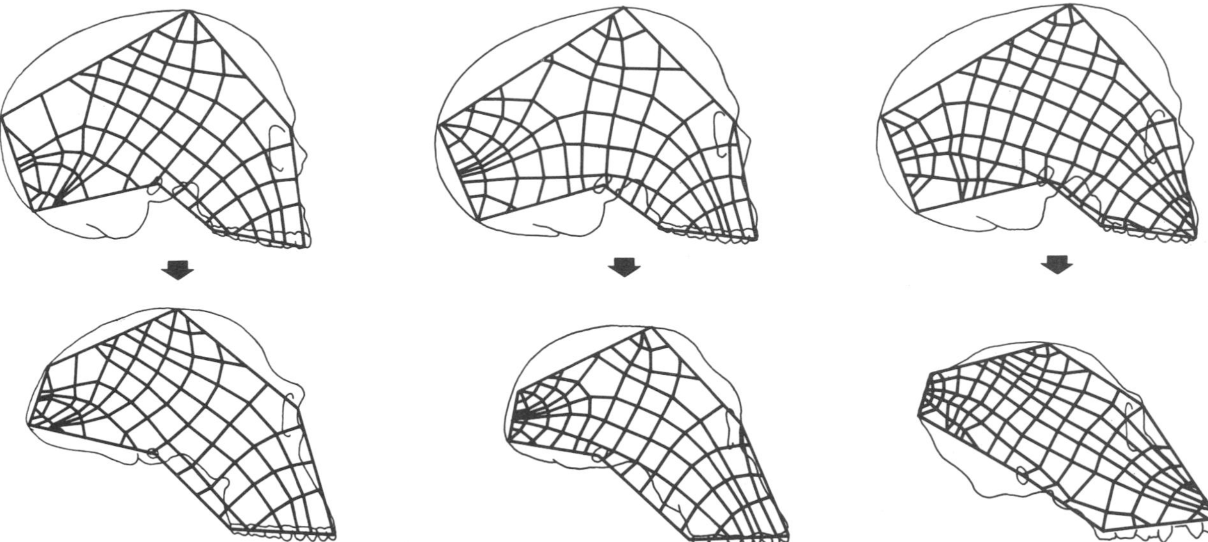

Fig. 3.55. The deformations between a human skull and, in turn, of Neanderthal, australopithecine and chimpanzee skulls using the method of bio-orthogonal grids as by Bookstein. The original starting grid for humans is different in each case because it is related to the actual deformation being examined. The amount of change appears to be least for the human to Neanderthal comparison and much more for the deformations into the other two skulls.

However, the method is not quite so easy to comprehend and, until comparisons are examined between each of the other pairs, Neanderthal to australopithecine, Neanderthal to chimpanzee and australopithecine to chimpanzee, it is less easy to see what is the overall meaning of the comparison. This method is yet more subtle than the previous ones. (Redrawn after Bookstein.)

'need' to be taken. Some investigators believe, for instance, that morphometric problems are not as complicated as presented here and that a single cleverly chosen measurement can supply the important information (e.g. Lovejoy, 1979). Others seem to think that adding extra measurements merely produces extra accuracy and that, though extra accuracy is a good thing in its own right, the extra work involved in gaining the additional accuracy may not be worthwhile (e.g. Day, 1978).

It is certainly true that some shapes can indeed be characterized by a very small number of measurements. A single measurement of a circle or sphere (if we know that the figure or object is a circle or sphere), for example, supplies us with the full information. Two measures completely define a rectangle and three a box (if we know that the shapes and structures are truly rectangles and boxes). But biological examples are, of course, more complicated — we may well not know to what shape they approximate — and more measurements are required.

It is also true that there are those investigators who believe that if two measurements are better than one, then 20 must be better than 10, 200 better than 100, and in a few cases very large numbers of measurements indeed have been proposed or taken upon biological objects with the notion that somehow this inevitably means a better study with increasingly more accurate estimations. This can be reduced to the absurdity mentioned by Howells (1969) who reminds us of the old study in which as many as 5,000 measurements were suggested as being necessary to describe, morphometrically, a single human skull. To a degree, a naïve use of statistics contributes to this problem because of misuse of the idea that increasingly large samples increasingly refine the accuracy of statistical estimates.

In fact, of course, the real situation is much more complicated. In some cases a small number of measurements may give a totally wrong answer simply because the numbers are so small that important elements of shape may be missed. But in other cases large numbers of measurements may not produce better estimates; they may actually produce worse estimates because it is possible that, beyond some optimum point, addition of extra measurements adds only a modicum of true information but a great deal of 'noise'.

Some studies from our own laboratories provide information with practical relevance to this matter.

Thus characterization of a shape such as that of the shoulder girdle has been undertaken using three, six, nine and seventeen variables. The analysis of these suites of measures using multivariate statistical methods makes it rather clear that for an object of this degree of macroscopic complexity, something like nine measures produces far more information than six or three, but scarcely any less than that provided by seventeen. Table 3.1 indicates that even for a restricted number of genera (restricted purely to keep the table short) the rank order of the genera becomes stabilized at about nine variables.

Another object of roughly comparable complexity, the talus, one of the bones of the ankle, has been examined through the study of three, six, eight and sixteen variables. In a similar way it has been found that eight measures provide a great deal of information not contained within three and six, but sixteen do not given any major information not already supplied by eight. Table 3.2 indicates that even for a restricted (restricted in this table for the same reason as before) number of genera the rank order becomes stabilized at eight variables.

Studies of the pelvis, on the other hand, show that this number, of about eight or nine measures, is inappropriate for this particular object. The pelvis is indeed a more complicated shape than the talus or scapula, and analysis of four, then another five, then nine, then seventeen and finally a total of twenty-five variables shows that information of new orders is added up to the analysis of seventeen dimensions. Thus, for an object as macroscopically complex as the pelvis, probably twice as many

Table 3.1

RANK ORDERS OF SOME GENERA IN PRINCIPAL AXES OF SHOULDER STUDIES: COMPARISONS OF DIFFERENT NUMBERS OF VARIABLES

3 variables	6 variables	9 variables	17 variables
baboon	patas monkey	patas monkey	patas monkey
patas monkey	squirrel monkey	baboon	baboon
mangabey	baboon	squirrel monkey	squirrel monkey
langur	capuchin	capuchin	capuchin
capuchin	mangabey	mangabey	mangabey
squirrel monkey	human	langur	langur
gorilla	langur	human	human
proboscis monkey	proboscis monkey	proboscis monkey	proboscis monkey
human	orang-utan	gorilla	gorilla
gibbon	gibbon	orang-utan	orang-utan
orang-utan	gorilla	gibbon	gibbon

Table 3.2

RANK ORDERS OF SOME GENERA IN PRINCIPAL AXES OF TALUS STUDIES: COMPARISONS OF DIFFERENT NUMBERS OF VARIABLES

3 variables	6 variables	8 variables	16 variables
orang-utan	macaque	macaque	macaque
Proconsul	*Proconsul*	*Proconsul*	*Proconsul*
macaque	*Limnopithecus*	*Limnopithecus*	*Limnopithecus*
Limnopithecus	orang-utan	orang-utan	orang-utan
gorilla	chimpanzee	human	human
chimpanzee	human	chimpanzee	chimpanzee
human	gorilla	gorilla	gorilla

measures are needed to obtain a reasonably full characterization of its external form (table 3.3 shows a selection of the generic rank orders varying in the different analyses but stabilizing at seventeen).

At yet another order of complexity is the human skull. Howells (1973) has shown that the skull may not be adequately described metrically until something of the order of fifty measurements are used. And this is confirmed through independent studies by Brown (1973).

Obviously, one of the things that matters is not how many measurements, but how many of what measurements. Attempts have been made in the past to prescribe the measurements that should be taken (e.g. for man, internationally agreed measures, Martin, 1957–66). Thus, often particular studies take measures necessary to compute special defining dimensions such as cranial (facial, nasal, crural, intermembral, etc.) indices. There are many good reasons for such standardization (e.g. in monitoring the growth and nutrition of human populations, in ergonomics and industrial design). But it must surely be the case that a best choice of measures for most evolutionary problems is determined by a creative approach to the particular question at hand. For even though a particular nine measures of the shoulder girdle may seem to be as good as seventeen, this is only the case when we are trying to define the overall macroscopic shape of the region. If we are interested, for instance, in the arrangement of the cortical bone within the scapula, the number of useful and necessary measures is likely to be considerably greater; and, if consideration of the grain of the cortical bone or the pattern of the cancellous network within the scapula is our aim, even more variables are likely to be required.

Problems: two dimensions versus three

Many of the new methods are currently employed for characterizing form and pattern within two dimensions. There are several reasons for this. First, some problems present themselves in two-dimensional form. Many studies of structure are exactly studies of two-dimensional sections of structures. One of the most extensive fields for morphometrics is cranio-facial growth and development because of the clinical importance of the applied information. The cranium, face, jaws and teeth are, of course, three-dimensional structures, but because a most important tool in such studies is radiography, the two-dimensional lateral radiograph is especially inviting. And many other problems in biology have been tackled because of two-dimensional simplicity. Thus, studies of objects as different as sepals, shells and shoulders exist exactly because these biological structures are often flat and easily examined through two-dimensional representations.

The two-dimensional limitation applies to a great many of the methods as they have so far been explicated. D'Arcy Thompson's original use of Cartesian coordinate deformations is demonstrated through two-dimensional examples and this has continued each time that method has been used (e.g. Richards and Riley, 1937, on growth in amphibian larvae; Medawar, 1944, on the stages of human growth; Lull and Gray, 1949, on growth in ceratopsians; Langston, 1953, on Permian amphibians; Smart, 1969, for looking at the formation of the avian egg; Chatterjee, 1974, on the form of rhynchosaurs; and, most recently perhaps, Appleby and Jones, 1976, and Gabrielson, 1977, in expositions that show how the method can be automated

Table 3.3

RANK ORDERS OF GENERA IN PRINCIPAL AXES OF PELVIC STUDIES: COMPARISONS OF DIFFERENT NUMBERS OF VARIABLES

4 variables	5 variables	9 variables	17 variables	25 variables
orang-utan	chimpanzee	chimpanzee	orang-utan	orang-utan
chimpanzee	orang-utan	orang-utan	chimpanzee	chimpanzee
gorilla	gorilla	gorilla	gorilla	gorilla
Australopithecus	*Australopithecus*	*Australopithecus*	human	human
human	human	human	*Australopithecus*	*Australopithecus*

using computer graphics).

Other applications that have stemmed at least in part from the method of deformed coordinates are also two-dimensionally confined (Sneath, 1967; Bookstein, 1977; Tobler, 1977). Many other uses of such methods have confined themselves, necessarily at this stage, to the two-dimensional example (e.g. the various usages of Fourier coefficients, in biology: Anstey and Delmet, 1972; Kaesler and Waters, 1972; in anthropology: Lu, 1965; Lestrel, 1976, Lestrel and colleagues, 1976, 1977); the few attempts to use central axis transformations: Blum, 1962, 1967; Oxnard, 1973b, Waddington, 1977; some of the applications of optical data analysis, Oxnard 1970, Scheibengraber, 1974. In order to make it clear that I am not making pejorative criticisms of others, I have included some of my own investigations in the above lists.

Almost every worker who has used these particular methods has been enormously aware of this two-dimensional limitation; almost every researcher has noted that, intellectually, extension to the three-dimensional situation is trivial; but almost every investigator also knows that such an extension is not only vitally important to raise these methods beyond mere examples, but also, logistically, most difficult. Yet an application in engineering (Gordon and Hall, 1973) shows us how the three-dimensional coordinate system can work.

It is in the realm of multivariate statistical and clustering analyses of measurements and the overall appreciation of shape through optical data analysis that the greatest strides towards three-dimensionality have been taken. Thus, it is usual to use, for multivariate statistical analysis, (a) measurements taken in three dimensions (and thus providing three-dimensional information about objects), (b) measurements made within planes aligned at known angles to one another (as in many of the studies of radiographs, sections and projected outlines) or (c) measurements in the form of three-dimensional coordinates of points. Likewise, the methods using optical equipment, though especially valuable for the study of two-dimensional patterns, are also being used to characterize three-dimensional arrays using methods capable of analysing three-dimensionally distorted patterns projected upon objects (e.g. skulls, Scheibengraber, 1977, and contour maps, Srivastava, 1977).

Problems: keeping geometry versus losing it

It may appear on the surface that geometric relationships are the important element for the study of shape. It is partly for this reason that many of the techniques using deformed grids have been employed so frequently. But, of course, it is also because there seem to be some very special reasons why geometrical relationships have biological importance.

Many studies of the face, jaws and teeth lead to practical applications in different branches of dentistry where the actual final geometrical adult shape may be of vital importance (e.g. Macgregor, Newton and Gilder, 1971). And many other biological studies try to maintain geometrical relationships, for example studies of change in shape during the movements of an organism (slime molds, Robertson, 1972; Robertson and Cohen, 1974) or during organismal growth and development (many examples: e.g. amphibian larvae, Richards and Riley, 1937; tobacco leaves, Richards and Kavenagh, 1945; living growing embryos, Prothero, Tamarin and Pickering, 1974) or even differences between shapes stemming from the actions of forces upon those shapes during either their biological formation (e.g. eggs passing through the oviduct, Smart, 1969) or their geological deformation (e.g. the shell of *Angelina* deformed during the process of fossilization, Appleby and Jones, 1976).

But the attractiveness of keeping geometrical relationships to the fore sometimes overshadows the importance of other relationships to which the geometry may speak. And it is probably too little realized that geometrical information is not lost merely because the results of the analyses are not presented as the relationships of compared pictures.

Those types of analyses which take a set of measurements of an object and analyse them by multivariate statistical methods or clustering techniques, or many forms of image transformation, do indeed seem to lose the geometry inherent in the original objects. But in general they do this in order to reveal a new dimensionality within the data that is extraordinarily important to the results. Thus, objects characterized by morphological data in three dimensions may be transformed into relationships between kinds of animals that may be in one, a few or even many dimensions. The vectors, axes and distances of statistical methods, the minimum spanning trees and dendrograms of cluster-finding techniques and the polar coordinate arrangements of the optical densities of image analyses may all contain the new information that is desired.

Yet in these studies the original geometrical arrangements are not lost. Techniques exist that can reveal those shapes once again given that this information is truly contained within the initial data. In multivariate statistical methods, for instance, particular differences in the biological forms may well be contained within individual canonical axes of particular discriminant functions, or in separate principal axes of particular factor analyses or even within oblique axes in such studies. This information may be demonstrable in a pictorial manner (Fig. 3.56).

The same thing applies to Fourier transformations. Although the analytical result may be contained within particular Fourier coefficients or within the flux of light across a transform plane (and biological meanings may well be posited for them), we have already seen that it is also possible to return to the original pictures and determine just what are the geometric differences between them.

Problems: special versus general morphometric points

Many studies, especially earlier ones that depend most upon measurement, also rest heavily upon the definition of special points on the bones. These may be of various kinds. Some are geometrical points that can be repeatedly defined such as most distal, most proximal, most medial, most lateral and so on. Such points do not necessarily have any special biological significance but are used frequently by anthropologists because of their practical simplicity.

Others are defined for *a priori* reasons, e.g. so-called homologous points. There is an enormous and complicated literature on the meaning of

Fig. 3.56. The use of Thompsonian deformed coordinates to display differences in shape of primate scapulae. In this case those scapulae are chosen for examination that are separated by only a single discriminant axis in a multivariate statistical analysis of shoulder data. In the case of each axis, a simple distortion is visualizable (as indicated by the arrows). These three simple trends truly are the trends that underlie the aspects of form picked out by the individual discriminant axes.

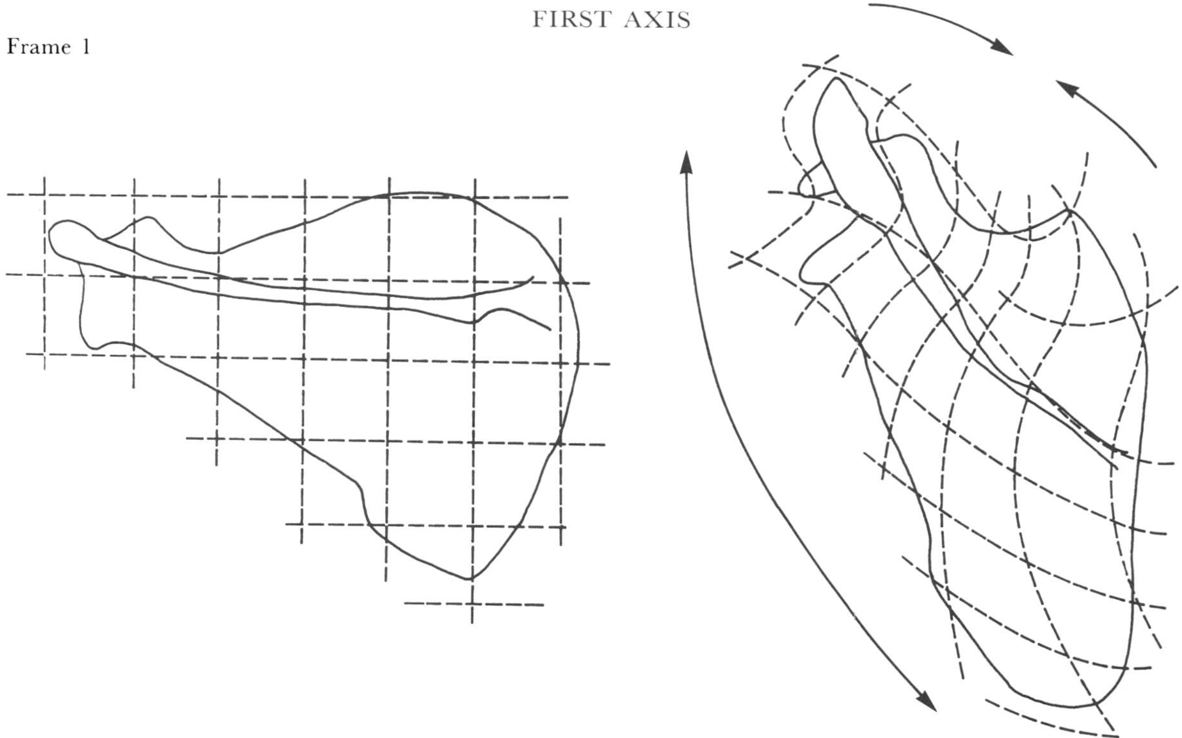

SECOND AXIS

Frame 2

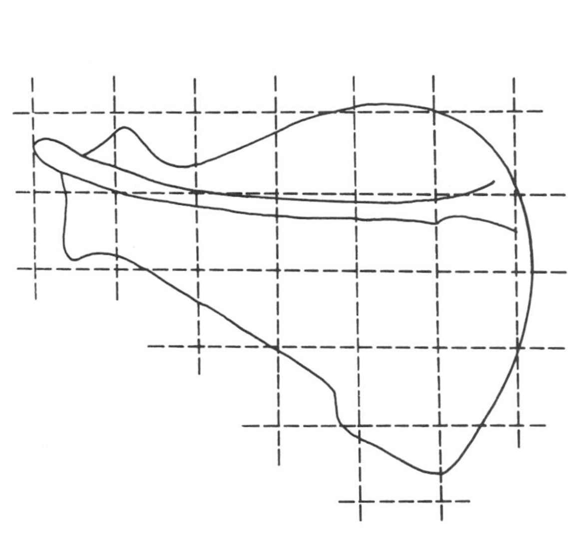

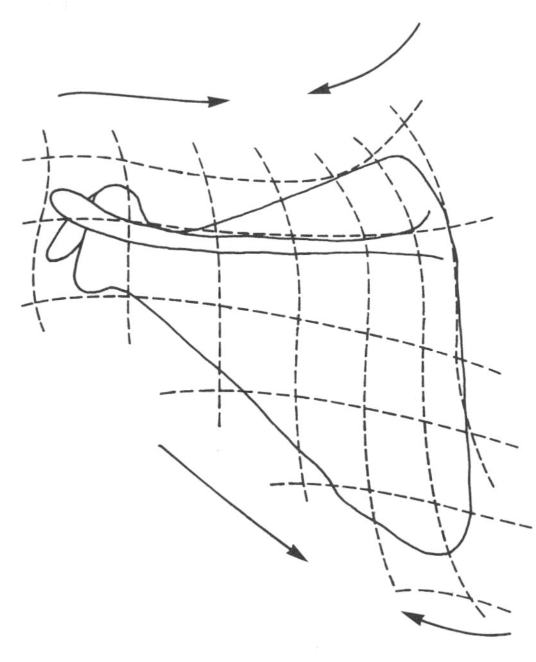

THIRD AXIS

Frame 3

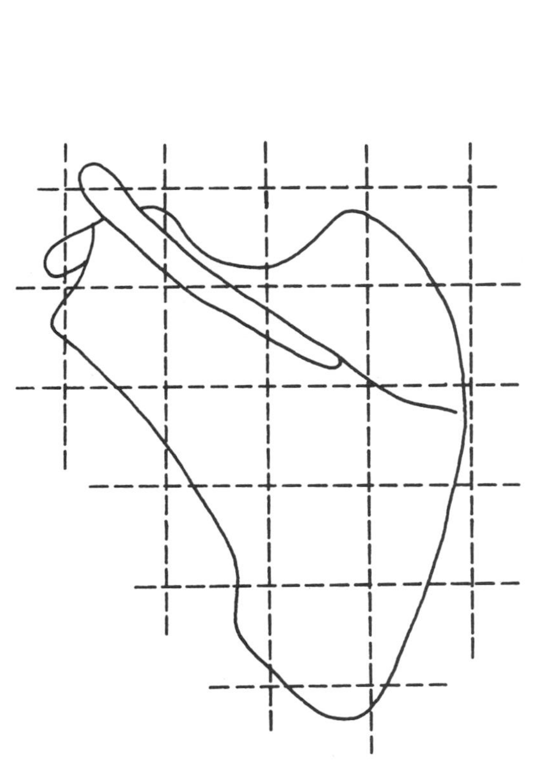

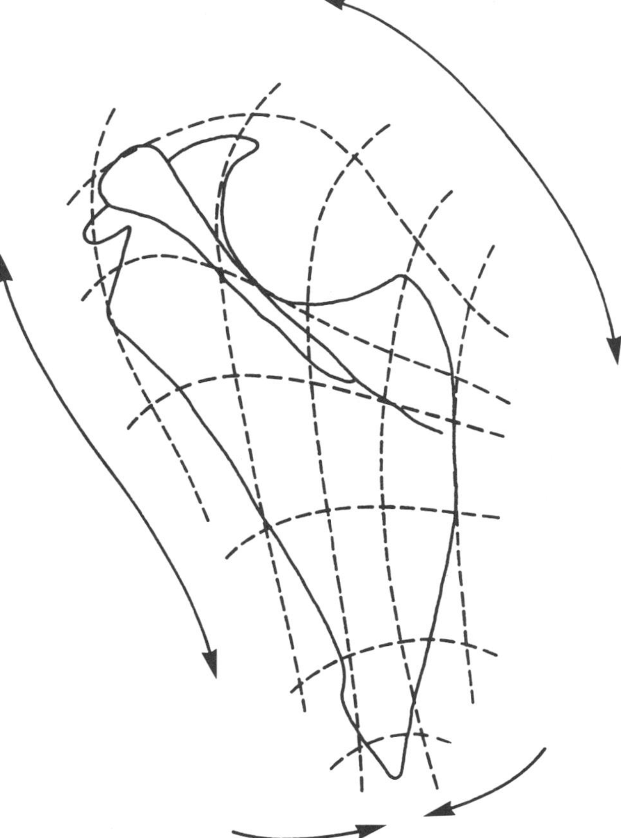

homology, but, without going into that matter, it is clear that useful biological points may be those that are thought to be produced through, for instance, the action of similar developmental processes during the ontogenies of the animals being compared. However, the definition of such points depends upon developmental information being truly available; when absent, and for most rare primates it is, such points may be defined through inferences, that, without care, can easily produce circular reasoning. It would seem obvious, for instance, that the styloid process is a useful homologous point for studies of skulls. But careful examination of comparative anatomy and development of the styloid process in higher primates shows that this is not the case (Zuckerman, Ashton, and Pearson, 1962).

Points can also be defined for other reasons. For example, our essays in recent years include attempts to characterize biomechanical (functional) aspects of various forms and patterns; the definition of points may be made using good biomechanical information. Again, however, this depends upon good biomechanical information being available and that is sometimes a pipe dream. For instance, a particular point may be chosen as defining a useful biomechanical element of a muscle especially important in walking (e.g. the function of the gluteus maximus muscle). New biomechanical information may tell us that this is spurious, gluteus maximus being actually of lesser importance in walking. The point means something else (see Chapter 4 for more complete discussion).

Some studies, however, are carried out without defining special points at all. This is automatically inherent in some of the methods already mentioned, for instance those of image analysis. And they may be deliberately sought by investigators in situations where we might expect that special points would be required. Sometimes a desire to avoid special points may relate to the need to characterize areas or regions that are relatively featureless (e.g. the vault of a skull, the form of an egg); at other times it may be a genuine desire to produce 'reference-free' characterizations. It is worth pointing out that specially defined points can easily be retained alongside general points in systems like the central axis transformations. This allows us to include studies of the relatively featureless vault of the skull alongside studies of special points on the base of the skull.

Problems: two specimens versus many

Although, technically, these methods speak to the problem of characterizing form and pattern, and even though this is exactly what is required in many of the non-evolutionary applications that have been noted, simply characterizing a form or pattern by itself is not very useful in evolution. Evolutionary problems require comparison.

Comparisons can certainly be achieved by inspecting two characterizations of forms or patterns placed side by side. Sometimes comparison is even carried out by superimposition of pictures. This is very common in studies of growth especially, for instance, in studies of cranio-facial growth. Comparison can also be carried out by metrical or optical subtraction of one specimen from another as indicated previously.

But the great majority of problems in evolution require the comparison of many different specimens belonging to a population. In this sense, many of the most attractive methods, especially the geometric or visual ones, have not been developed far enough. Coordinate transformations, trend-surface analyses, bio-orthogonal grids, central axis transforms can all, in theory, provide sample parameters. In their actual usages they have not yet been so developed. The intellectual task of improving these methods to take in this additional matter is trivial — the statistical background exists. But the practical work necessary to carry it out is by no means easy. Until it is done, these methods and many other similar ones not referenced will remain elegant, much quoted, but little used examples.

The univariate and multivariate approaches have, of course, been especially developed with this problem in mind. And a start has been made with optical data analysis where 'average' faces among 'samples' of faces can be obtained (Fig. 3.57). The average of a sample of measures is an especially useful concept in biology; so too is the average of a sample of pictures or patterns.

Everything suggested under the problem of characterizing the members of a single population applies *a fortiori* to comparing several populations. The only methods so far available for several populations are those that utilize the multivariate statistical approach or one of the modern offshoots of it. Yet many of the other methods could be so modified, probably by their combined usage with statistical methods.

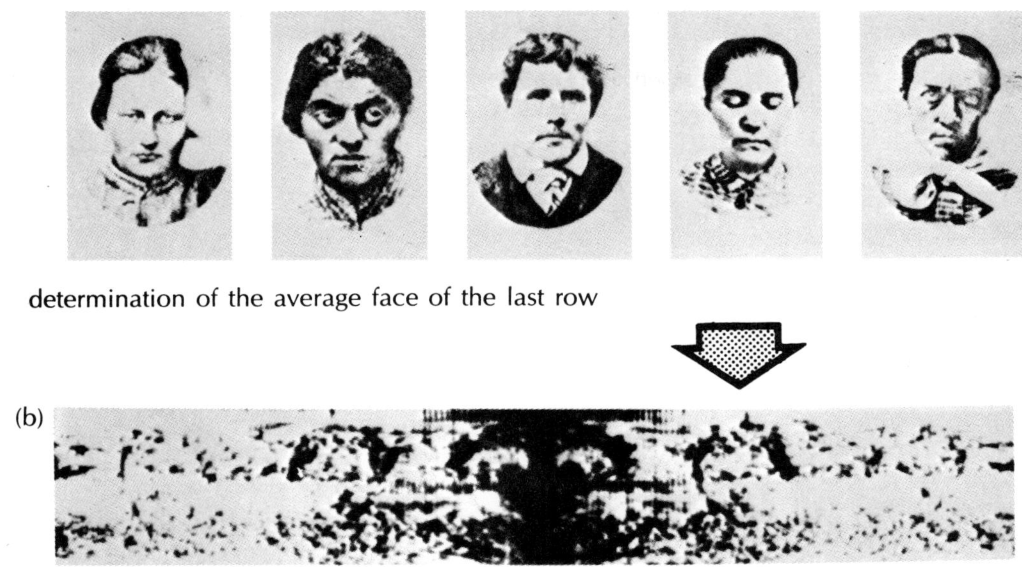

Fig. 3.57. Using the methods of optical data analysis, Vienot, Duvernoy, Tribillon and Tribillon are able to scan the row of faces (top) and determine which one is the one (the fourth) that is closest to the 'average' face in the row. It is the face that is most clearly seen in the optical transform at the bottom. Reproduced with permission.

Such studies assume the existence of different populations and determine the differences among them. This is not a major problem in many aspects of evolutionary biology where there are no difficulties in defining biological groups and where the problem is indeed the nature of relationships among known groups. For example, most biological questions already depend on a wide variety of older studies that render incontrovertible, on any objective basis, the prior existence of groups.

My own studies of primates are a case in point where limitations of materials often force work at the generic level. Although there is a great deal of debate concerning many primate entities, most primate genera are well-defined.

But in many investigations the point at issue may well be the very existence of lower level groupings. Thus, at the subspecific level, when possibilities lie between continuously varying clines, fairly distinct groups with narrow zones of hybridization, separate groups with narrow zones of sympatry, or separate allopatric groups, then the existence of already known objective groupings cannot be taken for granted. Under these circumstances group-finding procedures become of especial importance. A wide variety of these exist from those that utilize multivariate statistics in a group-finding mode, through to those depending directly upon cluster-finding procedures that are non-statistical in nature and including, of course, the combined use of these two major approaches (Fig. 3.58).

The problems raised by the placing of an unknown sample or unknown specimen within or among known specimens or populations are especially severe. Although not necessarily the simplest, placing an unknown sample is obviously the least difficult because statistical parameters can be derived from the unknown sample; it can therefore make its own genuine contribution to the determination of its position among a set of known groups. A good example here is our own investigation of the primate genus *Daubentonia*. Data from samples (albeit very small) of specimens of *Daubentonia* demonstrate their uniqueness from all other primate genera (see Chapter 8).

With single specimens, however, extra care is necessary, and this problem is especially marked in dealing with fossils. Thus, initial treatment using multivariate methods (Day and Wood, 1968) seemed to confirm an intuitive view, resulting from a study of the other remains of the Olduvai foot (Day and Napier, 1964), that the foot and thus the talus of the fossil are similar to that of humans and show evidence of bipedal walking of human type.

But further studies taking account of the problems of interpolation show first, that the Olduvai talus is not like that of man and in fact resembles somewhat more those of species able to climb in trees (Lisowski, Albrecht and Oxnard, 1974), and

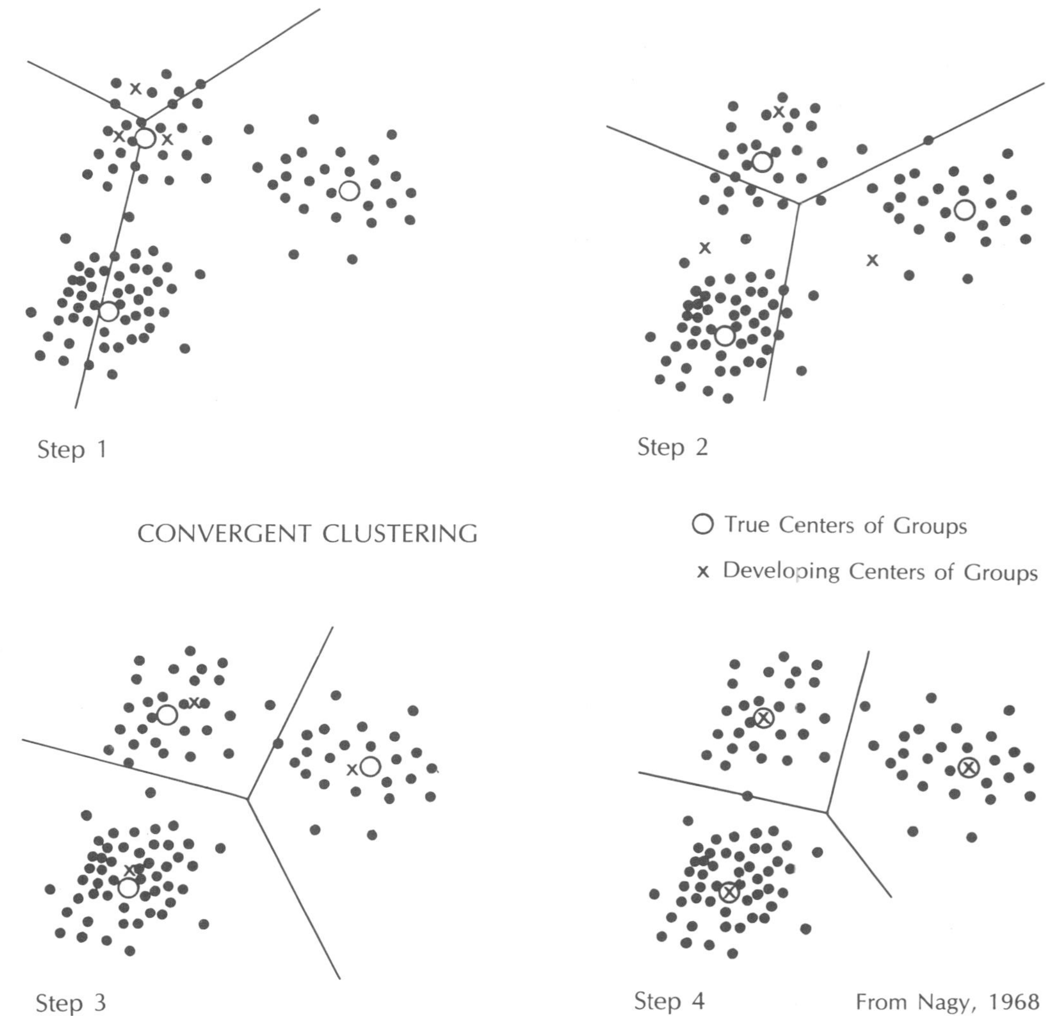

Fig. 3.58. This diagram demonstrates one of the many methods of discovering which clusters may exist among a number of data points. From step one to step four, the clustering process improves markedly. (Redrawn after Nagy.)

second, that the Olduvai talus is even more closely similar to those of several fossil apes and monkeys that are clearly not bipedal (Lisowski, Albrecht and Oxnard, 1976).

A return to the remaining elements of the Olduvai foot shows that they too confirm this new picture (Oxnard and Lisowski, 1980). The problem is that the initial rearticulation of the fossils into an arched foot resembling that of man was incorrect (Oxnard, 1980b).

Problems: testing

For this entire chapter the word 'test' is important. Many of these methods can be used as searching tools, where investigators look diligently at complex information in order to find, if they can, its inherent pattern. This usage is sometimes spoken of as being 'hypothesis free'. Many of the methods can also be used in a different way where some *a priori* idea is the underlying concept. This is often designated a 'hypothesis-testing' mode. There is a major difference between these two ways of using such methods and it has been spelt out nicely by Albrecht (1978).

However, there is another usage of the word 'test', a usage that applies to any, indeed every method of investigation. It therefore also applies here. This may be designated 'method testing'; it is not at the whim of the investigator whether it shall be used; unless appropriately tested, no method has demonstrated the 'right' to be used in science.

In the case of various observational methods, this type of testing is not carried out as a separate activity. It is inherent in the standards brought to the observational process by the observer. A novice may not have the experience to provide such testing. An expert has developed that experience over many years.

In the case of the various new methods that I have described, the nature of the approach allows this type of testing to be carried out as a separate, independent, overt act. In this case novices may fail to test because, perhaps, they may not realize that testing is required. The expert tests because that is the way he has learned to use the method. However, it often occurs that novices test extremely well, exactly because they are learning the methods and have become fascinated as much by the methods themselves as by the underlying biological problem. The question at issue here is: what is this testing?

Perusal of the literature reveals many papers written to criticize lack of testing of the methods. The reason for this is that, unfortunately, the mores of our subject are such that the tests and their results are rarely published. If they appear in print at all, they may be disguised within a sentence that states that the usual tests were carried out. In a book like this, however, it is probably well worth having a quick look at the various types of testing that there are; and it is probably worthwhile pointing out that such tests have been undertaken in every one of the studies that I have carried out (Oxnard, 1983a).

Testing starts at every level. First, in collecting the data we must make sure, as it were, that our rulers are steel and not rubber; that our bones are rigid and not flexible; that our eyes have the same degree of astigmatism as others; that our minds march to the same drummer. In other words, we must make sure that our data can be replicated in our materials, our instruments, ourselves and others. Let me emphasize that these somewhat flippant phrases cover a great deal of the business of research.

Second, in taking the primary data we must study each datum item separately; if looking at a population of items, we must study separately each sub-unit of that population; if interested in many populations, we must look at each population separately. In terms of the studies described earlier, we can be more specific; if we are interested in interactions of many variables, we should have some notion of how small numbers of variables interact; if interested in many groups of animals, we should understand something of how subsets of animals are related; if interested in complex anatomies, we should know something of how smaller anatomical regions are arranged. Tests of this type are undertaken not only so that we can know if there is anything wrong with the data — error of measurement, mislabelling of specimens and so on — but also so that, when we apply more complex methods of analysis we shall know when we have discovered a rogue result or when the data do not fulfil whatever underlying axioms are necessary for the method to work properly.

Third, when eventually we do apply the more complex analysis in the course of the investigation, we should look towards a series of tests that test the ultimate result. One of these is to use a different approach on the same data. Another is to use the same approach on different data. A third is to use both different data and a different approach. Sometimes such tests are provided by the accidents of the activities of different investigators (e.g. Feldesman's (1976) study on the arm and forearm was carried out at the same time as that by Ashton, Flinn, Oxnard and Spence (1976); they provide remarkable replication of each other). But one ought not to depend upon the accidents of science; it is good, obviously, if such tests are part of the planned study of the individual worker.

Even this short classification does not cover all of the types of testing that we do, nor all of the far larger battery of tests that are available. Table 3.4 attempts to group the various tests that we undertake, rather routinely, in our various studies (and see Oxnard, 1983a).

It is worth finishing this plea for adequate testing with a second plea for common sense in testing. Many tests are entirely appropriate for a particular task, but not for another. Many critiques about testing can be applied correctly to underscore the weakness of an investigation, but many others are applied to investigations to which they are not appropriate in order to cast doubt upon unpopular findings. Testing can be an enormous strength in an investigation, allowing the researcher to go to the very limit of the data, but testing can also produce an unthinking paralysis that prevents the investigator from moving at all, either because of the enormous bulk of work that it can involve, or because of fear of the method that it may engender. Indeed, it is entirely possible for an

Table 3.4

TESTS ON MULTIVARIATE MORPHOMETRIC ANALYSES

Simple tests

Errors of replication
Inter-observer errors
Inter-instrument errors
Differences between sex, age, sub-specific, pathological and other confounding groupings.
Normality of data, and if not normal, appropriate transformation
Equality of variance, and if not equal, appropriate transformation
Univariate, bivariate and multivariate searches for outliers

More complex tests

Principal components analysis on individual groups to study normality and homogeneity of dispersion
Principal components analysis on individual groups to identify hidden sub-groups of specimens
Principal components analysis to summarize information from group means alone

Tests internal to ultimate analysis

Significance of latent roots
Variance co-variance tests
Homogeneity of dispersion matrices
Significance and per cent of information with in individual axes
Tests of significance of positions of individual groups

Tests of ultimate analysis

Groupings of animals
- Based on all genera separately
- Based upon pooled groups from classifications, e.g. all Old World monkeys, all apes
- Based upon locomotion, e.g. all quadrupeds, all leapers
- Based upon combinations, e.g. all leaping prosimians

Groupings of anatomies
- Individual anatomical regions, e.g. pelvis, femur
- Individual functional regions, e.g. lower end of humerus, upper ends of radius and ulna = elbow
- Individual anatomical form of variable, e.g. all transverse variables, all longitudinal variables
- Individual metrical form of variable, e.g. all angles, all indices, all measures

Sample size and numbers of variables
- Comparison of groups with large samples only and groups with large and small samples together
- Comparison without and with interpolated data from single specimens
- Comparison without and with particular peculiar variables (e.g. tail length different, so tested separately)

Tests relating to overall bodily size

Tests comparing analysis of measures with analysis of other variables
Tests of extent to which individual discriminant axes represent animal size
Test of extent to which individual factor axes represent size
Tests of effect of manipulating size through regression adjustment (or other method)

Methods of display

Plots of single, paired or three (model) discriminant axes
Dendrograms of generalized distance connections
Minimum spanning trees of distance connections
Models, usually three-dimensional, of generalized distance connections
High-dimensional displays of patterns of discriminant or factor axes

Independent corroboration

Use of several appropriate multivariate methods
Use of methods with different axiomatic bases, e.g. neighbourhood limited classification
Combination of methods, e.g. neighbourhood limited classification and discriminant function analysis (Oxnard, 1967)

extremely useful method to be rejected outright simply because a theoretical test fails or cannot be carried out; yet the practical situation shouts loudly that the test is irrelevant.

A few examples are, perhaps, in order. It is important to try to discover if data are normally distributed or not in many situations where multivariate statistics are to be used. For normality of data is an important underlying axiom of this method, especially when used in an hypothesis-testing' mode where the distinctness (or otherwise) of the groups being investigated is at issue. This is the case if we are studying races or geographic varieties which may be so close that they are scarcely, in truth, distinguishable. But when we know that groups truly exist and that they are enormously different from one another — marmosets and mandrills, mice and men – then multivariate statistics may be used in a second manner, an hypothesis-free manner, where the information that is desired is not the separateness of the groups, but the pattern of their separation. Under this circumstance, it becomes unimportant if the data are not completely normally distributed. In fact, the requirement for normality is inversely proportional to the distinctness of the groups. That data are not completely normally distributed does not nullify a study of groups which are vastly different from one another. It is therefore appropriate to use the kinds of variables that sometimes depart from normal: ratios, angles, counts, even measures where samples are too small for normality to be determined. Of course the wise investigator will still desire to know if the data depart from normal and, if they do, will undertake appropriate transformations, if possible, to normalize them (Atchley, Gaskins and Anderson, 1976).

A second underlying axiom for multivariate statistical methods is that samples shall be approximately equal and that the numbers of variables shall be less than the numbers of specimens in the groups. Again, this is critical if the study involves a test of whether or not different groups are statistically separate from one another. But this is not critical if the groups are vastly different from one another and where the method is being applied to understand the pattern of relationships among them. Under such a circumstance, given that most of the groups are reasonably large, critical groups that are small for logistical reasons (that the animals are rare, that the fossils are unique) can still be entered without upsetting the analysis. Of course, a test of the effect of adding small groups must be undertaken, but this is easy. And, of course, if the group is small, though its relationships with others can be found, a greater degree of lability attaches to that relationship (but this is easily defined through the larger standard error of the mean position that applies to that group).

Let me give one final set of examples. It is rather important to test the particular statistical or other method that is used in comparison with others that may be available. For instance, some methods of discriminant function analysis normalize the variances of the different groups being studied so that distances are defined that are the same in all directions for all groups. Other methods of discriminant function analysis do not do this. Units differ from one discriminant axis to another. Obviously conclusions about the pattern of relationships between groups can be drawn from studies using the first method that are invalid using the second. One method produces groups that are circular (or hyper-circular) in the discriminant space, the other groups that are markedly elliptical (or hyper-elliptical).

Another example of the difference between individual computer programs relates to the matter of tables of classification or mis-classification. These tables are usually a part of canned discriminant function programs. In some cases such tables are appropriate and the mis-classifications that they provide can be readily correlated with the generalized distances upon which the tables are based. But other programs exist where tables of mis-classifications demonstrate very few specimens as mis-classified. Yet a glance at the generalized distance results indicates that this cannot be so, many more individual specimens are extreme than the table indicates. It is likely that there is an important statistical reason for this: a change-over within the study from the use of the generalized distance as providing information about the patterns among groups to its use as a statistical test of separation. Such tables should be used with extreme caution.

Finally, let me instance the matter of intermediate statistical manipulations. As I have said, if groups are enormously separate, as in our studies of primates at the generic or specific level, then some of the caveats about such features as normality become unimportant. But it is entirely possible, in carrying out a study at the generic level, to perform some intermediate manipulation that removes a large part of the information in the data (e.g. a multivariate regression subtraction) and then be left with such small differences between the groups that caveats about normality, and about statistical differences, once again become decisive. Indeed, it is not unknown in such studies that the manipulations actually remove everything that is of interest and leave little more than a residuum of statistical 'noise'.

There are many other problems in the usages of these methods to which I could point. But perhaps these few examples are sufficient to draw attention to the entire genre.

Some biological implications of the new methods

The main contribution to evolution of these new methods for studying form and pattern is mostly through the recent development and growth of the multivariate approaches. Given that we can now characterize and compare many biological forms, what relationships can we see among them? The resulting answers may suggest new arrangements of known forms, provide for better placement of unknown forms, and affect our assessment of phylogenetic relationships.

Such relationships may arise through investigations that look at aspects of form and pattern without regard to biological principles (e.g. some uses of numerical taxonomy); there is little doubt that much that is valuable may come from such approaches, although it will sometimes presumably be the case that some parts of the data will confuse the picture presented by other parts.

Such relationships may also be obtained by using more restricted data based upon prior hypotheses about their phylogenetic content (e.g. some usages of cladistic analysis); there is little doubt, again, that much useful information will result, although it will also presumably be the case that some of the results will be channelled by the prior hypotheses (which are not being tested and which may, of course, be wrong).

However, these methods may also be most powerful for evolutionary studies by testing biologically determined hypotheses. One obvious usage here relates to studies of geographical variation and racial affinity. Another may be through the understanding of developmental changes in structure during growth within individual groups of animals, for observations of structure in an animal group may reflect underlying developmental patterns or channels. In this way these methods may provide interpretative materials for systematic and evolution.

Yet another usage may help interpret functional adaptation of animal form. Methods using quantitative handling of form and pattern may well allow animal structure to tell for itself, as it were, something of its functional adaptation. The way in which particular morphologies are adapted for function may be revealed by the way in which such methods arrange different kinds of animals (functional divergences, convergences and parallels may become especially obvious, and so on) or by the way in which the structure itself is partitioned in providing separations of the animals (see later chapters).

Finally, it is worth remembering that these methods may be applied to structures other than the real three-dimensional structures of animals. The 'structures' of population interactions, of physiological variables, of ecological environmental parameters, of temporal paleontological sequences, of the arrangements of biological taxa are some of the 'structures' whose analysis by these methods contributes to knowledge of evolution. Certainly, describing anatomies can go a long way beyond observation and dissection. Some of the implications of the biological interpretation of structure is the subject of the next chapter.

Summary. For centuries, the main way of looking at animal structures has been through the efforts of the human eye and mind. Keen though these organs are, many aspects of the form of organisms are not readily obtained using such tools.

The discussion of this chapter has demonstrated a range of new techniques that we can bring to bear upon the problems of the analysis of form. These range from methods that use highly visual and geometric ways of studying form, to those that expand upon the relationships of form through algebraic and optical transformations. They span the spectrum from those that define form through discrete measurements, to those that attempt to analyse the information contained within an entire pattern. They include methods that discover empirical information and methods that attempt to provide explanatory models. For all of this, there are many other ways of viewing form that have not been dealt with in our discussion.

The discussion of this chapter also reminds us of constraints that may exist within ourselves. We must train ourselves, and our students, so that we are able to discover the advances made in other fields, and apply them, when appropriate, to problems in our own. It is vital for us to watch for new problem solving techniques in standard subjects as different as astronomy, metallurgy, geography, geology and so on. It is equally important for us to be aware of developments in disciplines new to the academic scene, subjects that have been born in recent years. Communications, image analysis, pattern recognition, language and picture processing, and many others, may all provide ideas and methods that can be grist to the biologist's mill. We have to be willing to learn something of mathematics, statistics, physics, engineering, electronics, computer programming and so on, so as to be able to apply, or more likely collaborate with those who can apply, these methods to our own problems.

Finally, it is worth remembering once again, that forms and patterns contain far more information than that available to the naked eye. A not inconsiderable part of the importance of the new analyses of biological form resides in the new insights that they can make available, the surprises that they may generate, and the fun, therefore, that they give us.

CHAPTER 4

Biological 'Meaning' of Structures

Abstract. In this chapter is discussed what we can learn from studies of structure. Of course, many studies of structure lead to the understanding of matters such as heredity and development. But, for the post-cranial skeleton, many investigations reflect the functions that the structures subserve. Though direct information about function is produced by the studies of animal mechanics, indirect insights can come from observing the way in which animals with different behaviours are arranged by structures. For this purpose the attempts to quantify and analyse structure, as in the last chapter, are so critical.

Arguments about the nature of the association between structure and function are discussed and lead to the concept of the 'structural-functional interface' and the 'average biomechanical situation' acting upon a structure. One form of average biomechanical situation is reflected in the classification of locomotor patterns: the clustering of animals according to similarities in locomotor patterns. In the past such ideas have been quite useful.

But consideration of the deficits of such classifications leads to the idea of the functional spectrum and to the notion that such spectra are limited to individual anatomical regions. This, in turn, leads to the idea that searching for relationships between structure and function can have a strict 'design of observations' that parallels the design of experiments which has long been known to be critical in direct experimental studies of animal mechanics.

The application of these concepts to the study of structures, the functions of which are unknown (i.e. the structures of fossils), has limits over and beyond those applying to the study of living forms. Of course, some of these limits stem from the fragmentary nature of much fossil data, but not inconsiderable parts stem from the problems of prediction.

Biological meanings of structure

Once differences in structure have been defined, as, for instance, by the methods of the last chapter, we can try to assess their biological (evolutionary) meaning. For some workers the fact and nature of structural differences are alone the result of the process of evolution, and speculation about evolution is therefore made directly. But for the great majority of biologists, discovering structural differences is merely the first step in making evolutionary assessments. As next steps, such biologists expect that variations in structure can be weighted according to whether or not there is evidence that they reflect specific biological information, such as that due to sexual dimorphism, or to environmental factors, or to individual variation and, as we shall see, to much more.

Adaptation of structure to function is exactly such a specific piece of biological information. It is believed to be especially important for evolution, and the recognition of functional parallels, convergences and divergences are traditional ways of trying to make this kind of judgement.

Information about the degree to which these structural differences are thought to be genetically determined is also of paramount importance. Such matters can be tested out readily in experimental species such as viruses, bacteria, peas, fruit flies and mice. But it is usually very difficult to obtain such information directly about natural populations of primates, whether living or fossil, that may be the object of evolutionary enquiry.

Information may even be sought about the extent to which features are 'primitive' or 'derived' in relation to a postulated evolutionary lineage, and this may help in evolutionary speculations. But, although such judgements are sometimes based upon biological thinking they also sometimes depend only on logical examination of the way in which differences are defined and distributed in various animal groups.

Interpretations like the above have often, in the past, been carried out in an intuitive manner. Some of the more recent studies, however, attempt to separate and examine these various possibilities. They do so while recognizing that complete separation is impossible, and indeed also unwarranted, because interactions among these many factors may be a most important part of evolution.

Although these many approaches are available for studying the structures of the skeleton of the living primates, and, of course, for making assessments of fossil fragments, different emphases have been placed upon them over the years. The shape of skulls, jaws and teeth have been especially important in the past because fragments of skull, jaw and tooth are far more frequently found as fossils than are any other parts of the skeleton. Such studies have mostly emphasized the overall resemblances of the individual parts; they especially have sought out and weighed certain characters in favour of others; they have placed especial importance upon the notion that these cranial features are genetically controlled and therefore reflect rather highly the genetic changes that have occurred in evolution. In particular, attempts have been made to use them in evolutionary assessments based upon estimates of their 'primitive' or 'derived' nature. It is uncommon indeed that function has been seen as an important element in these anatomical areas.

This is presumably fair enough for some information about skulls, jaws and teeth. Take, for instance, cusp patterns on teeth. Although functionally related to diet in a general way (few sharp cusps in shearing carnivores, many complex blunt cusps in grinding herbivores and so on), the precise and detailed patterns of cusps on teeth seem to be highly dependent upon the genetics of the situation, perhaps in a manner analogous to the genetic control of finger prints. In both cases (tooth cusps and fingerprint ridges) the general anatomical features are highly important for function (type of chewing, fingertip sensation and control, respectively). But in both cases, once an overall pattern is available for the functional adaptation, the precise local details (whether a particular cusp possesses a local cuspule, whether a particular fingerprint has a loop or a whorl) seem to be less important for function. It is presumed, and in man has been shown, that such local characters are indeed under genetic control. It is also presumed, but has not yet so far been shown, that these precise details are unimportant for function.

In like manner, it is often assumed that genetic control is paramount for many other detailed structural patterns of bone. Within the skull, for instance, arrangements of sutures in the orbit, at the cranial base and in the ear have all been treated in this way.

This seems not unreasonable. The orbit, the cranial base and the middle ear are all most important indeed for function in their respective anatomical areas. But there seems to be little functional importance attaching to precisely how a particular orbit, cranial base or middle ear has been achieved from the local bones that happen to have become available during development and evolution. Differences in arrangements of sutural patterns between bones in orbits, cranial bases and middle ears have, therefore, been extensively used in evolutionary assessments (e.g. Le Gros Clark, 1959).

The importance of internal developmental factors

Notwithstanding these ideas, which are very much in current vogue, it is beginning to be appreciated that this may not be the whole story. Let us take, as an example, the case of the infra-orbital foramen, the opening of a canal upon the face from which pass the nerves and vessels supplying much of the cheek and upper lip.

It has long been assumed that the number of infra-orbital foramina is under some kind of genetic control and a good measure of the relatedness of different species. Among the hominoids, for instance, man and australopithecines have generally a single foramen; in the great apes there are multiple channels through the face. This kind of feature is examined to discern genetic or racial links. And the particular evidence just quoted has been used to support the idea of the closeness of man and australopithecines, and their separation as hominines from various apes.

But as we learn more and more about developmental mechanisms, and especially as we come to realize the multiplicity of local developmental pathways (only loosely related to genetics, but closely associated with local factors of the internal environment during development), it becomes clearer and clearer that such features are not the special genetic markers that they were once believed to be.

For instance, the numbers and patterns of branching of the infra-orbital canals and foramina on the face (the subject of my Honours Bachelor thesis in 1955) have much more to do with (a) the functions of local structures in the face that are supplied by these nerves (the lips alone are a most important tactile organ), (b) the local growth of the face and especially its bones, (c) the local embryological development of the individual perineural sheaths determined by purely local factors during the development of the nerves, (d) the local development of particular vascular channels as they are picked out from among the meshwork of capillaries by minor haemodynamic differences during development, and especially (e) the rates of development of neural, vascular and bony elements, relative to one another, as they come to be anatomically adjacent.

Ideas such as these certainly fit the broader pattern of the arrangement of infra-orbital foramina in the primates. Thus, there is a single foramen not only in man and the australopithecines (both have shorter faces than the apes) but also in many of the relatively shorter faced lesser apes and prosimians. There are multiple foramina not only in the various great apes (with huge facial structures) but also in many other long faced New and Old World monkeys (and the longer the face, as in baboons, the greater the number of foramina).

However, we do not really know too much about such matters yet. For it is not yet fashionable to follow the later stages of developmental processes wherein may occur changes which we may not even have imagined.

A small piece of preliminary work gives a glimpse of such a developmental phenomenon. The deltoid muscle in the adult rhesus monkey possesses three primary parts forming an epaulette over the shoulder joint (as indeed it does in humans). Embryological texts describe the process of formation of this muscle in humans from three separate tissue clumps. In the embryo at this early stage can be detected the beginnings of three branches of the axillary nerve destined for each head of the muscle.

But the adult muscle in both humans and rhesus monkeys is supplied by only two branches of the axillary nerve, one branch going to the posterior head and a second to both the middle and anterior heads. What has occurred?

Dissections of a series of rhesus monkey foetuses (Oxnard, work in progress) provide an answer. At a stage far later than embryologists normally study, one of the muscle heads disappears and one of the remaining two splits into two new heads. In parallel with this, one of the three nerve branches disappears, and one of the others takes on the supply of the two 'new' heads.

And a second preliminary study of the temporal muscle and fascia in humans (Moore and Oxnard, work in progress) demonstrates an equivalent developmental/evolutionary phenomenon. The fleshy superficial head of the temporalis muscle found in so many primates is only occasionally present in humans; more usually it is absent, but evidence of its disappearance exists in the unusual form of the temporal fascia in humans; in humans this structure is not homologous with that of the non-human primates.

Thus are explained curiosities of the human adult nerve supply of deltoid, and the human adult structure of the temporal fascia. And thus are we reminded not to take developmental information for granted.

The all-pervasive effects of function

As we come to study more and more the post-cranial skeleton, the notion of similarity being based mainly upon direct genetic control becomes more and more tenuous. This is increasingly important in understanding the evolution of primates because it is the post-cranial part of the skeleton that is increasingly of interest today. Although for the post-cranial skeleton it is not doubted that genetic factors have their part to play, it is obvious that many features, especially most quantitative measures, are heavily dependent upon aspects of function during behaviour.

Let me repeat. It is not doubted that hereditary mechanisms play a part in the origin of such structures as the bone-joint-muscle complex and particularly, therefore, of bone (the principal material whose direct study allows us insight into fossils). But hereditary mechanisms are not necessarily implicated in a direct way in the development of the detailed form of those structures involved in biomechanisms.

Genetic mechanisms result in the appearance of *a* bone, *a* joint or *a* muscle, in the production of *an* epiphyseal plate or *a* sesamoid bone at a particular locality, in the formation of *the* basic shape of the femur or pelvis. But the part these hereditary mechanisms play in the emergence of ultimate detailed external form and internal pattern of bones is at the level of inheritance of adaptability towards mechanical stimuli during development and growth.

Even though it is not known how hereditary mechanisms are able, through evolution, to anticipate approximate adult size and shape, the plasticity of the entire mammalian organism is such that the anticipation is only fully realized when external functions no different from those in ancestors are placed upon the structure. Indeed the genetic control of adaptibility may be so loose in mammals that genetics may only rarely and grossly be the ultimate determining factor of bone shape in the natural situation. Placing functions upon descendants that are markedly different from those in ancestors can override structural patterns that might be thought to be determined by heredity.

This is supported by much experimental evidence over the years (summarized in Murray, 1936; Evans, 1957; and Hall, 1978). Should something occur which alters function (for instance, prevention of movement before birth through the use of drugs, or production of new forms of movement by experimental interference or surgery after birth) in a particular organism, then that organism's structure has no difficulty in changing to a new form or pattern in relation to the changed mechanical demands.

A rather complete expression of this idea is the widespread change that occurs in animals that are forced to be bipedal (either because of congenital anomalies or through experimental interference). Not only are the hind limbs radically altered to accommodate walking on two legs, but so also is every other part of the skeleton. The chest becomes broader and more barrel-shaped, the skull becomes shorter and rounder. The changes are so profound, indeed, that respiratory, heart and metabolic rates all alter completely (Lisowski, Van der Stelt and Vis, 1961).

There are two questions here. One is how natural adaptations that appear superficially Lamarkian become subsequently included within the hereditary mechanism in an evolutionary time scale. The other is how such changes in function actually affect structure during ontogeny. The second question is no less important than the first if we are to understand the adaptation of bone form to function. Partial answers are available.

Thus, one problem in understanding animal morphology is inherent not only in the structures that animals possess and the functions that they display but also in the association between the two. This structural-functional interface looms ever larger as our abilities to understand the complexities of both structure and function increase.

Structure, function and the structural-functional interface

Structure. Structure itself can be described rather simply by its appearance to the naked eye, both directly and through dissection procedures. From such descriptions are derived, for instance, the relationships of the primates outlined in the second chapter. However, the application of the new techniques, such as those outlined in Chapter 3, shows that much of the information contained within animal structure is far more complicated. Visual assessment of morphology may miss much that is important to function.

For instance, visual assessment of radiographs of bone suggests that, in general, the small bony spicules within some bones are arranged at right angles to one another; such patterns are very similar, apparently, to the right angle network of principal stresses that can be calculated for bone during function. This has lead to the theory (e.g. summarized in Murray, 1936; Evans, 1957) that the arrangements of the bony plates parallel the arrangements of the principal stresses (Fig. 4.1). Such a theory, if correct, would allow very direct predictions about principal stresses in fossils, given that their bony networks could be examined. Certainly some modern investigations (of architecture: Arnold, 1966; of stress: Kummer, 1966; of strain: Lanyon, 1974; and of stress, strain and architecture: Smith, 1962) appear to confirm such a theory.

But more detailed study of the internal structure of bones demonstrates that the minute bony spi-

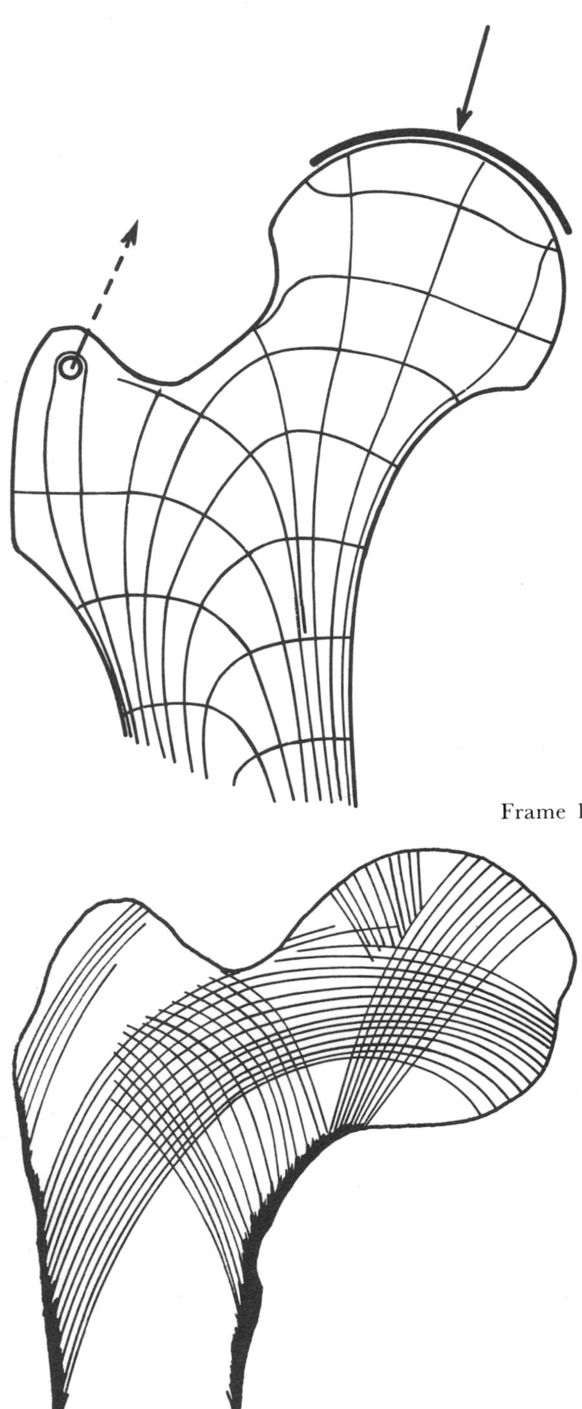

Fig. 4.1. The first frame shows the pattern of stress trajectories that exist in a two-dimensional model of the femoral head and neck when stressed as indicated by arrows. The second frame shows the general pattern of trabeculae found in an equivalent section of the femoral head and neck. The overall concordance between these two results seem clear.

cules are not arranged in a right-angle pattern, and do not correspond, therefore, in a one-to-one manner to the right-angle network of principal stresses (Fig. 4.2 and 4.3). While it is certainly not suggested here that there is *no* relationship between stress and architecture (there very clearly *is* such a relationship), the nature of the relationship is far more complicated (and probably far more interesting) than that implicit in any one-to-one association between orthogonal stresses and orthogonal architectures (Oxnard and Yang, 1981).

Many lines of evidence (e.g. theoretical stress analysis using finite elements: two-dimensional — Rybicki, Simonen and Weiss, 1972: three dimensional – Hakim and King, 1978; experimental stress studies of bone function: Kempson, Spivey, Swanson and Freeman, 1971; microscopical and scanning election microscope investigations of cancellous bone: Singh, 1978, Whitehouse and Dyson, 1974; animal experiments of trabecular bone loading: Radin, Paul and Rose, 1973; optical data analysis of radiographic patterns in bone: Oxnard and Yang, 1981) suggest nowadays that new relationships must be sought between mechanical stress and bone architecture (and see Chapter 3).

Function. It is also now becoming clear that the functions that organisms display cannot be observed in all their complexity through natural history description (the analogue for behaviour, of visual assessment for morphology). Although in general organismal function has not been investigated as fully as organismal structure, evidence abounds that its complexity is no less; in fact, it may even be very much greater.

If, for instance, we wish to know about the functions implied in the form of some post-cranial bone we have not only to study locomotion (that is, posture and movement). We have also to consider many other behaviours such as food-finding, avoidance of predators, and the construction of a variety of shelters, nests and burrows. These may all impact upon post-cranial function. We may have to consider a wide variety of behaviours that involve movement in less obvious ways; social behaviour, including courtship and care of the young, require complex interactions of many movements. Navigation, orientation, communication are all behaviours that are proving yet more complicated as organismal biology progresses. It is not suggested that all of these behaviours presuppose functions that impact biomechanically upon skeletal structure, but some of them may; and we cannot be sure that others will not until eliminated by study. As with structure described by dissection, function is more complicated than that defined by natural history description.

The structural-functional interface. When, finally, we come to define the nature of the interface between function and structure, we run into a further set of complexities. That association, too, may be studied at a number of different levels.

One of these is the understanding of the mechanisms that lead, from the genetic materials, through developmental processes within an individual, to a particular functional-structural association in that individual. Thus, a simple embryonic limb bud may become, through such mechanisms, a prehensile five-digit hand.

A *second* problem is the study of how the functional-structural bond changes over geological time in evolving organisms. A cursorial five-digit extremity, for instance, may evolve into a single digit associated with the evolution of galloping.

A *third* attempts to understand the direct impact upon structure of function through biomechanics.

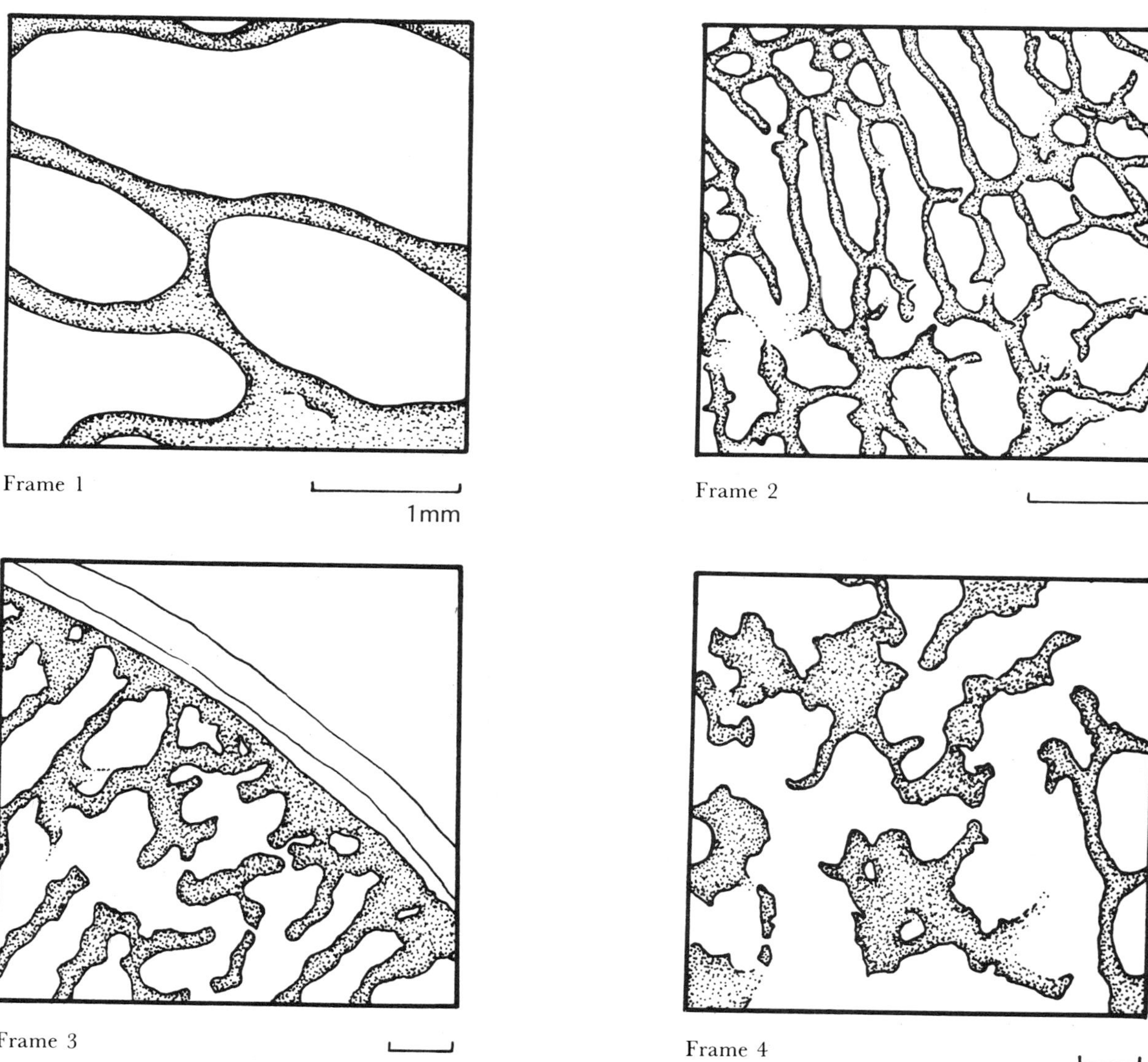

Fig. 4.2. That the picture of the previous figure is not totally correct can be seen from the four frames in this figure. Each of these represents actual drawings of trabecular intersections from different sites in the femoral head and neck. It is clear that the bony plates are not orthogonally arranged in their details even though the global view of Fig. 4.1 suggests they are.

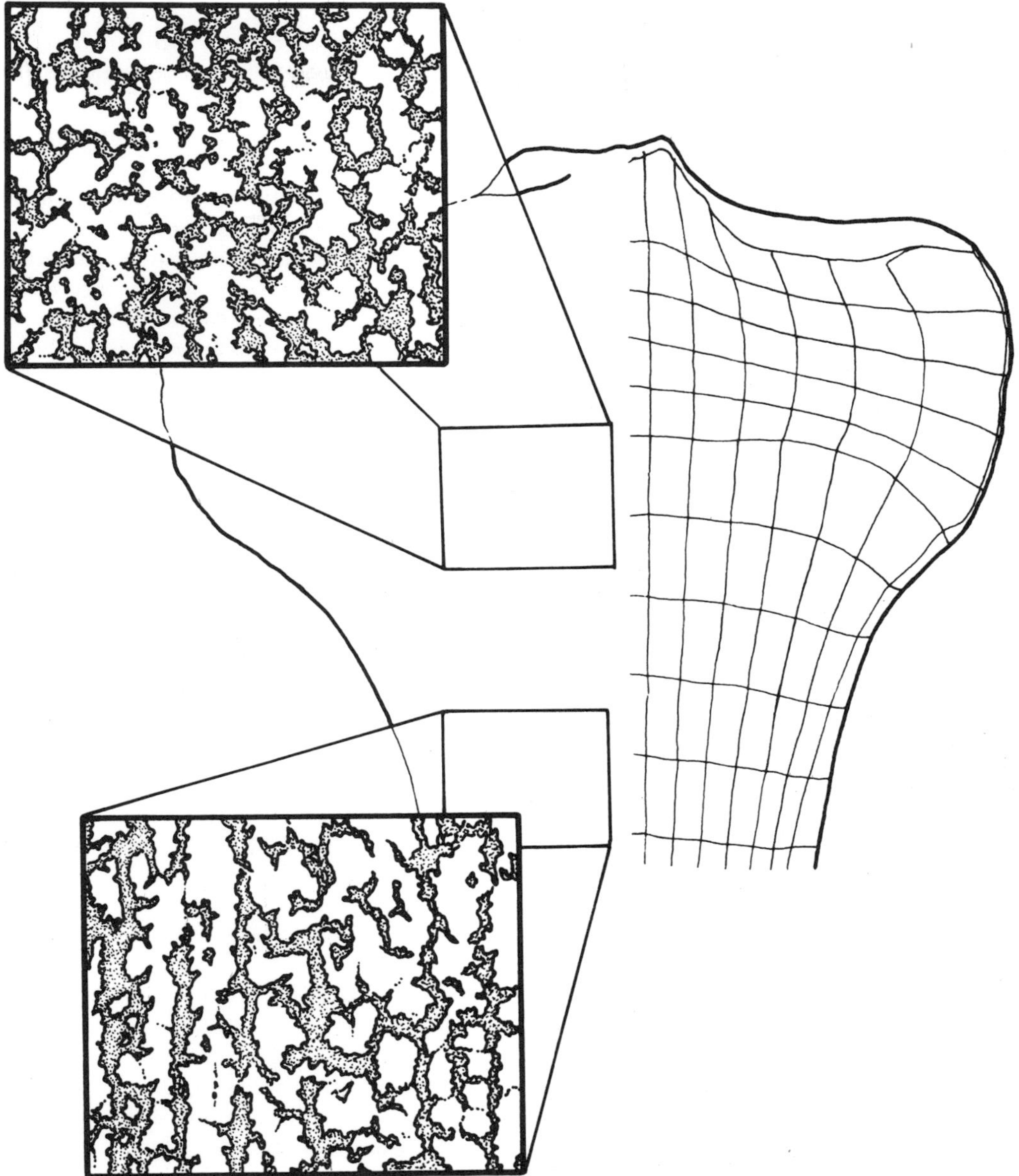

Fig. 4.3. This figure provides another example demonstrating that trabecular networks do not follow orthogonal patterns. In this case stress trajectories have been worked out for the upper end of the tibia. And, although in general it appears as though the trabecular network follows that pattern, careful examination of the different regions show that the bony materials are not orthogonally arranged. In both this and the prior figures it should be noted that even if the pattern of stress trajectories is not correct (for instance, because the loading is not sufficiently well modelled) the argument is not negated. For, although a different loading would indeed give a different set of stress trajectories, by definition these trajectories would still be orthogonal. It is the existence of structural orthogonality that these and many other data challenge.

Surgical interference, for example, may produce new mechanical situations that result in structural changes such as the formation of a new joint or a change in bony architecture in a healed fracture. Changes in behaviour during ontogeny may produce mechanical changes that result in the development of new bony buttresses (as the jaws develop in parallel with the development of chewing) or in the formation of new articular facets (as occurs in the ankle in people — Welsh miners, for instance — who frequently squat). Such changes may even reverse themselves as in the case of the ankle:

squatting facets that are found early in the human foetus (which squats in the womb) but that later disappear.

This third, causal relationship between impressed mechanical forces and architectural changes in tissues has recently received a major new impetus from the discovery of the various mechano-electric forces and implied feedback mechanisms that help relate stress and architecture (for reviews see Alexander, 1968; Wainright, Biggs, Curry and Gosline, 1976; Simmons and Kunin, 1979). Thus, it has been discovered that strain in biological materials of various types (especially bone) produces effects that are reflected in electrical disturbances. And electrical fields applied to these same biological materials produce changes in the alignments of their components. Though it is not certain whether the electrical effects produce or are merely evidence of the phenomena, and though there is still much argument about the details, it is clear that strain directly affects architecture in materials such as bone in a way that involves feedback to improve the efficiency of the architecture for strain-bearing.

None of the above attacks on the problems of evolutionary biology is isolated; each depends upon the others for full understanding. Yet disentangling the functional-structural relationship often starts with the study of animal mechanics as supplying one set of constraints within which both heredity and evolution must operate.

Animal mechanics: a direct approach

Biomechanics can be studied at a relatively simple level. It is easy to see, for instance, that the physical principles relating to size, load and properties of materials provide considerable constraints that determine some of the limits of biological form. As a result, it has long been known that land animals constructed of conventional biological materials cannot exist above a certain size.

But biomechanical studies have become more complex as a result of many modern developments such as equipment miniaturization and computer simulation. Experimental methods such as cineradiography, electromyography and experimental stress analysis are providing information about biomechanics that continually hone our understanding, and in many cases provide, entirely new views of traditional subjects. The application of information from the engineering world, as, for instance, that contained within finite element analysis, or that related to the properties of new materials such as fibre-glass, is also heavily involved in modern changes of this ancient biological study.

Let us took at some examples. Cine-radiography has shown that when chimpanzees walk upright, they do it by using a modification of their regular quadrupedal mechanism within the new upright position. They do not do it by adopting a biomechanically similar form of bipedalism to that of man (Jenkins, 1972). This finding has obvious implication for the study of bipedalism in possible human fossils.

Electromyography has suggested that considerable forces may be borne by anatomical structures even when muscles are electrically silent (Basmajian, 1967). This particularly emphasizes the importance of knowing about a generally much neglected topic, the existence of tension in biological structures, and the study, therefore, of ligaments and joint capsules. In turn, this provides new information about the workings of the limbs of fossil creatures.

Experimental stress analysis utilizing strain gauges *in vivo* has demonstrated, as a yet further illustration, that not only does the vertebral column help bear the obvious loads associated with posture and locomotion, but that each individual vertebra expands and contracts like a concertina as a result of hitherto generally neglected and apparently small cyclical forces such as those of the heartbeat and respiratory rhythm (Lanyon, 1971, 1972 and 1973). This new finding, too, may have an impact on the study of fossil vertebrae.

Many researches can be cited that today are providing new information about functional structural relationships in living species, and that, by analogy, may provide estimates about fossil forms. Thus, for instance, though we have no immediate extant biological equivalents in living species for the 'sails' of pelycosaurs or the 'crests' of hadrosaurs for studying their functions experimentally, in the case of many of the problems relating to human and primate evolution, reasonable inferences may indeed be made from models provided by study of closely related living species.

The models may be entirely theoretical. For example, utilizing theoretical stress analysis of both living and fossil species, it has been suggested that the forms of some hand bones in dryopithecine apes are well adapted to use in palmigrade postures and gaits. Models of digitigrade locomotion in these fossils, as in some terrestrial monkeys today, or of knuckle-walking locomotion as in the terrestrial living African great apes, would appear to lead to greater stresses than are efficient in portions of the fossil bones. This indicates the unlikelihood of digitigrade or knuckle-walking locomotor patterns in these particular dryopithecine apes; it implies that these fossil apes were arboreal rather than terrestrial (Preuschoft, 1973).

In a similar way, experimental stress-modelling has been applied to fossil shapes in comparison with the structures of extant species, the functions of which are known. It has been shown that the complicated architecture of the finger bones in the living apes relate fairly well to primary functions during locomotion and foraging, whether of the terrestrial knuckle-walking African apes, the gorilla and chimpanzee, or whether of the arboreal, hanging-climbing Asian great ape, the orang-utan. A fossil finger bone from Olduvai has been shown, in related comparisons, to be mechanically more efficient in the arboreal mode than the terrestrial and to be nothing like that of man (Oxnard, 1973a). Similar arguments would seem to apply to all of the fossil hand bones (from both Olduvai and Hadar) that display the kinds of curvatures found in that finger bone.

Some of the fossil bones from Olduvai have been removed from consideration as of human ancestry exactly because they display these curvatures with arboreal implications. And, as some of the hand bones of the fossils from the Afar Valley discovered by Johanson and Taieb (1979) are also said to show marked curvatures, we may wish to be very careful before we describe these hands, too, as from human ancestors equipped for primary activities of tool-using and -making; it is entirely possible that they were primarily adapted for use in some mode that involved finger-grasping (of branches?) with great power.

Both of these studies suffer from the deficiency that the bones modelled, whether theoretically or experimentally, are treated as though they were 'simple uniform elastic bodies'. Bone is not, of course, a 'homogeneous, isotropic material', operating as a 'uniform elastic body', under loading dependent upon 'infinite beam theory'. These were guiding ideas in most biomechanical studies that follow the mechanics of an earlier generation of engineers, ideas that are, of course, still completely appropriate to materials such as iron and steel. There is now every good reason for believing that most biological materials, especially bone, must be considered as 'anisotropic, poroelastic, materials', and that 'finite element theory' may be more appropriate, especially for complexly shaped anatomical parts (Rybicki, Simonen and Weiss, 1972). These ideas and methods include some of the newer engineering concepts of the last twenty years that have become necessary as engineers came to grapple with more complex materials such as fibre-glass (a useful analogue for bone).

Thus, theoretical ideas well-developed in engineering have marked implications for biology. We may hope that the inferences derived from simpler models as used above, whether theoretical or experimental, are not unreasonably inaccurate despite the simplifications adopted. Yet we must be careful because in studies of some regions, for instance the femoral neck and head, the inferences obtained from the simpler methods have proven to be wrong by almost an order of magnitude (Rybicki, Simonen and Weiss, 1972).

Functional morphology: an indirect evaluation

Indirect methods are also available for the study of animal form. These methods consist of allowing the structures to speak, as it were, for themselves. The approach is a good deal less directly aimed at biomechanics than the direct experimental methods. And because the structures do not always speak very clearly for themselves, we may run into different problems. Nevertheless, though the results must be inferred indirectly through comparison, rather than neatly displayed directly by experiment, the inferential approach to structure is decidedly superior to the experimental in some important respects.

It is better able to deal with populations rather than individuals, something that is usually prohibitive in experimental work. It can cope with different anatomical regions and a diversity of taxonomic groups of animals such as can scarcely be involved in carefully planned experimental studies with adequate controls. It is capable of dealing with fragmentary, incomplete and rare specimens in a manner difficult to arrange in experimental studies (particularly important for examinations of fossils). And it may require only study of museum materials already collected without the need to interfere with living species, many of which are in conspicuous danger at the present time.

It must be pointed out that the inferential methods are not at variance with the direct techniques. In fact, it is clear that the converse is true. The two approaches are complementary rather than competitive; concordance between them is an important strength in evaluating results; disagreements should make each investigator look to the beam in his own eye.

The inferential approach includes, of course, the old method of simply associating visually, functional observations with morphological information.

But a rough association between function and morphology readily springs to mind in most anatomical situations. How often have morphologists been able, glibly, to hypothesize a particular relationship between function and structure on the basis of some newly determined morphological fact? How often have they also known that if the morphological information had happened to be exactly the opposite, then they could still have hypothesized a superficially attractive biomechanical relationship? Examples abound in the literature.

Thus an early 'explanation' of the large gluteus maximus muscle of the human was that 'it produces the increased extension of the hip necessary in bipedal walking'. However, had the gluteus maximus muscle been smaller in man than in the other primates it could presumably have been as readily suggested that 'this smaller muscle in man is not required to produce strong extension of the hip as is necessary in the leaping and climbing of non-human primates'.

In fact, as is now well known, this problem has been worked out. The gluteus maximus muscle is large in man (fact) but it is likely that this has little to do with simple walking (old hypothesis) but with running and raising the trunk as in climbing rocks or steps and in rising from a sitting position (new hypothesis). The test does not come from the postulated simple association between morphology and function but from more complex studies of gluteus maximus in man in a variety of different activities using biomechanical and electromyographic techniques (e.g. Basmajian, 1967).

This example is relatively clear with hindsight. But it was not clear at the time; there are many useless words in the literature about the functions of the gluteus maximus in walking; and even today there is a good deal about the structure and function of this complex muscle block that deserves and is receiving (for example, Stern, 1971; Vangor, 1978) further study.

Notwithstanding this example, whatever the function and whatever the structure happen to be, once we become interested in shades of difference, the complex nature of bone form and pattern and the complicated series of developmental, evolutionary and biomechanical processes that are responsible for its precise expression deny the easy explanation.

The structural-functional association revisited. The causal relationship between structure and function seems not to be between primary elements of function and primary elements of structure. Each particular function, as observed, must be associated causally with several structures; each structure has impinging upon itself the effect of several functions; the association between structure and function must be far more complicated than any simple one-to-one relationship.

That this is likely to be so has always seemed intuitively obvious to those who have a '*gestalt*', whole-organism, view. And it has recently become more obvious as a deeper understanding of biological correlation reveals that any given primary functional feature is partially correlated with each of several pieces of structures. Likewise, the idea can be read in reverse. Particular structures can rarely be aligned totally with particular functions; structure has to be partitioned out, through correlation, with several, perhaps many aspects of function.

The development of ideas like correlation and the analytical methods that go with it provide tools for investigating the situation. They suggest that the morphological behavioural interface is unbelievably complicated. So complicated is it that we can never understand its precise, real nature. But do we have to? That may be almost like asking for average gas pressures from determinations of the actual position of each molecule in a gas. Quantum mechanics has shown us that we do not need the detailed information about each molecule in order to be able to understand the gas pressure laws. And in an analogous way we do not need the detailed information about behaviour; some simpler estimate of behaviour may be all that is necessary to understand particular biological structures. It may well not be necessary to know everything that an animal does in the field situation.

Much experimental work over many decades (e.g. Murray, 1936; Evans, 1957) suggests to me that the form of a bone is not related separately to each particular function that the bone has to perform. The relationship between a complex of functions and a complicated structure is through a narrow interface of an average or resultant biomechanical situation produced by functions upon structures. However, this average or resultant biomechanical situation is not easy to envisage. It is certainly not the simple average of all of the forces that function impinges upon structure. Presumably, it must mean a more complicated average that has to take into account some of the following factors.

First, this average must presumably reach a certain critical scalar threshold before it starts to have any effect at all in causing adaptational changes in structure. *Second*, the situation must presumably exist for some minimum length of time before it has adaptive effects, a temporal threshold. A *third* idea affecting the resultant or average biomechanical situation relates to the existence and effect of cyclicly changing functions such as those due to the heartbeat and respiratory rhythm, perhaps those of repetitive reflex movements before birth or locomotor cycles after birth. A *fourth* includes non-scalar elements such as the directions of the forces, a vectorial component. A *fifth* is the likelihood that the adaptation of bone form in relation to the average biomechanical situation is highly unlikely to be linear. A *sixth* is the possibility that the adaptation of bone in normal function may involve not only non-linear features (such as non-Newtonian stress-bearing) but also phenomena that are actually discontinuous or quasi-discontinuous (such as microfracture, Radin, Paul and Rose, 1973). A *seventh* is that each of these elements of the average of any single identifiable bit of function requires to be averaged against all those other functional features also acting during the period of time the morphology is being influenced. And *finally* it seems that, though the average is the important element for some features — for example, the amount of force — the sum may be the important factor for other features, for instance, the direction of force (Oxnard and Yang, 1981).

All this adds up to the final influencing factor being not any particular force and certainly not any particular function, but the resultant or average biomechanical situation that can take appropriate account of all.

Even though we have reduced the problem to an average it still seems unbelievably complicated. But let us press on.

Some practical examples. In a particular anatomical region where function is such that its biomechanical contributions to the average biomechanical situation are of approximately similar size, it is difficult to see what basis we would have for estimation. This might be the case if we were interested in understanding the association between function and structure in, say, the human hand. Though we can recognize power and precision grips in the human hand, the many different actions that the hand carries out in one individual over time — the enormous differences that exist between what one human hand does and the next — will prevent us from ever being able to estimate, in the present state of the art, the average or resultant biomechanical situation for the human hand (Fig. 4.4). But this disability does not apply to most anatomical regions.

In an anatomical region where most of the forces contributing to the average biomechanical situation are small, and a few are overwhelmingly large, then an estimate of the resultant or average can be made from the latter alone. This is because, unless the small forces are similar, they do not much affect the resultant or average biomechanical situation. The large forces are excessively dominant.

Thus, although, for example, both the gorilla and chimpanzee can use the fingers for a wide range of manipulative activities that cause many small forces to act upon the fingers, the two functions that impinge by far the biggest forces upon finger bones are those when the hand is used in heavy compression during knuckle-walking and in high tension during climbing. Examination, then, of these major functions alone may provide us with reasonable estimates of the resultant biomechanical situations; and it is to these resultants that the form and architecture of the finger bones in these apes will be most closely related (Fig. 4.4).

This form of estimation of the average is a very local phenomenon. It is easy to be completely wrong about which are the important elements involved in the adaptation of a structure. Thus, in the case of the hand of the African apes just cited, the example is true for *fingers* which are heavily involved in knuckle-walking and hanging. But in the case of *thumbs* the argument may well fail. For, though thumbs are undoubtedly involved in knuckle-walking and climbing activities to some degree, the lesser power of their participation is such that the locomotor forces in thumbs are quite small. These forces may be small enough to be similar to the small forces produced in the thumbs

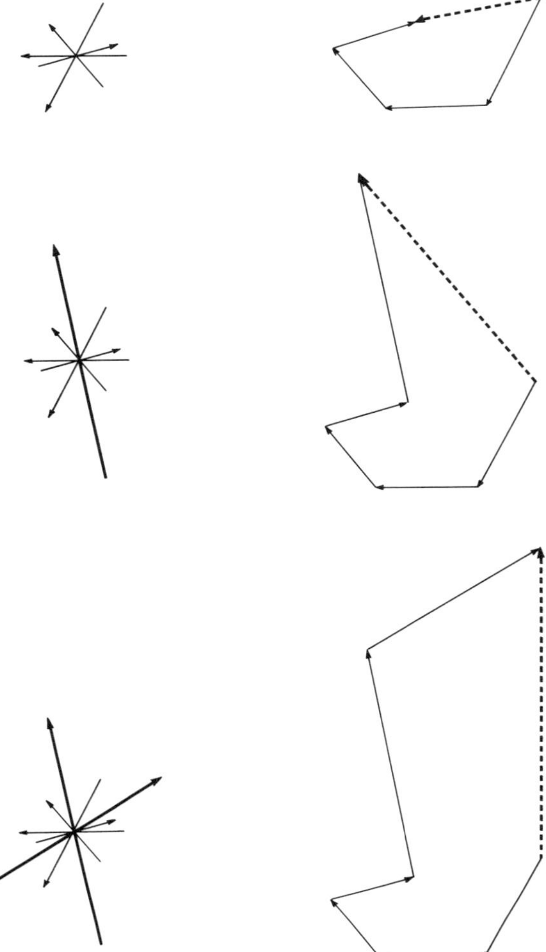

Fig. 4.4. This figure attempts to demonstrate how small and large forces respectively contribute in the formation of a resultant force. In the upper pair of figures it is assumed that many small forces are acting at the central point of the left-hand diagram. The right-hand upper frame shows (dashed line) the resultant of that set of small forces. Because the small forces are not aligned with one another, the resultant is also a small force. In order to know that resultant we would need to know each of the small forces contributing to it.

In the middle frame, in contrast, there is, in addition to a number of small forces, one large force acting at the central point in the diagram. The resultant force (again a dashed line) is obtained from the vector addition of all the forces. But it can also be seen that a reasonable estimate of the resultant force is given by consideration of the large force alone (dashed line). The solid line representing the largest force is very similar to the dashed line representing the resultant.

In the lower frame a similar picture is seen, but in this case there are two large forces. Again, although an absolute resultant (dashed line) stems from considering all the forces, a very fair estimate of the resultant comes from viewing only the contributions of the two large forces.

of apes during other manipulations of the hand. In this example, therefore, before a good functional-structural association can be worked out, more of these other forces would have to be included in the formation of the average.

Thus, though understanding the morphology of finger bones in apes may require rather little information about function involving only knowledge of knuckle-walking and climbing, understanding the morphology of thumb bones may involve a far more detailed study of manual function.

Another aspect of this localizing phenomenon can be seen through an additional example. Although the hip joint is undoubtedly subject to many different forces during the lives of humans and of orang-utans, it is clear that the biomechanics involved in the human hip when small boys hang upside down on horizontal bars are irrelevant to the estimation of the average biomechanical situation that is involved causally in human hip structure. In the case of the orang-utan, however, the opposite is likely to be the case; hanging upside down in orang-utans is a regular activity and may well be a major contributor to the average biomechanical situation that causally effects the shape of the hip in those animals.

Both of the average biomechanical situations here turn out to be locomotor averages because the non-locomotor things that orang-utans and humans do with their hips involve much smaller forces. Both orang-utans and humans, for instance, spend quite large amounts of time simply sitting and lying. Although the forces involved in sitting and lying act for long periods of time (it has often been shown that animals in the field spend large amounts of time in such postures as compared with movements), the forces involved in sitting and lying are so small that they are truly unlikely to have much effect upon the average biomechanical situation and therefore upon the bone adaptation.

In other words, knowledge of sitting and lying behaviours is unimportant in understanding the functional-structural associations of the hip joint; and field studies that attempt to discover functions pertinent to hip joint structure need not obtain information about sitting and lying, or indeed about most other behaviours.

But we should notice exactly how precise the arguments are. If we are interested in the structure of the bare surface of the ischial tuberosity (the sitting area of the pelvis) then the forces generated by the various locomotor activities described above may be of little importance. But forces involved in sitting may well have a major effect upon the morphology of the bare area of the ischial tuberosity. This is because, though small, they are the largest or even the only forces that operate in that particular anatomical locality. In this case knowledge about sitting postures may be very important in understanding the structural-functional association.

And a final example of this phenomenon relates to ranges of movement. It is highly likely that the extent of the edge of an articular surface is associated with the extreme range of movement performed by an animal irrespective of what that animal usually does. Knowledge of that extreme range of movement may thus be most important in deriving the average biomechanical situation for the structural-functional associations of the articular edge. But if the extreme movement is carried out only rarely (say during some vitally important but very infrequent escape behaviour) then its contribution to the average mechanical situation for some other nearby feature, say the size and form of the central part of the articular surface, may be very much less. Other movements and postures involving heavier and more frequent forces shape the centre of the articular surface.

In general, however, in the study of the non-human primates, except for special regions such as those cited above, the major anatomical features will be most clearly associated with average biomechanical situations that are influenced to the greatest degree by locomotor activities such as those involved in travelling, foraging, playing and escaping. Postures and other small movements such as those involved in behavioural and communicative interactions, and indeed any other movements involving small forces, however much time is devoted to them by the animal, are unlikely to have much effect upon major aspects of post-cranial morphology. They need not, therefore, be known; and field studies to obtain such wide-ranging and difficult data need not be undertaken for this purpose.

Average biomechanical situations: locomotor classifications

Thus, no fixed behavioural profile of a set of animals is necessary for all functional-structural studies. There are many investigations where information about postures is totally unimportant and where, in contrast, information about infrequent but powerful movements is critical. There are a few studies where posture is more important than movement. There are some studies where knowledge of the infrequent extremes of movement is useful; others where knowledge of usual ranges of movement is important. Just as, therefore, there can be no overall prescription as to what anatomy it is important to examine, so too there can be no overall recipe as to what behaviours are the most useful. There is no single plan of attack that will suit all problems.

It is argument like this that relates to our own studies of limbs and limb girdles in the primates. Although limbs are capable of many different activities, it is only necessary to consider the subset of activities involving the larger and more frequent forces in order to investigate form and function in limbs.

The shoulder girdle (Ashton, Flinn, Oxnard and Spence, 1971; Oxnard, 1973a) is a case in point. The form of the shoulder can readily be associated with those various activities relating to supporting and moving the entire body weight as in locomotion. Shoulder activities, such as those involved in resting, nursing an infant, even sleeping, produce forces that relative to those of locomotion are very small even though some of them may occur for rather long periods of time. Their effects upon the average biomechanical situation are very small; their associations with overall shoulder structure seem, accordingly, to be negligible. The average biomechanical situation for the shoulder is, then, one that pertains almost entirely to upper limb movements in locomotion and foraging, and to precious little else.

Similar arguments exist for the pelvic girdle (Zuckerman, Ashton, Flinn, Oxnard and Spence, 1973; Oxnard, 1973a, 1975a); the very large forces generated in muscles and bones during the movements of locomotion and foraging presumably have much greater associations with pelvic shape than movements and postures produced by such activities as sitting, lying, sleeping, or even the activities involved in the reproductive function of the pelvis. The average biomechanical situation is, again, one that relates most closely to pelvic function in locomotion.

In the same way, the major features of most other post-cranial bones are most closely associated with average biomechanical situations influenced most by relatively few locomotor activities involving large forces: behaviours such as travelling, foraging, and escaping.

It is in this context that classifications of the locomotor patterns of primates have been of value in earlier studies. Thus, different primates have been categorized as runners, leapers, climbers and so on. Such classifications have been used since the turn of the present century (e.g. Mollison, 1910; Priemel, 1937; Campbell, 1937; Erikson, 1963; Napier and Napier, 1967; Rose, 1974) to help understand the structure of the post-cranial skeleton. They have especially formed a useful background for functional assessments of fossils (as of *Dryopithecus*, e.g. Napier and Davis, 1959). But as the complexity of the functional-structural inter-relationship has become recognized, the utility of such classifications has waned.

The complexity of animal behaviour is so great that it is impossible to describe locomotor patterns of entire animals using such a device. The generalities expressed in terms such as 'runner' or 'leaper' about the entire behaviour of animals may be unacceptable to many who study field behaviour of living primates (e.g. Ripley, 1967).

Furthermore, even when, for a given animal, the more detailed spectrum of behaviours is known from field studies (and this is gradually becoming the case for increasing numbers of primate species) the very complexity of the behaviours is such as to defy defining their entire locomotion.

Finally such classifications, although resting in part upon behavioural realities, do also sometimes depend upon prior knowledge of morphological features; the resulting circularity in argument detracts even further from their value in evaluating fossil and living forms.

However, classifications of function in a restricted anatomical region (within the demands of the behaviour of entire animals) can be much closer to the average biomechanical situation for that region. This regional restriction makes it easier to see what may be occurring during the function of the part. It especially allows identification of equivalent functional situations even in animals whose entire locomotor patterns may differ totally. It allows us to develop parallels for use in evolutionary studies.

Thus, although the overall locomotion of uakaris, howler monkeys and orang-utans differ totally from one another, the movements of the hip in these three species display certain similarities in terms of its extreme mobility and its special use in support and suspension during climbing and foraging in the trees. These functional similarities contribute uniquely to the average biomechanical situation in the hips of these animals and may account for some of the otherwise inexplicable resemblances that exist (Stern, 1971; Oxnard, 1973a, 1975a). In this example, the average biomechanical situation results from the addition of hind limb tension-bearing. The regional functional parallel more easily allows the discovery of structural implications.

In exactly the same way, although there is no similarity between the manner in which orang-utans, angwantibos and sloths (to choose a non-primate example) move, there are considerable similarities in the way in which the shoulder may be used in suspension within their vastly different locomotor patterns. Each of these species is capable of hanging by the upper limbs and of bearing, therefore, tensile forces in the shoulder. Again, therefore, in spite of the many different things that these animals do, the average biomechanical situation is one that pertains to the addition of tension-bearing in the shoulder. Again, the regional functional parallel accounts for some of the similarities of shoulder structure found in each, and differing in most other related animals (e.g. Miller, 1932, 1943; Oxnard, 1968a).

Such classifications or definitions of 'average biomechanical situations' within localized anatomical regions have been used by a few authors (e.g. Ashton and Oxnard, 1964a, b; Ashton, Healy, Oxnard and Spence, 1965; Ashton, Flinn, Oxnard and Spence, 1971; Zuckerman, Ashton, Flinn, Oxnard and Spence, 1973; Feldesman, 1976). However, the controversies surrounding the problem of the classifications of the locomotion of entire animals have so confused the picture that the usefulness of even these regional groupings is masked. And in any case advances in our ideas about 'average biomechanical situations' suggest that better ways are available for handling the problem.

Alternative ways should not, of course, attempt impossible tasks: that is, they should not attempt to obtain and meld the very large amount of information inherent in describing the entire life patterns of individual primates. In fact, they should only summarize those aspects of function big enough to be main contributors to the average biomechanical situation. They must be flexible enough to be able to take into account locomotor variations that truly exist within particular primate groups. In this way the locomotor plasticity that is an essential characteristic of many primates (as compared with some other animals) is built into the mode of characterization. Finally, and perhaps most importantly of all, they should not force a grouping or classification in a situation where function is patently so continuous that groups do not exist.

An alternative: the regional functional spectrum

The alternative that suggests itself is viewing the average biomechanical situation as a continuum, or spectrum or trend (Oxnard, 1975c). It is easier to see what is meant if we apply the idea to specific anatomical regions.

A spectrum of upper limb function. Although most non-human primates are able to swing by their arms and walk bipedally, nevertheless they are primarily quadrupedal in habit. In quadrupedal walking and running on a level substrate, the upper limbs act mainly as propulsive levers and struts; they move generally in a low quadrant relative to head position, and they bear primarily compressive forces. In the arboreal milieu, locomotion also occurs along inclined surfaces and to this degree the upper limbs may be required to operate more and more in front of the head in a somewhat higher quadrant, and, to a somewhat greater degree, they may be required to bear tensile forces. The trend is further accentuated in the three-dimensional network of the small branch environment. The concept reaches its acme in those species that often hang by their upper limbs. Clearly, there is a trend, or continuum, or spectrum of upper limb function.

Because of the problem of deciding how much emphasis to place on the different expressions of a given animal's movements, on the variety of environments that it may occupy, on differences that may occur during its ontogeny, on variations in size of different bodily parts, on the importance of rare but critical escape movements, on the effects of a wide variety of other non-locomotor movements and many other complicating factors, it may be difficult indeed to place any given species at a specific locus within the spectrum. However, the very idea of a spectrum allows us to visualize a fuzzy or indeterminate description for any given form. The spectrum is a descriptive tool capable of assimilating new information about any individual species without causing a breakdown of the entire model.

Thus, the model of upper limb function may be thought of as a broad spectrum or band of function (Fig. 4.5) running from those species (e.g. patas monkeys, baboons) in which the average biomechanical function in the upper limb is usage in a cranio-caudal, two-dimensional arc within a lower quadrant under compression. These species may be compared with others (e.g. gibbons, spider monkeys and orang-utans) in which the average biomechanical situation in the upper limbs is usage in a three-dimensional, highly mobile cone within a raised quadrant under tension. Within such a spectrum, some species (e.g. woolly monkeys, proboscis monkeys) are placed at intermediate positions because they possess, whatever their overall locomotor patterns, average biomechanical situations in the upper limbs that lie somewhere between the extremes.

Humans are not, of course, placed anywhere within such a spectrum, for they do not use the upper limb for locomotion at all.

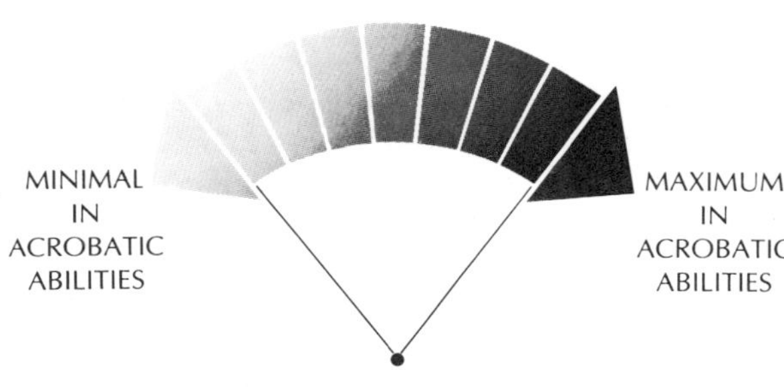

Fig. 4.5. A possible model of function in the upper limb in primates: a band-shaped arrangement. Humans, because of the non-locomotor activities of their upper limbs, do not fall in the spectrum.

A spectrum of lower limb function. When we come to view the functions of the lower limb, we are much closer to locomotion than is the case for the upper limb which also has manipulative, communicative and other functions. Yet the range of locomotor mechanisms for the lower limb is actually more complicated than for the upper limb. Thus, although in quadrupedal walking and running the lower limb acts as a propulsive lever and strut (as previously described for the upper limb), it also has a series of additional functions.

In some species, for instance, there are marked accentuations upon leaping. This demands powerful and often synchronous retraction and extension of the lower limbs from highly flexed to highly extended positions. There are a variety of mechanical ways in which leaping may be staged. Leaping may depend upon overall lower limb length and especially hip extension (e.g. colobines, indriids); it may be related to overall limb length together with special foot mechanisms (e.g. bush-babies and tarsiers); it may even be markedly asymmetrical (e.g. Allen's bush-baby; Jouffroy and Gasc, 1974).

Another locomotor function exists in those species, rather fewer in number, which possess abilities to suspend the body in part or even totally by the lower limbs. This too may be carried out in different ways, e.g. the lower limb suspension of some lorisines when passing under branches is different from that of uakari monkeys when utilizing lower limb suspension in foraging (this is not infrequent), and different again from the frequent and acrobatic lower limb suspension of orang-utans.

Yet another activity of the lower limb is the more specialized, somewhat cursorial quadrupedalism of the terrestrial habitat. This special type of quadrupedalism is found, for instance, in ground-living cercopitheques as compared with most of the more arboreal ones; it is rather more obvious in the terrestrial behaviour of baboons as compared with closely related more arboreal mangabeys; it is perhaps most clearly characterized in the highly cursorial gait of patas monkeys in full flight.

And a different form of terrestrial locomotion is associated with yet another set of mechanisms for lower limb function in the African great apes. This is a lower limb correlate of the upper limb descriptive form of locomotion: knuckle-walking.

In yet other primates the lower limbs play a comparatively smaller role in locomotion as in brachiation proper (e.g. gibbons) or tail-assisted brachiation (e.g. spider monkeys); although of course, in these same species, other activities involve the lower limb fully.

The model of lower limb function that we may derive from these descriptions is considerably more complicated than for the upper limb. A single broad spectrum is too simple a descriptive device. A good analogy is a star-shaped spectrum (Fig. 4.6).

The nucleus or body of the star may be thought of as containing the generalized arboreal forms in which the average biomechanical situation in the lower limbs is that stemming from regular quadrupedalism. The various rays or arms of the star may be thought of as mini-spectra emanating from

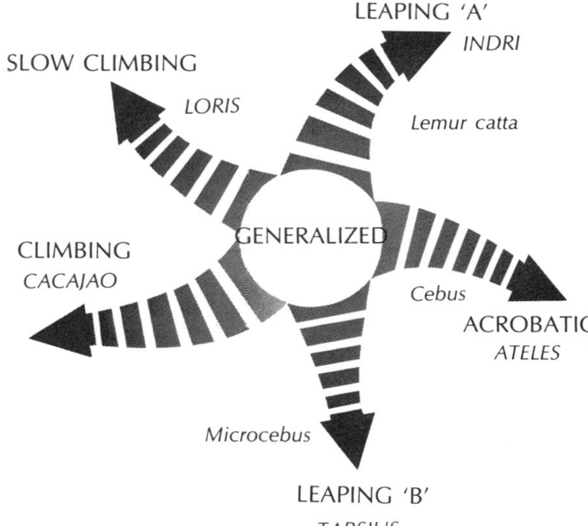

Fig. 4.6. A possible model of function in the lower limb in primates: a star-shaped arrangement. Though not shown in the diagram, it is reasonable to place humans as entirely outside the star-shaped arrangement because of their complete difference in lower limb function.

the nucleus for each of the qualitatively different biomechanical situations resulting from the more extreme functional modes of the lower limb.

Thus, a ray towards average biomechanical situations resulting from regular terrestrial constraints on lower limb activities contains at its extreme the patas monkeys; half-way along the ray are baboons; lying near the root of the ray with close relationships to regular quadrupedal monkeys are the semi-terrestrial green monkeys. This description is achieved without defining any locomotor groups.

Another ray of this star may pass towards average biomechanical situations that involve the constraints imposed by the curious slow climbing activities of the various lorisines: most extreme for the potto at the point of the ray, least so for the loris at the ray's root.

Yet other separate rays may incorporate various biomechanical modes of the different groups of leapers; the leaping of bush-babies and tarsiers might be expected to be located in one ray, that of indriids in another. These 'leaping rays' may possibly be, in turn, separated from leaping of a less specialized kind such as is found in mangabeys and colobines; and we may expect to find each species at its own approximate place within each ray.

As with the model of upper limb function, humans do not fall anywhere within the model of the star. For there are no species, among extant forms, that provide appropriate links with such a bizarre form of movement as human bipedalism.

And again, as with the simpler analogy posited for upper limb function, the model is flexible enough to provide a locus for any given species as a relatively fuzzy, indeterminate position at some place within the body of the star, or at some location along one of the rays of the star. The model is not so inflexible as to require that any particular animal belong, unequivocally, to a 'locomotor group' within a 'locomotor classification'.

A summary: the classification versus the spectrum

Thus, to summarize, we may note that, though useful in times past when structures and behaviours were understood at only relatively superficial levels, classifications of entire locomotor patterns are today less than useful. It is not possible to provide groups within classifications of entire locomotor patterns that satisfy behaviourists because of the more detailed knowledge of locomotion that exists. What any one animal does is different enough from that of any other that the classifications become so split as to include only a single animal in each group. And in any case much of the information in such classifications was partly based upon structural realities; the resulting circularity in argument is obvious.

Classifications based upon comparisons of functions in specific anatomical regions, in contrast, are still most useful. They get around the problem of every animal becoming its own locomotor group as it were, because, though indeed what every animal does is different, the comparative activities of anatomical regions may be similar. Thus, however different may be the entire locomotions of the various apes and some prehensile-tailed New World monkeys, the functions of their upper limbs compared to the upper limbs of other primates are similar due to tension-bearing in acrobatic situations. However different may be the running patterns of baboons and patas monkeys, there are similarities in hind limb functions related to their terrestriality as compared with hind limb functions in the arboreal activities of other monkeys.

But it is the concept of a spectrum of function that is the most useful now that both behavioural and structural studies have become more detailed and quantitative. The notion of a spectrum of function allows a given animal to have a fuzzy position within the spectrum. This can cope with differences in function due to different local ecologies, changes in views about function as more becomes known from behaviour and existing variations in function that are part of the biological reality.

Thus, instead of implying a spurious similarity in leaping ability inherent in the idea of classifying thick-tailed bush-babies and Senegalensis bush-babies in a group of leapers, we can imply a useful similarity *and* difference by placing, within a spectrum of leaping, the thick-tailed bush-baby at the lesser leaping end and the Senegalensis bush-baby at the greater leaping end. Conversely, instead of implying a spurious difference between *Lemur catta* placed in a group of quadrupeds and *Propithecus* in a group of leapers, we can recognize their similarity *and* difference through their placement within another spectrum of leaping. In this case, again, one species occupies the lesser leaping end and the other the greater. In both of these examples, much of the subsequently determined structural data fits the concepts.

The 'design' of observations

The above discussion shows that, although function and structure in a particular anatomical region are complicated, we can truly hope to obtain some assessment of the functional-structural spectrum in a finite study. But a second reason why we can study the association between structure and function (in animals upon which we cannot easily do experiments) relates to the use of the comparative method. Thus, although we may not know the totality of the variety of forces that contribute to the absolute value of the resultant biomechanical situation in a particular animal, we may be able to see that the major difference between this species and the next is fairly well-confined to a relatively simple activity. Two animals may well walk, run, leap and in general cavort about in the trees in a way which makes any estimate of the resultant in absolute terms difficult. If the major difference is (say) that one animal does a somewhat greater degree of leaping than the other, then studying leaping alone may give us the difference between the average biomechanical situations in the two animals. Knowledge of this behavioural difference may be enough to allow us to envisage the structural difference between the two species and to pinpoint, therefore, this particular part of the functional-structural association.

We must be careful in using this comparative method of course. There is not actually overly much value in comparing only one group of animals with only one other group. For differences can always be found between two groups, and these differences may easily be spuriously associated with the difference in behaviour rather than with some other, unrecognized factor.

In order to avoid this problem there are several steps that can be taken. *First*, the comparison may be made upon several groups of animals (rather than a pair) that differ in their functions in a consistent way; this allows the determination of whether or not the proposed functional-structural associations change in a regular and predictable manner. A spectrum of change over several groups, in series, helps prevent spurious or accidental associations from wrongly channelling our ideas.

Second, and even more powerfully, comparisons may include several parallel sets of animals, each set displaying the particular functional spectrum in which we are interested. This further magnifies the likelihood that the associations that we draw between differences in function and structure will be real.

A *third* way of attempting to prevent a wrong diagnosis in this type of study is, of course, to look at the nature of the morphological difference and see if it bears a reasonable mechanical relationship to the particular function at issue. This, too, helps distinguish spurious or accidental associations from those that are functionally meaningful. Unfortunately, it is almost always possible to postulate a biomechanical meaning for whatever morphological difference is found. But making this method more rigorous by using experimental biomechanical methods helps prevent the accidental or spurious association.

Considerations such as these can have a profound effect upon the conduct of functional-structural studies. Thus, from being merely a comparison of two groups of organisms, such studies can have a detailed internal design exactly aimed at increasing the information flowing from them and designed especially to detect spurious results. Such 'observational design' is the equivalent of 'experimental design' in other biological situations.

Thus, a simple attempt to discover the association between, say, climbing and upper limb structure might be tackled by comparing two animals, say, *Cercopithecus aethiops* which, being more terrestrial climbs less, and *C. diana* which, being more arboreal, climbs more.

But how much more powerful is a study examining the *linear* sequence afforded by comparison of:

(a_1) terrestrial,
(a_2) semi-terrestrial,
(a_3) arboreal main branch and
(a_4) arboreal fine branch

cercopitheques? Any morphological finding linearly consistent within such a detailed series stands considerably greater chance of being truly related to the functional characteristics implied by the habitat descriptions.

A similar linear design to discover associations between leaping and hind limb structure could involve a sequence such as:

(a_1) more quadrupedal, less leaping thick-tailed bush-babies,
(a_2) more leaping needle-nailed bush-babies,
(a_3) even more highly leaping Allen's bush-babies,
(a_4) most highly leaping Senegal bush-babies.

Again, any morphological finding consistent within this *linear* design is highly likely to be related to the functional characteristics implied in the locomotor descriptions of the species.

The second level of observational design with even further power involves *parallel* series. Studies of differences in climbing in Old World monkeys might thus include comparison of:

(a_1) more and
(a_2) less terrestrial cercopitheques,
(b_1) more and
(b_2) less terrestrial mangabeys,
(c_1) more and
(c_2) less terrestrial macaques and
(d_1) more and
(d_2) less terrestrial langurs.

This *parallel* design is thus such, that differences in structure associated with differences between habitats may be detected.

In each of these examples, the observational designs do the best that they can, given the species that actually exist, to hold constant, or *constantly varying* features not related to the elements under investigation. In this sense these designs truly mirror the 'design of experiments'.

Observational design: some precautions

Now we have to be careful with the 'design of observations'; because of the accidents of evolution they are not, and cannot be, as elegantly arranged as is possible in the design of experiments. It may well be that when activities and structures are compared in animals of different taxonomic categories, say guenons and colobs, then the activities may not be associated with structures in the same way in each exactly because one is a cercopithecine and the other a colobine. It may also be that when activities and structures are compared in animals with different underlying locomotions, say again, guenons and colobs, that the activities may not be associated with the structures in the same way exactly because in the case of guenons they are superimposed upon a basically quadrupedal structure, and in the case of colobs upon a basically leaping one.

These possibilities add to the richness of the functional adaptation of structure. But because of them it may well be important to make one further addition to our observational design.

That is, it is most useful to look at functional relationships in sequences of animals completely outside the desired comparison. In the Old World monkey example, therefore, we might look towards:

(e_1) more terrestrial and
(e_2) more arboreal

quadrupedal prosimians, say *Lemur variegatus*, which is more terrestrial, and other lemurs which are more arboreal. We may even look toward:

(f_1) certain more terrestrial and
(f_2) more arboreal

quadrupedal mammals, say ground squirrels as compared with tree squirrels.

This extended discussion of more complex observational design does not mean that it is improper to carry out a study where one has only one sequence and only a single pair of forms in that sequence. Many good investigations with this structure exist in the literature (e.g. Fleagle, 1976, 1978; Mittermeier, 1978; Rodman, 1979; Jungers and Fleagle, 1980). But this discussion does indicate that it is possible to avoid pitfalls, to greatly increase sensitivity and to make the fullest use of the data if studies can be designed in this more detailed manner.

It is not common that attention is drawn to this form of investigational design in physical anthropology. It is, however, my opinion that, as we come to study in greater and greater detail more refined groupings of animals, then differences will become of so fine a grade that design of this type will be necessary to produce results. And as such studies progress, design of this type will become more and more necessary to help separate associations that are relevant from those that are accidental.

Certainly an investigational design of this type makes the application of analytical methods easier. It is on this basis, for instance, that the studies presented later in this book have isolated morphological differences apparently related to function in groups of animals — the Old World monkeys — which were described only about a decade ago as being comparatively uniform and with little variability (Schultz, 1970).

Applications to fossils

Study of the morphology of fossil specimens may utilize information from living forms, thus providing assessments of the resultant biomechanical situation that may have acted upon the fossil specimens during life. But because so many behaviours can result in similar resultant biomechanical situations, it is not possible to make the further leap to assess behaviour itself in the fossil. That the general form of fossil post-cranial bones will speak to this level of function, rather than to actual behaviour or to direct genetic propinquity, is the idea that is becoming so much more clearly understood at the present day.

'Primitive' and 'derived' features. One of the modes of analysis of morphological features much in vogue at the present time and distinctly useful in evolutionary studies is based upon the notion of the evolution of features from 'primitive' to 'derived' states. Of course, in examining extant forms we cannot see directly what is primitive and what derived.

But if we examine differences between features in several groups of organisms, it may be possible to suggest indirectly which features (primitive) may not have changed over a period of time (in the groups under discussion) and which (derived) may have changed in that period. It is necessary to go to animal groups outside the ones being studied (outgroups) in order to obtain information bearing upon the matter of primitive and derived.

This method has been mainly applied outside the primates, although Luckett (1975) has used it for the primate reproductive system, and Szalay (1975b) for the primate skull base. In these, and a small but now gradually increasing number of studies, it seems as though information useful for the study of primate phylogeny is arising (e.g. Delson, 1977).

In the case of Luckett's reproductive investigations, it really seems possible to say that a placenta with a fairly strong barrier between mother and foetus (a diffuse epithelio-chorial placenta) is primitive with respect to primates, and that a placenta with a much closer link between mother and foetus (a discoidal haemochorial placenta) is derived. The first is found in most prosimian primates (but not *Tarsius*) and these prosimians share it with a number of outgroups whose representatives include such forms as disparate as cows and horses, pangolins and whales. The detailed resemblances in the structure of the placenta among these diverse extant groups suggest to Luckett that it is likely that their presence is due to retention of a primitive ancestral condition in very distantly related descendants.

The second, haemochorial placenta is found in monkeys, apes and humans together with *Tarsius*. It is also found in the animal groups represented by elephant shrews, hedgehogs, armadillos, 'flying lemurs' (not primates but belonging to a small, special group, the dermopterans) and rodents. In this case, detailed internal differences between all of these species in the internal structure of this type of placenta suggest to Luckett that they are convergently evolved but that, within the extant primates, a haemochorial placenta is uniformly derived.

In a similar manner, Szalay (1975b) uses this method to look at skeletal features of the skull and has available, therefore, information about fossils (i.e. some structures that are truly earlier than others). Szalay believes that this method demonstrates clearly the origins of the Order Primates from a single stock. And he suggests that the omomyids (a problematical group) are near that origin. It should again be pointed out that Szalay himself notes the difficulties about being certain about some of the 'primitivenesses' and 'derivednesses' involved in the cranial base.

Although undoubtedly much that is useful can be garnered by this method, it is not foolproof because the determination of what is primitive and what is derived can be so easily, and falsely, channelled by our prior thoughts about evolution.

But quite apart from any methodological pitfall that may result from the method, the use of the terms: 'primitive' and 'derived' to characterize many quantitative post-cranial structures of the primates may be inapplicable for good biological reasons. These reasons come, as we shall see, directly from the nature of the evolution of post-cranial features and the special form of the features' measurements.

Let us consider first the problem in relation to evolution: the case of a feature changing over evolutionary time. Fig. 4.7 attempts to show that the feature possesses variation at any one time; some of the variation at that given time is more like that in ancestors; some other part of the variation is more like that in descendants; some variation exists that resembles neither. From a diagram like Fig. 4.7 it is clear that there has existed, for this hypothetical character, an earlier primitive condition, and that it has changed, in the lineage figured, to a later derived condition; it is even clear that some of the variation in the derived feature points back towards the primitive feature.

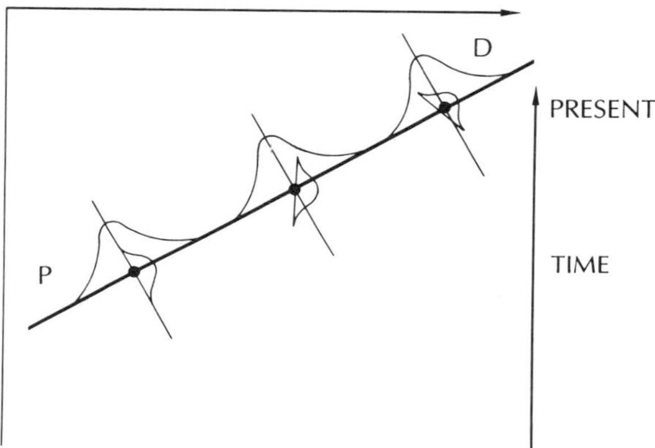

Fig. 4.7. This diagram shows the change in a theoretical morphological feature over time. The vertical axis represents time, the horizontal axis morphological change. The small bell-shaped curves erected on the line of change of the feature represent portions of the variability of the feature at each particular point in time. Part of the variability at a given time is similar to the average situation at some prior time; and another part of it is similar to the average situation at a later time. The bell-shaped curves drawn on axes at some angle to the main line indicate that, of course, other aspects of variability may well not be related to the direction in which the feature happens to be evolving at the given time. (This diagram should not be read to imply gradual rather than punctuate evolutionary modes – the same point exists for each).

Fig. 4.8 shows that this description might truly apply to a quantitative feature of the human post-cranial skeleton that is closely related to func-

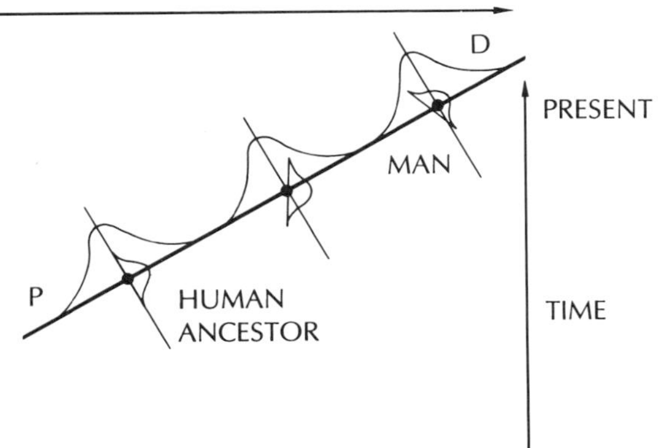

Fig. 4.8. This figure replicates Fig. 4.7, but with a particular anatomical feature in mind, the changing length of the lower limb during evolution from arboreal to terrestrial human movement. For the sake of discussion, it is assumed that a shorter hind limb is evolving towards a longer one, and that this is associated with an evolutionary change from an arboreal quadrupedal to a terrestrial bipedal habitat. All other conventions are as in Fig. 4.7.

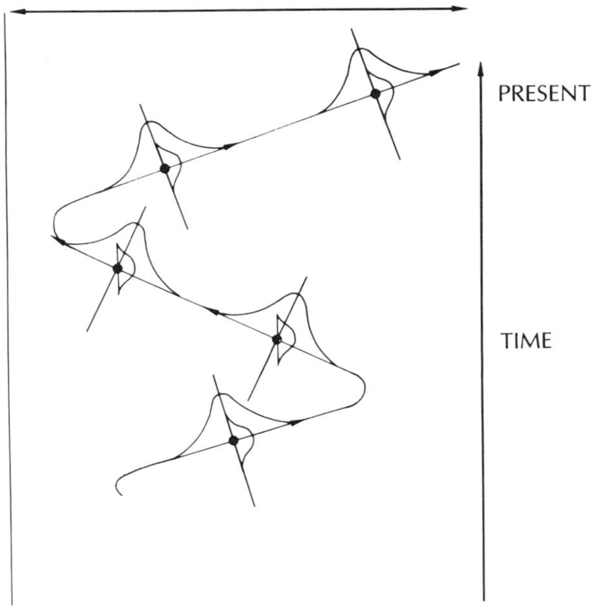

Fig. 4.9. This figure also uses the theoretical discussion of Fig. 4.7 in order to look at a morphological change that we might find among the primates, a more supporting or more grasping hand. Such a morphological difference might be associated with a particular evolving function during a terrestrial-arboreal change. In this case, however, it is not at all unlikely that a particular group of animals may have moved from a terrestrial habit to arboreal and back, and that this swing may have occurred more than once. Longterm changes in inundation or in arridity, for instance, may well have channelled such effects. Under these circumstances, earlier features may well also look like later features; and who is to say what is primitive and what derived without actual prior knowledge as in Fig. 4.7.

tion. For simplicity, it has been assumed that the feature is a relatively long lower limb (as in present-day humans) and that this is derived from a relatively shorter lower limb in some much earlier ancestor. The functional association is that, not only has the structure evolved from primitive to derived, it has also evolved from a function associated with that in, say, an arboreal creature (where a shorter lower limb similar in length to the upper limb may be biomechanically appropriate) to a function in a terrestrial bipedal creature (where a longer lower limb is functionally better). So far, then, there is no difference between our functional post-cranial charcter of Fig. 4.8 and the theoretical character of Fig. 4.7. Both mirror those features used by Luckett and Szalay and others who employ this technology generally.

However, Fig. 4.9 demonstrates what, for primates, must be a usual pattern of evolution for post-cranial features. In this case it is assumed that the structure that is evolving is some feature related to evolving quadrupedal functions stemming from differences between arboreal and terrestrial quadrupedal environments. And it is assumed that over a long period of evolutionary time several reversals have occurred; at some times animals have been quadrupedal in the trees; at others they have been quadrupedal on the ground; and we do not know how often these reversals may have occurred nor in what order.

Under this circumstance, a feature on the left of Fig. 4.9, like those on the left of Figs. 4.7 and 4.8, is primitive relative to the extant condition at a particular time. Equally, however, a feature like that on the upper right of the diagram, although occurring later in time, has also occurred earlier (lower right); such a feature would be both primitive and derived in comparison with the first primitive character. But although primitive from the point of view of time to the extant character, the earliest feature would apparently show no difference from it.

The enormous complexity of whole animals and the existence of the prior history speaks against complete reversals ever occurring. But in individual quantitative features of small anatomical regions this is by no means the case; the existence, in many quite different extant species, of similar functional measures supports this idea. It may just be impossible to determine what is primitive and what derived in this type of functional situation.

Finally, we may look at Fig. 4.10 to demonstrate just how complex this line of thinking becomes when we view an evolving lineage in which splitting has occurred. In this case a single lineage, already alternating over time between arboreal and terrestrial environments, has given rise to two daughter lines that also continue to oscillate in this manner, although not harmoniously. The diagram makes it clear that there is no way that we can decide which of such functional character states is primitive or derived especially when we may not know to which line the fossil fragment bearing the feature belongs.

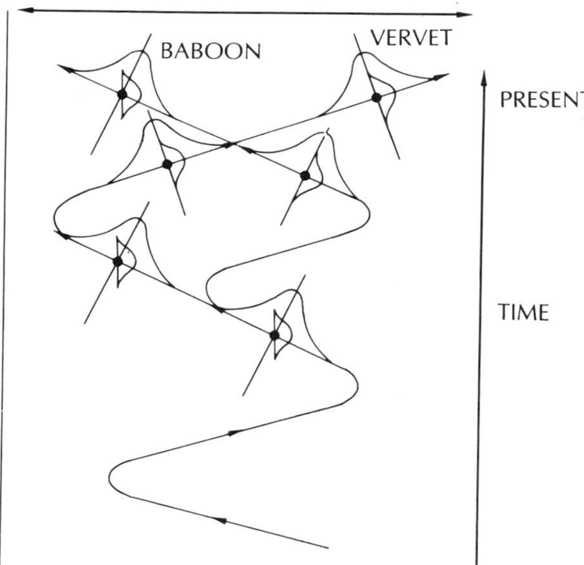

Fig. 4.10. This figure demonstrates just how complex the problems of Fig. 4.9 may become if one species, already demonstrating a cyclical change, splits during the evolutionary process into two. Both descendant groups may continue to evolve in the same direction for a short time; both may evolve in the reverse direction at a later time; the two could easily become decoupled so that at any single time (e.g. the top of the diagram, the present) each might actually be in the opposite adaptive mode. Conventions as in Fig. 4.7.

These arguments do not deny the use of primitive and derived features in situations like placental or skull evolution. It is highly unlikely that, with such fundamentally complex anatomical regions, several reversals would occur as suggested here. But we may certainly ask ourselves if all this is merely of theoretical interest or if it has enough practical importance that it may truly apply in studies of quantitative features of the post-cranial skeleton.

In fact, it is very likely that such reversals have indeed occurred in the sorts of features that relate the internal structures of animals to external functions. Cyclical phenomena, swings, reversals and so on, are the one feature of the physical world upon which organisms can count. Lakes come and go; forests advance and recede; temperatures rise and fall; glaciers freeze and melt; droughts precede and follow floods. Arboreality and terrestriality must have been entered many times by different primates; both arboreal-terrestrial and terrestrial-arboreal trends must have occurred many times throughout the evolution of the primates.

Examinations of some actual ecological and anatomical situations demonstrate this. The more extreme adaptation of terrestriality in baboons suggests that it occurred a rather long time ago in relation to the evolution of drier and less wooded parts of Africa, as compared with the lesser adaptations to terrestriality in the Hanuman langurs, which may well have been a later phenomenon related perhaps to the development of human communities (and, therefore, of scavenging activities in these monkeys living in the shadow of man). The adaptation of the patas monkey to terrestriality seems so extreme that it may well have occurred an even longer time ago. And it seems entirely possible that the anatomical features of some of the (today arboreal) cercopitheques may stem from at least one prior period of terrestrial adaptation.

There is evidence of the opposite. The inundations that seem to have come and gone in Malaysia in the past must have meant that, although not infrequently terrestrial at the present time (McKinnon, 1974), the orang-utan has been prevented from moving upon the ground in some prior periods.

Finally, however, the very complexity of some of the functional-structural adaptations prevent us from deciding what may have been primitive or derived. For instance, in Figs. 4.7 and 4.8 it is assumed that the morphological feature shows a uni-directional change from terestrial to arboreal in otherwise similar animals. An actual example is the robusticity of limbs (Manaster, 1979) as seen from measures of cross-sections of bones. Such cross-sections are greatest in terrestrial forms such as *Cercopithecus aethiops* and show a gradual decrease in arboreal forms (*C. mitis* to *C. diana*) as the mode of locomotion occurs more and more on finer branches, the springiness of which reduces and spreads the dynamic forces acting on the limbs in running and leaping.

But in other features, careful anatomical study shows that trends may not be uni-directional. For instance, in intermediately arboreal monkeys (e.g. *C. mitis*) the lengths of the scapula and the radius are short compared with longer lengths in fully terrestrial forms (e.g. *C. aethiops*) which are more actively quadrupedal on the ground. But the lengths of the scapula and radius are also short in intermediate *C. mitis* compared with longer elements in the most highly arboreal forms (e.g. *C. diana*) which are more actively acrobatic, living as they do to a greater degree in the small-branch milieu (Manaster, 1975, 1979). In both of these latter functional milieux, but for different reasons, it can readily be seen that a longer scapula and radius might be biomechanically more adaptive than in the intermediate species (*C. mitis*).

Which of these conditions is primitive or derived depends upon whether the more terrestrial, the more arboreal or the intermediate species is the one most similar to the ancestor; it is tempting to suggest that in this case it is the middle form, but it is not impossible that it was a form like either of the others; and to suggest that we already know the form of the ancestor by assuming one type of character to be primitive and the others derived begs the question.

In situations like these, then, guesses about primitive and derived states do not help us; it is the nature of the biomechanical adaptation that may possibly allow some estimate to be made of the evolution of function.

Let us now consider the second set of problems with the terms 'primitive' and 'derived': the relationship with measurement. Although in biometrical shorthand we may refer to particular measures of a bone as 'features' of the bone, in fact such primary measures (or secondary data such as ratio, angular, logarithmic or other transformations of measures) are not 'features' in the technical sense required by cladistic methods of data analysis.

For any measurement of a bone is merely the distance on that bone between two defined points. By definition, therefore, it is not a measurement of a 'feature'. It is a measurement that includes within itself, partial measures of many features. A measure might include 20 per cent of a first feature, 10 per cent of a second, 25 per cent of a third and so on. Methods have not yet been devised for deciding just how much of what features any particular measure represents. But it is certainly the case that such measures are complexes of the parts of many features.

How then can the terms 'primitive' or 'derived' be applied to such a measure. For some of the component features of that measure may be truly primitive, others truly derived, yet others, possibly, something else. The final mix may be complex. If that same measure taken upon different species is identical with the first, this would not mean in any way that the mix of component parts was also identical. Conversely, a measure might be quite different in two animals, yet contain identical proportions of each of the primitive and derived features.

And the terms 'primitive' and 'derived' must be even more inapplicable to simple derived dimensions such as the ratios, indices and angles that have been so useful in these studies, or to more complex compound dimensions that involve additions of variables, regression-adjusted treatments of two or several variables, multivariate statistical reductions of many variables and so on. Such complex dimensions can never represent individual characters and must contain within themselves the interactions of many individual characters in completely different ratios in each dimension.

Yet another line of argument stems from even further consideration of just what is represented by these mensurational characters of bones. Thus, in addition to the individual value of any single measurement of a bone being a compound of partial influences from many character states of that bone (some presumably primitive, some derived as expressed above), that value also contains information about (a) other, non-bony parts (such as the various soft tissues, muscles, ligaments and joints, the form of which influence measures of bones). It contains information that is not merely about structure (whether of hard or soft tissues) but is also about (b) function (a measure of a bone contains information about locomotion). Indeed, a measure of a bone even contains information about

(c) the nervous system, responsible in part for locomotion, (d) diet as associated with locomotion, (e) behaviour as allowed by locomotion, (f) ecology within which locomotion occurs, even (g) social structure (which has interrelationships with animal locomotion). Thus, we are forced to recognize just how great an amount of biological information may be reflected within a suite of measurements of the skeleton.

It could be said that this means that measures are of much less value exactly because they contain a mish-mash of biological information that cannot be disentangled. And, this is certainly one reason why their examination one by one in a univariate manner, is not of much use in evolutionary studies. But the multivariate-approach, with its ability to partition, at least in part, the different contributions of biological information to the values of many such measures when they are taken together, is able to provide insights into the complexity.

In parenthesis, it should be noted that, although we think of the more traditional qualitative data as comprising individual "characters", each of these is, too, likely to be just as complex as any individual measurement; it is most unlikely to be the single heritable character uninfluenced by function or anything else, that the cladistic approach assumes it to be in the diagnosis of primitive or derived. But the character possesses the disadvantage that its non-quantitative nature does not allow the disentangling of the complexity. In other words, in the same way as for quantitative data, many of these qualitative data cannot really be described as simply primitive or derived. That is not a difficulty for the analysis of quantitative data as in this book. It is, however, a major problem for those who use such qualitative data through the methods of cladistics.

Be that last as it may, it is surely clear from this discussion that a measure is not a 'feature' that can be 'primitive' or 'derived'. It is entirely wrong to treat measures by this methodology (although other quantitative descriptors — e.g. the presence or absence of a foramen in a bone, or the numbers of scales on an operculum) may possibly still be investigated in this way (Oxnard, 1982).

Limits, for fossils, to the functional argument. Realization of all the previous arguments limits, rather suprisingly, what we can expect to learn from fossils. In the examination of living species, logical arguments may progress from the detailed behaviours of the living forms, through the resultant biomechanical situations, to the detailed structures of muscles, bones and joints. But, when we make the opposite assessment about the equivalent anatomical regions of fossils, we cannot draw the entire reverse sequence. We can legitimately progress only from the detailed structures of fossil bones to resultant biomechanical situations. We cannot pass further to the estimation of the behaviours of the fossil forms. This is because, as we have seen, many sets of behaviours may be associated with the rather narrow interface or bottleneck, the resultant biomechanical situation, that lies between behaviour and structure.

Assessments of anatomies that provide information about fossil behaviours or, even more widely, fossil ecologies, should be viewed exactly for what they are: guesses that cannot be tested without new fossil information. But estimations that confine themselves to descriptions of the resultant biomechanical situations are better than guesses. They can indeed be tested through examination of the forms and functions of other, functionally parallel, living species.

Evaluations of fossil skeletal remnants (of the post-cranium at any rate) that attempt to move directly to genetic relationships or taxonomic positions without passing through stages related to assessments of biomechanical situations may have little validity. The all-pervasive effects of function and structure, especially within the post-cranial skeleton, are such that it is highly unlikely that any major quantitative features are entirely hereditary in the primary sense.

With these ideas about the relevance for evolution of bone form in the forefront of our minds we may pass to more detailed study of individual anatomical regions. Let us go first to upper limbs.

Summary. Though factors such as heredity and development are closely related to structure, the form of the post-cranial skeleton is especially related to the functions that it subserves during the lifestyles of the different animals. The relationships between structure and function involve the functional-structural interface. In this interface, particular subsets of structure are associated with special subsets of function. The subsets of structure may possibly be obtained by the methods of the last chapter. The subsets of function may relate to average biomechanical situations, although these averages are undoubtedly far more complex than simple arithmetic averages.

One way of studying the structural-functional interface is, of course, through the direct approaches of animal mechanics which produce exemplars resulting from study of particular experimental species. But another way is through the indirect study of the associations between the two descriptions: of structure and of function. A traditional form of functional description for the primates applicable to the locomotor system (the post-cranium) has been the classification of locomotor patterns. Nowadays, we recognize in such classifications clear deficits.

An attempt has been made to improve these functional descriptions (a) through the usage of regional anatomical classifications of the average biomechanical situations resulting from locomotion and (b) by departing from the very idea of a classification of groups and turning instead to the notion of spectra based upon average biomechanical situations resulting from locomotion.

These ideas have lead in turn to the concept of a design of observations that is parallel in some ways to the design of experiments long known to an experimental biological community. Though such a design of experiments must be used cautiously, it has far greater powers than the traditional simple pairwise comparison of animals for revealing the structural-functional relationships.

The application of these ideas to fossils requires certain precautions and has certain limitations. These stem partly from the fact that fossil data are fragmentary in a number of ways compared to evidences from living forms, and partly because, of course, fossil data lead to predictions that cannot be tested directly. Such testing as is possible depends upon independent parallel assessments of other fossils.

CHAPTER 5

Upper Limbs and Tension

Abstract. In this chapter we discuss form and function in primate upper limbs. From early simpler views of upper limb function, we move towards new natural history observations of primates, so much more extensive now than even only two decades ago. These have many implications for the function of upper limbs and are related to biomechanical ideas such as tension-bearing. We thus recognize the wide number of activities (not just brachiation) that give rise to tensile forces in upper limbs; we see the great mix of activities undertaken by individual primates (though, of course, different in each); we discuss their implications for biomechanics. This summary leads to a definition of the average biomechanical spectrum for primate upper limbs following the general discussion of the last chapter.

We then examine structural descriptions of various upper limb parts using the multivariate statistical methods described in the third chapter. The anatomical parts are the shoulder, upper arm, forearm and overall proportions of the entire limb including some measures of the hand; in addition, various combinations of parts are examined. But a detailed view of the hand is not presented because this work has not yet been undertaken. Notwithstanding different individual functions of specific upper limb parts, the overriding finding relates to the total functional spectrum of the entire upper limb.

The degree of concordance between the functional concept and the anatomical description is very close. The overall structure of the various parts of the upper limb, when viewed quantitatively, speaks most clearly about overall function.

Nevertheless, we end with a caveat about the possible systematic content of the morphological data; it is a part of the story to which we shall return in a later chapter.

The functions of fore limbs

Studies of fore limb function in vertebrates, especially mammals, are many fewer than for the hind limb and have usually been aimed at quadrupedalism. Thus, in his consideration of quadrupedal structure using the model of a table, Gray (1968) treats fore limbs essentially in the same way as hind limbs, although it is true that he differentiates the propulsive lever from the propulsive strut and indicates that the former is more generally associated with hind limb function and the latter with fore limb activities.

And though there have been other studies of fore limbs (for example, the fore limbs of pandas and of sloths; Davis, 1964; Wislocki and Straus, 1932; and Miller, 1943), their functions have not been examined in the same detailed way that has been achieved for hind limbs. It is not that we believe that the functions of fore and hind limbs are the same, although obviously they do share in many of the same activities. It is just that, among the vertebrates generally, no special functional activity, such as leaping for the hind limbs, has been extensively studied on fore limbs.

We are thus unable to undertake a summary of

the biomechanics of fore limb functions in vertebrates as is done in the case of the hind limb (Chapter 6). And we are thus equally unable to discuss the varieties of vertebrate fore limb function in any terms other than those described for the hind limb.

In contrast to this situation among vertebrates in general, studies of upper limb structure and function have been more detailed in the primates; this relates quite simply to the fact that it is among the primates that a remarkable form of upper limb dependent locomotion, brachiation, is found.

(The terms 'upper' and 'lower' limbs, rather than 'fore' and 'hind' limb, are used here because so many primates maintain orthograde positions of the trunk, and also because, of course, we would have to change the terms when we came to refer to humans. Rather than separate humans artificially in this way, I have chosen to use the terms 'upper' and 'lower' limbs almost exclusively throughout the rest of this book.)

Early views of upper limb function in primates

Although we are readily able to associate rather specific functions of the lower limbs with leaping (as in Chapter 6) and of all four limbs together with quadrupedalism (as in Chapter 7), it is rather less obvious what are the major functions of upper limbs considered alone. This is because the biomechanical function of the upper limbs in non-human primates has for so many years been overshadowed by the remarkable form of locomotion practised by gibbons and siamangs swinging by alternate arms beneath branches — brachiation.

Brachiation is, of course, an activity that involves more than just the upper limbs; witness the well-known actions of the lower limbs during brachiation in which they may be flexed and drawn upwards as much as possible, thus shortening the pendulum of the swinging body. This is an old concept (e.g. Carpenter, 1940) that is being investigated in more detail at the present day (e.g. Carpenter, 1976; Fleagle, 1977).

However, even as early as 1923, Frey inspected the anatomy of primate upper limbs, not only from the viewpoint of the special adaptations for brachiation that are possessed by the lesser apes, but also to take into account the fact that the other apes can perform somewhat similar activities.

And this was followed by the very detailed study by Miller (1932) that attempted not only to discover the structure of the upper limb in the primates as a whole, but also to relate it to a broader notion of upper limb function. Thus she saw rather clearly the idea of increasing degrees of mobility in upper limb functions. This was a new concept that described what was judged to be extreme mobility in the upper limbs of 'the ape' (through brachiation and other acrobatic activities in trees, together with a different kind of mobility in humans) and related it to lesser degrees of mobility in Old World monkeys, lesser still in New World monkeys and least of all in prosimians.

This concept marked a considerable advance in thinking, although hidden within it was the notion that the main taxonomic grades of the primates represented somewhat equivalent grades of functional activity. Miller did not overtly recognize the possibility that perhaps functional activity is not especially related to systematic position.

Miller's study was followed by that of Inman, Saunders and Abbott (1944) who, though they dissected both human and non-human primates, undertook experimental (electromyographic) studies only on humans. They, too, were able to see the general relationship, within the primates, of increasing degrees of mobility as expressed by Miller. And, of course, they were also able to postulate and, by analogy through human electromyography, test ideas about the real activities of muscles, joints and bones. The concepts that they obtained did not actually determine what goes on in the different primates because they could only use these methods on humans. But for the times it was a ground-breaking study.

These studies were then followed by the reviews of locomotion and the dissectional and osteological investigations of Ashton and Oxnard (1963, 1964a, b). These authors used such meagre functional information as was then available, mainly the study by Inman, Saunders and Abbott together with anatomical common sense, to help interpret the structural differences that they found in dissecting and measuring a large number of primates. Although it is clear that their study, too, was made very much in the 'shadow' of brachiation, it did make certain advances. Thus, Ashton and Oxnard pointed to two new ideas in our understanding of upper limb function and structure in primates.

The *first* of these is the notion that, included among the movements of which primate upper limbs are capable, are many that reflect something that is also basic to the biomechanical requirements of brachiation. Thus, they postulated that the extreme tension-bearing and the extreme raised positions required in brachiation proper were mirrored to lesser degrees in many other activities of upper limbs. And they further suggested that the tension-bearing activities and raised positions of upper limbs were, in turn, found far more widely among the primates than in the brachiating gibbons and siamangs alone (Fig. 5.1).

They demonstrated that a spectrum of such activities could be discerned by 'dissecting' the functions of upper limbs. And they acknowledged that their functional spectrum of tension-bearing activities and elevated positions rested fairly heavily upon the prior idea of a spectrum of mobility (Ashton and Oxnard, 1964a) as put forward by Miller (1932), and Inman, Saunders and Abbott (1944).

The *second* improvement in our understanding made by Ashton and Oxnard was the idea that this spectrum of function was not related to systematic position. Thus, they described activities in prosimian species whose upper limbs function in such a manner that they are more involved in bearing tensile forces and in elevated positions. These were contrasted with other prosimians whose upper limbs bear tensile forces to minimal degrees and operate in mainly lowered positions under compression. Likewise, Ashton and Oxnard noted New World monkeys displaying a similar diversity. They recognized that such a diversity also existed, though to a somewhat lesser extent, among the Old World monkeys. The completion of the diversity of Old World forms in this regard was

Fig. 5.1. Tension-bearing and raised upper limbs in primates (after Oxnard, 1963).

made up by consideration of upper limb abilities and capabilities in the great apes as well as the lesser (Ashton and Oxnard, 1964a).

The investigational weapons available at that time to these authors (little more than univariate statistics such as could be computed on a mechanical calculator) were meagre in the extreme. This forced Ashton and Oxnard to simplify the complexities of the spectrum of upper limb function by grouping them broadly into a number of catgories: those most involved in tension-bearing and raised positions, those involved in such functions to the least degree and those involved to intermediate degrees.

Unfortunately, for some years this grouping of upper limb function became embroiled in controversies of other workers who were attempting a different and more problematical task: the classification of locomotor patterns overall (e.g. especially Napier, 1964; Prost, 1965; Napier and Walker, 1967; Ripley, 1967; Walker 1967; Rose, 1974). However, some of the problems of locomotor classifications overall have now been realized; the separation of that intellectual activity from consideration of basic functions within anatomical regions has been clarified by more recent work (perhaps especially, Stern and Oxnard, 1973; and Oxnard, 1975c and 1979a, b).

This history of the development of our ideas about the functions of upper limbs in primates has been followed in the last few years by a large number of new studies. Some of these attempt to discover what is going on when particular animals move in the field. Others attempt to find out what is really happening in experimental animals in the laboratory. And yet others attempt to synthesize the information of both prior groups of workers in order to apply it to the structure of the primates across the board, as it were, in ways that do not involve field or experimental work with a few species, but the study of structure in large numbers of specimens and species already existing in museum collections.

The activities of upper limbs

Tensile forces in brachiation. Many field studies have now been carried out since that review of upper limb function by Ashton and Oxnard. Although the new studies do support the basic concepts that were put forward at that time, the notion of a spectrum of tension-bearing and raised upper limb positions, those ideas can now be taken very much further. Thus, although it was early recognized (e.g. Coolidge, 1933; Carpenter, 1940) that brachiation proper (Fig. 5.2) involves tension

Fig. 5.2. Tension-bearing and raised upper limbs: brachiation proper.

in upper limbs together with a markedly raised upper limb position, as in gibbons and siamangs, and, in somewhat different ways, in woolly spider monkeys, spider monkeys, orang-utans, and chimpanzees and gorillas on occasion, many other activities have been explicitly noted as also providing degrees of tension-bearing and raised upper limb postures.

Tensile forces and raised upper limbs. Many field and laboratory studies document how primates that can brachiate have mid-air positions with upper limbs raised preparatory to landing by grasping branches above the head. This results in the upper limb bearing very high tension although for a short but important period of time (Fig. 5.3, lesser apes; e.g. Carpenter, 1976).

Fig. 5.3. Tension-bearing and raised upper limbs: three lesser apes in mid-air postures associated with raised upper limbs and tension-bearing on landing.

Frame 3

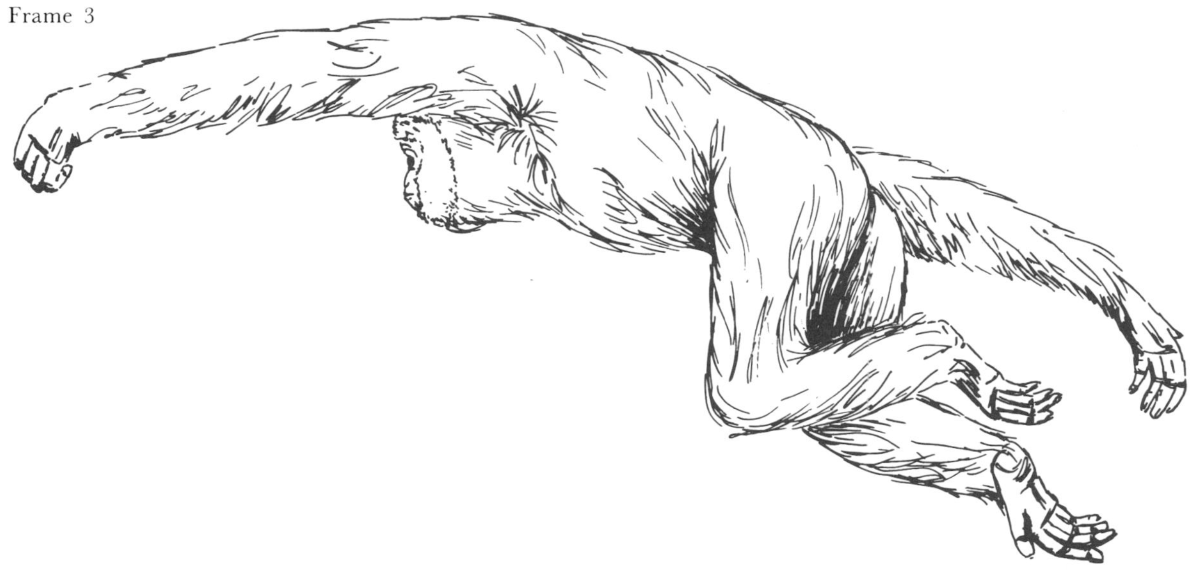

A number of other primates that do not brachiate also impel themselves into the air, land among the branches with the upper limbs above the head and support the body weight in tension during the critical moments of landing. Such abilities can be seen in the mid-air positions of these primates where the upper limbs are raised and lead the way. Fig. 5.4 shows examples of this in monkeys of both the New and Old Worlds (e.g. Napier and Napier, 1967; Fleagle, 1976). Fig. 5.5 gives examples among leaping prosimians (e.g. indriids, Walker, 1967).

Tensile forces in support from above. Upper limbs may be used during changes between locomotor activities to support, steady and secure the body from above. This is found most markedly in all of the species mentioned above, but it occurs frequently in some New and Old World monkeys such as capuchins and howler monkeys, colobus monkeys and langurs (see frontispiece: Chinese Golden Monkey). And it is certainly not unknown in many of the other primates that live in the small-branch milieu (Fig. 5.6 and Morbeck 1976, 1979; Ripley, 1976a, b; Fleagle, 1976).

Upper limbs are also used in tension-bearing modes, though perhaps with less elevation, by animals capable of progressing quadrupedally along the under-surfaces of branches. Again, this *can* be done by all primates, but field studies make it clear that it is somewhat more frequently carried out by some species than others, e.g. orang-utans, howler monkeys, uakari monkeys among the anthropoids and perhaps rather especially by the lorisines among the prosimians, each to differing degrees, of course (Figs. 5.7, 5.8, 5.9 and 5.10).

It is now also well-documented that upper limbs may be used in modes involving tensile forces and raised positions when animals suspend themselves momentarily, often prior to a downward drop. This, again, *can* occur in any primate. But those that are most quadrupedal and most terrestrial are least likely to do it. And it is again especially documented as being used by uakari monkeys, capuchins, howler monkeys, woolly monkeys, spider monkeys and woolly spider monkeys of the New World, by colobus monkeys, langurs and other colobines of the Old World, and especially by the indriids among the prosimians. Of course, it goes without saying that such activities form a part of the spectrum of actions of all apes (for example, Tuttle, 1977).

Many other actions are also related to the bearing of tension in upper limbs, and, to lesser degrees, to the propensity for raised positions. These include body-lifting actions of limbs in raising the trunk on to higher ledges or branches. This is obviously rather more frequent in those species whose environments more freely offer such ledges and branches. It is particularly described for colobines among the Old World monkeys, and for uakaris, sakis, capuchins and howler monkeys among the New World (it is, of course, a *sine qua non* for the atelines, hylobatines and great apes: MacKinnon, 1974; Baldwin and Teleki, 1976; Tuttle, 1977).

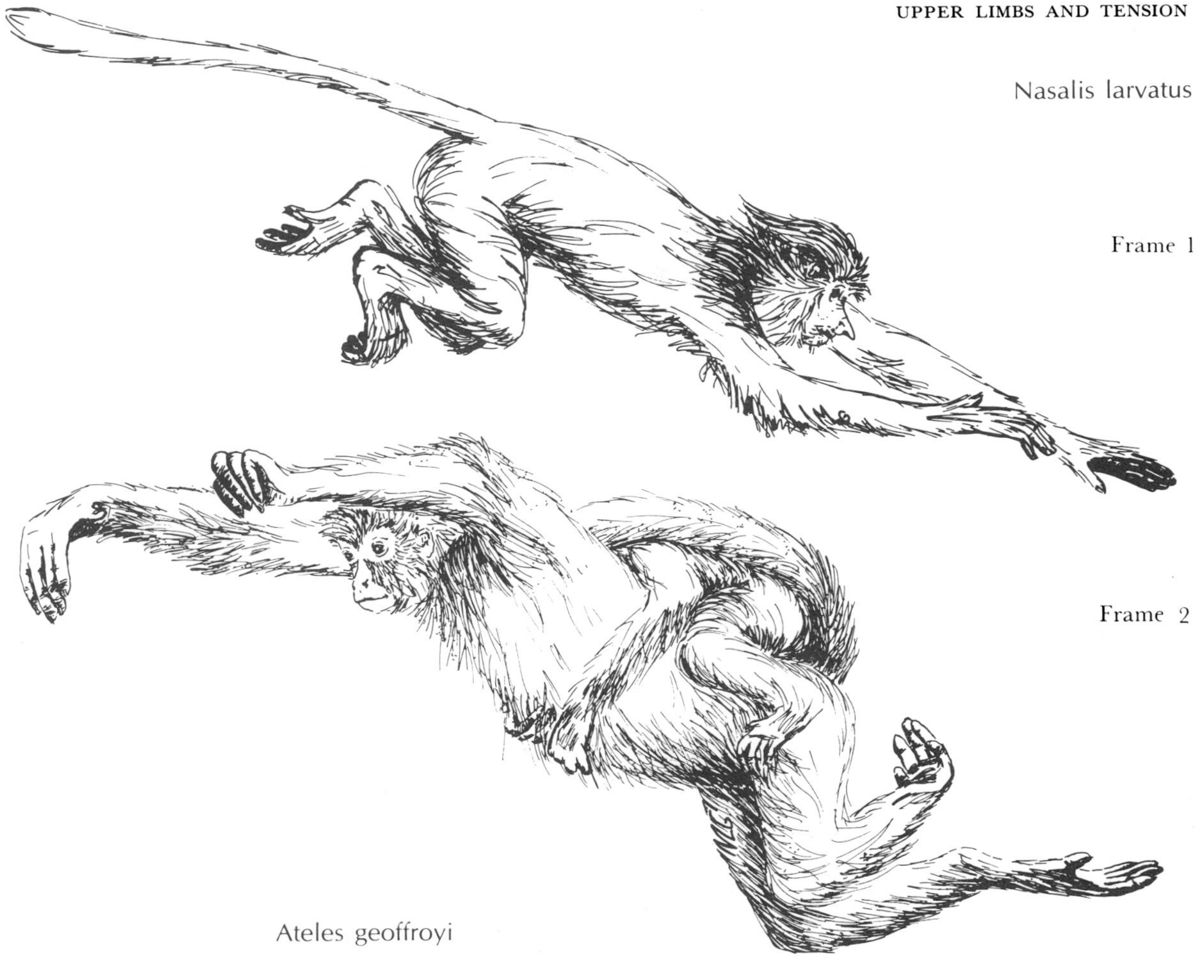

Nasalis larvatus
Frame 1
Frame 2
Ateles geoffroyi

Fig. 5.4. **Tension-bearing and raised upper limbs: two monkeys in mid-air postures associated with raised upper limbs and tension-bearing on landing. Compare with Fig. 5.3.**

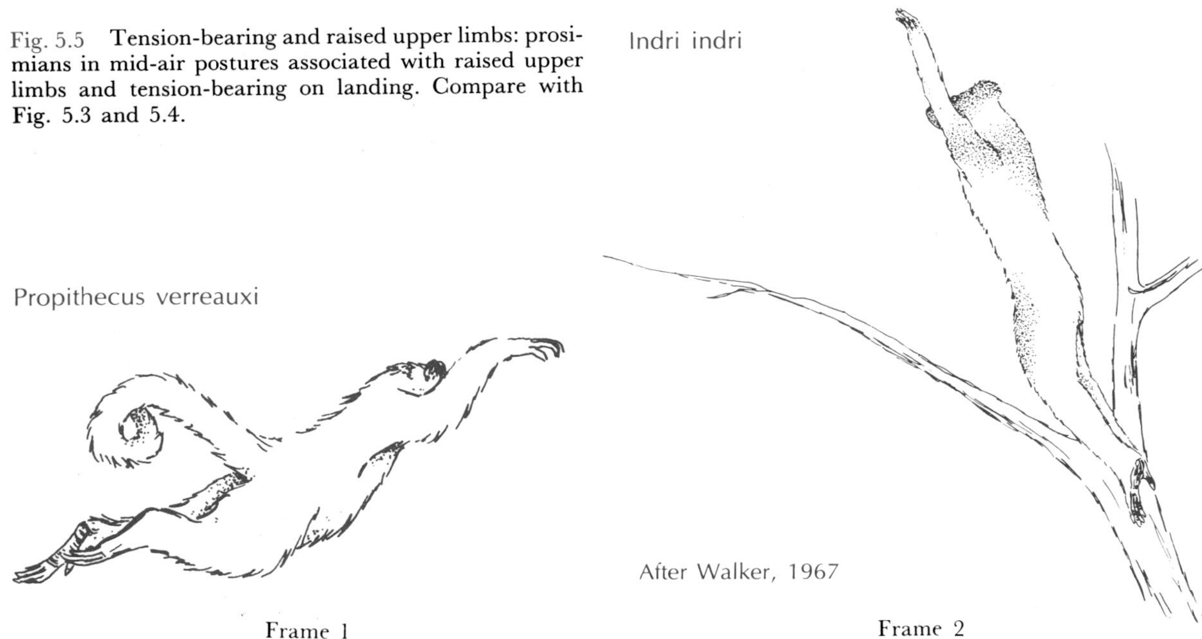

Fig. 5.5 **Tension-bearing and raised upper limbs: prosimians in mid-air postures associated with raised upper limbs and tension-bearing on landing. Compare with Fig. 5.3 and 5.4.**

Propithecus verreauxi
Frame 1

Indri indri
After Walker, 1967
Frame 2

Fig. 5.6. Tension-bearing and raised upper limbs: support from above (potto).

Fig. 5.7. Tension-bearing and raised upper limbs: underbranch progression (woolly monkey).

Fig. 5.8. Tension-bearing and raised upper limbs: suspension (howler monkey).

Fig. 5.9. Tension-bearing and raised upper limbs: more suspension (chimpanzee). Compare Fig. 5.18.

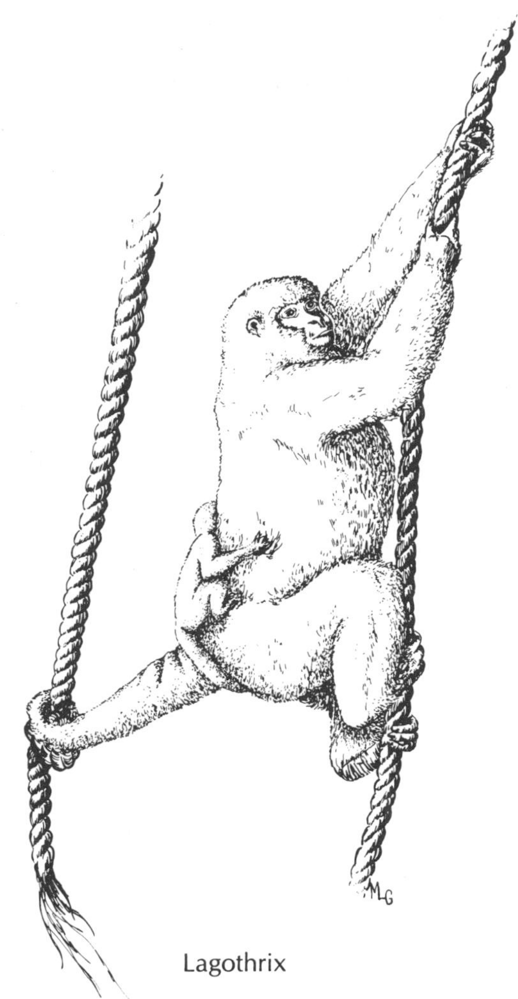

Fig. 5.10. Tension-bearing and raised upper limbs: vertical climbing (woolly monkey).

Tensile-bearing activities also include propulsive actions of upper limbs in the rather special vertical climbing modes that can be adopted by all primates. Such vertical climbing can be done in a number of ways. One involves climbing upwards (or downwards) using alternate movements of limbs similar to regular quadrupedal movement. Another involves simultaneous movements of pairs of limbs as in 'inch-worming'. But, because the upper limbs are above the centre of gravity, all such vertical climbing modes imply the existence of tensile forces together with some degree of elevation. Activities such as these are also found in any primate on occasion. But they are undoubtedly more frequently noted in particular groups such as indriids and lorisines, uakaris and saki monkeys, capuchins and howler monkeys, colobs and langurs and, of course, all species capable of the more arm-swinging activities, the ateline New World monkeys and the lesser and greater apes.

Finally, tension-bearing occurs in yet other situations. One is during those many postures of primates in which the body is pressed against a vertical tree trunk and the upper limb used in front of the face or above the head to secure the animal's position (Figs. 5.11 and 5.12). Another is in the actions of upper limbs in front of or above the head when used to steady the trunk when the animal is sitting atop the branches with the trunk in a relatively upright position (Figs. 5.13 and 5.14). A third is when animals are standing and walking bipedally upon branches and at the same time holding on with the hands to branches above the head. Again, all primates, even man, *can* do these things; but those more frequently adopting such truncally vertical postures do this to greater degrees, e.g. indriids do it more than most other prosimians, prehensile-tailed species more than the other New World monkeys, colobus monkeys and langurs more than most cercopithecine monkeys and, of course, it is habitual in the repertoire of the apes.

Frame 1

Fig. 5.11. Tension-bearing and raised upper limbs: vertical postures (tarsier, sportive lemur and slow loris).

Tensile forces in non-locomotor activities. Finally a wide variety of movements are now being noted that do not specifically come into the category of locomotion and posture at all, but that are really associated with other kinds of activities. For instance, some species feed by drawing terminal branches towards themselves with the upper limbs. Obviously this occurs more in those species that feed on leaves and fruit especially in the small-branch milieu (e.g. the various leaf-eating species and the various canopy-living monkeys). Such activities clearly generate tensile forces in upper limbs because of the elasticity of plant materials. Other feeding activities involve tension in the

Frame 2

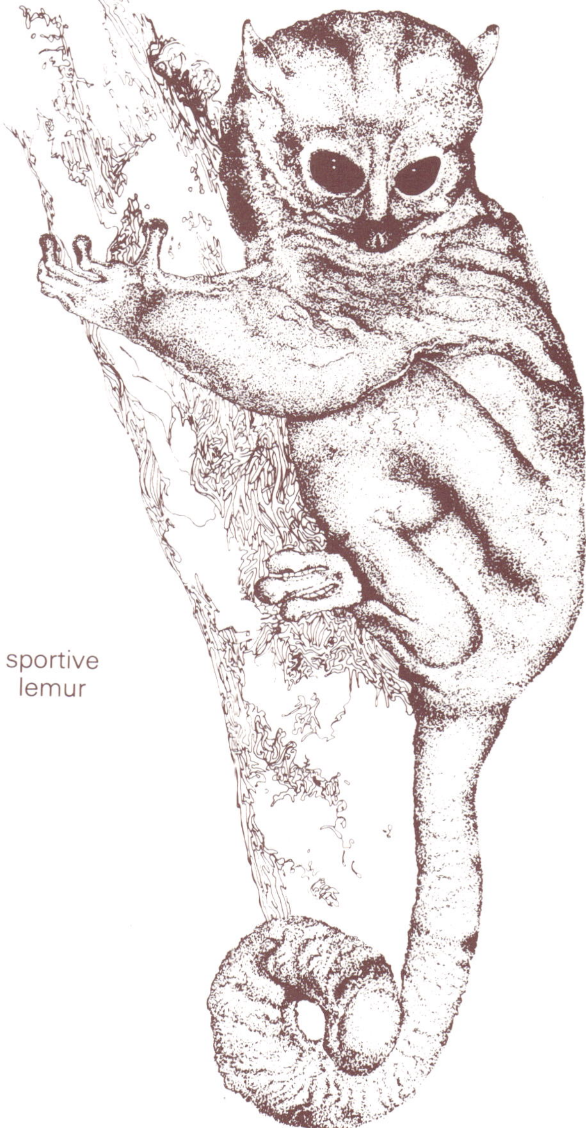

Frame 3

Fig. 5.12. Tension-bearing and raised upper limbs: yet another type of vertical posture (dwarf lemur).

Fig. 5.13. Tension-bearing and raised upper limbs in truncally erect positions (indriids).

Fig. 5.14. Tension-bearing and raised upper limbs in truncally erect positions (*retroussé*-nosed langur).

other, non-foraging upper limb as the animal steadies itself with a secure hold. They all may involve raised positions depending upon particular postures during foraging (Figs. 5.15, 5.16, 5.17, 5.18 and 5.19 and Stern and Oxnard, 1973).

Another example, also less readily identified as locomotion *per se*, includes movements involved in nest-making. Here, again, materials that are strongly elastic are seized with the arms extended and pulled in towards a nest. This also produces tensile forces in upper limbs and is found primarily in the great apes.

A final example that I will cite, but there are surely many more, includes the various play activities indulged in by all species and all ages. Such activities occur, obviously, more in juveniles than in adults, more in species that live in the small-branch milieu than in those existing on the large branches or even on the ground, and more in those capable of brachiation proper than in those who rarely do this. Thus, young bonobos (by inference), chimpanzees while young and even young gorillas may be involved in far greater degrees of tension-bearing activities with upper limbs raised during play than characterize the movements of the adults of these same species. And as the final adult form may be especially moulded by some of these juvenile activities, this may be a prominent part of adaptation as seen in the adult.

The spectrum of activities. Such a catalogue of upper limb functions results in a spectrum of tension-bearing and upper limb raising. This spectrum forms an average or resultant biomechanical situation that displays similarities and parallels among many primates that are important in understanding functional adaptation in the upper limb.

This does not, of course, imply that the same movements are present to greater or lesser degrees in all these animals. It is clear, for instance, that the use of the upper limb during these many behavioural activities may be different in the various animal species. It is therefore important to study the mix of upper limb functions in each species.

The mix of upper limb functions in individual species

The prior section has attempted to analyse the mix of activities that upper limbs may undertake. It is also important to look at the mix of activities that pertain to each individual species. It is not, of course, possible in this book to cover every animal, but a summary of a selection of animals supplies information that confirms the initial idea of a basic spectrum. And such a summary can also be used to point to special species where there is argument about what they actually do in the field.

The most extreme species, apes and atelines. We can quickly pass by the lesser apes and the ateline monkeys of the New World. There is no doubt that these species carry out a very wide range of the items mentioned above, even though, in addition, some of them have prehensile tails for other types of hanging activities. These species must be recognized as having the highest level of tensile forces and the greatest degrees of raised positions for upper limbs among the primates (Carpenter, 1940; Ellefson, 1974; Fleagle, 1976).

A second set of animals that we should especially investigate are the African great apes. These are usually thought of as terrestrial, although recent studies suggest that among them the bonobo is least so adapted. The amount of actual brachiation that they carry out is probably quite small. Bonobos brachiate a little more, chimpanzees and gorillas somewhat less (e.g. Nissen, 1931; Coolidge, 1933; Goodall, 1965; Patterson, 1979; Susman, Badrian and Badrian, 1980). All perform slower forms of brachiation (slow, that is, compared with gibbons) and in a non-richochetal manner (given that brachiation in gibbons is usually richochetal).

It is nowdays frequently overlooked that, in fact, these animals are very accomplished in the trees, especially when young. They can and do perform every activity mentioned above in an excellent manner, although they are much more cautious than are gibbons and siamangs. There is no doubt whatsoever that, notwithstanding their primarily terrestrial habitat, all the African great apes are able to utilize the upper limbs in modes that involve a great deal of tension and elevation. And, when we realize that they do these things with shoulders more fully abducted, with elbows more

Fig. 5.15. **Tension-bearing and raised upper limbs in an acrobatic position (orang-utan).**

Fig. 5.16. **Tension-bearing and raised upper limbs in foraging (New World monkey).**

Frame 1

Pan troglodytes

Fig. 5.17. **Tension-bearing and raised upper limbs in foraging (great apes: chimpanzee and orang-utan) next page.**

Frame 2

Fig. 5.18. Tension-bearing and raised upper limbs in foraging (another great ape).

completely extended, with hands more powerfully grasping and with heavier bodies increasing forces in all these joint systems than any other primates, we realize that they are near one extreme of the tension-bearing, upper limb elevated spectrum.

Recent attention paid to knuckle-walking and terrestrial activities in African apes has tended to obscure these facts. The more newly recognized activities of the bonobo, which seems to be considerably more arboreal (e.g. Patterson 1979; Susman, Badrian and Badrian, 1980), may help to draw this back into correct focus.

The Asiatic great ape, the orang-utan, is far more effective in these tension-bearing, upper limb raised modes and undoubtedly rivals the gibbon even though it moves in a quite different way as a cautious acrobatic climber that uses richochetal brachiation only occasionally and mostly when young.

The spectrum in Old World monkeys. Among the various Old World monkeys even the terrestrial baboons are not ineffective in trees. The semi-arboreal, semi-terrestrial species such as the vervet monkeys obviously are reasonably accomplished in the trees. The mainly arboreal species undertake even more of the activities mentioned above. But almost all of the cercopithecines perform most of

Fig. 5.19. Tension-bearing and raised upper limbs: in a variety of activities (orang-utans).

these activities in modes that resemble what a truly quadrupedal animal does when it merely changes its substrate to the smaller-branch milieu within the trees. The limbs still do not reach far in front of the body and the arms are thus only raised to smaller degrees. When they are so raised, it is not in a manner with full abduction of the shoulder and full extension of the elbow. And the amount of tension born by the limbs in such positions is not so great.

However, there is no doubt that tension-bearing activities in raised positions occur in upper limbs in cercopithecine species to some degree. As a generality, degrees of this are increasingly great in species that spend a little more time in the trees, or a little more time in the small-branch part of the arboreal environment, as compared with others which do not.

It is in the evaluation of the various colobines that a problem has arisen, although I would say that the general consensus today is that even that matter is fairly well solved. The problem stems in part because of early reports that some of these animals swing by their arms (e.g. Tennent, 1861). This has not been confirmed by primary field studies and is therefore probably only an accidental or occasional occurrence in play activities (as for any primate).

The problem has also arisen because some of the earliest langurs that were studied in detail were the Hanuman langurs (e.g. Ripley, 1967). They spend a great deal of time on the ground and are, to some considerable degree, habituated to human contacts and provisioning. It is now well-documented that the terrestrial habitat is not representative of how most langurs live. Even terrestrial langurs, when in the trees, are able to replicate fairly well many of the activities of other more generally arboreal langurs (Ripley, 1979; Morbeck, 1976, 1979). In these latter, the upper limbs bear fairly heavy tensile forces, not as great as in the apes and New World monkeys just discussed, but considerably greater than in the cercopithecine Old World monkeys just mentioned.

It is well known that these animals hang before drops more than do most Old World monkeys. Lifting actions are particularly conspicuous in colobus monkeys (Tuttle, 1975) especially when they are in the small-branch milieu. Pulling actions also occur not infrequently. Though Tuttle never saw colobus monkeys hanging by their upper limbs while feeding, he documents very well how they pull in, with the upper limbs, foods on springy branches well above their heads while sitting atop the larger boughs. And it is also well documented that, during feeding, colobus monkeys frequently employ their hands above their heads to steady themselves with grips on overhanging limbs (e.g. Morbeck, 1979). Such modes are less common in the cercopithecine monkeys whose propensities for full shoulder abduction and full elbow extension are considerably less.

The spectrum in New World monkeys. As we view the various New World monkeys we can identify similar upper limb activities (e.g. Erikson, 1963; Fleagle, 1977). It is well documented that the various atelines possess tension-bearing facilities and raised upper limb positions to large degrees. Of the atelines, the woolly monkey is the one least so endowed. But there is still no difficulty documenting that even this species does these things to very considerable degrees.

There is equally little difficulty in demonstrating that the various other New World monkeys, such as the marmosets and tamarins, owl monkeys and douroucoulis, undertake such activities to minor degrees only. And there is even not too much difficulty in showing that uakaris and saki monkeys are somewhat acrobatic in the trees (certainly more than marmosets and tamarins, equally certainly much less so than the atelines: e.g. Erikson, 1963). It is when we come to the howler monkey that there are problems somewhat similar to those with the colobines. It is entirely possible that the problems stem in part from somewhat similar historical causes.

Thus, the anatomical similarity between howler monkeys and atelines was recognized early so that a degree of similarity in tension-bearing and raised positions in upper limbs was assumed. But some of the earlier studies examined the activities of howler monkeys in milieux where they could be easily viewed — not, as in the case of the langurs, because they were upon the ground, but for a somewhat similar reason, that they were clearly visible upon open and relatively large branches. Such an environment militates against the more acrobatic activities predicted by the anatomy, and that might have been seen had the animals been studied in a smaller-branch milieu. In contrast, more recent investigations examine the movements of the animals in the close-branch environment; this allows the animals more scope for acrobatic movements.

This problem was further accentuated because, again as with the langurs, different groups of howler monkeys actually do inhabitat slightly different micro-habitats and actually do somewhat different things. (This is now well known for many other primates). Again, therefore, different patterns of movement were observed.

Thus, when we come to study in detail the many different activities described for howler monkeys, we gradually realize that they do use the upper limbs in modes that involve considerably more tensile forces and considerably more elevated postures than do other more quadrupedal New World monkeys. This is evident from the works of a number of investigators who describe various activities — hanging before drops, lifting, pulling, steadying with limbs above the head, reaching out for feeding materials and various acrobatic activities during play in the young.

Indeed, though a few workers (e.g. Schon, 1968) are among those who most deny tension-bearing in the limbs of howler monkeys, it is study of their own publications that easily reveals evidence of these activities. What Schon is denying, correctly, is the absence of arm-swinging, not tension-bearing.

The spectrum in prosimians. Even when we come to study prosimians, such differences can be found. Those prosimians that are most nearly regular quadrupeds with the most pronograde bodily habits do the various things leading to tension-bearing in raised positions to the least degrees, although even in this group these tension-bearing activities do occur because all move in the small-branch habitat for at least part of the time.

But among the lorisines a variety of slow climbing activities are carried out that lead, in comparison to other prosimians, to increasing degrees of such activities. This is most so in the potto and angwantibo, less so in the loris and slow loris.

And in the indriids, though their locomotion is characterized by long leaps, the various ancillary movements of which their upper limbs are especially capable (certainly more so than for most other prosimians) make it clear that they too bear tensile forces and have raised upper limbs more than do most other primates.

The spectrum in primates. It is thus rather clear that within both the anthropoids and the prosimians there are many different activities of primate upper limbs which, when summarized in this rather anecdotal way, demonstrate that there is a spectrum that consists of an underlying biomechanical mode, an average or resultant biomechanical situation, that involves tension-bearing in raised positions.

These many descriptions are not too dissimilar from what was postulated by Ashton and Oxnard (1964a) more than two decades ago. But they now rest upon a far wider range of primary data from field studies of a large number of investigators. And they also rest, far more than previously expressed, upon the knowledge that many movements, not mere arm swinging, result in tensile forces in the upper limb. Finally, they can be seen, far more clearly than at that earlier time, to be represented by a spectrum or a continuum, rather than by five separate artificial groups. With such a description it is possible for us to confirm the model of upper limb function that was postulated briefly in Chapter 4, Fig. 4.5.

Of course, this picture of a single spectrum of function in the upper limb does not imply a single evolutionary trend during the evolution of this most complex group, the primates. The various activities that result in this trend have undoubtedly arisen many times within the group; indeed, these directions are the natural new directions that might be taken by any group whenever it enters (or re-enters) the trees.

And the existence of an apparently unitary trend certainly should not suggest that each time it has occurred during primate evolution it has occurred in the same way, or with the same mechanical parameters, or producing the same anatomical features; clearly to assume the unitary position would be very much of an oversimplification.

Indeed, one of the problems of the evolution of upper limb function is exactly to disentangle all the various ways in which it may have occurred, and, from the viewpoint of humans, to make some assessment from just which part of that enormous functional complex the human upper limb may have derived.

The new biomechanics

The concept of the use of the upper limbs in tension and the question of the raised position of the upper limb seemed relatively simple in the earlier days of Inman, Saunders and Abbott (1944). But many new biomechanical studies make us realize that some of the early concepts and questions were naïve (although in fairness to the earlier investigators concerned, this naïvety could not be removed without the new techniques of today).

The question of tension in upper limbs. Today, the concept of the biomechanics of the limbs in tension is far more complex. One biomechanical matter relates to the question as to whether or not the *bones* of the limbs actually bear tensile forces during the various activities described in prior sections.

The behavioural descriptions of upper limb function make it certain that there are many times when the entire limb is bearing tensile forces. But although the limb as a whole may be under tension, the bones themselves are mainly under compression. In part we know this because it is the task of the various ties (ligaments, tendons and muscles, joint capsules, interosseous membranes, collagen and elastin and so on) to bear the tensile forces, and the task of the struts (bones and cartilage) to bear the compressive. In part also we know that there are compressive forces in the bones because the animals do indeed move during the support of these tensile forces; movement implies passage of joint surfaces over one another; that in turn implies compressive forces across joint surfaces during function.

Yet there are special peaks of loading when the bones within the limbs bear tensile forces. This, too, follows from considerations of bending in relatively rigid structures. If there were ever any doubts about these matters, they have now been removed by those studies using strain gauges *in vivo* which have demonstrated exactly that tensile forces exist in bones at some particular times (for instance, in primates in the clavicle: Stern and Fleagle, personal communication; and in non-primate mammals in a variety of sites: Lanyon and Smith, 1970; Lanyon, 1973, Lanyon *et al*, 1975).

Muscles and ligaments are, thus, not always available to spare tension-bearing by bones. It has long been known that in static, motionless standing postures in humans muscles are not electrically active. Newer findings (Tuttle and Basmajian, 1976) indicate that the situation is no different in suspensory postures, e.g. simple hanging in apes. Here, too, muscles may be electrically inactive; in such cases tensile forces exist though they must be taken up by ligamentous, interosseous and capsular mechanisms, but must also be borne by localized regions of the bones themselves.

Yet the *major* task of bone and bones *is* the bearing of compressive forces. Whatever may be the particular situation from moment to moment, it seems clear that the overall, or average or resultant biomechanical situation for most bones is one of compression. In the same way, the overall or average or resultant biomechanical situation for muscles, tendons, ligaments and capsules is tension.

This has been shown in certain special situations. The first is the consideration of forces in linear crests of bone with muscles arising from either side of the crests, such as the spine of the shoulder blade and the crests found on the skulls of some large primates. The second is consideration of forces in sheets or tables of bone with muscles arising from both sides, such as in the flat blades of bones like the shoulder and pelvic girdles. The third includes consideration of forces in blocks of bone with muscles or tendons arising from both ends of the blocks, such as in sesamoid bones and some cubical bones in hands and feet (Oxnard, 1971). Work in progress demonstrates that this is also true in more localized, microscopical situations, such as the precise points of bony attachment of tendons and ligaments into pits and onto tuberosities and in the elements of bone that develop embedded in some muscles and tendons.

It is rather likely, in fact almost certain, that though there are special peaks of load-bearing when bone bears tension, the average situation within bone itself is compressive.

Questions of upper limb muscle activities. We must also look at the matter of the raised positions of the upper limb for this, too, is no longer the simple activity it was once thought to be. The original studies of Inman, Saunders and Abott (1944) in particular, and the work of many other anatomists over the years, have supplied some view as to how muscles work in upper limb raising. These views were actually overly simplistic, although there is no doubt whatsoever that they were a most important part of the development of our general understanding of these matters and should by no means be underestimated.

They revolved around some relatively simple ideas, such as that during the raised position some muscles act primarily as stabilizers and others mainly as movers. They involve concepts like the existence of protagonists — muscles, the actions of which produce particular movements and antagonists — muscles, the actions of which produce movements opposite to protagonists but which may very well be actively helping to 'guide' protagonistic functions. These ideas also involve more complex notions of stability using terms such as fixators — muscles fixing various parts to allow other movements to occur, and synergists — muscles acting to remove unwanted actions of other muscles. These ideas postulated the existence of rather simple muscular couples and rotations as well as angular swings.

There has always been considerable discussion as to the precise meaning of these various terms. But none are especially wrong. New studies of exactly how the movements occur show that they are all just more complex; and in particular they show that individual parts of the earlier ideas require modification, elaboration and even some correction.

Thus more recent studies have elaborated shoulder function in relation to the raised position of the upper limb. We should clearly understand that many of these new studies confirm previously existing ideas based only upon anatomical inference and electromyographic studies in humans. One of these is, for instance, the notion that the supraspinatus muscle alone among the so-called 'rotator cuff' muscles (the short muscles of the shoulder blade believed to have mainly stabilizing functions) is also especially important in working with the deltoid muscle during arm-raising. This is a long held concept in humans. It has recently been confirmed by Tuttle and Basmajian (1978a, b) in some apes.

On the other hand, some of the new information from electromyography does not confirm the old speculations, and, indeed, offers new evidences of its own that are most important. Thus, based upon the usual kind of anatomical inference, some investigators suggested that the muscles of the forearm that are responsible for flexing the fingers also help in the special finger postures that are adopted by the African great apes during knuckle-walking. In this case electromyographic studies of Susman and Stern (1980a) have exactly demonstrated that these muscles do not perform this function actively.

In contrast, these authors are able to show that the finger flexing muscles are heavily recruited in climbing and suspension. As a result, it seems as though climbing and suspension may have played a major role in the function of the Olduvai hominid hand. This is further confirmed by new studies of the form of these finger bones (Susman and Stern, 1980b). And all of these support the results of earlier investigations of the mechanical efficiency of curved phalanges of the Olduvai hand in which it is demonstrated that it must have operated in a milieu involving very powerful grasping (Oxnard, 1973a).

But the major new findings resulting from these studies are of an order different from confirming individual older facts (such as Tuttle and Basmajian above) or suggesting individual new ones (such as Stern and colleagues above). They relate to two completely new concepts.

The first of these is the idea that particular muscular structures may be important for a variety of different functions. An example here also stems from the work of Stern and colleagues. Thus, they have shown that the clavicular head of the pectoral muscle mass is electrically active in a number of different movements in the prehensile-tailed New World monkeys and in gibbons. The muscle is important when these species engage in regular quadrupedal activities; it is also important when these animals brachiate, a movement that would seem almost entirely the opposite of quadrupedalism; and it is clearly effective when the animals employ vertical climbing activities, yet a third set of movements that are important in the life styles of these animals. Such a finding has implications for the evolution of the human upper limb (Stern, Wells, Jungers, Vangor and Fleagle, 1980).

Stern and colleagues have made a number of such findings relating to muscles in various regions of the body (e.g. Stern and Wells, 1976; Stern, Wells, Jungers, Vangor and Fleagle, 1980; Jungers, Jouffroy and Stern, 1980). One of their examples relates to the function of the serratus anterior muscle. This muscle passes from the trunk to the shoulder blade and is responsible for both the positioning and movements of that bone. The structure of that muscle as it is found in the great apes, large (especially in its caudal portion) and extending far down the chest, is closely related to both the propulsive effort of climbing with the limbs in raised positions and in brachiating with the limbs also elevated; but it does not seem to be related, in these animals, to the actual movement of raising the arm itself.

Again this has implications for human evolution, because the human muscle, although rather different from that in the great apes, is somewhat large; its importance *is* in arm-raising, and hence the possession of a somewhat large muscle in humans is not necessarily positive evidence for an arboreal past (Stern, Wells, Jungers and Vangor, 1980). Of course, it is always possible that the condition in man did derive from arborealism but has become modified as found in man exactly because suspension and climbing in the trees are no longer carried out.

However, the general theme resulting from these various new studies (and I have merely cited examples from quite a large body of work) is that many muscular developments may well relate to a variety of quite different activities, not merely a most obvious single activity. Hence evolutionary speculation from structure alone must be additionally careful.

There is a second general concept that arises from the new work. Again, examples from Stern and colleagues are most clear. This is the notion that when different animals do the same thing, they do not necessarily do it in the same way. This idea also arises in Chapter 6 where we realize that, though many animals leap, the structures for leaping vary rather enormously.

In the same way, many animals carry out arm-swinging movements. But electromyographic studies in prehensile-tailed New World monkeys and in gibbons have shown most effectively that arm-swinging is not a homogeneous behaviour. Thus, in gibbons almost all muscles are active in pull-up phases of activity just after mid-swing. In spider monkeys this is not the case and may be related to a role in support for the prehensile tail (Jungers and Stern, 1980; see also Jenkins, Dombrowski and Gordon, 1978). And Jungers and Stern (1980) also believe that their work shows that the difference does not reside in an already well-known anatomical difference: the existence of muscle 'chains' in gibbons and their absence in atelines. These muscle chains do seem to be present anatomically in gibbons, but electromyographic studies suggest that whatever their function is it is not in acting as a continuous muscle chain in arm swinging.

Another example of such a phenomenon, pertaining in this case to the lower limb, is the finding already mentioned for the hip and thigh (Jenkins, 1972). He shows that when chimpanzees and men walk bipedally they do not do it in the same way. Of course, this was somewhat obvious even before the study: the bipedalism of the chimpanzee is so clearly behaviourally different even to the naked eye. But Jenkins demonstrates the difference in detail and he further provides the most important information that the reason for the difference is that, when chimpanzees are bipedal, they use muscle activity patterns that are the same as when they are quadrupedal.

This, then, is the second major finding: that apparently similar behavioural activities in different animals must be examined in detail because it is highly likely that they are not truly biomechanically similar. Such findings have grave implications for the 'indirect' studies where the structures 'attempt', as it were, to 'tell about themselves'; and an understanding of these phenomena is thus most important before such studies can be used as a basis for looking at the overall relationships of the primates. We have particularly discovered this for the structures involved in leaping (Chapter 6).

Let us summarize this discussion. The main importance of all these studies is to demonstrate that

(a) many muscles are employed precisely as anatomists have judged them to be employed, using as evidence the results of experimental studies in man and the estimates of reasonable anatomical inference.

But these studies also demonstrate that

(b) some muscles are actually used rather differently from the conventional wisdom; these new departures from anatomical inference or from the human condition need to be known. They are very important in making judgements about the evolution of function.

The new studies more particularly, however, demonstrate two new concepts. The first of these is that

(c) structural findings that seem to relate to predominantly single actions, may actually be very important for a range of activities that is so wide that we would almost think them to be contradictory.

Who would have thought that the structure of the clavicular head of the pectoral muscle as found in atelines and gibbons was useful in three activities as widely different as quadrupedalism, vertical climbing and brachiation?

The second of these is that

(d) activities that appear to be the same in different animals may actually involve quite different functional relationships among the structures.

Again, who would have thought that the actions of muscles in arm swinging could be so different in arm-swinging atelines and arm-swinging gibbons?

It is thus clear that major new ideas have arisen since those early simpler studies based purely upon the human condition together with reasonable anatomical inference.

Questions of upper limb joint function. Finally, the new biomechanics allow us to 'dissect' shoulder function in a different way: one involving joint mobility.

What was previously interpreted as increased mobility in the shoulder in passing along the spectrum of shoulder function is now seen to include not only actual mobility in raising the upper limb at the shoulder, but also, and this is more important, potential mobility, as it were, the shoulder being already cranially placed and cranio-laterally twisted in appropriate species. This was implicitly inherent in the older descriptions of Inman, Saunders and Abbott (1944) and in Ashton and Oxnard (1964a); it was rather more overtly expressed by Erikson (1963); but it has required the more recent studies (especially of Stern and colleagues) to make it perfectly clear.

Relatively less work has been done at the elbow. We understand rather clearly the increased degrees of tension-bearing by the upper limb as a whole. But it is much less easy to associate this with features of the elbow region, which is perhaps adapted to the compressive forces in a lowered position by being less completely habitually extendable and to tensile forces in a raised position by being more completely habitually extendable. Such anatomical differences have effects upon the positions of attachment of the biceps and brachialis muscles, providing for their different leverages in flexion. Similarly, they affect the form of the olecranon process on the ulna giving different attachments for the triceps muscle that extends the elbow. In the case of the elbow joint, however, we still await those studies that will dissect it functionally in more detail.

Again, for the radio-ulnar joint, the joint allowing rotation of the forearm along a longitudinal axis, the inferences that can be made from structure suggest that the attachments of the pronator and supinator muscles of that joint are different in those species in which upper limbs bear tensile forces. Thus the biceps and pronator teres muscles show differences that relate to (a) how circumferentially peripheral and how angulated from the longitudinal axis of the radio-ulnar joint are their lever arms as flexors. These same two muscles also show differences that relate to (b) how circumferentially peripheral and how angulated from the longitudinal axis of the radio ulnar joint are their insertions on this same bowed bone: differences that relate in this case to their functions as pronators and supinators.

Preliminary work in my own laboratory suggests that the bowed form of the radius and ulna, together with the structure of the interosseous membrane that links these two, may provide a very special set of constraints for the radio-ulnar joint. Most uni-axial joints have two so-called close-packed positions in which the joint ligaments are rendered taut. Often there is one such position at each end of the range of movements. The radio-ulnar joint, however, may well have three such positions. Two of these are in the positions that we would expect, at extreme pronation and extreme supination. But it seems as though in man a third nearly close-packed position may exist so that somewhere in the region of the mid-prone posture the interosseous membrane moves into a somewhat tighter stress-bearing position. More needs to be done to confirm this preliminary finding. But it would certainly fit with the importance, in man, of the mid-prone position being especially stable: it is the working position for powerful action; we punch most heartily, we use a battle-axe most viciously, with the forearm bones in the mid-prone relationship to one another.

My suspicion, also, is that the arrangement of this third close-packed position may differ markedly among the monkeys and apes according to the degrees of pronation and supination that are important. In quadrupedalism, for instance, there is a premium on prone positions because during quadrupedal weight-bearing the hand must be placed palm down. In feeding and drinking there is a premium upon supine positions because in most primates the hands are used for conveying food to the mouth. In brachiation there is a premium upon both positions because both prone and supine positions occur regularly in arm-swinging. Other possibilities may be linked with other forms of upper limb activity. This matter certainly needs further study.

Finally, at the wrist, some inferential studies have been done suggesting biomechanical parameters, but in the main this region remains yet to be studied. Those aspects of the wrist that pertain to pronation and supination have also not been studied biomechanically, although recent investigations of the positions and structures of the semilunar discs have bearing upon this matter and have resulted in some controversy (Lewis, 1965, 1969, 1971, 1974; Tuttle, 1975; Cartmill and Milton, 1977, Schon and Ziemer, 1973).

When, therefore, we carry out multivariate morphometric statistical studies of upper limbs, we have very much less in the way of prior biomechanical studies to go on than we have in the case of investigations of leaping. But we nevertheless possess considerable information. Some of this stems from the mix of activities that are possible in upper limbs as a result of locomotion. Others depend upon the mix that occurs in individual animals. Yet others depend upon the precise biomechanical ways that the various movements are carried out. Finally, some depend upon the functions of individual joint units within upper limbs. This is a complex picture indeed. Paradoxically, the upper limb provides us with a greater range of studies in which indirect, morphometric studies have been used to assess trends of adaptation within the entire Order.

Morphometric studies of the primate upper limb

The multivariate morphometric approach has been the basis of a series of our own investigations on the shoulder, the arm, the forearm, the shoulder-arm-forearm complex and the proportions of the entire upper limb. Like the studies of other chapters they depend, first, upon an assessment of function by reviews of literature descriptions of the movements of the animals, by study of films and by discussions with field workers studying the animals directly (although not usually to investigate locomotion but to study other matters such as social and sexual structures). These ideas have produced the notion of the spectrum of tension-bearing in raised positions that we have already discussed.

The ideas were checked against the arrangements of muscles, joints and bones obtained through soft tissue dissection in these anatomical areas. The information that was gained was about such functional features as the relative extent of muscular leverages, the relative orientation of muscular fibres and joint facets, the relative dispositions of bony attachments of tendons and other similar data presumably of biomechanical import in the function of various joints.

Upper limbs: materials and methods. For the shoulder this involved dissection of 52 specimens representing 28 genera of the primates. For the arm and forearm, this phase was more extensive involving dissection of 145 specimens representing 27 genera of the primates. All of this first-hand information was backed by review of the extensive dissections of a host of prior workers, many in the last century, but with special note being made of the work of Miller (1932). More recently these data have been increased through investigations by workers such as Tuttle (1972) and Roberts (1974).

Based upon such part of this wealth of information as was available at the time, the various bony features of these anatomical regions were examined, both through visual observation of the bones and, more particularly, through examination of osteological measurements. Such data were examined first in a univariate manner. But it is their multivariate combination that provides the main picture with full information, elimination of redundancy and remarkably greater sensitivity.

These studies include multivariate statistical assessments (canonical variate analysis) of 17 variable measured on each of 376 sets of shoulder bones representing 39 genera (table 5.1); 19 variables measured on each of 525 sets of arm and forearm bones representing 41 primate genera (table 5.2); and 36 variables measured on each of 241 sets of shoulder-arm-forearm bones in 37 genera (table 5.1 and 5.2 combined and genera not common to both studies being removed).

In addition, these studies also include 9 variables of the proportions of the upper limb as a whole in 34 genera of primates (table 5.3); and 13 variables taken upon the upper limb as a whole in 12 genera of prosimians in a separate study (also table 5.3).

The variables on the localized anatomical parts are those taken by Mr. T.F. Spence and I, and reported in collaborative studies with the Birmingham, U.K., laboratory (Ashton, Healy, Oxnard and Spence, 1965; Ashton, Flinn, Oxnard and Spence, 1976). Those variables taken upon the overall proportions of the upper limbs of primates are a subset of the dimensions taken by the late Professor A.H. Schultz and studied in a multivariate manner by Ashton, Flinn and Oxnard (1975) and by Oxnard (1975a, 1977). Those variables taken upon the overall proportions of the upper limbs of prosimians alone are a subset of those taken by the Paris laboratory (Dr. Jouffroy and the late Dr. J. Lessertisseur) and studied in a multivariate manner by Oxnard, German, Jouffroy and Lessertisseur (1981).

The variables and species are fully described in the original publications referenced, but are here summarized briefly in tables 5.1 through 5.3. The various tests of technique applied to these studies are outlined in table 3.4 of Chapter 3.

Upper limbs: results. The result of these various studies show that, indeed, the morphometric arrangements of the various primates correspond well with notions of the biomechanical function of the upper limbs as discussed in this chapter. Indeed these studies suggest further hypotheses about structural functional relationships in the upper limb that remain to be tested, both by further biomechanical studies on individual species in the laboratory and by further investigations of actual behaviours in the field.

The picture is best summarized, perhaps, by the results of the analysis of the overall proportions of the entire upper limb in the whole Order. Fig. 5.20 demonstrates, diagrammatically, the actual measures involved. And Fig. 5.21 shows that the primates are strung out in a relatively linear sequence. This sequence may be thought of as paralleling the linear sequences of Chapter 2 which shows, through many different pieces of information, the systematic arrangements of the primates: prosimians, New World monkeys, Old World

Table 5.1

DIMENSIONS AND ANIMALS IN MORPHOMETRICS OF THE SHOULDER

Anatomical features

Locomotor	*Residual*
Medial extent of insertion of trapezius	Relative heights of supra and infraspinous fossae
Angulation of insertion of trapezius	Cranial displacement of superior scapular angle
Caudal prolongation of insertion of serratus magnus	Form of superior scapular border
Length of couple between trapezius and serratus magnus	Degree of projection of acromion process
Angle between fibres of trapezius and serratus magnus	Shape of acromion process
Distal extent of insertion of deltoid	Form of supraspinous fossa
Orientation of shoulder joint cavity	Orientation of scapular spine relative to lateral border
Torsion of clavicle	Orientation of scapular spine relative to cranio-caudal axis
Lateral projection of shoulder joint	

Genera and numbers of specimens studied

PROSIMIANS

Hapalemur, 2; *Lemur*, 18; *Lepilemur*, 3; *Propithecus*, 4; *Daubentonia*, 2; *Loris*, 4; *Nycticebus*, 3; *Arctocebus*, 5; *Perodicticus*, 17; *Galago*, 9; *Euoticus*, 11

NEW WORLD MONKEYS

Aotus, 7; *Callicebus*, 7; *Cacajao*, 4; *Pithecia*, 7; *Alouatta*, 6; *Cebus*, 10; *Saimiri*, 7; *Ateles*, 8; *Brachyteles*, 1; *Lagothrix*, 8; *Callithrix*, 6; *Leontocebus*, 8

OLD WORLD MONKEYS

Macaca, 19; *Cynopithecus*, 2; *Cercocebus*, 8; *Papio*, 9; *Mandrillus*, 2; *Cercopithecus*, 36; *Erythrocebus*, 6; *Presbytis*, 16; *Rhinopithecus*, 5; *Nasalis*, 1; *Colobus*, 16

HOMINOIDS

Hylobates, 16; *Pongo*, 15; *Pan*, 20; *Gorilla*, 14; *Homo*, 40

Table 5.2

DIMENSIONS AND ANIMALS IN MORPHOMETRICS OF ARM AND FOREARM

Anatomical features

Locomotor	*Residual*
Extent of humeral trochlear joint facet: elbow movement	Relative position of radial tuberosity: flexion of elbow
Extent of insertion of triceps: mechanical advantage	Angle of ulnar joint facet: flexion of elbow
Projection of ulnar styloid: ulnar deviation at wrist	Angle of long axis of ulnar joint: elbow flexion, extension
Relative sizes of styloids: both deviations at wrist	Relative projection of epicondyles: elbow flexion, extension
Projection of styloids: both deviations at wrist	Projection of radial styloid: radial deviation at wrist
Distal insertion of pronator: efficacy of pronation	Relative widths of radius and ulna: radial and ulnar deviation
Angulation of insertion of biceps: efficacy of supination	Position of lateral bowing of radius: related to pronation
Bowing of radius: efficacy of pronation and supination	Position of maximum bowing of radius: pronation and supination
	Angle of radial interosseous ridge: pronation and supination
	Angle of radial notch: stability of radio-ulnar joint
	Angle of ulnar head: movement at radio-ulnar joint

Genera and numbers of specimens studied

PROSIMIANS

Tupaia, 6; *Hapalemur*, 4; *Lemur*, 18; *Lepilemur*, 4; *Cheirogaleus*, 1; *Microcebus*, 2; *Propithecus*, 2; *Daubentonia*, 2; *Loris*, 11; *Nycticebus*, 7; *Arctocebus*, 5; *Perodicticus*, 19; *Galago*, 10; *Euoticus*, 13

NEW WORLD MONKEYS

Aotus, 15; *Callicebus*, 5; *Cacajao*, 11; *Pithecia*, 13; *Alouatta*, 12; *Cebus*, 15; *Saimiri*, 17; *Ateles*, 10; *Lagothrix*, 14; *Callithrix*, 17; *Leontocebus*, 18

OLD WORLD MONKEYS

Macaca, 20; *Cercocebus*, 14; *Papio*, 23; *Mandrillus*, 7; *Cercopithecus*, 35; *Erythrocebus*, 13; *Presbytis*, 24; *Rhinopithecus*, 5; *Nasalis*, 2; *Colobus*, 18

HOMINOIDS

Hylobates, 23; *Symphalangus*, 2; *Pongo*, 16; *Pan*, 24; *Gorilla*, 20; *Homo*, 163

Table 5.3

DIMENSIONS AND ANIMALS IN MORPHOMETRICS OF THE UPPER LIMB

Anatomical features

Schultz's data on upper limbs	*Jouffroy and Lessertisseur's data on upper limbs*
Relative chest circumference	Relative length of humerus
Relative shoulder breadth	Relative length of radius
Relative upper limb length	Relative length of hand
Intermembral index	Relative length of total upper limb
Brachial index	Relative length of carpus
Relative hand length	Relative lengths of metacarpals
Relative hand breadth	Relative lengths of fingers
Relative thumb length	

Genera and numbers of specimens studied: Schultz's data

PROSIMIANS

Tupaia, 5; *Lemur*, 3; *Microcebus*, 3; *Lichanotus*, 1; *Propithecus*, 1; *Indri*, 1; *Daubentonia*, 1; *Nycticebus*, 9; *Perodicticus*, 9; *Galago*, 10; *Tarsius*, 8

NEW WORLD MONKEYS

Leontocebus, 28; *Aotus*, 9; *Cacajao*, 2; *Pithecia*, 1; *Cebus*, 25; *Saimiri*, 49; *Alouatta*, 4; *Ateles*, 74; *Lagothrix*, 2

OLD WORLD MONKEYS

Macaca, 27; *Cercocebus*, 3; *Papio*, 5; *Cercopithecus*, 3; *Erythrocebus*, 2; *Presbytis*, 14; *Nasalis*, 26; *Colobus*, 2

HOMINOIDS

Hylobates, 78; *Pongo*, 13; *Pan*, 26; *Gorilla*, 6; *Homo*, 25

Genera and numbers of specimens studied: Jouffroy and Lessertisseur's data

PROSIMIANS

Hapalemur, 7; *Lemur*, 46; *Lepilemur*, 9; *Cheirogaleus major*, 6; *Cheirogaleus medius*, 3; *Microcebus*, 15; *Avahi*, 6; *Indri*, 6; *Propithecus*, 7; *Galago*, 39; *Euoticus*, 9; *Tarsius*, 8

monkeys, apes and man. But, as can be seen from the disposition of these major groups (the brackets in Fig. 5.21), the one thing that the upper limb linear pattern does *not* resemble is the systematic arrangement of the primates. All the taxonomic groups are mixed up with one another; even the apes are not entirely separate.

An even simpler view of this idea was much earlier evident from the study of the shoulder alone (Oxnard, 1963). Fig. 5.22 shows that the result of that investigation could be summarized in a single discriminant axis alone. It shows, of course, a linear sequence; but it is not one related to systematics or taxonomy.

The prior discussion about upper limb function within all of the primates presents an answer immediately. The species are linearly related to one another in a way that mirrors upper limb function. Species such as terrestrial baboons and patas monkeys, and leaping bush-babies and tarsiers lie at the left-hand end of the entire separation of the primates. They have in common the func-

Fig. 5.20. A skeletal diagram demonstrating the elements (shaded dark) defined in the studies of the upper limb as a whole.

tional element previously enunciated: that their upper limbs bear, to the greatest degree among the primates, compressive forces acting in a lower position. At the other end of the linear separation of the primates are the brachiating lesser apes, the orang-utan and some of the atelines; and these, it will be remembered, are the primates whose upper limbs bear to the greatest degrees tensile forces in raised positions. In between these extremes, at not unreasonable positions within the spectrum, are a variety of species which, from left to right, demonstrate increasing usages of tensile force in increasingly raised upper limbs: regular quadrupedal New and Old world monkeys (marmosets and macaques, for example), more acrobatic New and Old World monkeys (howler and woolly monkeys, and colobs, for instance), slow climbing acrobatic prosimians (such as lorisines). This arrangement is quite different from the taxonomic one of Chapter 2, but it closely resembles the functional spectrum described earlier in Chapter 4 (Fig. 4.5).

In order to be certain that this set of relationships is indeed correct, we must also view the results of other studies. One of these is analysis of the proportions of upper limbs taken on prosimians alone. In this case the dimensions are those of Drs. Jouffroy and Lessertisseur; they are not identical to those of Professor Schultz; they are more detailed and they do not have the complicating factor of being related to the trunk length as do those of Professor Schultz. The range of dimensions is also considerably wider, involving lengths of

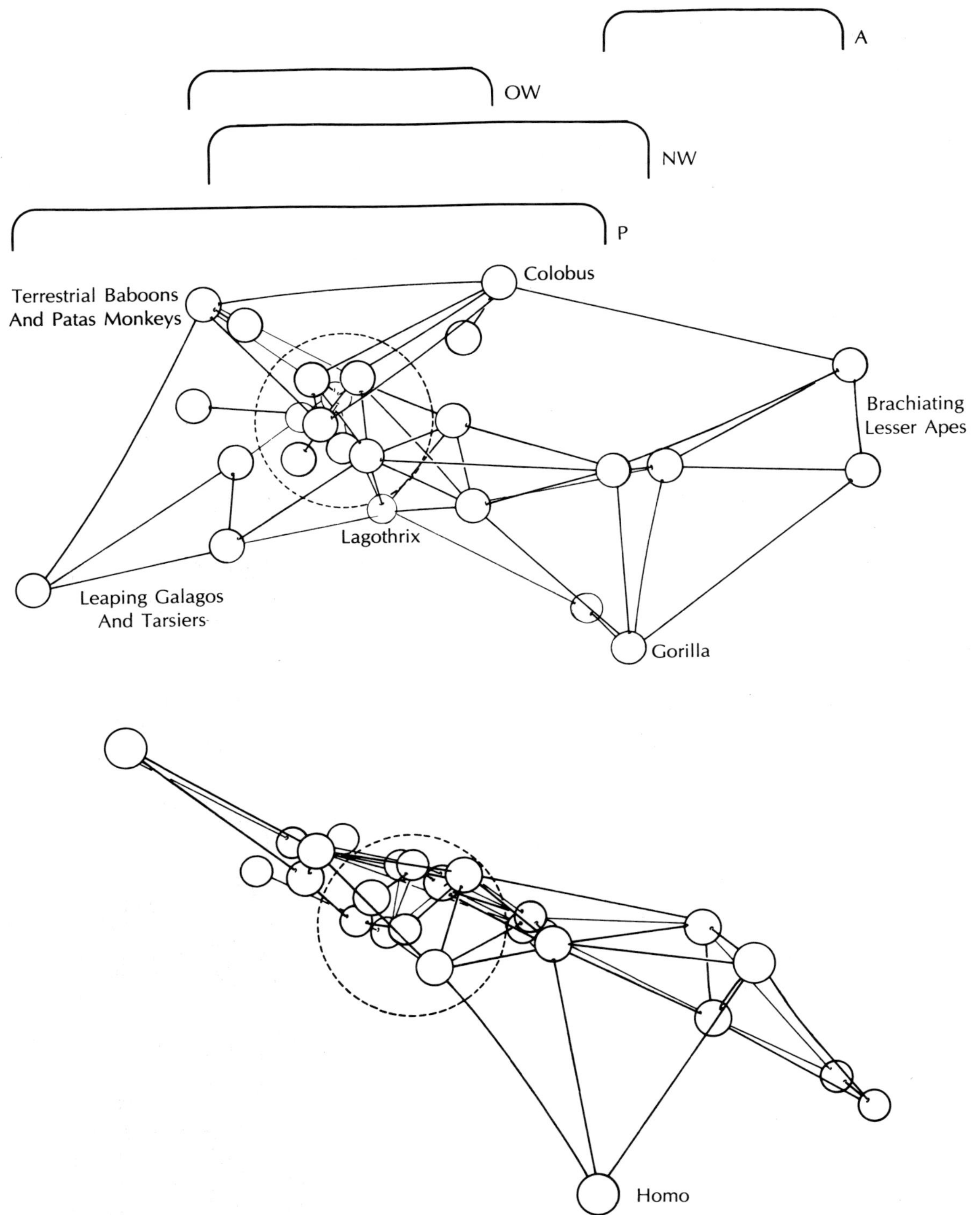

Fig. 5.21. A multivariate morphometric study of the data resulting from the measures of Fig. 5.20 displayed as a three-dimensional model of generalized distances. The overall scale of the diagram is some 25 distance units.

The primates are strung out in a linear array, but consideration of taxonomy (P = prosimians, NW = New World monkeys, OW = Old World monkeys, A = apes) shows that this is not an array that coincides with taxonomy as was the case for most variables in Chapter 2. There is enormous overlap of taxonomic groups.

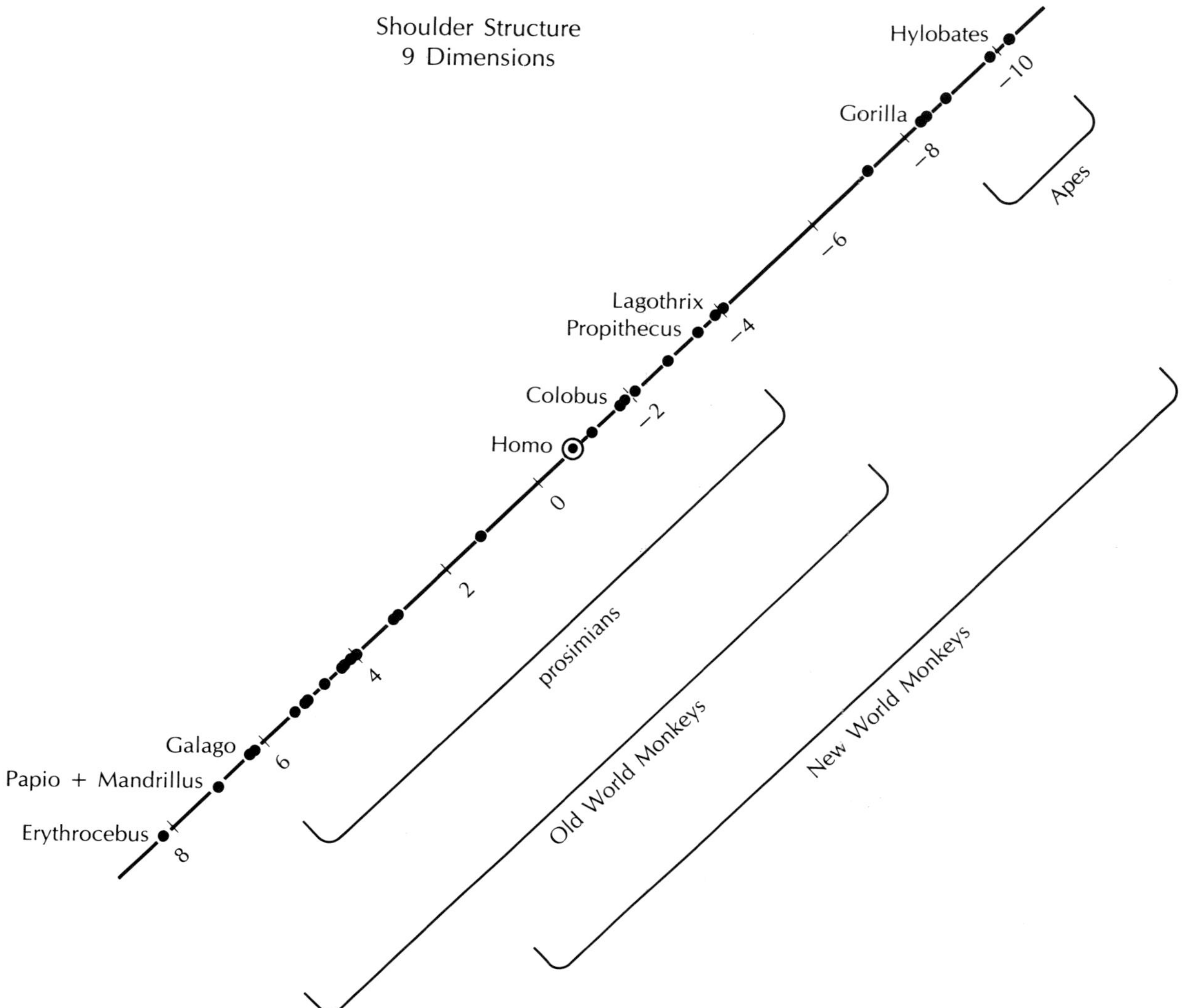

Fig 5.22. A multivariate morphometric study of data from the shoulder displayed as a single axis. The primates are strung out in the linear array to be expected from a single discriminant axis, but consideration of taxonomy shows, again, that the array does not coincide with the taxonomy. There is a very big overlap of taxonomic groups.

individual digits rather than the simple segment lengths of Professor Schultz. Fig. 5.23 presents these new dimensions in diagrammatic form.

The result of analysing these dimensions using multivariate statistical methods is almost identical to what we have obtained from Professor Schultz's data. The three-dimensional model that it is possible to construct is shown in Fig. 5.24, and the most important finding is that there is a similar linear spectrum. Of course, in this case the spectrum is confined to prosimians because these are the only species examined in this study. But the positions of these prosimians relative to one another is exactly the same as their relationships in the prior study.

At the left-hand end of the linear separation of the species are the bush-babies and tarsiers; these species have upper limbs that bear mainly compressive forces as a result both of the way that they land on their upper limbs after the very long leaps of which they are capable and of their otherwise generally quadrupedal habits. At intermediate positions are the more regular quadrupedal forms (e.g. lemurs); these species, leaping and climbing in trees, are not overly different from the general group of quadrupedal monkeys in that the upper limbs bear tensile forces in raised positions to degrees somewhat greater than of the bush-babies and tarsiers. At the other extreme are located the indriids whose upper limbs bear even more tensile forces in raised positions as a result of the arboreal activities that they carry out. (The upper limbs, in contrast to the situation in the leaping bush-babies, do not bear much compression in landing after leaps for these forces are mainly borne by a forward movement of the lower limbs just before landing).

When we view the results of the more localized anatomical investigations (the shoulder, arm and forearm separately) the findings are similar. The general picture is a linear spectrum of species with animals such as baboons and bush-babies at one end, woolly monkeys and colobus monkeys, lorisines and indriids intermediate, and spider monkeys, orang-utans and gibbons at the other extreme. These results mirror what we have already seen foı

Fig. 5.23. A skeletal diagram demonstrating the limb elements (shaded dark) defined in the study of the upper limbs of prosimians alone.

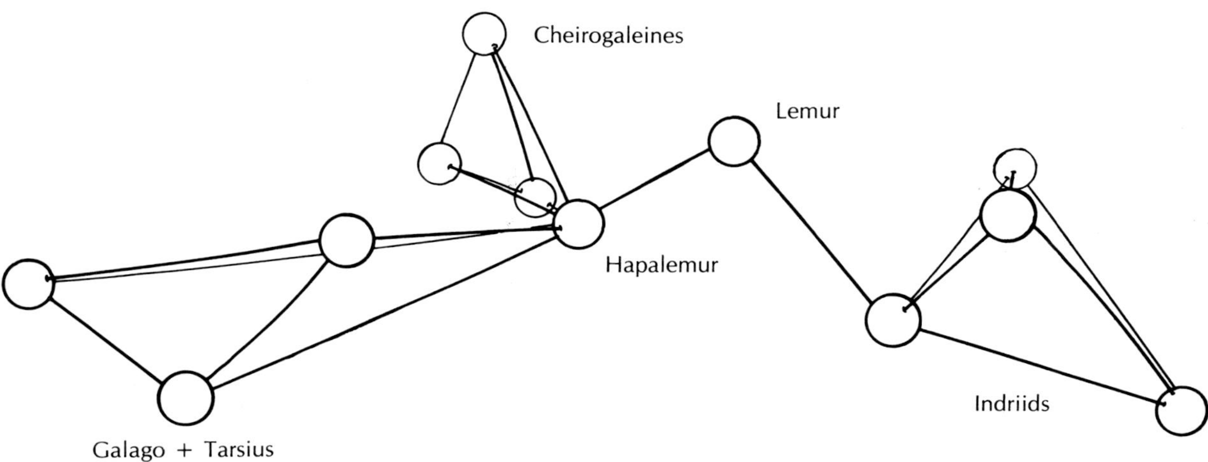

Fig. 5.24. A multivariate morphometric study of the data from the prosimian upper limb in Fig. 5.23 displayed as a three-dimensional model based upon generalized distance connections. The prosimians are strung out in a linear array. Comparison with Figs. 5.21 and 5.22 demonstrates a marked similarity.

the upper limb as a whole. These similarities are most evident in comparing the rank orders of the various genera on the most important (first) discriminant axis; despite some rather interesting differences (the positions of humans and baboons, for instance) the overall similarity cannot be denied (Fig. 5.25).

Although these various studies confirm one another remarkably well, it is still necessary to question them because they have come from the laboratory of a single group of workers. Luckily, several other investigators from independent laboratories have either replicated these studies or carried out investigations that parallel them partially, and in each case the results are corroborated. Thus, replicate studies of the arm and forearm by Feldesman (1976) and of the shoulder by Corruccini and McHenry (1978) provide rather similar confirming pictures (Figs. 5.26 and 5.27; and Oxnard, 1979a, b).

Upper limbs: explanations. If, then, these results are valid, what can be the explanation for them? It must reside in the idea that, in characterizing a particular anatomical area in this way, we are characterizing a morphology that is associated with the average or resultant biomechanical situation found in that anatomical area. Viewing the skull in this way does not give a single coherent picture because the skull itself is not a single functional region but a compound of several, some of which are among the most complicated to be found in the animal body. But visualizing the upper limb in this way probably does provide some kind of overall average view of the functional-structural relationships within that anatomical member.

It is thus not unexpected that gibbons, siamangs, orang-utans, spider monkeys and woolly-spider monkeys all fall near one another in all of these studies. These are all animals in which, irrespective of the details of the actual behaviours, upper limbs operate more in raised positions and bearing tensile forces.

It is not unexpected that a large number of regular four-footed animals are near the other end of the spectrum (e.g. baboons and patas monkeys from the Old World, squirrel monkeys and douroucoulis from the New World, lemurs of various kinds from among the prosimians). These are all creatures in which, however wide the range of activity that they carry out, and however different

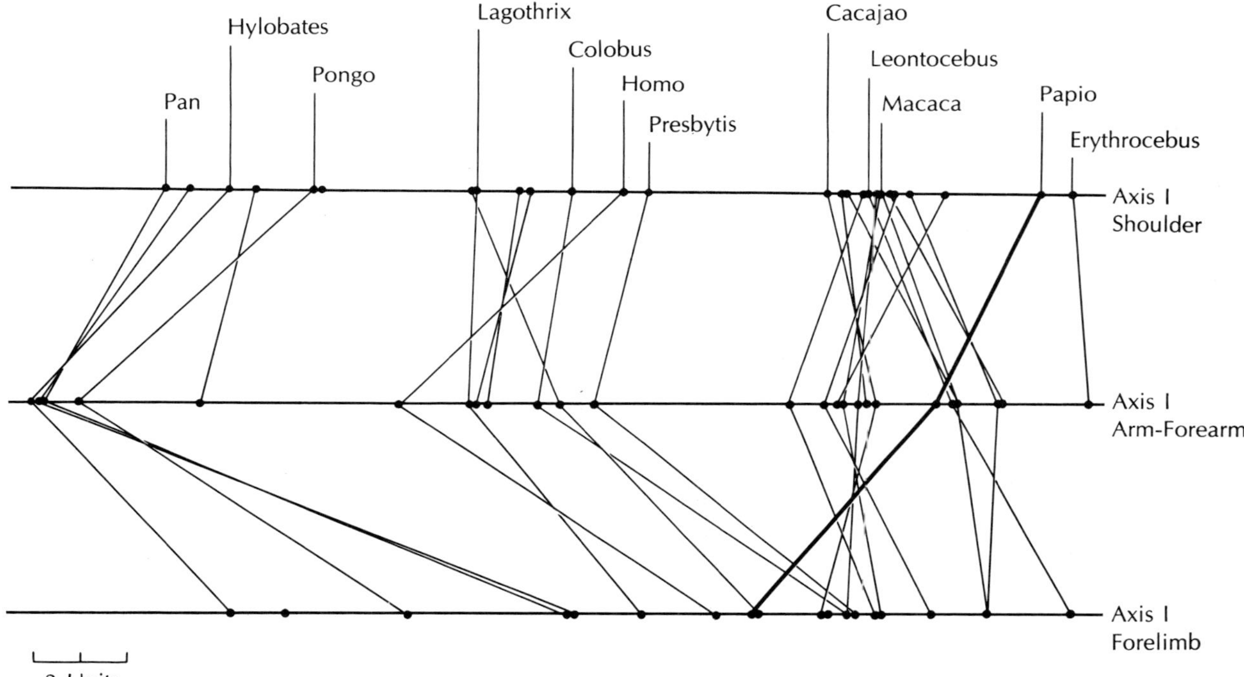

Fig. 5.25. The general similarities that exist between the results of the more restricted studies, e.g. those of the shoulder, arm and forearm, and overall form of the upper limb. The rank orders of most genera are rather similar in the first axis of each study.

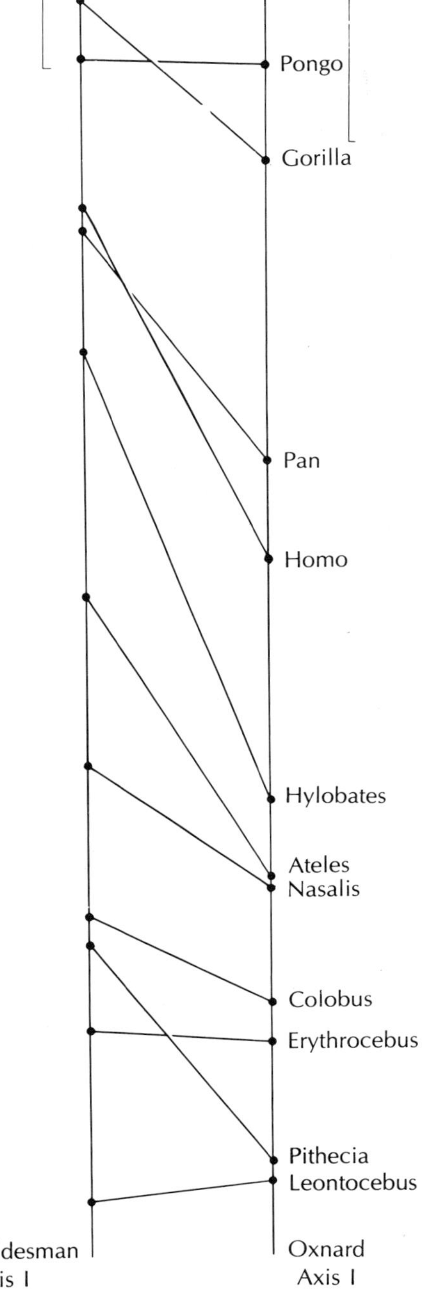

Fig. 5.26. General similarities that exist for such replicate morphometric studies as have been carried out for the arm and forearm.

the details of their locomotor patterns, upper limbs are used more in lowered positions and bearing more compressive forces as compared with the species just mentioned.

Predictably, we find yet other species at intermediate points along the spectrum (colobus monkeys, langurs, proboscis monkeys from the Old World, woolly monkeys, howler monkeys from the New World, lorisines and indriids from among the prosimians). These, as we have seen, carry out many very different sets of locomotor activities but have certain fundamental similarities in possessing intermediate degrees of tension-bearing and intermediate degrees of raised upper limb postures between the two aforementioned groups.

It is not at all unreasonable to expect that different behaviours might produce similar morphometric positions for individual species. For different behaviours may well produce similar average biomechanical situations. This is presumably why certain aspects of upper limb structure seem to be somewhat similar between, say, sifakas and proboscis monkeys, between, for instance, pottos and woolly monkeys or between, for example, squirrel monkeys, macaques and lemurs. Though none of the members of these sets of genera move in similar ways, their various movements may well result in similar average biomechanical situations. Certainly structural analysis demonstrates that there *are* morphological similarities.

Knowing what we do of the ways their upper limbs function, we can only suppose that the morphological resemblances are due to the parallels in function. Different as the overall behaviours are, they are similar enough to produce (through development and evolution) somewhat similar morphologies. Thus, sifakas and proboscis monkeys, though progressing in markedly different ways, both employ their upper limbs in a relatively three-dimensional range of activities, with much use of raised positions and bearing tensile forces to some considerable degrees as compared with those closer systematic relatives of each, such as macaques and lemurs. Pottos and woolly monkeys move quite differently, yet both employ their upper limbs in such a way that they may be under some resultant forces of tension during the various underbranch and hanging activities of which both are capable. Finally, squirrel monkeys, macaques and lemurs all move in very different ways and could not be mistaken for one another when compared through locomotor profiles; indeed only a momentary glance at each of these animals moving serves to distinguish each from the next. Yet the functions of their upper limbs are such that, compared with

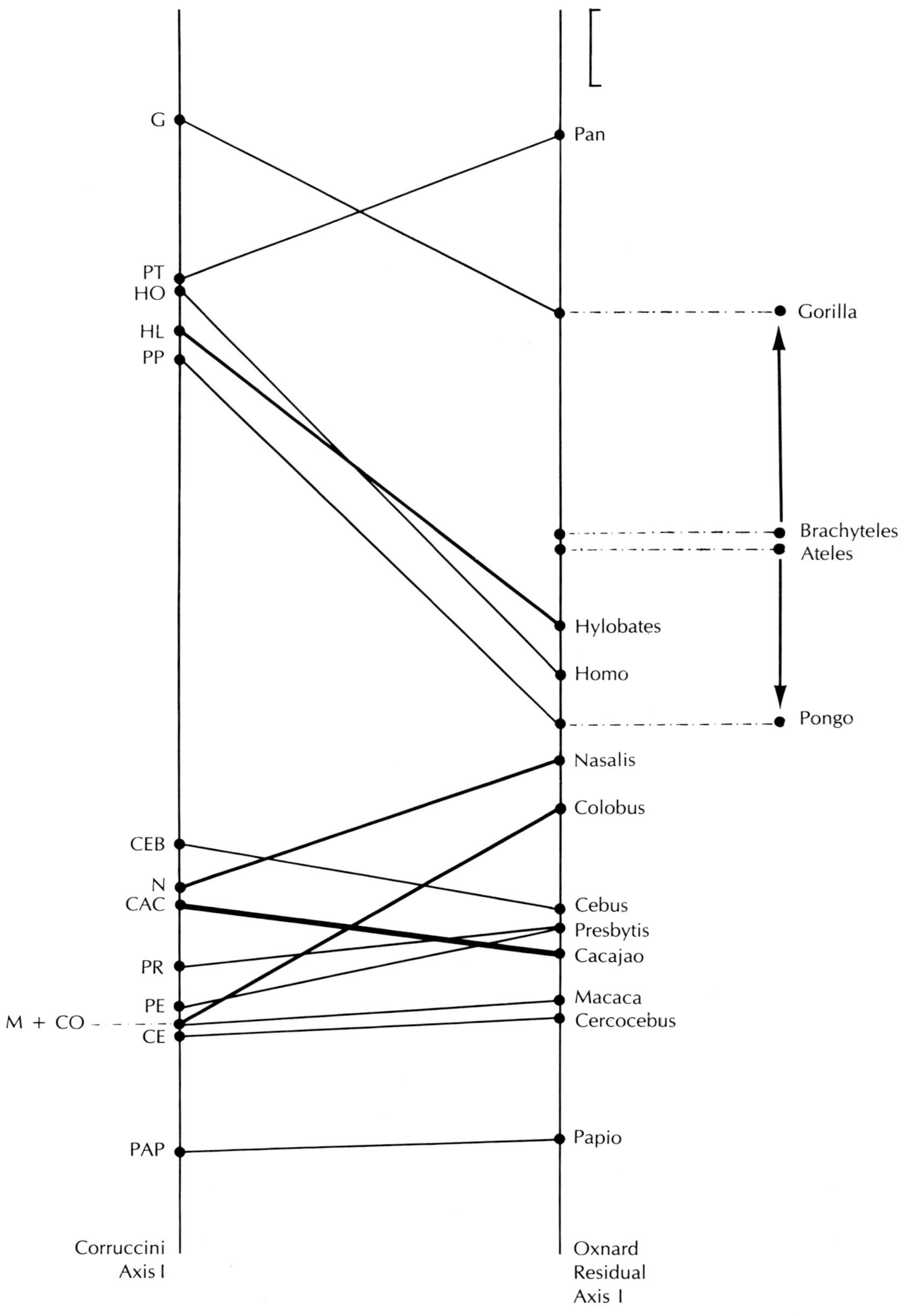

Fig. 5.27. General similarities that exist for such replicate studies as have been carried out for the shoulder.

other primates, they presumably function with overall smaller ranges of movement, more in lower rather than raised positions and bearing compressive forces rather than tensile.

Upper limbs: humans. It is worth looking for a moment or two at the position of *Homo* within these various studies. This is best given by the additional view of the plots and models represented by Figs. 5.28 and 5.29. In each case humans appear to occupy a somewhat intermediate position along the linear spectrum. But in each case consideration of additional information demonstrates that they actually lie in markedly offset positions. They are not like any of the non-human primates; their position in the analyses is unique; their morphology must be correspondingly unique; and this implies uniqueness for the average biomechanical situation in their upper limbs. This last we know they have; the functions of human upper limbs are, indeed, uniquely different from those of any other primates.

Upper limbs: fossils. It is worth also thinking for a moment or two about fossils. For this study makes it clear that the structure of a fossil, if analysed in this way, might allow us to say a very great deal about the upper limb of that fossil. Depending upon where the fossil fell within the linear spectrum, we might be justified in being quite detailed about (a) the degree to which that fossil upper limb was capable of acting in two-dimensional or three-dimensional modes; (b) the degree to which that upper limb was capable of being used in raised as compared with lowered positions; and (c) the degree to which that upper limb was adapted for bearing tensile rather than compressive forces.

All of this seems remarkably detailed. Yet a reminder of the ways in which sifakas and proboscis monkeys are not clearly differentiated from one another, or the ways in which pottos and woolly monkeys are similar or even the ways in which squirrel monkeys, macaques and lemurs are alike, makes it quite clear that we cannot hope to have a picture of the locomotion of the fossil from such a study. The morphology is not related to locomotion *per se* but to the average or resultant biomechanical situation of which locomotion is the major, but not the only determinant. It is accordingly only the average biomechanical situation that can be pinpointed for the fossil.

It is also worth thinking about the position of presumed human progenitors. If not actually human, yet on the lineage to *Homo*, we would also expect progenitors to be unique from the non-human primates. But as ancestors stemmed from further and further back in time we would expect to see them lying more and more towards some part or other of the linear spectrum of non-human forms. For it is presumably from one part or another of that spectrum that prior human ancestors belonged. Fossils that do not lie on such a lineage may well also be morphologically unique; but this would be because they represent their own particular offshoots leading to their own curious and presumably extinct functional uniquenesses. This is a topic to which we will return in a later chapter.

Upper limbs: taxonomy. There is thus no doubt that the overall form of the upper limb is related to the biomechanical demands of its function. It is, however, worth asking if any aspect of the form of the upper limb seems related to the overall relationships of the primates. For it is well known that individual meristic features are so related; thus, a metacromion, a special process on the shoulder girdle to which is attached special musculature, is confined, among primates, to the Prosimii; an entepicondylar foramen (a foramen at the lower end of the humerus transmitting a major limb artery), is likewise found only in certain prosimii. Such features have long been used to place fossil fragments within the primate sequence.

Equivalent features are indeed found among the various quantitative measures used in the studies described above. Thus, in the shoulder investigation, a set of residual features (features that do not seem to have much relationship to the locomotor function of the shoulder and summarized in table 5.1) is definable. Similar sets of residual features are distinguishable for the arm and forearm (and are summarized in table 5.2). Study of these dimensions alone by both univariate and multivariate statistical methods provides some separations apparently related to taxonomic classification. Even these features, however, contribute markedly to the locomotor separations (see Oxnard, 1967; Ashton, Flinn, Oxnard and Spence, 1971; Ashton, Flinn and Oxnard, 1975). Yet the taxonomic information, however small, that seems to be contained within these and other dimensions is a facet of these results to which we shall return in Chapter 8.

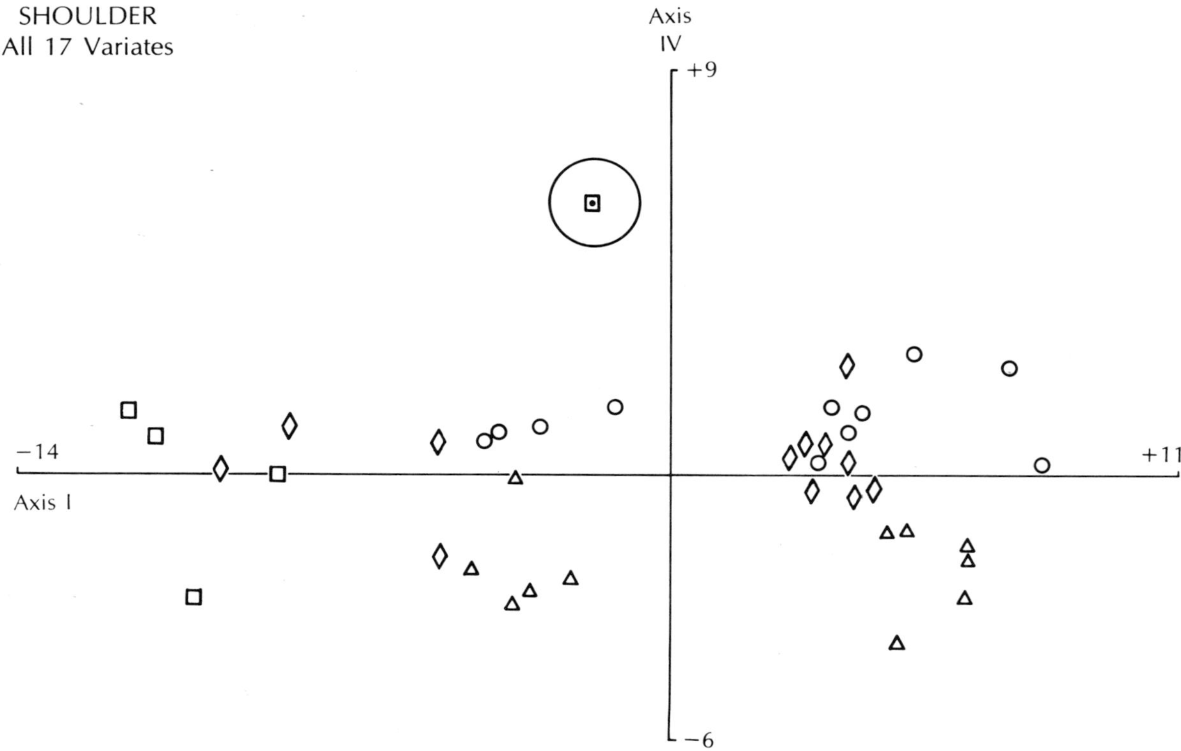

Fig. 5.28. The unique position of humans in the morphometrics of the shoulder alone: bivariate plot of discriminant axes one and four (the remaining axes do not provide any separation of humans from the rest). The position of the human genus is represented by the small square surrounded by a circle of one standard deviation radius to help provide a marker. Humans are separated markedly in a position above the plane of all other primates.

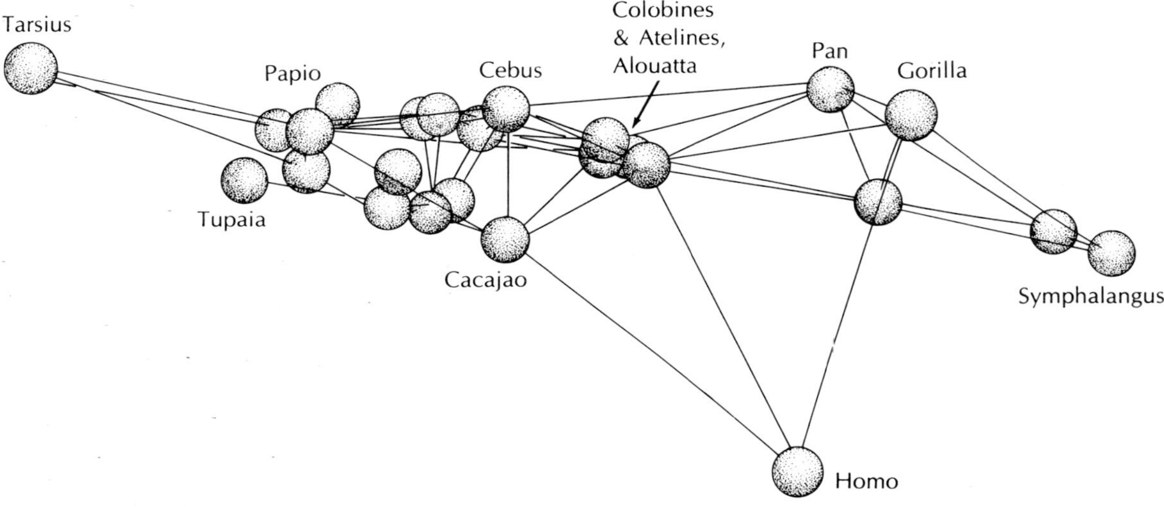

Fig. 5.29. The unique position of humans in both the functions of the upper limb and its morphometrics: model of generalized distance connections. Humans lie below the plane that defines all other primates. Compare Fig. 5.28.

Summary. This review of structural-functional associations within primate upper limbs started with the old notion of grades of function related to taxonomic grades. The idea that mobility of upper limbs was least in prosimians, greater in monkeys, greater still in apes, and most in humans was a useful idea for viewing the primates in earlier times. But new information and ideas about primate locomotion now reveal that upper limb function is better described in other ways.

We thus recognize the underlying difference between basically arboreal primates and terrestrial mammals: that is, in comparison to compressive forces in the forelimbs of terrestrial mammals, tension-bearing by upper limbs is a phenomenon of most activities that primates undertake. Thus, though tension-bearing is most easily seen in brachiation and arm-swinging, as practised by some apes and monkeys, it is also a part of most locomotor actions that involve upper limb raising, body support from above, climbing and other acrobatic movements, together with a wide variety of foraging and other activities only marginally within the concept of locomotion.

We also recognize rather more clearly that individual primates are less characterized by specific activities (e.g. brachiation in lesser apes — this is the old concept stemming from the idea of a locomotor classification) than they are by the amazing mix of activities, different in different primates, which they undertake. Thus, though all primates can do almost all things (even a baboon *can* brachiate after a fashion), the mix of activities, a mix producing variable degrees of tension-bearing, is what distinguishes the various species from each other.

And finally, in looking at the mix of activities and their implications for biomechanics, we can see that new ideas about biomechanics must be taken into account. These involve not only matters such as tension and compression bearing by bones, joints and muscles, but also new ideas about movements and postures, and about joint function. Much of this is coming from direct biomechanical experimentation involving methods such as stress analysis, electromyography, cinematography and radiocinematography and motion analysis among many others.

All these ideas end in an overall description of an average biomechanical situation for upper limbs that forms a broad spectrum or band relating to tension, position and movement. When we examine the structures of the various upper limb parts — shoulder, arm, forearm, upper limb proportions — and various combinations of these, we find separations of the non-human primates that also form a broad, band-shaped spectrum. Humans are uniquely placed outside the spectrum, and even this fits with the reality that the human upper limb, alone among the primates, is not involved directly in human locomotion.

The degree of concordance between the structural and functional spectra is extraordinarily high. It suggests that the morphometrics of upper limbs and upper limb parts speak most strongly to function. And this, in turn, implies that the examination of upper limbs and upper limb parts in fossils is most likely to give information about function in fossils.

CHAPTER 6

Lower Limbs and Leaping

Abstract. In this chapter we discuss form and function in primate lower limbs. Because leaping is so obvious an adaptation of many primates, and because leaping has, therefore, been very well studied, we first concentrate upon leaping species. The most marked leapers among the primates are all prosimians.

As in the case of the upper limb, early simpler views of leaping functions in lower limbs did not differentiate between the possibility of the existence of different forms of leaping. New natural history studies are starting to suggest that leaping is carried out in quite different ways by different animals. As a result we can envisage the average biomechanical situation (of the fourth chapter) for lower limbs as a rather complexly shaped spectrum of function.

We then obtain structural descriptions of various lower limb parts (hip, thigh, overall proportions of lower limb) using the multivariate statistical method of the third chapter. This also provides arrangements of the prosimians that fall within a complex spectrum. A detailed view of the leg and foot is not available because, apart from a few leg and foot dimensions within the study of overall proportions of the lower limb, these areas have not been studied.

The degree of concordance between the functional and structural spectra is examined; the two are remarkably similar. The overall structure of the various parts of the prosimian lower limb, when viewed quantitatively, speaks most clearly about overall function.

We also carry out the equivalent procedure for the primates as a whole. In this case the functional spectrum is even more complex because it includes not only those species that leap in their different ways, but also a series of species that do quite different things with their lower limbs (e.g. hang upside down on occasion: saki monkeys and orang-utans; walk on two legs habitually: humans; run terrestrially for most of the time: baboons and patas monkeys). Again, the structural spectra are determined using multivariate statistical studies of the hip, thigh, ankle, foot, and overall proportions of the lower limb as a whole. It is once again remarkably concordant with the spectrum derived from considerations of function.

The chapter ends with a caveat about the apparent systematic content of the morphological data; it does indeed exist though not as clearly as the functional. This is, again, a story to which we shall return in a later chapter.

What is leaping? Essays into biomechanics

Leaping seems, intuitively, a very obvious behaviour. An animal, whether at rest or during movement, propels itself into the air, moves in a ballistic flight pattern and then lands. Although leaps may be upward, horizontal or downwards, although they may end in running, flight or swimming, as well as in a posture on a solid substrate, and even though some animals progress almost entirely by leaping, in most animals this is a behaviour that is additional to walking and running. The walking and running movements are usually basic; leaping is superimposed upon them. However, exactly because leaping is a simple activity it has been very well studied; some of the background mechanics were elucidated years ago. It is to D'Arcy Thompson (1917) that we may turn for a summary of the older results and an excellent dissertation of his own accomplishments in this area.

Some of Thompson's essays into the biomechanics of leaping are aimed at implications of magnitude for this biological behaviour. He provides that famous description of leaping in a flea. He discusses such matters as the work done in leaping and demonstrates on theoretical grounds that, given equivalent anatomical proportions and certain other assumptions, the height to which different types of animals should be able to leap should be much of a muchness.

This particular interest in the biomechanics of leaping has been further elaborated. Alexander (e.g. 1968, 1974, 1977, 1980; Alexander and Vernon, 1975) provides information about the jump of the locust and gives data showing the high and long jump records for a number of different creatures. He, too, shows that within certain size ranges the heights to which different animals can jump are rather similar. A grasshopper, for instance, can jump about as high as a locust of at least twice its body length; a jerboa which is about as big as a rabbit can jump just about as high as a kangaroo of far greater size. This is explicable on the notion that, if animals have an equal ratio of jumping muscle to total body weight to be moved, and equivalent lever systems for moving the body weight, then they should be able to jump approximately equal absolute heights. Grasshoppers have bodies approximately in proportion to those of locusts, jerboas to kangaroos.

This way of analysing animal movement has been taken very much further nowadays through the use of modern equipment devised for studying animal mechanics.

Thus — the development of force plates and strain gauges, slow-motion cinematography and cineradiography, electromyography and, of course, accurate timing devices now allow far more detailed studies of animal locomotion. And, because leaping is a movement that can be set up in the laboratory with a fixed place of take-off and a regulated landing, these methods have been applied most and taken farthest in studies of leaping.

Alexander himself has carried out many of the most definitive recent investigations, having looked at leaping in the frog, dog and kangaroo (Calow and Alexander, 1973; Alexander, 1974; Alexander and Vernon, 1975). He has computed the stresses involved in leaping, studied changes in the angulations of the various joints, measured external forces against the environment and correlated these functional parameters with the architectural arrangements of the muscles, joints and bones. On the basis of such work he has been able to present succinct discussions of a matter such as the optimal working conditions for a muscle subject to inertial loading and of the notion that a good deal of what happens in muscles during leaping is due to storage of elastic energy in their tendons. In particular he has provided a mathematical model of leaping that gives good agreement with experimentally determined parameters, such as forces, directions, elasticity and power.

Anatomy of vertebrate leaping

Some knowledge of the anatomical implications of leaping in vertebrates has also been available for many years, although it does not stem from the biomechanical concepts mentioned above. Most striking, of course, is relative elongation of the hind limbs, often, especially, of their more distal segments, as compared with regular quadrupedal species. With this goes a more proximal positioning of the hind limb musculature than in non-leaping species. Such adaptations provide both the levers and the engines that are responsible for the enormous increases in acceleration that allow animals to achieve take-off velocity and direction. But adaptations are also found in many other anatomical regions.

Thus, there is frequently a greatly increased and exceedingly muscular tail. This can provide a movable counterweight so that major and important modifications can be made in the leap itself. For, though once an animal has left the ground its centre of gravity should move in a predetermined parabolic path until it touches the ground again, many animals can change their orientation in the air through movements of different bodily parts; rotation of a heavy tail is one major resource. A kangaroo rat can, for instance, completely reverse its position in the air through manipulations of its tail so that it is ready to bounce back in the opposite direction immediately upon landing.

There are sometimes found certain modifications of bodily parts so as to take advantage of the pressure of the air during the flight phase. This is most clearly seen in gliding, where an aerofoil in the form of a patagium or skin fold between limbs and trunk or limbs and tail may be very obvious. But it also seems likely that some leaping animals are able to capitalize on aerodynamic factors, not so much for true gliding, but rather to change by minor degrees what would otherwise be the parabolic path of the centre of gravity. There seems little doubt, for instance, that a spread-eagle position of the limbs and limb folds and a careful positioning of a bushy tail may allow some aerodynamic manipulations of leaps beyond the parabolic limitation inherent in a straightfoward ballistic path.

Finally, many leaping animals have a shortened trunk with a marked distinction between a robust lumbar region and a slighter thoracic segment. Undoubtedly this helps concentrate much of the body weight over and into the enlarged hind limb engines that are so heavily involved in the initial propulsive effort.

Leaping in primates

Most studies of vertebrate locomotion have been carried out on animals other than primates. But because there have evolved among the primates several creatures highly adapted for leaping, there are a few excellent studies of leaping in which various primates are the subject materials. Thus, Hall-Craggs (1965a, b) supplies an analysis of the leap of the lesser bush-baby utilizing high speed cinematography. This allows him to provide estimates of the tensions that different muscular groups are required to attain and the distances over which they need to shorten. Jouffroy and Gasc (1974) present a study of leaping in Allen's bush-baby using cine-radiography, and this permits them to determine the precise relative positions of the bones and joints and the mechanical levers and axes of the process of leaping.

Such studies as these provide fascinating information totally unreachable save through the mechanisms of the technology. For instance, Hall-Craggs (1965a, b) shows that the force developed in the limb applied to the ground is in the region of the tarsometatarsal joint and not at the distal ends of the metatarsal bones as has been assumed by most of the prior workers who depended upon anatomy alone and who did not have slow-motion photography. Jouffroy and Gasc (1974) show that the great majority of the jumps made by Allen's bush-baby are markedly asymmetrical, the body weight being shifted to a single foot, the active foot, immediately at the commencement of the jump. Again, this is something that might not have been easily guessed on the basis of anatomy alone. And a number of other biomechanical studies of leaping in primates are in progress in laboratories around the world.

The general insights provided by current studies of leaping (e.g. a vertical leap in a cercopitheque, Wells, Wood and Tebbetts, 1977) do indeed give us some information that, when used inferentially with knowledge of the behaviour of the animals, allows more detailed assessments. Thus, investigations of many primates from the viewpoint of defining leaping propensities in terms of actual or presumed behaviours and in terms of anatomical adaptations have been carried out over many years. Although almost every primate can and does leap, attention has primarily been focused on those that do it most frequently or to the greatest height or length relative to their body size. They are all prosimians.

The structural variety of prosimian leapers

Long before biomechanical investigations existed, it was known that many primates, especially many prosimians, are excellent leapers. And, for almost as long, investigators have known that these leaping primates possess special anatomical adaptations mechanically efficient for leaping. Some of these structural features, such as elongated lower limbs, parallel those already noted in other vertebrate species capable of leaping. Other architectural features are obviously related to special leaping mechanisms.

Thus, half a century ago, Morton (1924) understood well the importance of elongation of the hind part of the foot for leaping in the bush-baby and the spectral tarsier. Earlier, Volkov (1903, 1904) was aware of the specialized foot structure and understood, if in a less detailed manner, a relationship with leaping in these species. Even more than a century ago, Mivart (1867, 1873) and Fitzinger (cited by Brehm, 1868, and brought to my attention by Jouffroy and Gasc, 1974), although not directly expressing views about functional implications, nevertheless had an understanding of the morphological specializations in the foot that distinguish these leaping prosimians from many others.

Since those early days, however, the most overt expression of the idea that perhaps a large part of the overall form of the animals (not merely specific elements of the lower limb, such as the foot) may be associated with so extreme a form of locomotion as leaping is provided by Napier and Walker (1967). In examining a series of structural features of many primates, these workers assert that certain characteristics appear to be common among a variety of prosimian species of which some — tarsiers, bushbabies and the various indriids — are heavily adapted for leaping, and others — sportive and gentle lemurs — are possibly adapted for lesser degrees of leaping. Napier and Walker (1967) categorize this entire group of prosimians as 'vertical clingers and leapers' believing that, in addition to leaping *per se*, the vertical position of the body at rest when pressed against a branch or trunk is important, as also is the notion that the leaps are often from one vertical support to another (Figs. 6.1, 6.2 and 6.3). Parts of Napier and Walker's descriptions make it clear that they include other prosimians that are able to leap, as tending towards being incipient members of their group of vertical clingers and leapers (table 6.1). And, they provide information about individual descriptive and biometric features of the skeletons of these prosimians that seem to be associated with leaping behaviour. These include such quantitative measures as the intermembral index (longer lower limb in relation to upper limb in leaping forms), brachial index (longer arm in relation to forearm in leaping animals) and crural index (longer thigh in relation to leg in leaping species). And they include

Fig. 6.1. A drawing made from a photograph of vertical clinging (sifaka).

Fig. 6.2. A drawing made from a photograph of leaping (sifaka).

Fig. 6.3. Drawings, made from two series of cinematographic frames, of vertical clinging and leaping (sportive lemur and tarsier).

a number of observational features, such as the possession of a more cylindrical femoral head and a deeper patellar groove (both presumably in association with an increased accentuation on powerful extensor movements of the hip and knee) as diagnostic of anatomical mechanisms for leaping.

Shortly afterwards, two groups of studies take some issue with the conclusions of Napier and Walker (Lessertisseur, 1970, and Lessertisseur and Jouffroy, 1973; Cartmill, 1972). Each investigation notes earlier evidence of major anatomical differences in the lower limb between some of the leaping prosimians (tarsiers and bush-babies) which possess special hip and foot modifications associated with leaping and others (such as indris and sifakas) which do not. Both groups of scholars suggest that even in terms of the more general anatomical features, such as overall limb and body proportions, grouping these two sets of leaping prosimians into a single category is artificial.

Table 6.1

PRIMATE LOCOMOTOR CLASSIFICATION AFTER NAPIER AND NAPIER

Locomotor category	Activity	Genera
1. Vertical clinging and leaping	Leaping in trees and hopping on the ground	*Avahi, Galago, Hapalemur, Lepilemur, Propithecus, Indri, Tarsius*
2. Quadrupedalism	Moving on four limbs Many sub-categories	Most remaining prosimians, all monkeys
3. Brachiation	Arm-swinging Two sub-categories	Apes
4. Bipedalism	Striding	Humans

At about the same time, and working independently from these authors, Oxnard (1973c) and Stern and Oxnard (1973) also note both the very early evidence of Morton, Volkov, Mivart and others, and the more recent difficulties of understanding the arrangements of the various prosimians produced by examining both detailed facets of the structure of the hip and thigh and broader aspects of overall bodily proportions. They show that there is an element of truth in both descriptions. The various individual measures separate leaping forms from non-leaping forms only in part. The degree of separation of the mean values for the various animals by any individual measure is too little; the degree of overlap due to variation of the position of individual groups in different measures is too great; the large number of measures, undoubtedly correlated by different amounts, produce a series of confusingly different arrangements of the animals (tables 6.2 and 6.3).

But exactly because of these problems, Oxnard and co-workers take the investigations further by assessing measures of several different anatomical regions in the various prosimians, using the multivariate statistical approach (canonical variate analysis) in order to demonstrate the overall picture. These new studies show that, whether one looks at the pelvic or shoulder girdle, the proportions of the lower or upper limb or even the proportions of the entire limbs, trunk and head, those prosimians which are capable of extreme leaping are divided into two clear morphological modes (Figs. 6.4 and 6.5). One of these includes the tarsiers and bush-babies, the other all three indriid genera, sifakas, indris and avahis; each of these groups lies on opposite sides of a centrally located cluster of species that includes not only certain prosimians but also many Old and New

Table 6.2

NON-VERTICAL CLINGING AND LEAPING PRIMATES INCLUDED BY DEFINING FEATURES SO BROADLY AS TO INCLUDE ALL VERTICAL CLINGING AND LEAPING PROSIMIANS

Anatomical features defining all vertical clingers and leapers	Other genera also with these features			
	Lemur	*Daubentonia*	*Callimico*	*Loris*
Intermembral index < 72	x	x	x	
Brachial index > 99	x	x	x	x
Hand length index > 27	x	x	x	
Finger length index > 54			x	x
Lower limb length index > 125		x		x
Thumb length index > 70				x

Table 6.3

VERTICAL CLINGING PROSIMIANS EXCLUDED BY DEFINING FEATURES SO AS TO NARROWLY EXCLUDE ALL NON-VERTICAL CLINING AND LEAPING PRIMATES

Anatomical feature	'Vertical Clingers' in which these features may be absent			
	Lepilemur	*Galago*	*Tarsius*	*Indri*
Intermembral index < 68		x		
Brachial index > 117	x	x		x
Hand length index > 33	x	x		x
Finger length index > 64	x	x	x	x

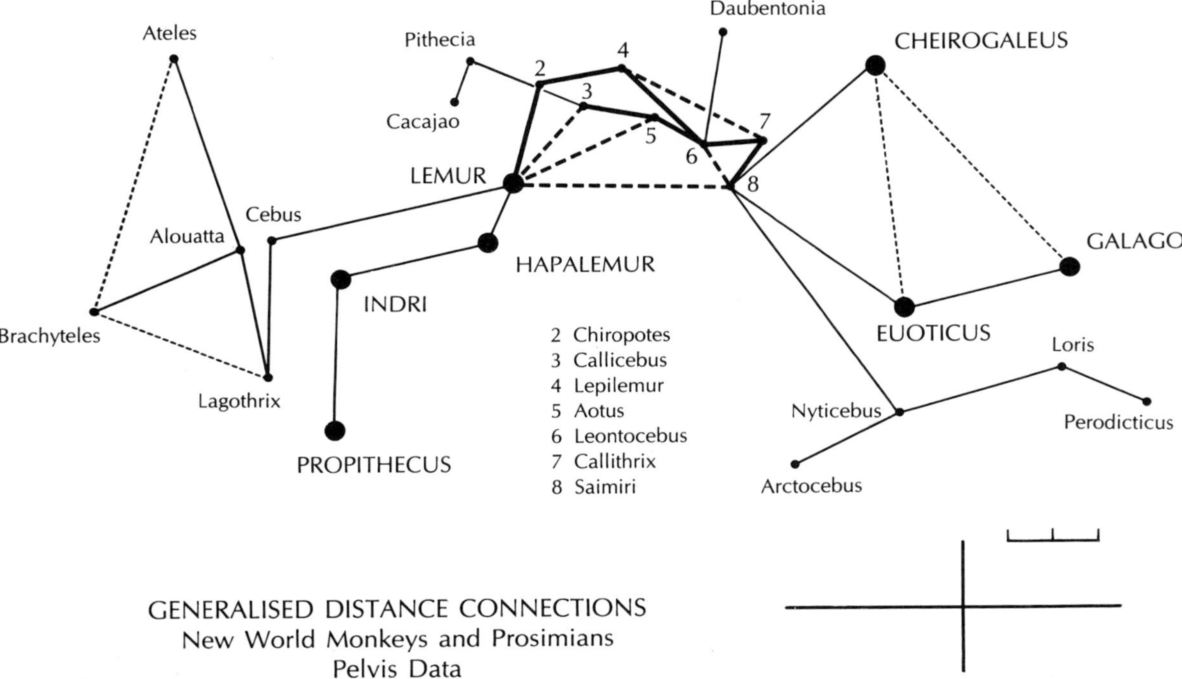

Fig. 6.4. The separations achieved by multivariate statistical studies of dimensions of the pelvis. The minimum and near minimum spanning tree of generalized distance connections is laid out within the framework provided by the plot of the first two discriminant axes. The marker is two standard deviation units but of course is only applicable to the lengths of lines connecting genera in the tree. Other distances in the diagram cannot be measured by the marker because of the multi-dimensional nature of the information.

Those leaping genera that are represented in the study are clearly divided into two groups with indriids (*Propithecus* and *Indri*) and galagines (*Galago* and *Euoticus*) lying each on opposite sides of a mixed group of prosimian and New World genera that have in common only that they are generally quadrupedal. Similar separations, but less marked, exist in studies of the upper limb girdle (not shown).

World monkeys all having in common that, though well capable of leaping, they are generally more quadrupedal in locomotor habit.

Further, these new investigations are sensitive enough to suggest that several other prosimian genera, believed to practise leaping to some intermediate degree, are also intermediate in relation to their morphological adaptations (for example, in the different studies of Oxnard, mouse lemurs and dwarf lemurs seem to be somewhat intermediate

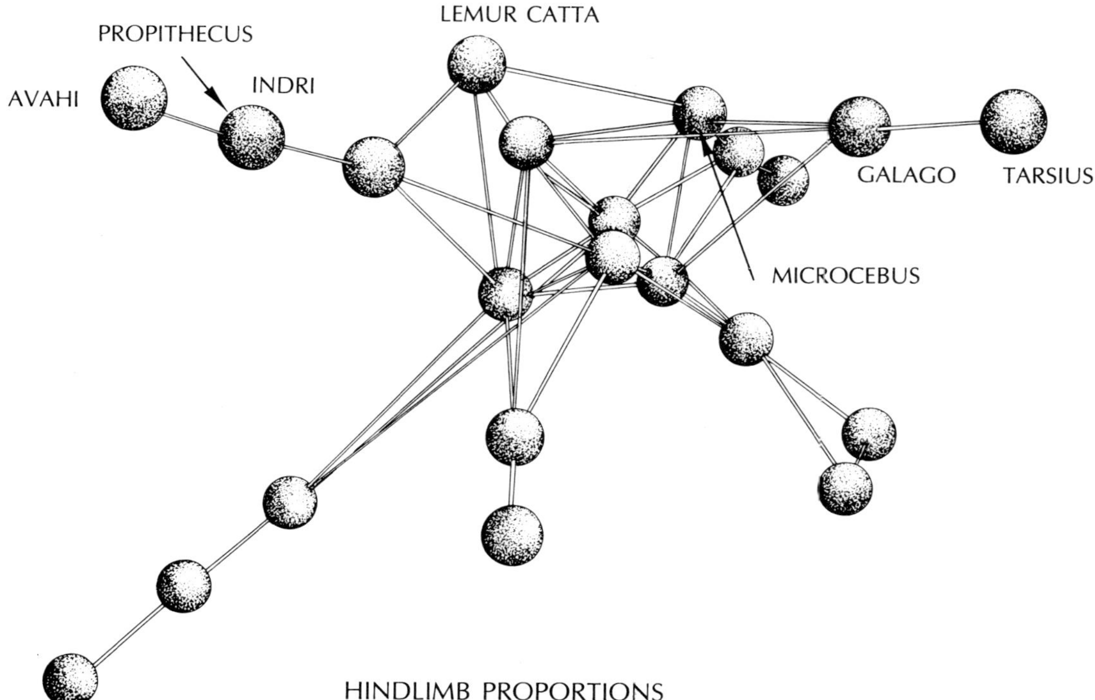

Fig. 6.5. The separations achieved by a similar study of dimensions of the entire lower limb. In this case the minimum spanning tree of generalized distance connections (together with a small number of other links to maintain stability of the model) is drawn within the three-dimensional space of the first three axes. The general scale of the diagram is thirty standard deviation units from side to side.

As in the previous figure, those leaping genera of prosimians that are represented in the study are divided into two groups, indriids on the one hand and bush-babies and tarsiers on the other, each separated by species (in the centre of the model, unlabelled) that are generally quadrupedal in locomotor habit. Similar separations, but less marked, exist for the dimensions of the upper limb (not shown).

towards the galagine-tarsier mode, and ring-tailed lemurs and mongoose lemurs seem to show some tendencies towards the indriid group, table 6.4). However, those investigations still leave as equivocal the morphological placement of sportive and gentle lemurs, genera which seem also to be reasonably able leapers. And Oxnard (1973c) suggests that yet other morphological modes adapted for leaping may be demonstrated on further investigation.

Is prosimian leaping even more complex?

Napier and Walker (1967) made it clear, in their original study, that the notion of a single group of 'vertical clingers and leapers' was a preliminary idea based upon what they knew at that time. And though in 1974 Walker still presented a more extended description of that same idea, it is highly likely that the results described above were not yet available to him. Yet even now, although there is no doubt whatsoever about the existence of two major morphological adaptations associated with different forms of leaping in prosimians, the work of Oxnard and colleagues must be further tested. For these studies leave a number of fascinating questions unanswered.

Are there yet other such adaptive modes for leaping (Oxnard, 1973c, leaves that matter open)? Are there really variously intermediately adapted species? And are there any relationships that speak to locomotor differences *within* each of the modes? Within each of these modes not all the species of each genus move in an identical way; some leap considerably more than do others.

The need for further testing of this original work of Oxnard and colleagues also relates to some limitations which, they themselves point out, stem from paucity of materials in those original studies.

First, although in each separate investigation

(whether of limb girdles, limb segments, whole limbs or overall bodily proportions) some representatives of genera from each morphological mode are analysed, in no single study are all genera represented.

For instance, for the shoulder girdle, information is presented for only one indriid form: the sifaka. For the pelvic girdle there are data for both sifakas and indris. In both shoulder and pelvic studies, regular and needle-nailed bush-babies alone represent what is subsequently discovered to be a mode including all bush-babies and tarsiers. It is only in the investigations of the limbs as a whole and of overall bodily proportions that data for tarsiers is included. And it is only in the limb and overall bodily proportion studies that avahis are represented and found to lie with indris and sifakas. Thus the picture presented by Oxnard and colleagues is necessarily built up in a patchwork fashion by assessing the results of several studies using somewhat different materials and with notable gaps in those materials.

The *second* limitation recognized by Oxnard and colleagues lies in the small number of specimens representing some of the genera. Thus, although the position of the bush-baby can be defined relatively accurately because data from many specimens are available, knowledge of the relationship of some other genera (e.g. dwarf and mouse lemurs) is restricted by data from only two or three specimens.

Third, although Oxnard and colleagues are well aware that in a group like the bush-babies, different species move in different ways, they nevertheless are forced to lump all such species into the single group. A similar lumping is necessary for lemurs. Such lumping simplifies and generalizes the true behavioural situation; thus increasing the representative materials allows better investigation at specific and subspecific levels.

A *fourth* qualification of that work is that many of the original data relate to studies of locomotor adaptations in the anthropoidea. The prosimians are included because of the possibility, since proven, that parallels among them exist that help elucidate anthropoid locomotor adaptations. It is only retrospectively that the data can be seen to be useful for studying prosimian leaping in its own right, as it were.

Because of all these qualifications a whole new series of investigations is needed to study leaping and its morphological associations in the prosimians.

New studies of leaping anatomy in prosimians

A new study (McArdle, 1978, 1981) of the muscles, joints and bones of the hip and thigh in prosimians examines, through dissection and observation, 55 cadavers and whole skeletons and 289 femoral and pelvic sets representing a total of 20 prosimian species. On this basis, osteological features that seem to be related to the soft tissue contrasts are identified and examined. And from these in turn a series of 15 quantitative osteological variables is defined and measured. Study of these various qualitative and quantitative parameters one at a time demonstrates a number of features common to most of the leaping species.

For instance, among the important soft tissue elements are the hamstring muscles, the extensor muscles of the hip which also flex the knee. These are relatively smaller in most of the leaping species where powerful hip extension in required to occur together with more powerful knee *extension* than in more regular quadrupedal animals.

One of the related osteological features is the ischium, the origin of the hamstring muscles. This appears to be shorter and stubbier in leaping species. And when the ischium is characterized through its length relative to the length of the pelvis, its shortness is confirmed quantitatively in leaping as compared with quadrupedal species.

Again, a new study has been undertaken (Jouffroy and Lessertisseur, 1979) of individual overall proportions of the limbs of prosimians. This study examines a total of 26 dimensions of the upper and lower limbs in 161 specimens representing 12 species of prosimians. In this case, of course, the dimensions are taken not because they relate to anatomical features of the soft tissues but simply because they are there. The study thus replicates, to a degree, many prior studies of overall proportions (such as those of Schultz, Napier, Erikson, Mollison and so on) but is different because of the more detailed nature of the measurements, the larger number of features measured and the greater number of specimens and species represented. The result of examining these features one by one also demonstrates that there are many characteristics that are common to most of the leaping prosimians. For instance, prosimians that leap have relatively longer lower limb segments (thigh and leg) than those that do not (Jouffroy and Lessertisseur, 1979).

However, notwithstanding the more detailed

nature of the anatomical findings provided by these two new univariate investigations, they do not show the clear evidence of two separate adaptive modes that is found in the earlier work of Oxnard and colleagues (table 6.4). In the main this is, once again, because the sensitivity of any single anatomical feature in producing separations is not great enough. An attempt to assess all the information together by eye does not readily distinguish the various genera from one from another because there are many different anatomical features and a great deal of confusing overlap.

Accordingly, then, the data resulting from these two new studies have been used in a series of new morphometric investigations. These studies eliminate most of the self-criticisms of Oxnard's earlier study listed above. Sample sizes, although still necessarily small because these animals are quite rare, are considerably greater than in the prior studies; almost every leaping prosimian is represented; where possible, genera are divided into the species or species groups that accord with the more detailed behavioural information about leaping; and, finally, the studies are aimed at understanding both very detailed anatomical arrangements within a localized region important for leaping, the hip and thigh, and at understanding overall proportions of entire upper and lower limbs.

Morphometric studies of overall limb form in prosimians

The first of the new investigations, carried out upon data supplied by Françoise Jouffroy and the late Jacques Lessertisseur, include a large number of prosimian species, and the numbers of dimensions studied are also far larger than in prior studies (summarized briefly in table 6.5 and figured diagrammatically in Fig. 6.6). The full materials and methods are provided in Oxnard, German, Jouffroy and Lessertisseur (1981).

First, canonical variate analysis of the raw measurements displays most of the significant separation among the genera in a single discriminant axis. Not surprisingly, these separations seem to relate to little more than differences in the overall sizes of the animals: most of the indriids are large, most of the bush-babies and the tarsier are small (Fig. 6.7).

Second, the measurements were rearranged and analysed as a set of nonsense indices: indices concocted on no good biological basis and using deliberately curious combinations of measurements, such as the ratio between the length of the big toe and the radius, or the length of the palm of the hand compared with that of the tibia. This analysis tests the notion that it does indeed matter what indices are studied. As with the analysis of measurements, this procedure produces only a low level of separation of the species, almost entirely within a single discriminant axis and apparently related to little more than the differences in sizes of the species (Fig. 6.8).

Third. In complete contrast are the results of analysing the series of dimensions outlined in tables 5.3 and 6.5 and figured diagrammatically in Figs. 5.23 and 6.6.

Thus, study of the entire suite of dimensions of both limbs provides a very clear result. It is (Fig. 6.9,) that lying around a centrally located group comprising lemurs, sportive lemurs and gentle lemurs, are three outlying groups, a first containing the indriids with indris themselves as the most extreme

Table 6.4

PROSIMIAN LEAPING AFTER STERN AND OXNARD

Locomotor category	Activity	Genera
Leaping indriid type	Leaping with stretched body posture	*Avahi, Propithecus, Indri*
Incipient indriid type leaping	Tending towards outstretched postures	*Lemur catta, Lemur mongoz, Lepilemur*
Leaping galagine type	Leaping in curled posture	*Galago, Euoticus, Tarsius*
Incipient galagine type leaping	Tending towards curled posture	*Microcebus, Hapalemur*

Table 6.5

DIMENSIONS AND ANIMALS IN MORPHOMETRICS OF PROSIMIAN LOWER LIMB

Anatomical features		
Measures	*Lower limb ratios*	*Foot ratios*
Length of femur	Femur to rest of limb	Tarsus to rest of foot
Length of tibia	Tibia to rest of limb	Each metatarsal to rest of foot
Length of foot	Foot to rest of limb	Each toe to rest of foot
Length of tarsus	Tarsals to rest of limb	Cross comparisons among metatarsals plus toes
Length of each metatarsal	Metatarsals to rest of limb	
Length of each toe	Toes to rest of limb	

Groups of animals and numbers of specimens studied

Hapalemur, 7; *Lemur*, 46; *Lepilemur*, 9; *Cheirogaleus major*, 6; *Cheirogaleus medius*, 3; *Avahi*, 6; *Indri*, 6; *Propithecus*, 7; *Galago*, 39; *Euoticus*, 9; *Tarsius*, 8

Fig. 6.6. A lemur in full flight demonstrates the skeletal elements (shown in black) measured by Jouffroy and Lessertisseur and used in this analysis.

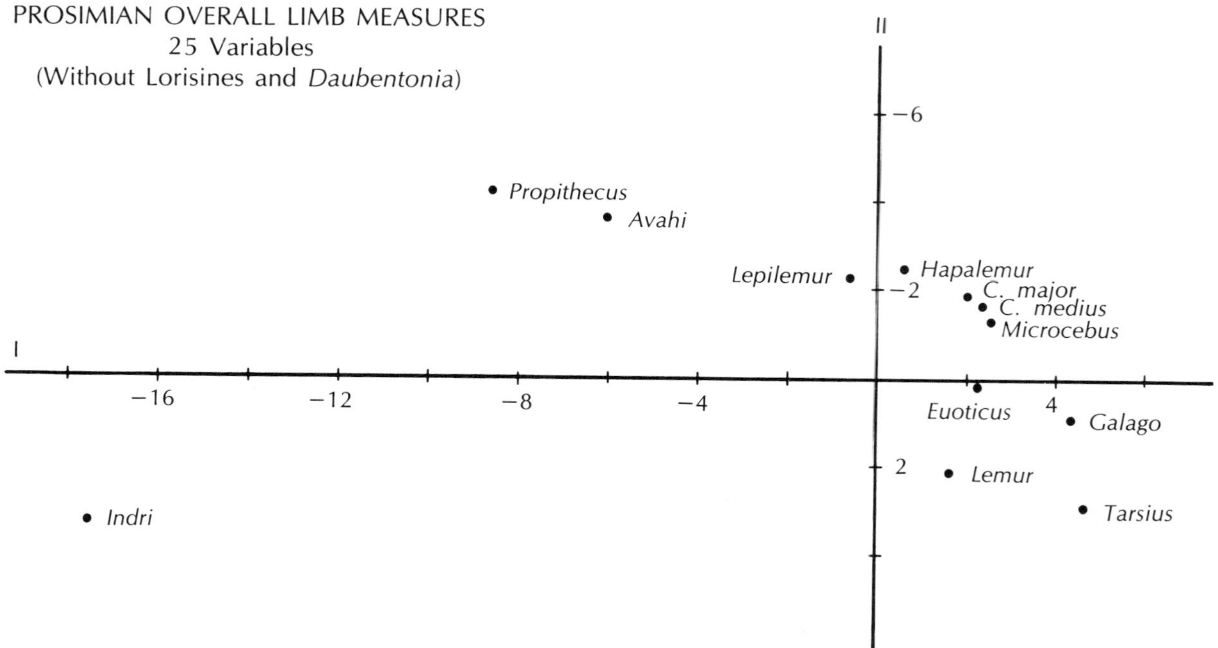

Fig. 6.7. Analysis of measurements of the limbs of leaping prosimians provides an arrangement of genera that is mostly contained with a single discriminant axis. This axis seems related mostly to the simple size differences among the species.

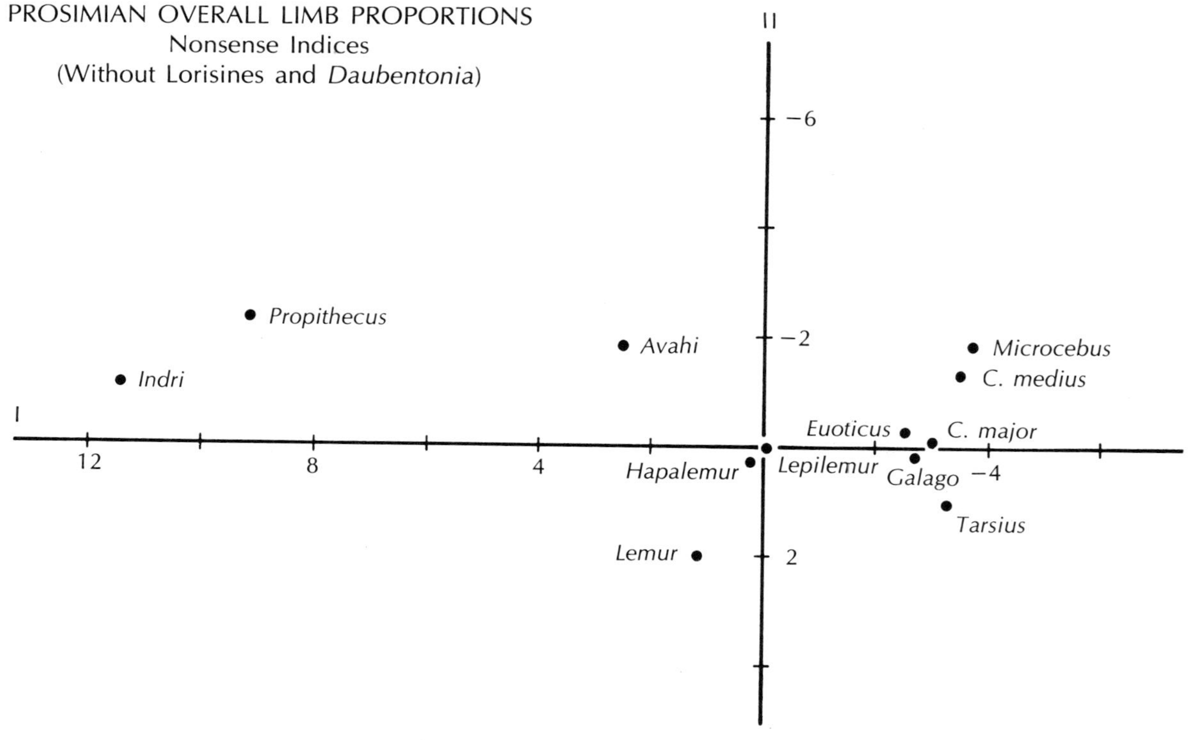

Fig. 6.8. Analysis of a set of nonsense indices provides a picture very little different from Fig. 6.7.

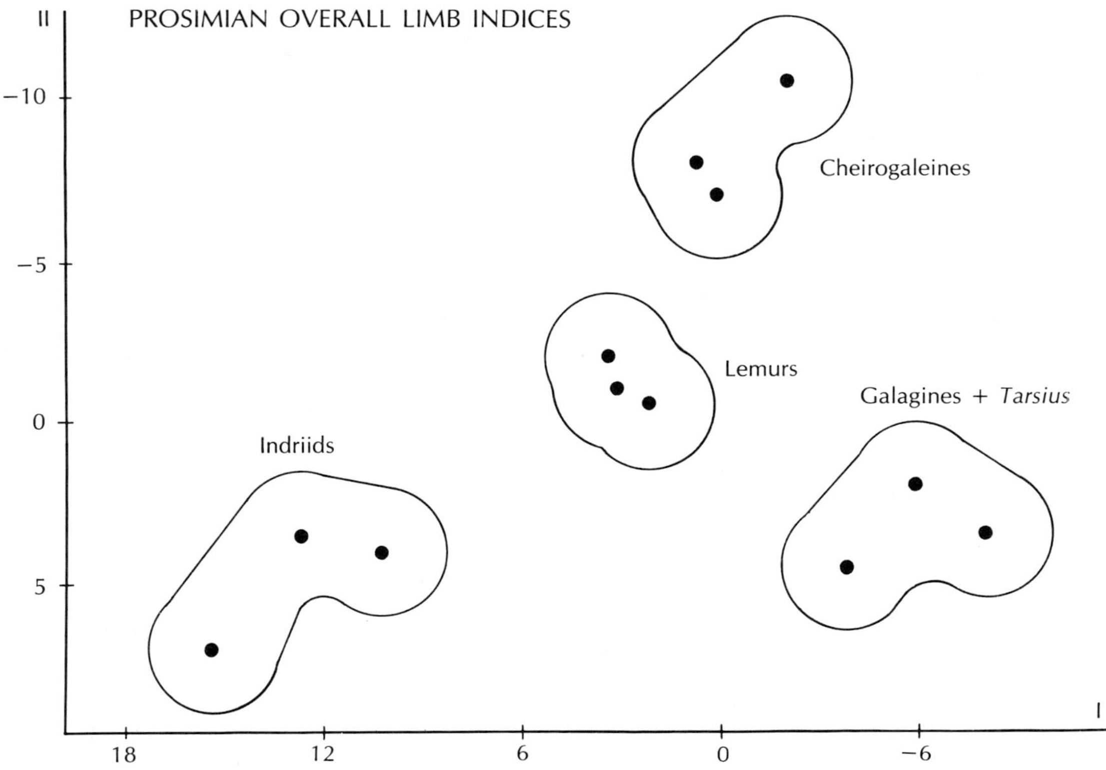

Fig. 6.9. Analysis of the entire suite of upper and lower limb indices. The contours drawn around individual means of groups are one standard deviation unit in radius.
The three peripheral groups of genera are markedly separated from a central group of lemurs.

species, a second containing bush-babies and tarsiers with tarsiers as the most extreme and a third containing greater and lesser dwarf lemurs and mouse lemurs with mouse lemurs as the most extreme. Whether we view these groups through their mean positions (Fig. 6.10) or through the placement of the individual specimens (Fig. 6.11) the same information obtains. Although not unsuspected in prior studies (see Oxnard, 1973c), this confirms the existence of three groups and is highly suggestive of a fourth.

Before we can discuss the functional implications of the results we must also examine those stemming from the study of detailed osteometric dimensions of the hip and thigh.

Morphometric studies of the prosimian hip and thigh

These investigations, carried out upon the data supplied by McArdle (1978) include an even greaer number of prosimian species, and the number of dimensions is large enough to take in a great deal of the morphological complexity of the hip and thigh (summarized briefly in table 6.6). Fig. 6.12 demonstrates in pictorial form the points from among which the various measures were taken. Again, the full materials and methods are provided in an original publication (Oxnard, German and McArdle, 1981).

The canonical variate analysis of the data expressed as measurements provides rather small separations among the species, and separations, moreover, that seem to point to nothing much more exciting than the size difference among the species. This is the same result as obtains from the analysis of measures of limbs.

But analysis of the data as a series of ratios (the dimensionless numbers of Alexander, 1977) chosen to reflect various functional notions about the positions of major bone struts, of muscle attachments and of joint facets provides far greater separations among the species. As with the examination of the combination of upper and lower limbs of these animals, the total picture comprises

Table 6.6

DIMENSIONS AND ANIMALS IN MORPHOMETRICS OF PROSIMIAN HIP AND THIGH

Anatomical features

Verbal definition	*Verbal definition*
Relative length of anterior ilium: position of hip flexors	Length of anterior ilium relative to pubis
Relative length of posterior ilium: hip extensors	Angle between pubis and ischium
Relative length of ischium: ham-string muscles	Angle between ischium and ilium
Relative length of pubis: adductor musculature	Position of lesser trochanter on femur: hip flexors
Relative length of pelvis in front of sacrum	Position of third trochanter on femur: hip extensors
Relative length of pubic symphysis	Ratio of lateral and medial femoral lengths: extensors and abductors
Length of ischium relative to ilium	Shape of femoral head
Length of ischium relative to pubis	

Groups of animals and numbers of specimens studied

Hapalemur, 14; *Lemur catta*, 20; *Lemur variegatus*, 15; *Lemur macaco*, 9; *Lemur mongoz*, 18; *Lemur fulvus*, 27; *Lepilemur*, 16; *Cheirogaleus major*, 7; *Cheirogaleus medius*, 7; *Microcebus*, 18; *Propithecus verrauxi*, 14; *Propithecus diadema*, 7; *Avahi*, 15; *Indri*, 7; *Galago crassicaudatus*, 32; *Galago demidovii*, 10; *Galago senegalensis*, 26; *Galago alleni*, 6; *Euoticus*, 14; *Tarsius*, 7

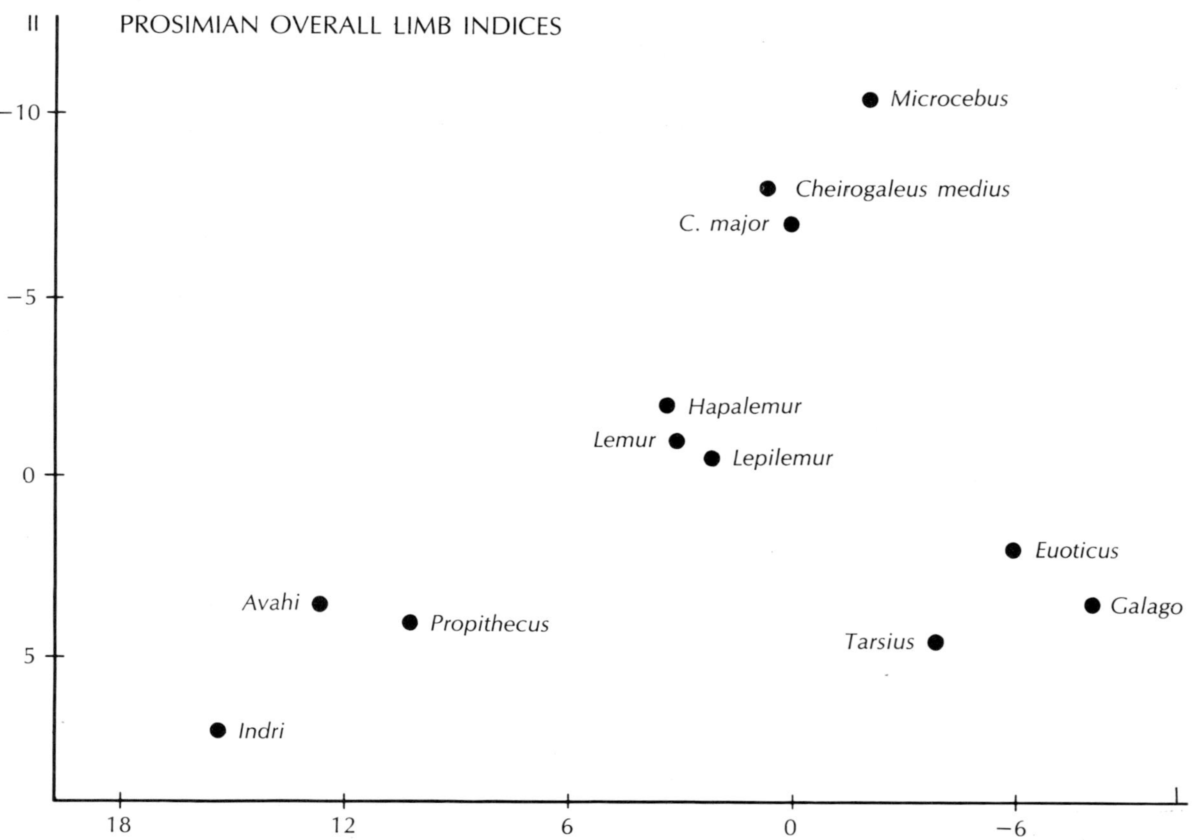

Fig. 6.10. The same data as Fig. 6.9 but with each individual species or genus identified.

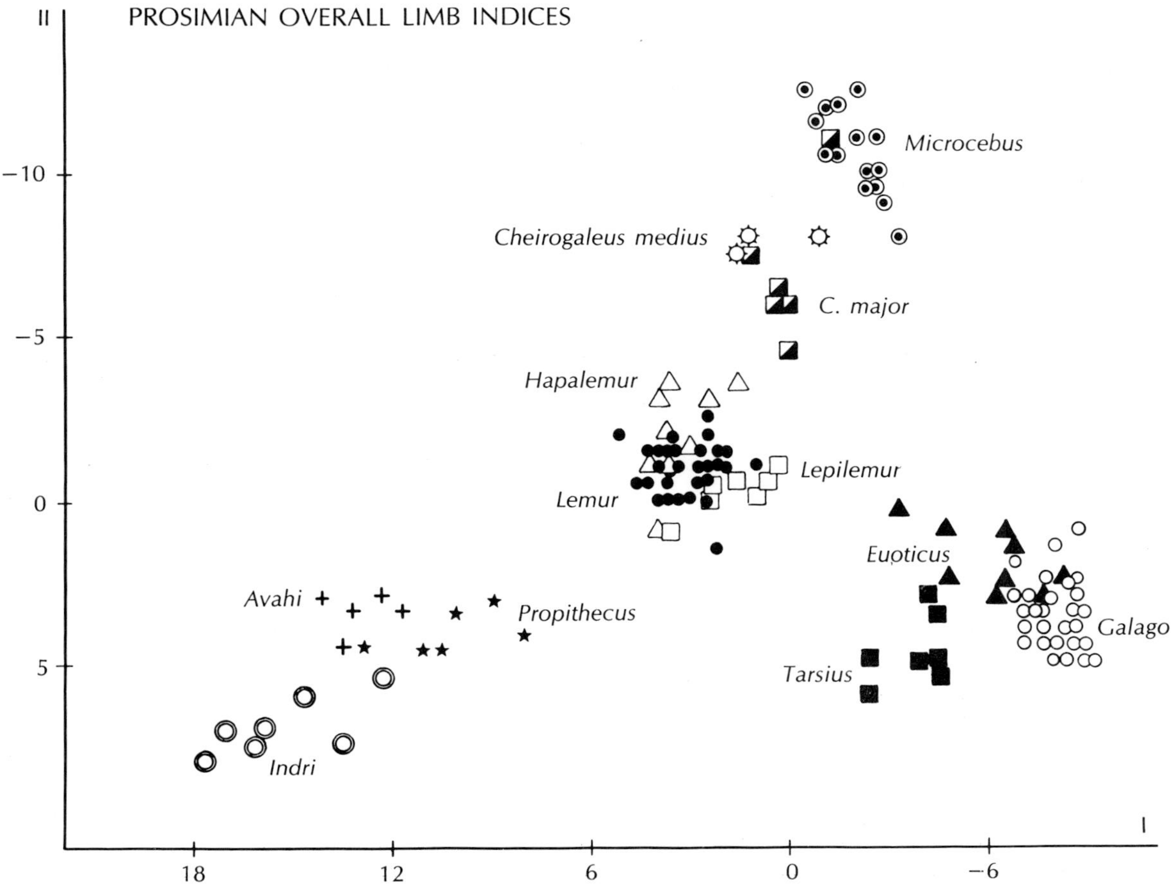

Fig. 6.11. The same analysis as Figs. 6.9 and 6.10 but with the positions of each individual specimen located. This demonstrates that the various groups are indeed tightly arranged with only a little overlap of specimens.

a relatively centrally lying group containing lemurs, sportive and gentle lemurs, with three outlying groups. One contains the various indriids, a second the bush-babies and tarsiers, a third the dwarf and mouse lemurs (Fig. 6.13).

As a further refinement, however, this new picture demonstrates something only hinted at in the prior results from the study of overall proportions: that is, that the species in the various groups are arranged in a special way. Of the central group, the lemurs lie most centrally of all and the sportive lemurs lie towards the outlying group comprising indris, sifakas and avahis. In contrast, the gentle lemurs, also lying centrally overall, tend towards both the other two groups, the bush-babies and tarsiers, and the dwarf and mouse lemurs, respectively.

Moreover, within each of the peripheral groups, although study of overall proportions only *suggests* that each is linearly arrayed, study of the hip and thigh *indicates strongly* that the arrays are linear. Within the bush-babies and tarsiers the linear arrangement is from the fat-tailed bush-baby lying nearest the gentle lemur in the central group, through Allen's bush-babies and needle-nailed bush-babies lying more distally, to tarsiers and Senegalensis bush-babies lying most distally of all. Similarly, the group of dwarf and mouse lemurs is linearly arranged with the greater dwarf lemurs lying nearest to the gentle lemurs in the central group, with the lesser dwarf lemurs more distally and with the mouse lemurs in the most extreme position of all (Fig. 6.14).

A diversion into testing

One of the most important considerations in multivariate statistical studies can relate to the question of outlying values in the data. We have always

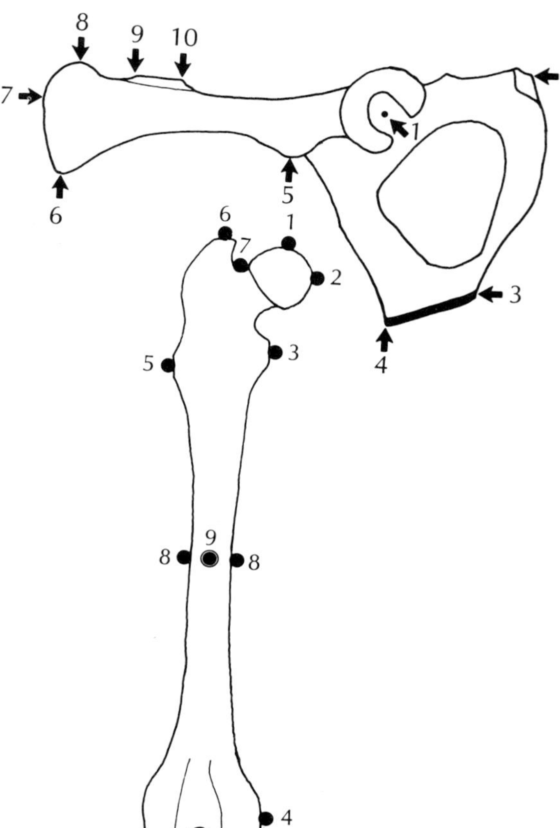

Fig. 6.12. Outline drawings of a pelvis and femur indicating the various points from which measurements were taken in obtaining the basic data of McArdle.

carried out a number of special tests to seek the presence of such outliers (see Chapter 3 and table 3.4). Such tests obviously include scanning the univariate raw data for outliers. Such tests also include plotting pairs of raw variables because, though the univariate scan may not detect outliers, the combination of two variables may do so very well in cases in which the association between two values for a single specimen is the reverse of that usually found. However, even this may not detect all outliers, for it is entirely possible that it is only in the multivariate combination of variables that certain outliers may be found.

Two examples of this occurred in the data on the hip and thigh, and the outlying specimens were detected by principal components analysis. Thus, Figs. 6.15 and 6.16 demonstrate principal components analyses in which the first two factors place particular specimens of bush-baby (G) and

Fig. 6.13. The full analysis of the data for the hip and thigh of all the prosimians available shows three very clear groups, the indriids, the cheirogaleines and the bush-babies plus tarsiers. The lemurs together with sportive lemurs (*Lepilemur*) and gentle lemurs (*Hapalemur*), all represented by single unlabelled dots, lie relatively centrally. The curious slowly moving lorisines also form a separate group well away from the leaping species.

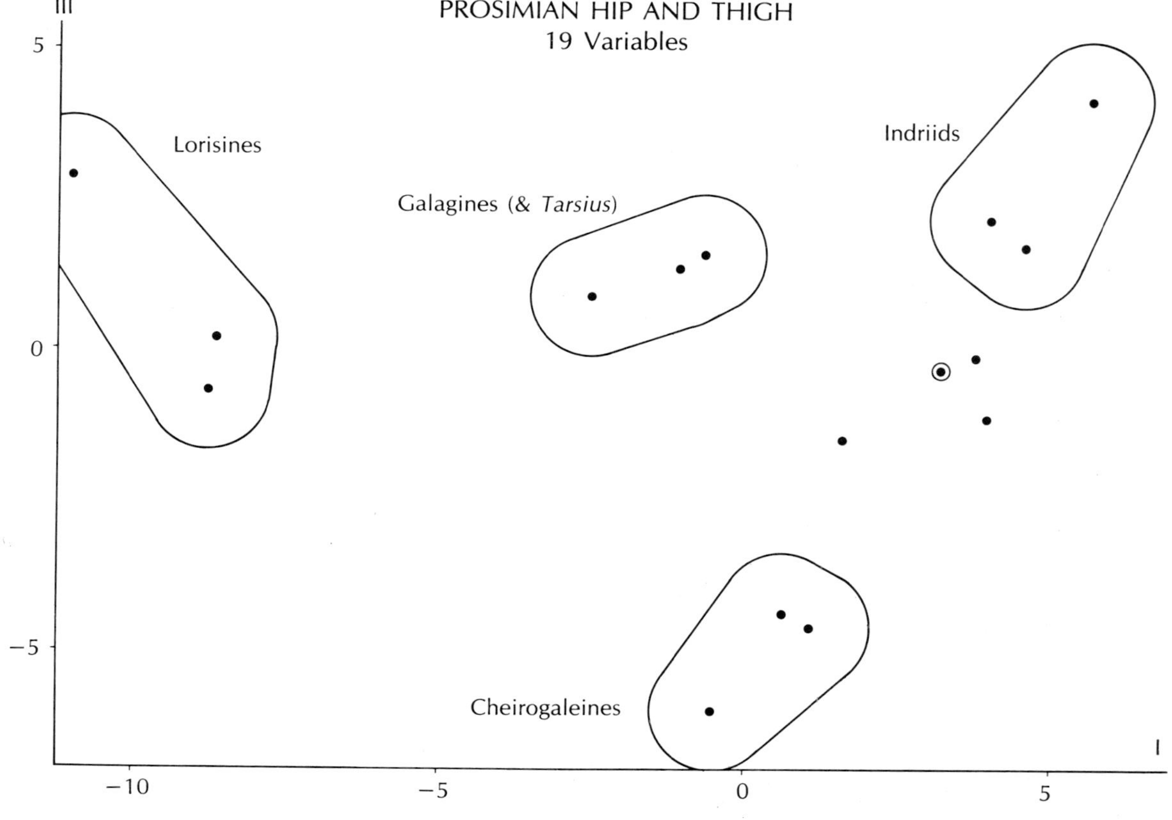

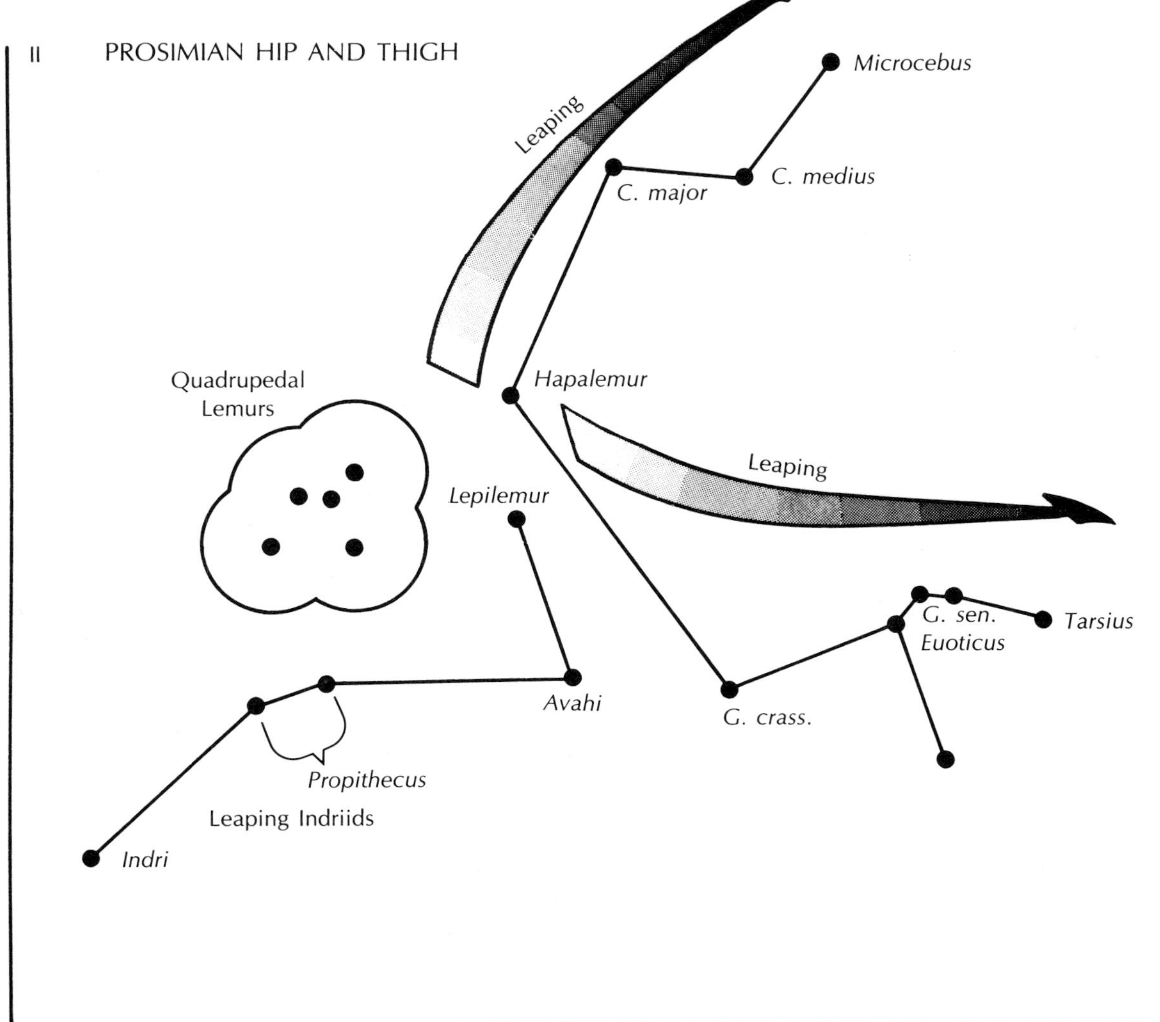

Fig. 6.14. Analysis of the hip and thigh for leaping forms alone. This figure shows that the separations of the previous figure include some minimum links in higher dimensions that make the nearest links of sportive lemurs with the indriids, and of gentle lemurs with the rest. It also shows that the spectrum of morphology in the cheirogaleine and bush-baby groups follows the spectrum of leaping in those groups.

mouse lemur (M) squarely in the centre of the distributions of the specimens. However, further scanning of the other principal component axes revealed one (axis 6, Fig. 6.17) in which these two specimens were markedly outlying. They were subsequently discovered to be errors of recording that involved all of the variables.

Yet another example of the importance of testing appeared in the study of the overall proportions of the limbs. In this case the specimen (*Indri*) was noticed to be allied with the wrong group (*Propithecus*) when a plot was made of the first two canonical variates, specimen by specimen (compare Fig. 6.18 with Fig. 6.11).

In this case it was subsequently found, by going back over the original specimens, that this particular specimen was markedly sub-adult. A sub-adult indri is, of course, smaller than an adult; it is small enough, in fact, to be similar in size to adult *Propithecus*. This immediately suggested to us that the difference between *Indri* and *Propithecus* in this analysis might be merely one of size only. It would have been easy to accept such an assessment.

However, closer study revealed that though this

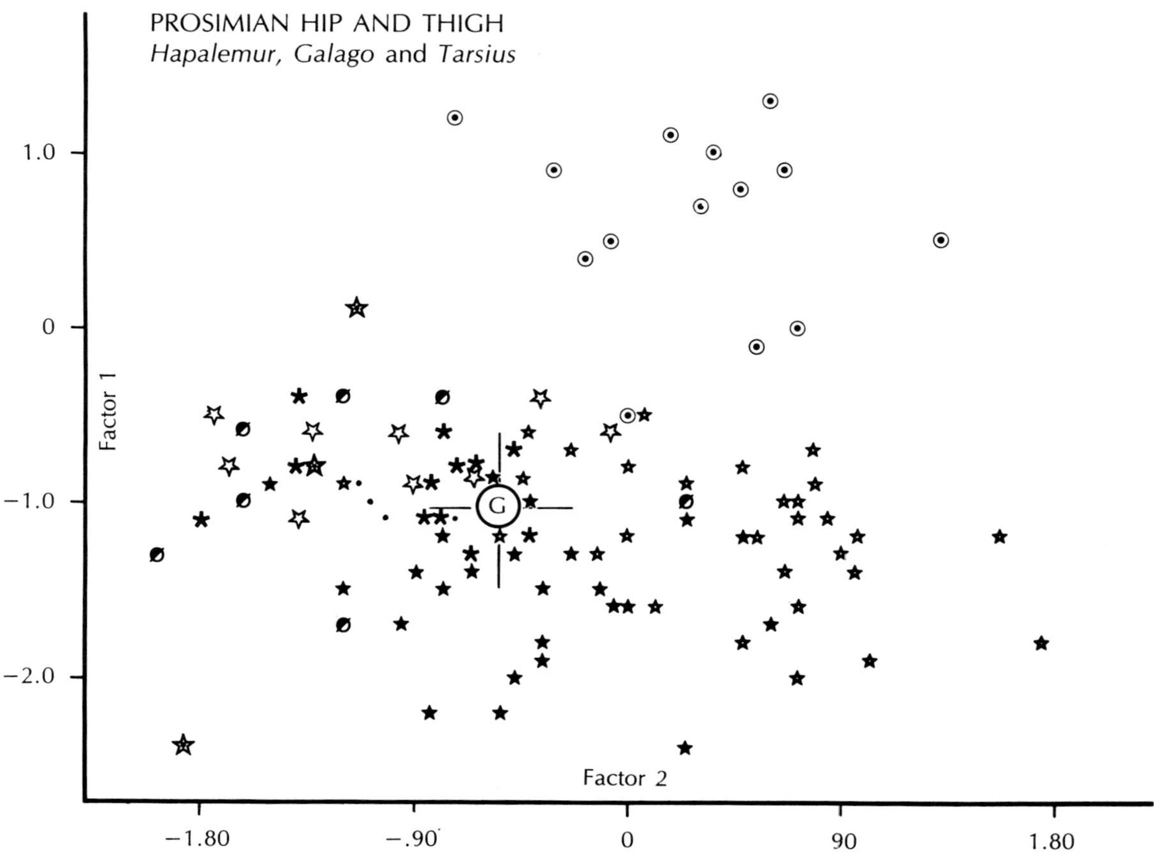

Fig. 6.15. Principal component analysis of the hip and thigh data for the genera *Hapalemur*, *Galago* and *Tarsius*, plot of the first two axes. One particular specimen of *Galago* occupies an entirely appropriate position near the middle of the plot.

Fig. 6.16. Principal component analysis of the hip and thigh data for the genus *Hapalemur* and the cheirogaleines, plot of the first two axes. A particular specimen of *Microcebus* lies near the middle of the plot.

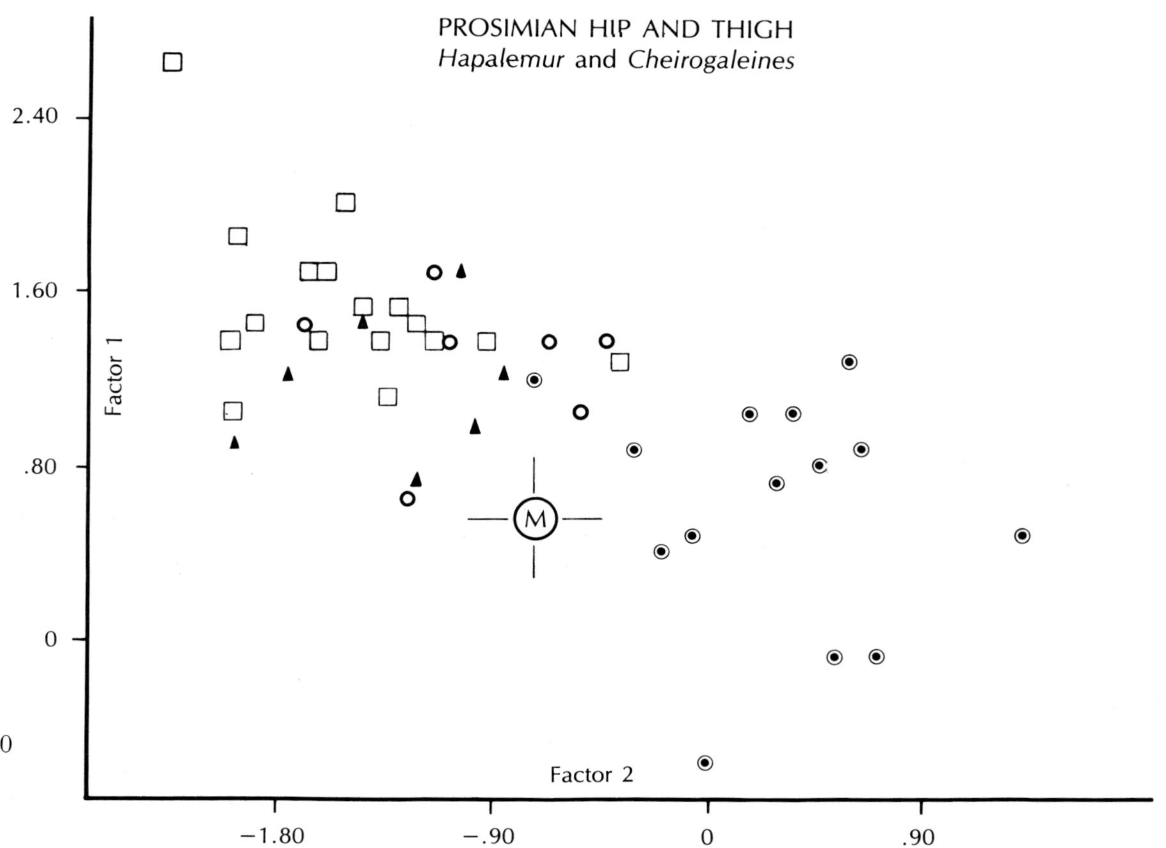

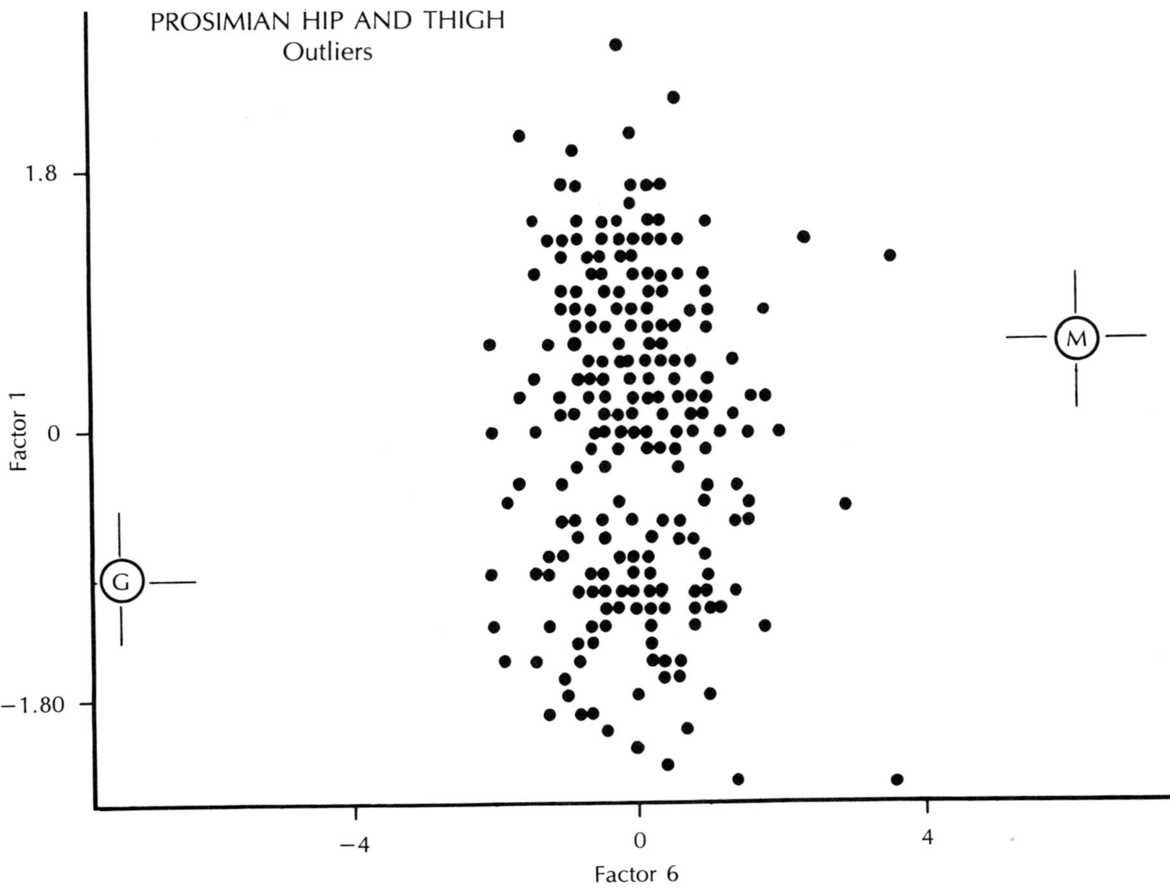

Fig. 6.17. Principal component analysis of the hip and thigh data for the combination of genera in Figs. 6.15 and 6.16. In this case, the first few axes do not separate any special specimens, but the sixth axis, plotted here, separates in different directions the two particular specimens of *Galago* and *Microcebus* noted in the two previous figures. These two specimens proved to be outliers due to problems in the collection of the data (see text).

specimen of *Indri* was similar to *Propithecus* in the two axes plotted, it still lay seven generalized distance units away from any individual *Propithecus* in the overall analysis. Accordingly, a simple size relationship between the species was not the cause of the finding. The specimen was eliminated from the data because we are studying only adult forms. But, *obiter dictum*, it is entirely possible that some fascinating shape-changes with size occur during development in these species. This should certainly be further studied should the appropriately aged materials ever become available.

A diversion into size

With the hindsight of the earlier discussion, the functional implications of these results seem obvious. Surely it relates to leaping, and by degrees. However, before we can look at that matter in detail there is another that must be settled. Analysis of measurements alone arranges the animals in order of size. How could it be otherwise? Whatever are the differences in structure of these various animals, overall size is the biggest difference.

In the studies just outlined no special attempts have been made to eliminate size through regression adjustment, multivariate allometric correction or any other such method. And arguments have been put forward elsewhere (Oxnard, 1979a) as to why this is not always a good thing to attempt. Thus, Pedley (1977) provides many examples of the part that size plays in animal movement; McMahon (1975) presents a particular example of the importance for quadrupedal locomotion of retaining information about size. We have here used dimensionless numbers such as ratios which allow for the non-allometric element

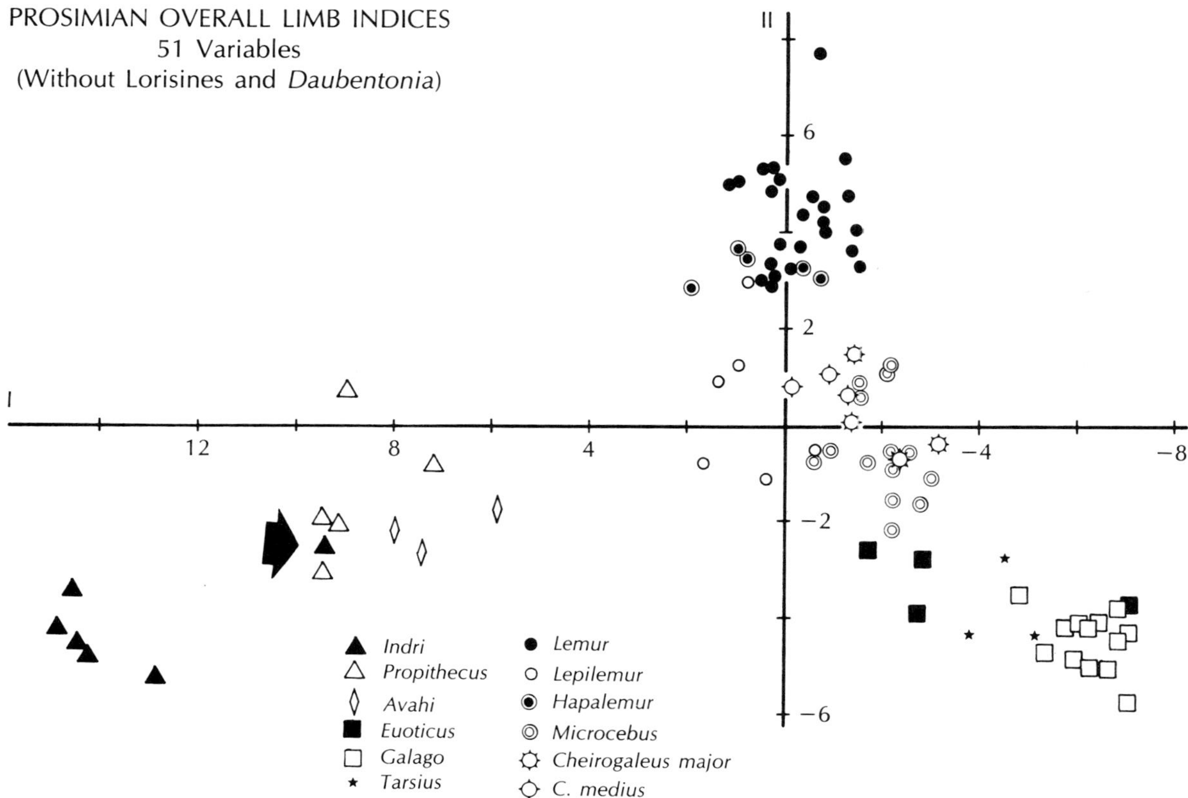

Fig. 6.18. Canonical variate analysis when first carried out upon the data of the prosimian overall limb proportions placed a single specimen of *Indri* as close to the group of *Propithecus*, as shown here. Contrast with Fig. 6.11. This specimen proved to be an outlier; the measurements on it were taken correctly but re-examination showed that it was a markedly immature individual that had somehow crept into the study (see text).

of size in an approximate manner; all allometric components of size, however, have been retained.

But the main reason for using dimensionless numbers such as ratios is to provide data that are of biomechanical import (Alexander, 1971). Because statistical significance is not at issue here, the problem relating to the non-normal distribution of some ratios (Oxnard, 1973a; Atchley, Gaskins and Anderson, 1976) is not of importance. Yet we must be careful to see whether the results of analysing ratios provide anything other than a measure of a simple size difference. This is examined in Figs. 6.19, 6.20 and 6.21.

We already know that the first discriminant axis resulting from analysis of measurements is closely tied to size. This is shown by Fig. 6.19 in which a regression between size and value in the first axis for the study of measurements confirms the close relationship between the two.

That the first axis resulting from the study of indices is not size is clear from Fig. 6.20.

However, the picture is yet more complicated. Study of the equivalent regression between size and the first axis of the study of indices can be viewed in several ways. First, Fig. 6.21 suggests that the overall regression line shows *no* relationship between size and position in the first axis. Yet in frame 1, Fig. 6.21, ellipses drawn around particular species suggest that there is a size relationship. In frame 2, other ellipses around other groups of the species deny any relationship with size at all.

The final evaluation here must be that size has little to do with the overall arrangement of the species in the first axis, but the final evaluation must also be that we must be careful in such examinations; cursory inspection may easily give misleading results (Fig. 6.22).

However, returning to the study of prosimians, even though overall size is not the primary factor in these separations, it would be wrong to suppose size is absent. In fact, allowing for size in a simple manner (i.e. not taking account of the allometric

Fig. 6.19. The close relationship between the size differences of the species and their location in the first statistical axis derived from study of measures. The first two frames compare the rank orders of the groups by size and by position in the first axis. The third frame provides the regression of size upon rank order in the first axis: the relationship is clear.

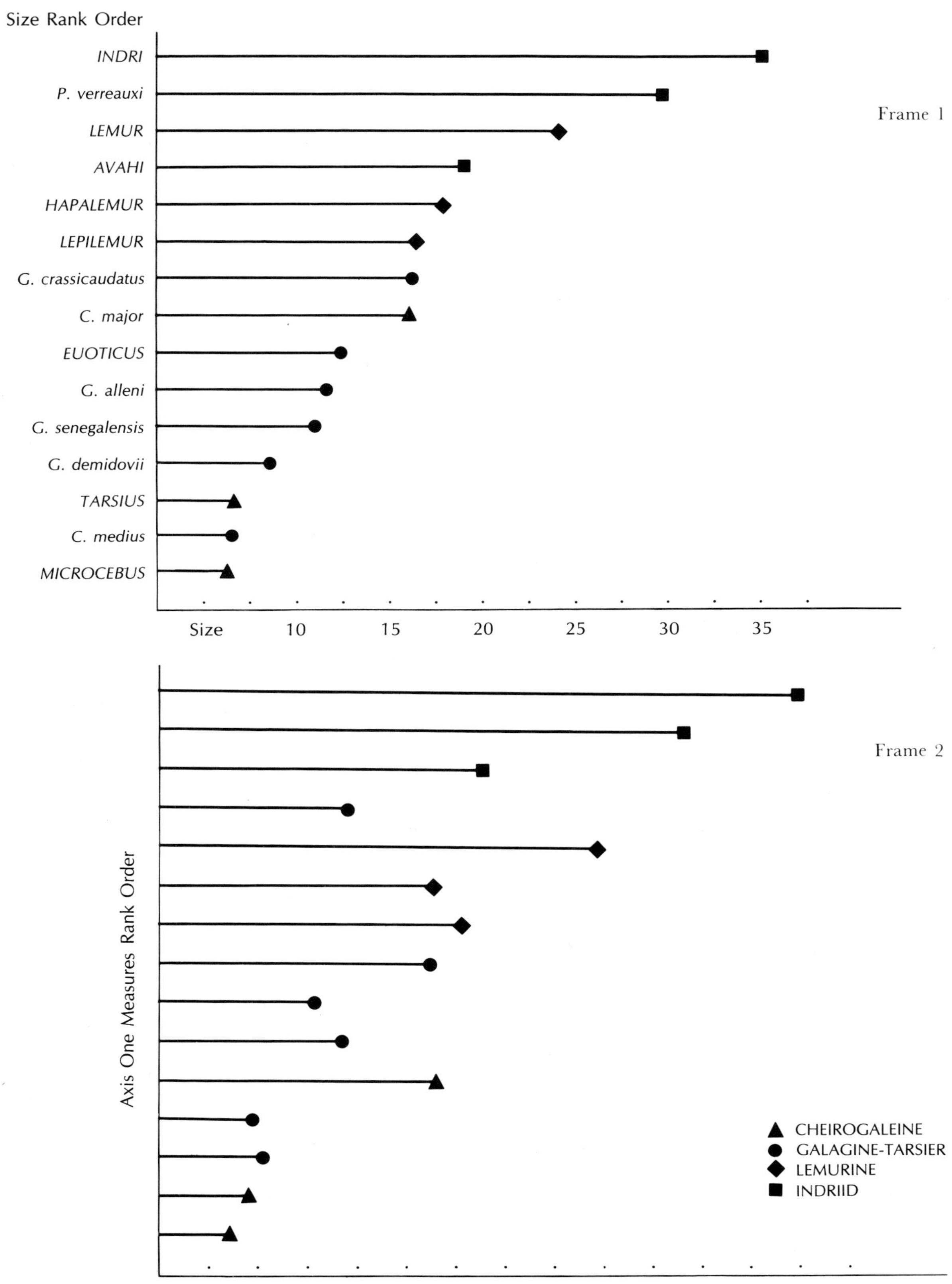

Frame 3

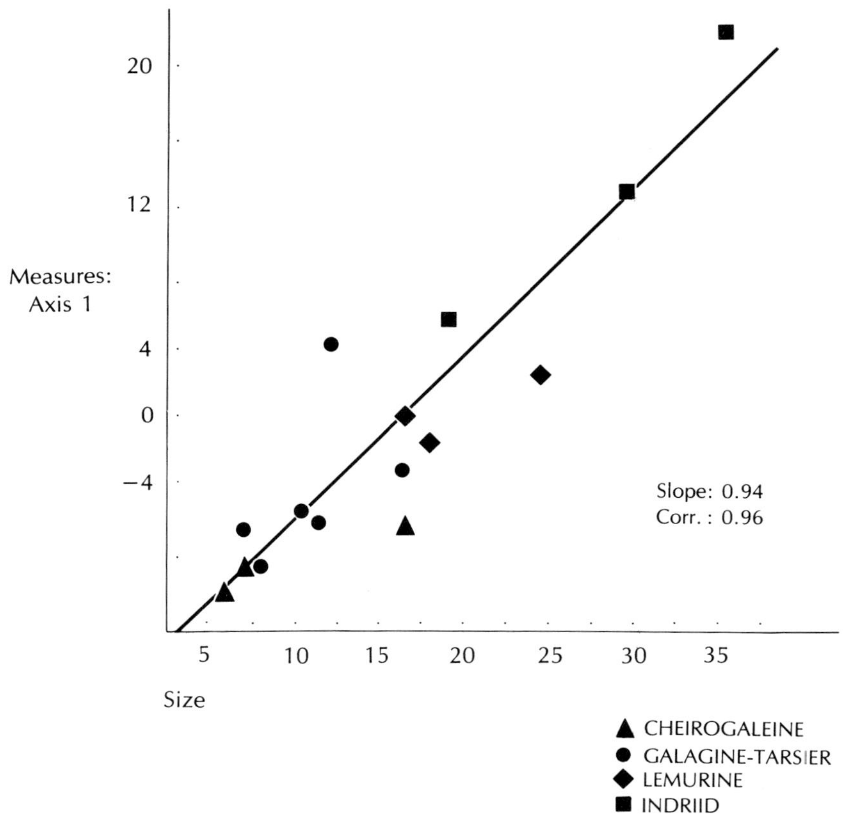

component of size by using only indices) may have allowed the results to reveal more subtle size effects that are closely tied in with functional aspects of the structure of these animals.

Thus, although no single axis across the entire group of animals reflects size, within each group there seems to be an expression of size gradients. From centrally to peripherally the gradient is from larger lemurs through greater dwarf lemurs, lesser dwarf lemurs to mouse lemurs. In a direction pretty much at right angles to this is another size gradient passing from large lemurs through thick-tailed bush-babies, Allen's bush-babies to Senegal bush-babies and tarsiers. And finally, among the indris, although a gradient is harder to see because the species do not form a neat linear array as do the others, there is at least some appearance of a *reverse* size gradient from the smallest (sportive lemurs) through intermediate species to the largest (indris).

That these three size gradients differ is clear because they are aligned at different angles and with different polarities within the plots. It appears that simple size is not a single predominant factor in these results, but that, in some more complicated way, size is differently implicated among the different groups. The implication becomes clearer as we study further the interpretation of the different morphological associations.

Functional implications of these results

The morphological linear array of the galago-tarsier spectrum seems clearly related to the different abilities of the creatures in leaping. All do of course leap, and compared with the central species (lemurs, sportive and gentle lemurs) they leap extensively. But recent more detailed field studies make it clear that there is a spectrum of leaping. The thick-tailed bush-babies spend considerable periods of time in quadrupedal movement. Allen's bush-babies and needle-nailed bush-babies leap considerably more. And of the bush-babies, Sene-

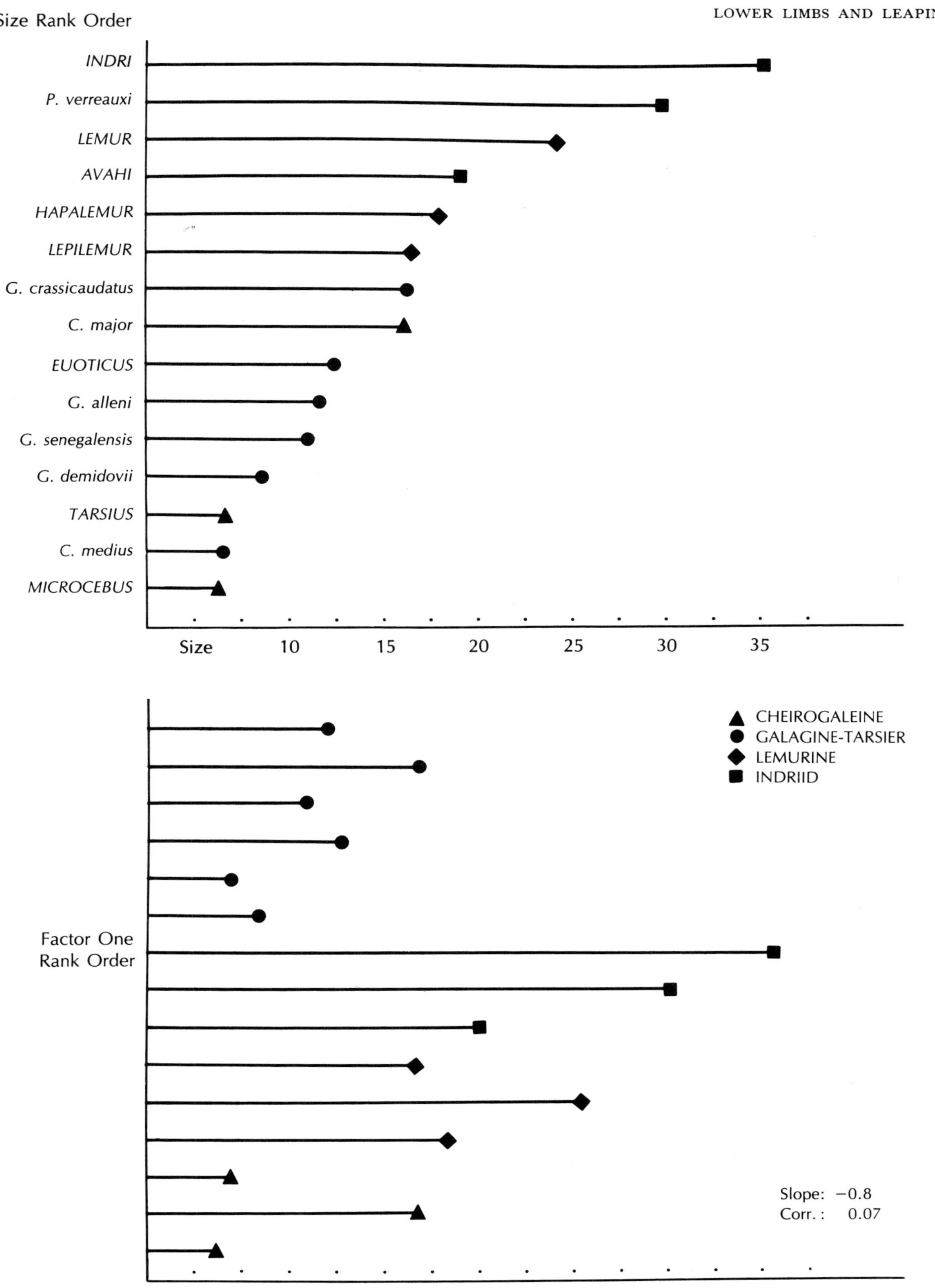

Fig. 6.20. In contrast to Fig. 6.19, and with the same conventions, the relationship between the first axis of a study of indices demonstrates that size is much less involved in the result.

gal bush-babies leap most of all (Bearder, 1974; Bearder and Doyle, 1974; Charles-Dominique, 1971, 1972, 1977; Kingdon, 1971; reviewed in McArdle, 1978). Tarsiers also leap as much as many bush-babies (e.g. Niemitz, 1974). Gentle lemurs also leap less but tend towards the bush-babies and tarsiers.

In the same way, the morphological linear arrangement within the gentle lemur — dwarf lemur — mouse lemur spectrum also seems related to leaping. Again, more recent field reports make it clear that these species leap a good deal more than previously thought. Thus, though usually grouped as quadrupeds, the suspicions of Napier and Walker (1967) and of Stern and Oxnard (1973) that these creatures have morphologies indicating that they might be doing some leaping is now borne out (but see later, this chapter). However, it is also apparent that they do not all leap to the same degree. Among them, greater dwarf lemurs probably leap least, lesser dwarf lemurs a little more, mouse lemurs perhaps more still (Petter, 1962; Walker, 1967; Petter, Albignac and Rumpler, 1977; Jouffroy and Lessertisseur, 1979). Again, gentle lemurs which do leap also tend towards this group.

When we view the indriids we do not see a morphological linear arrangement. The species are

Fig. 6.21. The regression of size upon position in the first axis of the analysis of indices shows no relationship throughout the entire group. The ellipses drawn around specific sub-groups in the first frame suggest, however, that there is a relationship with size in two different groups within the analysis. The ellipses and circles drawn around different groups in the second frame demonstrate that if we look at every individual group in the analysis there is, again, no relationship with size.

It is clear that we must not choose one of these results for emphasis over another without having good reasons. It is all too easy to find the result that is 'wanted' if the whole picture is not presented. Undoubtedly, size is implicated in these results in some way, but as suggested in the text, in a most complicated manner.

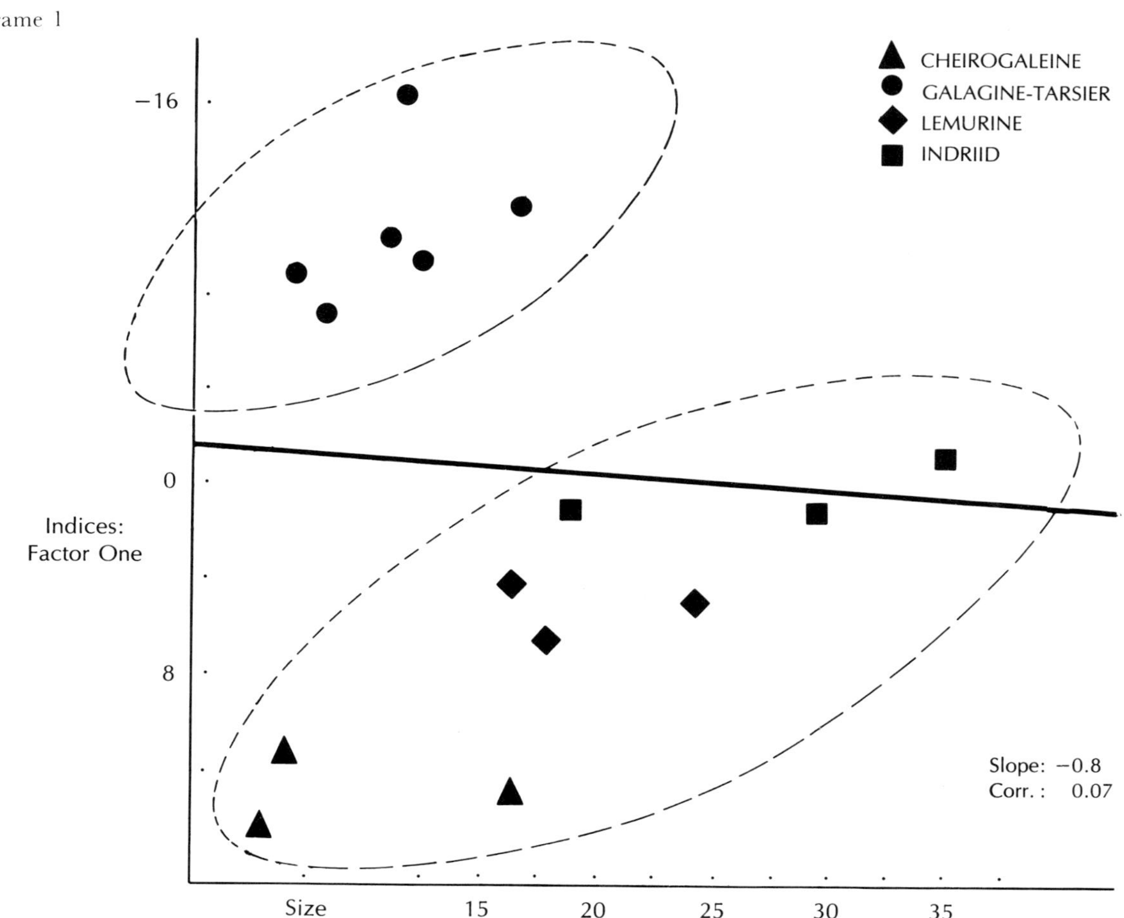

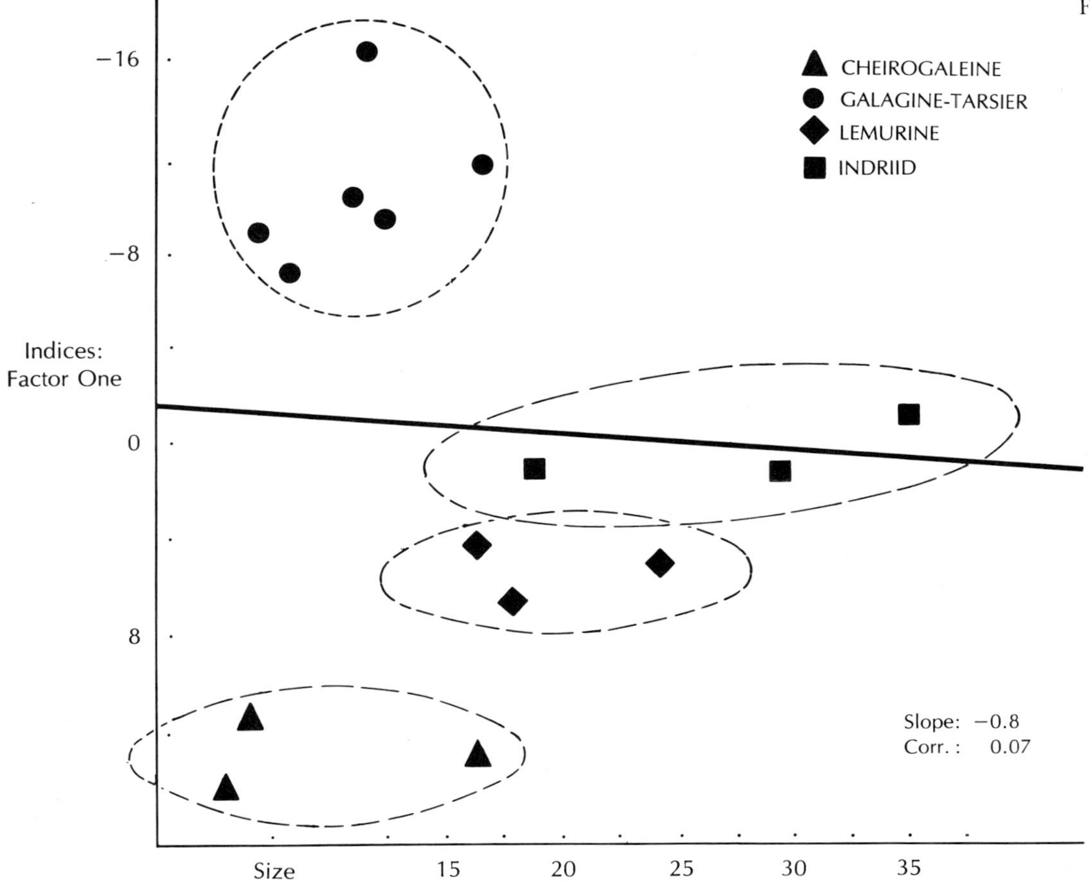

spread out in a way which does not especially mark any given species as less or more extreme. And such reports as are available about their leaping abilities do not seem to especially differentiate any single genus as most effective leapers (Petter, 1962; Petter, Albignac and Rumpler, 1977). In their different ways all are very good leapers. Only sportive lemurs leap less; and they, though tending towards the various indriids, are not as extreme morphologically.

Finally, ring-tailed lemurs, mongoose lemurs, black lemurs and ruffed lemurs are also reasonably able leapers though they are also more quadrupedal than, for instance, the indriids and the galagines. Of the various lemur species, *L. variegatus* leaps to perhaps the least degree of all, it being said to be quite clumsy in the trees (Walker, 1967). Certainly, some lemurs possess propensities for leaping not totally unlike those of the indriids (though much less in degree).

It thus seems clear that we now have at least three morphological adaptations (table 6.7) for leaping, in addition to the leaping superimposed upon the quadrupedalism of most *Lemur* species. We are thus left with a series of possibilities that complicate and extend those already mentioned briefly.

First, it is still possible that the leaping adaptation is a single adaptation and that these morphological differences are merely due to the fact that the adaptations have been evolved in parallel in species that started from different morphological and genetic bases. However, consideration of the following factors make this much less likely. The genera in each morphological mode come from different taxonomic groups. The most obvious example of this, is of course, the juxtaposition of tarsiers and bush-babies in the same mode; it is highly unlikely that animals in two separate infraorders can have started with the same morphological and genetic equipment. However, to lesser degrees, this same argument holds for each of the other groups. Although the indriids are a unitary taxonomic group, sportive lemurs and some regu-

Table 6.7

LOCOMOTOR TRENDS IN PROSIMIAN LEAPING AFTER OXNARD (OXNARD, GERMAN AND McARDLE; OXNARD, GERMAN, JOUFFROY AND LESSERTISSEUR)

Maximum (left)		Minimum (right)
Trend of indriid-type leaping from maximum (left) to minimum (right) Large upward leaps, stretched-out body posture during leap		
Avahi *Propithecus* *Indri*	*Lepilemur* *Lemur catta*	*Lemur variegatus*
Trend of galagine-type leaping from maximum (left) to minimum (right) Large upward leaps, curled-up body posture during leaps		
Galago senegalensis *Euoticus,* *Tarsius*	*G. demidovii*	*G. crassicaudatus* *? Hapalemur ?*
Trend of cheirogaleine-type leaping from maximum (left) to minimum (right) Leaps mostly downwards, limbs hanging down often begin and end in running		
Microcebus	*Cheirogaleus medius*	*C. major* *? Hapalemur ?*

Note: Positions ascribed to genera are only approximate; there is no comparability in degrees of leaping between the three major groups.

lar lemurs (lemurines) also manifest indications of this particular morphological adaptation. In the same way, although the cheirogaleines are a single taxonomic group, the gentle lemurs (lemurines) tend toward their morphological mode. In any case, the fact that the linear arrangement within each group associates with abilities for leaping renders this criticism even weaker.

Second, it is possible that these morphological modes reflect not leaping but some other aspect of the lives of these species. Given the three sets of parallel associations with leaping this seems unlikely. Given, furthermore, the fact that it is in the lower limb that the associations are most strong renders it highly unlikely. Yet it is a not impossible idea to which we shall return shortly.

Third, we are therefore left with the possibility that there really are several different modes of leaping and several different morphological adaptations in these animals. Is there any evidence of this? The increased numbers of field studies that have been performed on these animals in recent years have not been aimed at this particular hypothesis. But they have provided considerable evidence that impacts on these ideas.

For instance, when the various indriids leap, and

Fig. 6.22 As a comment on the previous figure let us look at a study in the literature in which an attempt has been made to look at size, in this case to investigate the association between body weight and femur size for the prediction of morphological parameters in fossils.

The first frame shows the analysis that was presented (McHenry, 1976) in which predictions about the named fossils were made upon the basis of the regression line calculated for humans alone. The close (apparently) relationship of the apes to the regression for humans is suggested by that author as extra evidence supporting the case.

His picture should be compared with the view presented in frame two. This indicates rather clearly that the real regressions may be totally different. It especially shows that the regressions in the various apes are not only different from each other, but also from that of humans. Any attempt to predict measures from unknown forms should certainly take all this information into account.

Given that we do not know what the pattern is for any of the unknown (fossil) forms, it may well be vitally important to make our prediction very cautiously. Indeed, once we have seen that different patterns are truly the case for each living species, it is not at all unlikely that for the unknowns yet other distinct patterns exist. ▶

Frame 1

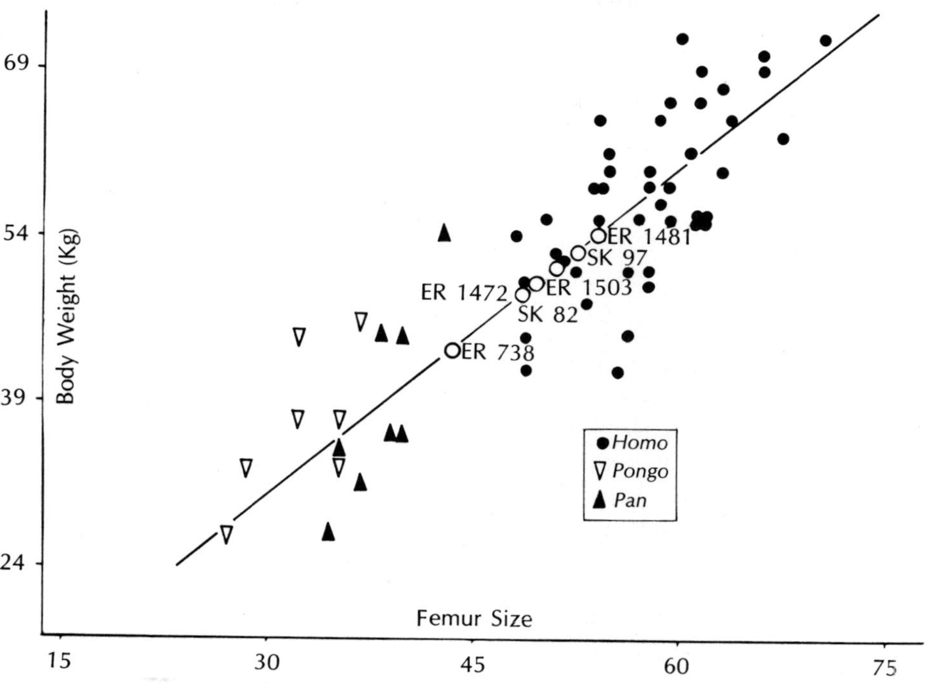

From McHenry, 1976

Frame 2

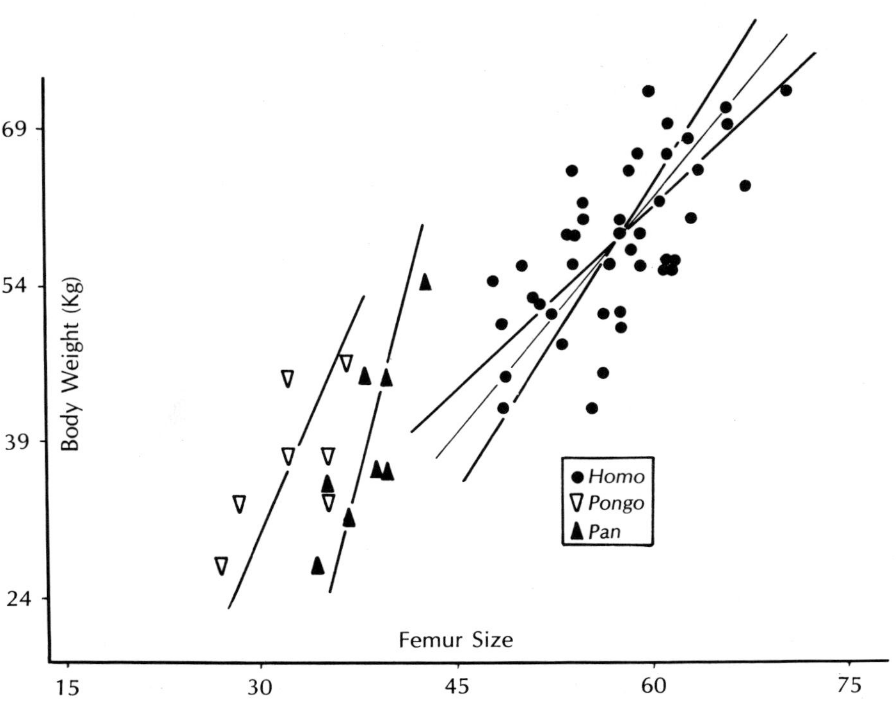

especially when they are leaping to their greatest extent, it seems as though they often employ stretched out postures during their flight path through the air. It is easy to find single frames of movies that show this for the different indriids (Fig. 6.23). Such postures during 'flight' also exist, although to lesser degrees, both in the genus *Lepilemur* (Fig. 6.24) and among specimens of the genus *Lemur* itself (Fig. 6.25). Although such positions are not exhibited so frequently as in the indriids proper, nor, one surmises, for such long distances in the air, it may well be that these phenomena also exist in an incipient manner in these other leaping species.

In complete contrast is the situation in bush-babies and tarsiers. Although, of course, there is a phase during the leap immediately after take-off when these animals are stretched out, it is apparent that the primary body posture during flight (especially during long leaps) is curled up. Many individual frames of these different species leaping demonstrate this (Fig. 6.26).

The rather fewer descriptions that are available for cheirogaleines suggest to me that they leap in yet a third manner, one more similar to regular primate leaping (Jouffroy and Lessertisseur, 1978, 1979; Jouffroy, personal communication). They leap with the body in a semi-flexed, semi-extended position, with the limbs neither curled up nor stretched out but hanging down. Certainly the only picture I have available of a cheirogaleine in mid-leap is in exactly this posture (Fig. 6.27). It can be compared with similar mid-air positions for *Hapalemur* and *Lepilemur* during smaller leaps. More information is needed to know for certain how dwarf and mouse lemurs leap and whether what they do really represents a bio-mechanically different mode from those of indriids on the one hand and bush-babies and tarsiers on the other.

Let us return to the *second* possibility: that these morphological modes depend upon associations with other aspects of the lives of these animals. A careful search of the literature has not revealed environmental and ecological associations that fall into the groups represented by the morphological descriptions. But a wider view of the literature on posture and locomotion does give some information that may bear upon this question. For instance, although the leaps that these animals carry out do seem to differ in the ways that I have indicated above, it turns out that there are also associated movements and postures that differ in the same ways.

Thus, the various indriids are similar to one another (and differ from bush-babies and tarsiers) because of the wide variety of movements, in addition to leaping, of which they are capable. These species maintain a generally vertical body

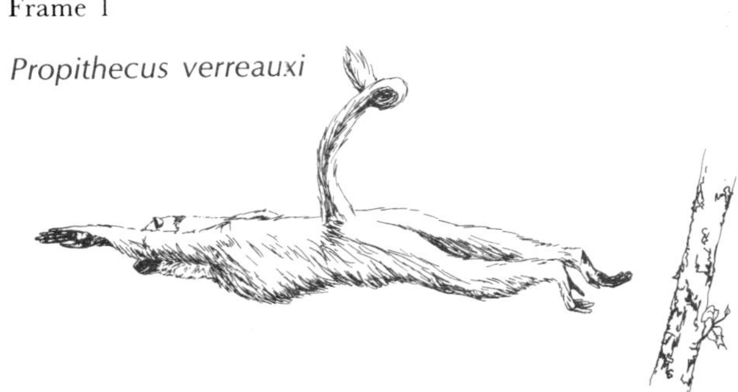

Frame 2

Propithecus verreauxi

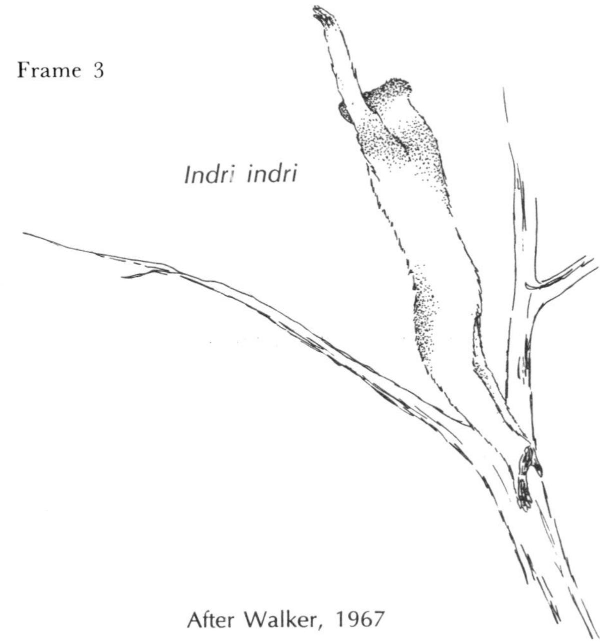

Fig. 6.23. **Various indriids in mid-leap.**

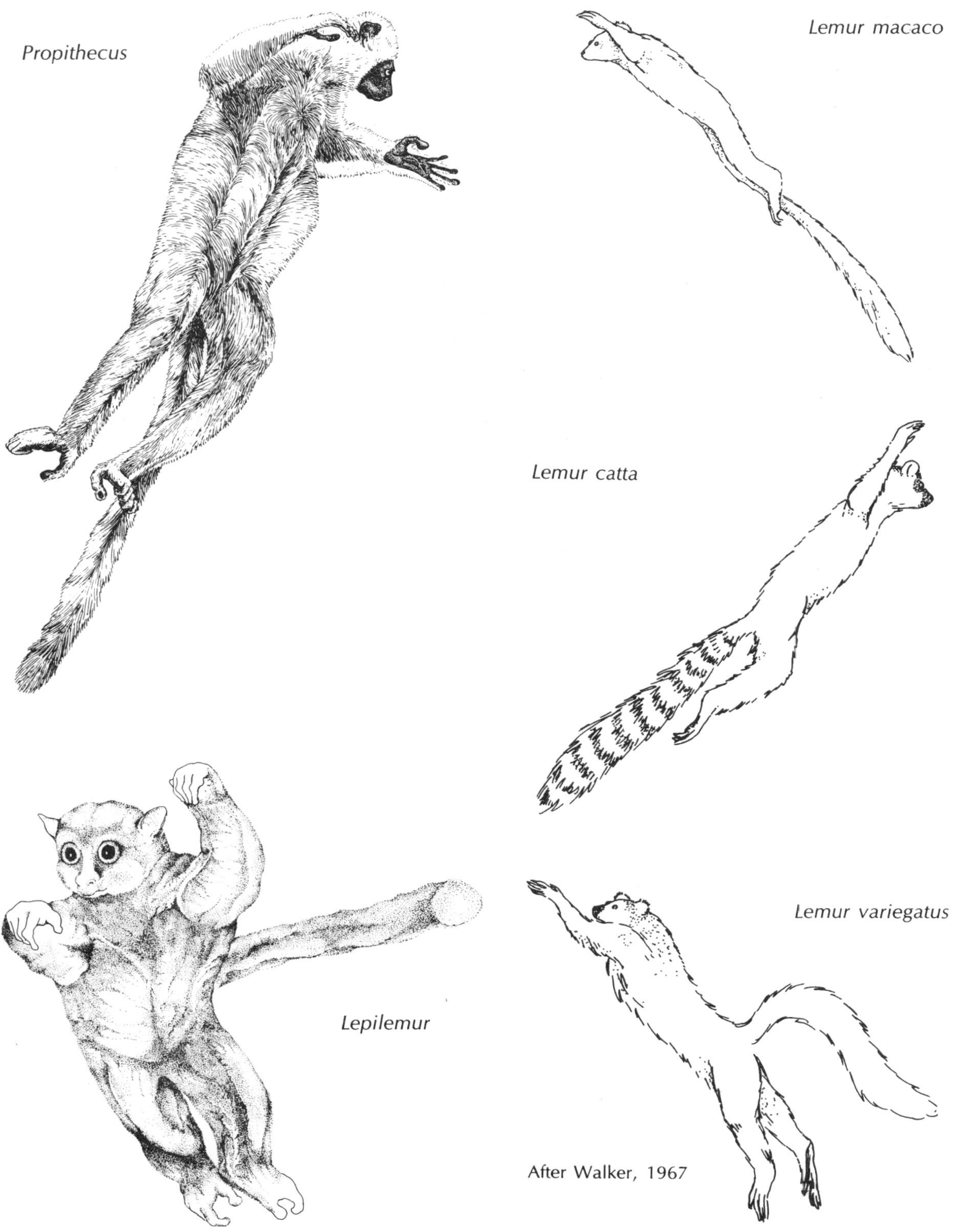

Fig. 6.24. A comparison between the outstretched posture of *Propithecus* and *Lepilemur* in mid-leap.

Fig. 6.25. Some lemurs in mid-leap: *L. catta*, *L. macaco* and *L. variegatus*.

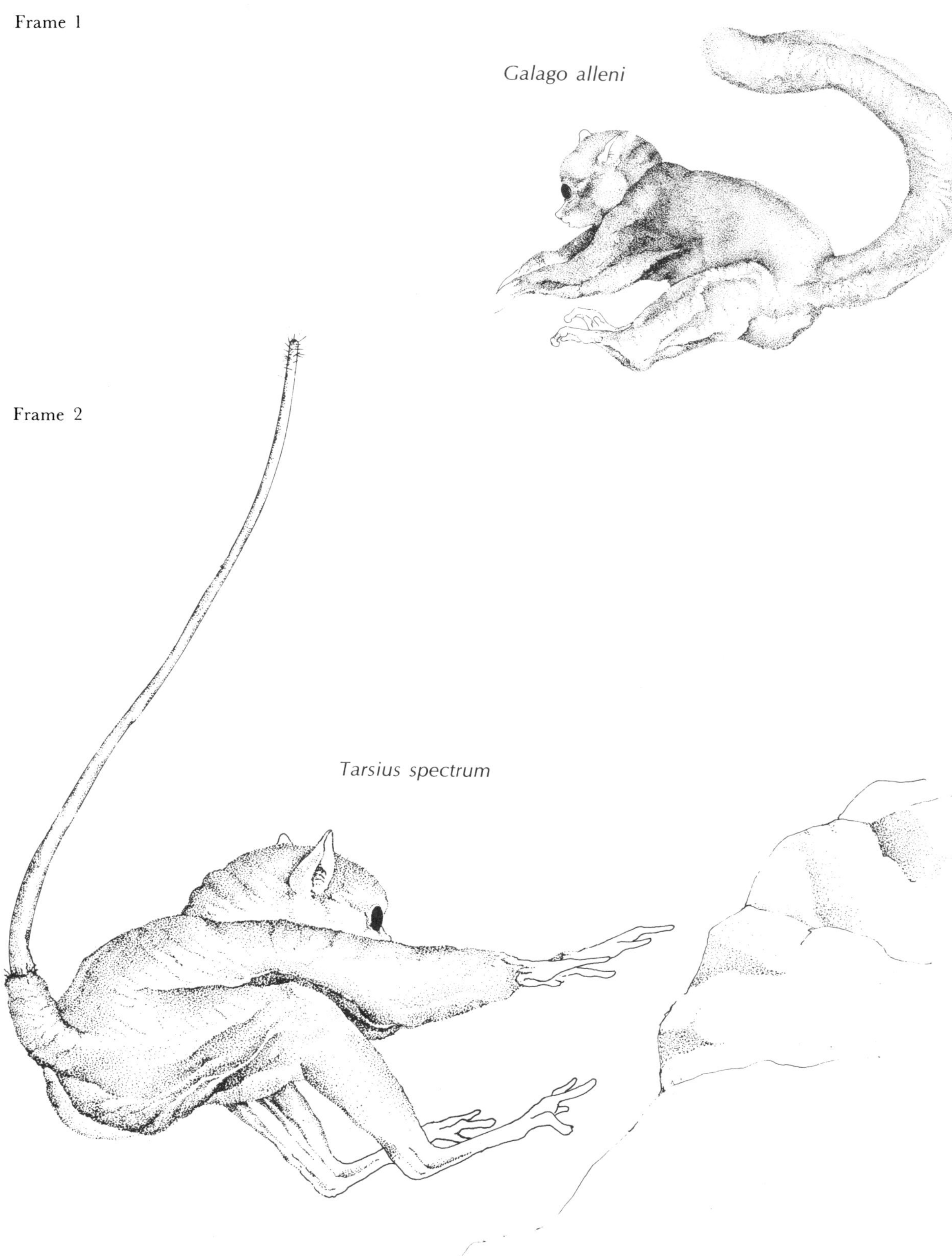

Fig. 6.26. **Mid-air postures during the leaps of a bush-baby and a tarsier.**

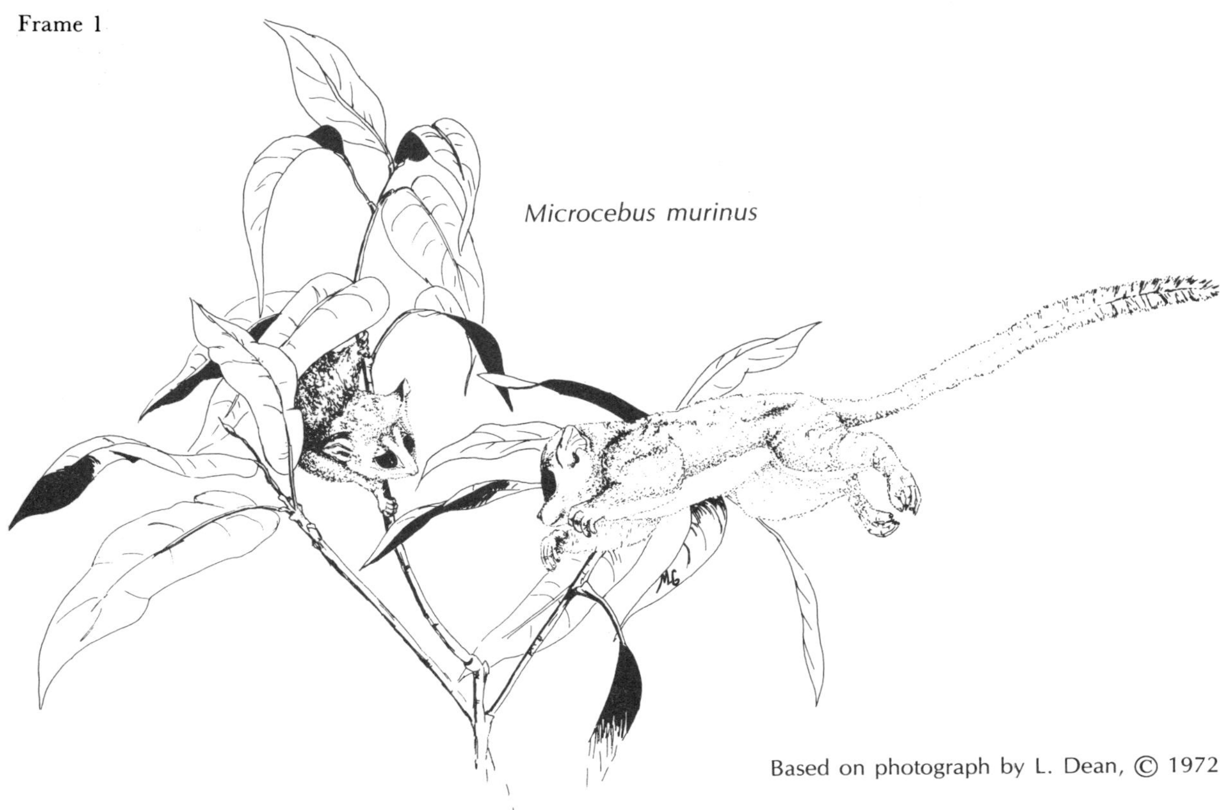

Fig. 6.27. A diagram made from a photograph of a cheirogaleine in mid-leap can be compared with those for a sportive lemur and a gentle lemur.

posture; but in supporting that body, in helping it to move through mechanisms other than leaping (such as climbing and descending) and in a variety of activities of play, foraging, escaping and so on, these animals demonstrate extremely versatile limb movements. Upper and lower limbs may be widely splayed. Upper limbs are frequently held above the head. Limbs frequently adopt markedly rotated positions. And these positions and movements are readily seen from inspection of films made both in the field and in captivity (Figs. 6.28 and 6.29).

In contrast, although it is true that bush-babies and tarsiers also maintain vertical body positions during vertical clinging, their upper limb move-

Fig. 6.28. An example of the wide repertoire of other limb movements and postures in indriids.

Fig. 6.29. Another example of the wide repertoire of limb movements and postures of indriids.

ments are generally more restricted. The upper limbs are much less often raised above the head, and much less often widely splayed and rotated. Both limbs are more frequently used either in more or less horizontal supporting positions associated with vertical clinging or in dependent positions associated with regular quadrupedal locomotion (Fig. 6.30).

Less is known about the ranges of postures and movements of the limbs associated with locomotion, play, foraging and escaping in cheirogaleines (e.g. Petter, 1962; Jouffroy and Lessertisseur, 1979; Jouffroy, personal communication). That it is different from the two previous groups is clear, however. Cheirogaleine leaping, such as it is, seems to be more frequently associated with prior quadrupedal walking or running; it often seems to occur as leaps that are mainly outwards and downwards from running starts and continuing into running; in their general postures these creatures seem to be considerably more pronograde, the body far less often vertical, than either of the other two groups.

It seems unlikely that these groups of accessory movements (accessory, that is, to leaping) would be the major reason why the animals fall into these three morphological spectra. However, it does seem likely that these additional postures and movements may be parts of the different leaping adaptations. The adaptations could thus be viewed as considerably wider than merely leaping itself. Pending further precise behavioural and biomechanical understanding of these species, adding this wider view to the picture of leaping may be the more cautious assessment.

Higher primate locomotion

It is perhaps also worth asking questions about locomotion in higher primates. Much attention has been paid to this in recent years (e.g. Prost, 1965; Napier and Napier, 1967; Walker, 1967; Rose, 1974) so that leaping is well documented, for instance, as a major mode of locomotion in the various colobines (e.g. Morbeck, 1976; Ripley, 1976; Fleagle, 1976, 1978). And even more recently we have been reminded that, indeed, almost all primates leap, even some of the most unlikely ones, such as gibbons (Fleagle, 1979).

But more detailed study shows that, among higher primates, most leaping involves descent from a higher to a lower branch; in contrast, most of the prosimians of which we have just been speaking are more capable of jumping from lower branches to higher ones.

And consideration of even these leaps in higher primates demonstrates biomechanical modes that are very far from the highly developed ones that we

Frame 1

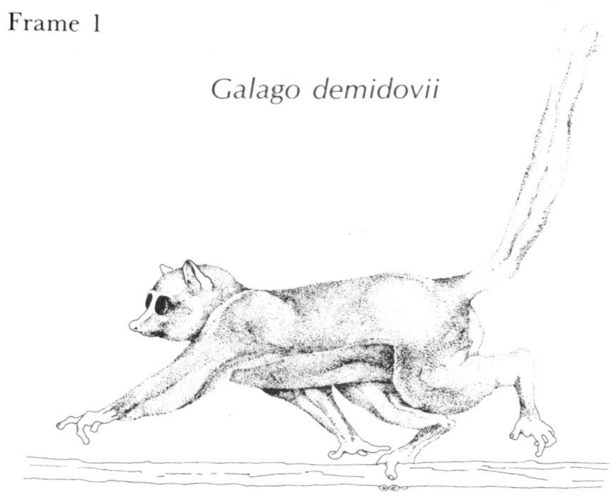

Frame 2

Fig. 6.30. This shows how, in contrast to those of indriids, the general pattern of limb postures and movements in bush-babies and tarsiers is much less rich. Limb positions and movements relate most to quadrupedal positions (first frame) and the special vertical clinging mode (second frame).

have just been considering. Figs. 6.31 and 6.32 show leaps in a series of higher primates well-known for their leaping propensities. Comparison of these leaps, their downward (generally) direction, their spreadeagled (generally) mid-air postures, and their nondescript (generally) modes of landing, show something far less well-developed than we have seen among the leaping prosimians.

And when we consider leaping in those other primates not usually thought of as leaping animals at all, we see mid-air positions in which limbs cycle or even wave wildly; it appears at first sight as though no good biomechanical mechanisms are at work at all. Thus, curious mid-air positions are found in forms such as macaques, mangabeys, gibbons and spider monkeys (Fig. 6.33). And the most unlikely mid-air positions of all are found in the most unlikely leapers of all, such species as spider monkeys, gibbons and siamangs. Indeed, though these last animals are extremely adept at mani-pulating their bodies on landing because of the prehensility of their hands (and tails when appropriate), their mid-air positions are most peculiar (Figs. 6.34 and 6.35).

Even other aspects of leaping are informative. Mid-air postures during upward leaps of some higher primates parallel what has been already described for indriids (Fig. 6.36). It may well be that there is a fascinating problem in the biomechanics of leaping even here.

And yet other locomotor possibilities may be important. Thus, it is well known that the apes move on two legs on occasion. It is somewhat less well known, perhaps, just how frequently most other primates also do this (Fig. 6.37). Again, this is an area for biomechanical study that has already been started by such investigators as Farish Jenkins (1972) and Ishida, Kimura and Okada (1974). Likewise, interest has recently been brought to bear upon acrobatic (tension-bearing) aspects of lower limbs (sometimes combined with prehensile tails, (Fig. 6.38). Stern (1971) has presented some of the first information here by recognizing anatomical correlates of hind-limb hanging in species such as uakari monkeys and orang-utans.

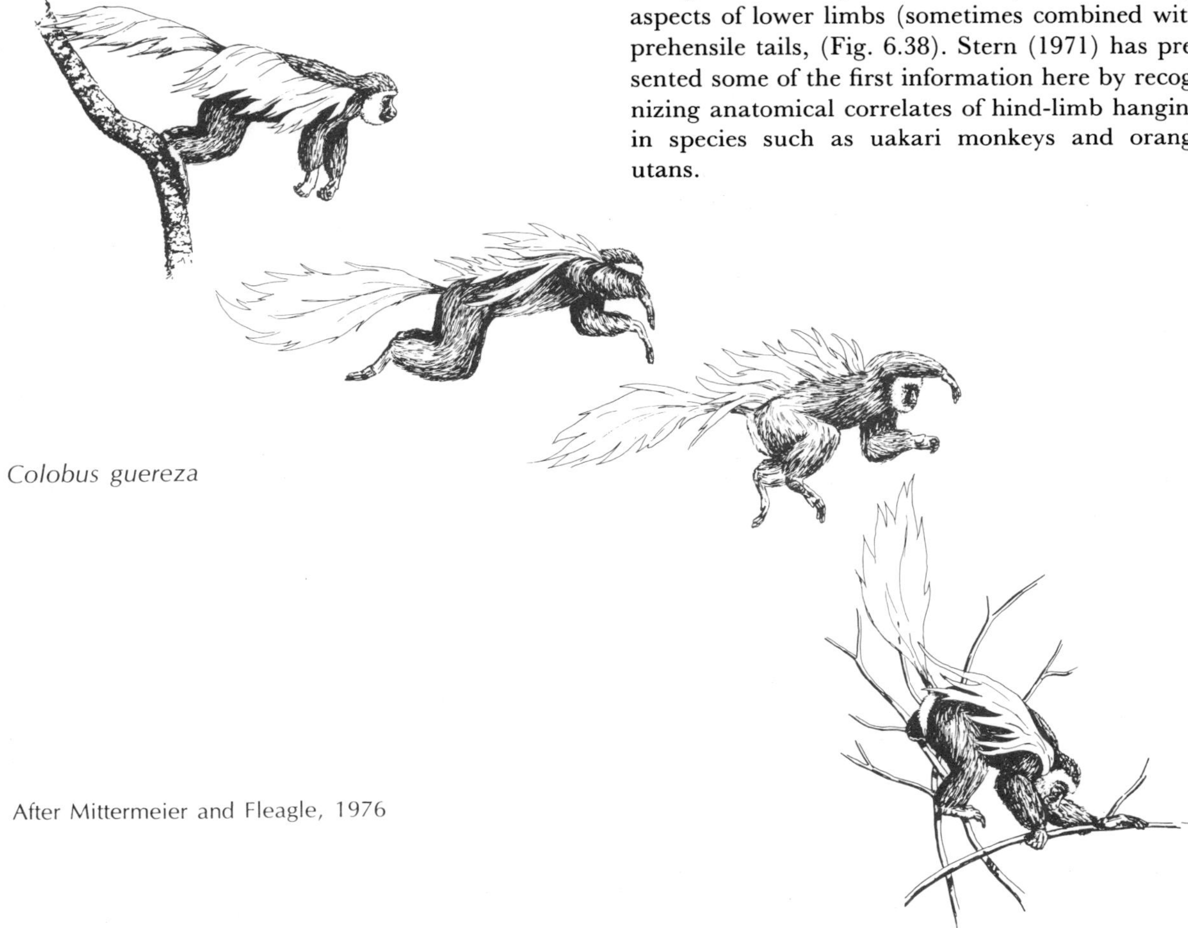

Fig. 6.31. A complete leap in a higher primate that is good at leaping.

Fig. 6.32. **Mid-air positions in higher primate leapers.**

Fig. 6.33. **Mid-air positions in higher primates less adept at leaping.**

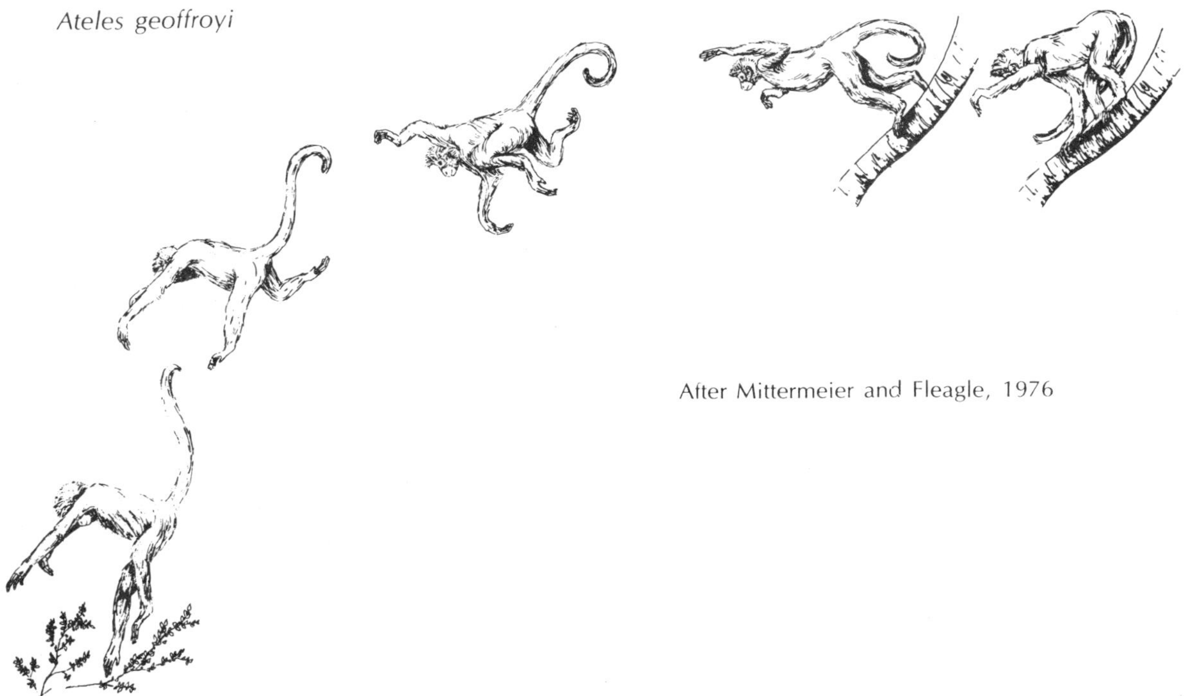

Fig. 6.34. A complete leap in a species usually assumed to leap scarcely at all. Note the fascinating mid-air posture.

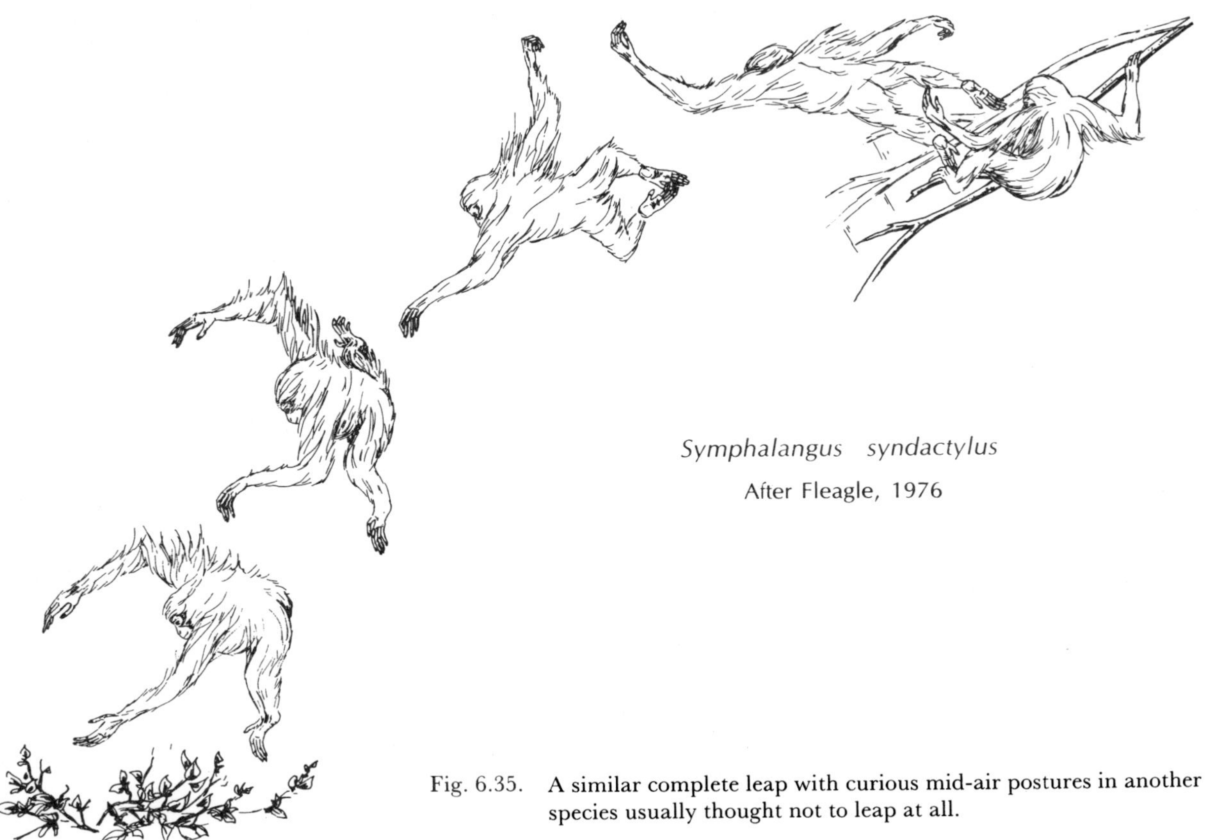

Fig. 6.35. A similar complete leap with curious mid-air postures in another species usually thought not to leap at all.

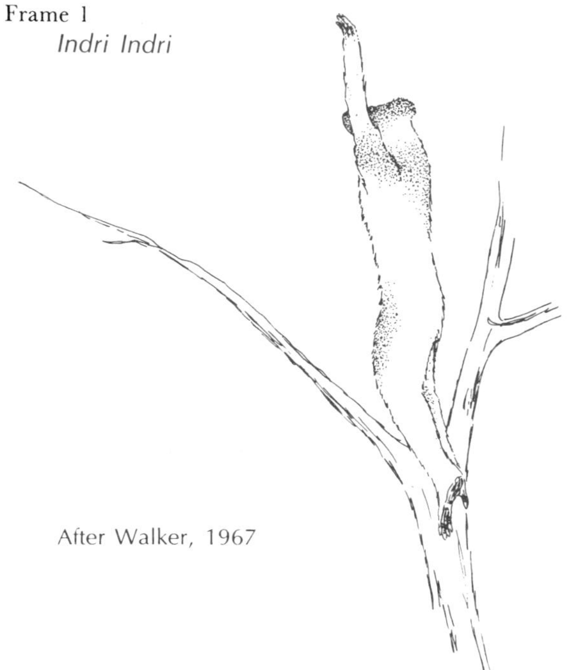

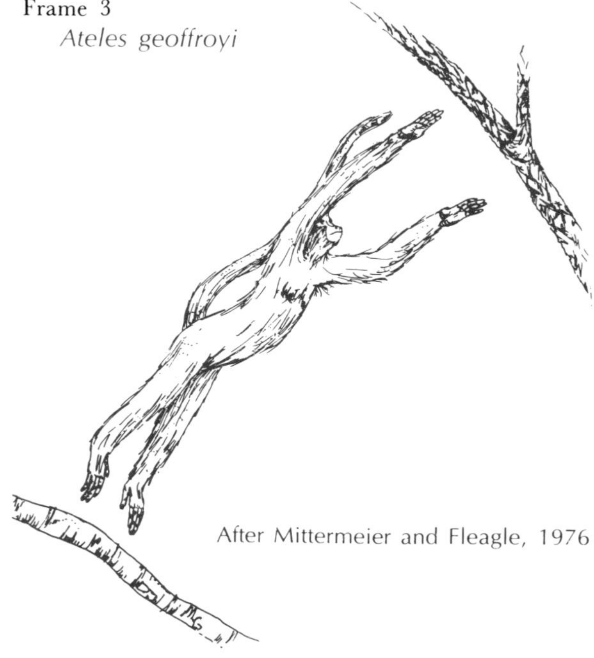

Fig. 6.36. Mid-air postures in upward leaps. A comparison of an extreme leaper, a good leaper and a (supposed) non-leaper.

Morphological modes in higher primates

Although leaping is indeed the activity most likely to be related to the form and architecture of the lower limb, leaping is by no means the only major lower limb activity in primates. It is, therefore, worth examining the relationships between lower limb structure and lower limb function in the entire range of primates.

Here we have so far carried out fewer investigations. Overall lower limb proportions and detailed pelvic structure have not been so fully studied throughout the entire primate order as in the more detailed prosimian studies of Oxnard, German and McArdle (1981) and Oxnard, German, Jouffroy and Lessertisseur (1981). The studies that are available (Zuckerman, Ashton, Flinn, Oxnard and Spence, 1973; Ashton, Flinn and Oxnard, 1975; Ashton, Flinn, Moore, Oxnard and Spence, 1981) cover the primates only at the generic level and with considerably smaller samples in each genus. Nevertheless, data do exist at this level and the results are available.

Univariate results: pelvis, talus, lower limb. The first essay is into the structure of the pelvis (Zuckerman, Ashton, Flinn, Oxnard and Spence, 1973) and, as with the more recent and detailed

Fig. 6.37. Two sequences and eight frames of different primates in bipedal postures and gaits. The last frame is of an animal raised in captivity and having seen only humans walking. (From a photograph taken at Hong Kong Zoological Gardens, courtesy Professor Lisowski and Mr. K.S. Lee.)

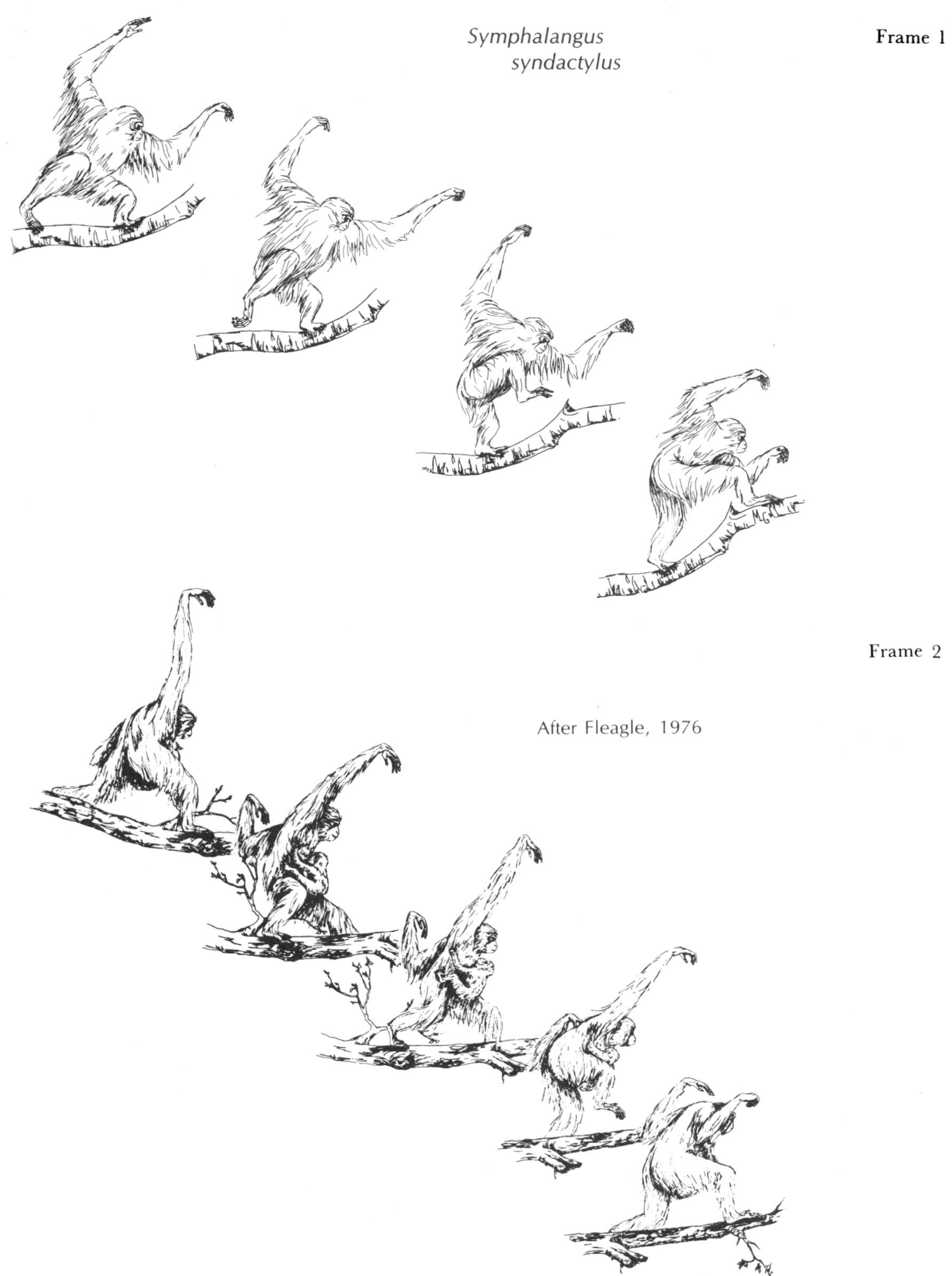

Fig. 6.37. Several different primates in bipedal posture and gaits. The last frame is of an animal raised in captivity and having seen only humans walkings. (From a photograph taken at Hong Kong Zoological Gardens, courtesy Professor Lisowski and Mr. K.S. Lee.)

Fig. 6.37. Several different primates in bipedal postures and gaits. The last frame is of an animal raised in captivity and having seen only humans walking. (From a photograph taken at Hong Kong Zoological Gardens, courtesy Professor Lisowski and Mr. K.S. Lee.)

Fig. 6.38. Five different primates in hind limb hanging positions One species is using a special 'hind limb' alone.

Frame 2

Frame 1

Alouatta

Macaca silenus

Frame 3

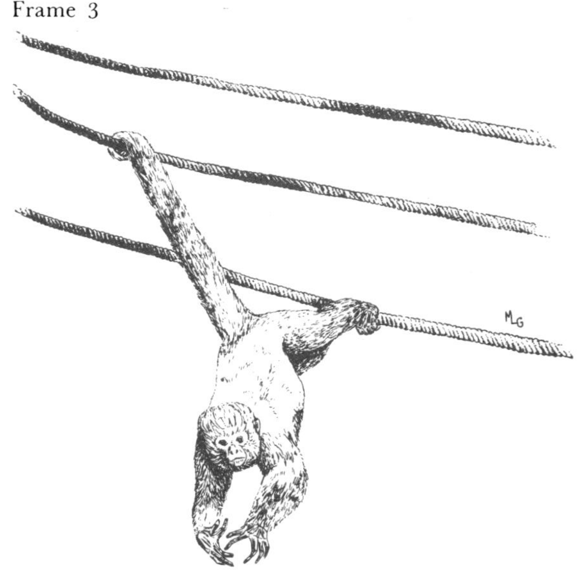

Lagothrix

Frame 5

Brachyteles arachnoides

Frame 4

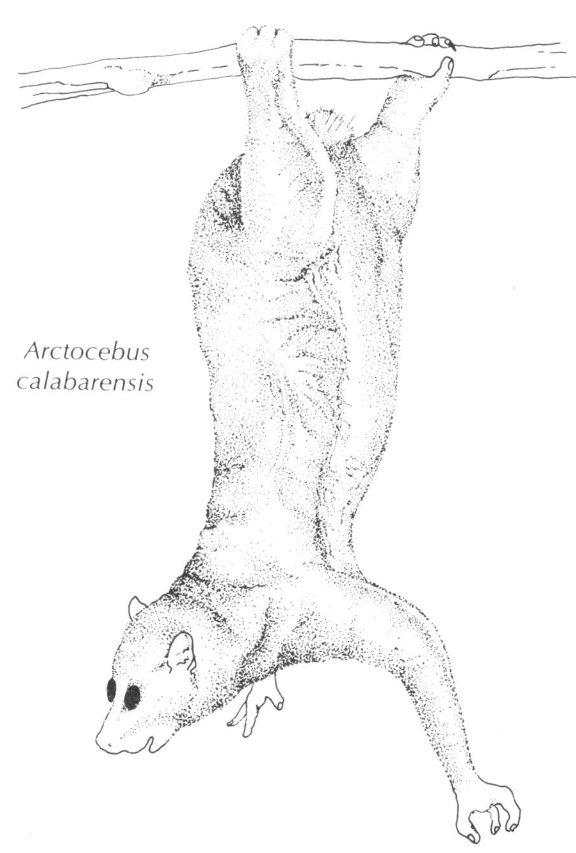

Arctocebus calabarensis

investigations of the prosimian hip and thigh, the initial examination is dissection. An extended series of 167 dissections of the hip and thigh in 33 genera of primates yields information about muscles, ligaments and joints pertaining to the various functions of the pelvis in primates.

For instance, the muscles around the hip joint in man are organized in four approximately equal blocks: two fore and aft and two on each side. Thus, in humans they act somewhat as guy ropes supporting a mast in every direction, useful when humans are supported on one leg in particular phases of walking and running. In non-human primates, where support upon a single leg is much less important in a four-footed animal, these muscles are arranged primarily in two blocks, fore and aft, with little emphasis on side to side components. They are more useful, therefore, in the craniocaudal swinging movements of the lower limb that occur during quadrupedal locomotion.

This knowledge of the soft tissue anatomy allows important bony landmarks relating to muscular leverages and joint positions to be defined. From these many elements of pelvic shape can be measured. Thus a series of nine osteometric features (Zuckerman, Ashton, Flinn, Oxnard and Spence,

1973), later augmented to a total of 25 osteometric features (Ashton, Flinn, Moore, Oxnard and Spence, 1981), were obtained from pelvic girdles of 430 extant primate specimens representing a total of 41 genera of the primates (summarized briefly in table 6.8). Individual bony dimensions give some useful information about the form of the pelvis in various primates. In most instances, naturally, the principal pattern of separation is between bipedal humans on the one hand and generally quadrupedal monkeys and apes on the other. This is especially evident in, for instance, the dimension measuring the dorso-ventral position of the auricular facet, the joint between the pelvis and the vertebral column. This is more ventral in man and hence closer to the special line of gravity in upright humans than it is in any non-human primate. But there are also other bony features in which the biggest differences are among the non-human primates. Thus, the relative distance between the sacro-iliac and hip joints in the cranio-caudal direction is greater in all the acrobatic forms (prosimian lorisines, New World prehensile-tailed monkeys, Old World apes) than in other more regular quadrupedal species.

Investigations of the primate talus are also available. These are not based upon personal dissection of the soft tissues of the foot, but upon what was

Table 6.8

DIMENSIONS AND ANIMALS IN MORPHOMETRICS OF THE PRIMATE PELVIS

Anatomical features

Locomotor dimensions	*Potential residual dimensions*
Measures of joint positions	Relative length, curvilinear, of iliac crest
Four measures of the hip joint	Medio-lateral inclination of iliac blade
One measure of the sacro-iliac joint	Angulation of hip joint
Two measures of both joints together	Orientation of ischio-pubic plane
Measures of muscular attachments	Orientation of obturator foramen
Two measures of hamstring muscles	Width of obturator foramen
Four measures relating to muscles attached to the ilium	Lateral separation of hip joints
Four measures relating to adductor muscles	Lateral separation of ischial tuberosities

Genera and numbers of specimens studied

PROSIMIANS

Tupaia, 6; *Hapalemur*, 2; *Lemur*, 22; *Lepilemur*, 5; *Cheirogaleus*, 1; *Galago*, 11; *Euoticus*, 9; *Propithecus*, 4; *Indri*, 1; *Loris*, 10; *Nycticebus*, 7; *Arctocebus*, 5; *Perodicticus*, 13; *Daubentonia*, 2

NEW WORLD MONKEYS

Callithrix, 13; *Leontocebus*, 17; *Aotus*, 10; *Callicebus*, 10; *Callimico*, 2; *Pithecia*, 9; *Cacajao*, 10; *Chiropotes*, 1; *Cebus*, 13; *Saimiri*, 18; *Alouatta*, 11; *Ateles*, 12; *Lagothrix*, 11; *Brachyteles*, 1

OLD WORLD MONKEYS

Macaca, 25; *Cercocebus*, 13; *Papio*, 15; *Mandrillus*, 3; *Cercopithecus*, 27; *Erythrocebus*, 12; *Presbytis*, 21; *Rhinopithecus*, 4; *Nasalis*, 1; *Colobus*, 19

HOMINOIDS

Hylobates, 15; *Symphalangus*, 3; *Pongo*, 18; *Pan*, 24; *Gorilla*, 20; *Homo*, 141

already known from the literature. (Since then, however, a small series of 24 dissections of the primate foot have been undertaken in order to check on some of the osteological alignments of the bones: Oxnard and Lisowski, 1980, and work in progress). Sixteen dimensions of the talus were measured (and these are summarized briefly in table 6.9). Again, when these data are examined in a univariate manner, the principal finding is of individual features, such as the angle between the neck and the body of the bone, and the angle of the head upon the neck of the bone, in which humans are sharply distinguished from non-human forms, presumably in relation to bipedality. The human values for these angles are related to the high arches of the human foot. However, other features separate different non-human primates from one another. For example, measures of overall talar height are least in those species whose foot participates less in weight-bearing because their locomotion is so heavily fore limb dependent (spider monkeys and gibbons).

A third investigation covering the entire primate order is the study of overall lower limb proportions. These are a subset of Schultz's data (summarized in 1969), and the examination of lower limb proportions was carried out by Ashton, Flinn, Oxnard and Spence (1975) and Oxnard (1975a, c). Seven dimension were taken upon the lower limbs of 472 specimens representing 34 primate genera (summarized briefly in table 6.10). The measurements, when examined one by one, separate the various primates in a number of different ways, a major basis for separation being the different locomotor habits. Thus, as before, one big separation is that of humans from all other primates; the relative lower limb length is at a maximum in humans as befits their bipedal status. In contrast, the relative hip breadth distinguishes from among the non-human primates those that have special abilities for acrobatic movement in the trees: the prehensile-tailed New World monkeys and the lesser and greater apes.

Multivariate results: pelvis, talus, lower limb. Each of these studies has been examined using the multivariate statistical technique (canonical variate analysis). The results of all three parallel one another remarkably well. It is to the analysis of Professor Schultz's overall proportions of the lower limb that we may turn for an overall summary.

Thus, Fig. 6.39 demonstrates the arrangement of the primates produced by discriminant function and generalized distance analysis of the data on overall lower limb proportions. What is first obvious is that this arrangement in no way resembles that produced by the three approaches outlined in Chapter 3. Rather than the linear sequence paralleling systematics discussed in that chapter, the arrangement of species here is generally star-

Table 6.9

DIMENSIONS AND ANIMALS IN MORPHOMETRICS OF THE PRIMATE TALUS

Anatomical features

Locomotor dimensions	*Residual dimensions*
Maximum medial height	Maximum length of talus
Maximum lateral height	Maximum breadth
Curvature of trochlear joint facet	Maximum median height
Transverse width of trochlear facet	Maximum trochlear height
Posterior trochlear breadth	Width of medial joint facet
Angle between neck and body of talus	Total head neck length of talus
Torsion between head and body	Maximum diameter of neck
	Minimum diameter of neck

Groups of Animals and Subjects, and numbers of specimens studied

Ateles, 9; *Macaca*, 29; *Papio*, 14; *Colobus*, 12; *Hylobates*, 3; *Pongo*, 13; *Pan*, 98; *Gorilla*, 98; *Homo* (Jericho) 16; *Homo* (mixed) 14; *Homo* (Australian) 7; *Homo* (Vedda) 4; *Homo* (Andamanese) 5

shaped, with many genera embedded in the nucleus of the star and with numbers of others lying in different outlying rays of the star. Studies of the hip and the talus replicate this star-shaped pattern. In no way is current taxonomy reproduced in this arrangement.

What is the meaning of such a star-shaped format? This becomes immediately obvious as we note just where each primate genus lies in this structural array. The centrally located genera include tree-shrews, some lemurs, marmosets and tamarins, owl monkeys and saki monkeys, macaques and cercopitheques, among many others. They thus include many prosimians and both New and Old World monkeys. The only common factor in such a wide range of animals is that they are all capable of generalized arboreal quadrupedalism, irrespective of taxonomic group.

The individual genera within different outlying rays of the star also form interesting aggregations. One ray includes, for example, vervets, patas monkeys and baboons. Each of these is more closely related taxonomically to other species enclosed within the centre of the star (baboons to

Table 6.10

DIMENSIONS AND ANIMALS IN MORPHOMETRICS OF THE LOWER LIMB

Anatomical features

Schultz's data on lower limbs	*Jouffroy and Lessertisseur's data on lower limbs*
Relative hip breadth	Relative length of femur
Relative lower limb length	Relative length of tibia
Intermembral index	Relative length of foot
Crural index	Relative length of total lower limb
Relative foot length	Relative length of tarsus
Foot length relative to lower limb	Relative lengths of metatarsals
Relative foot breadth	Relative lengths of toes

Genera and numbers of specimens studied: Schultz's data

PROSIMIANS

Tupaia, 5; *Lemur*, 3; *Microcebus*, 3; *Lichanotus*, 1; *Propithecus*, 1; *Indri*, 1; *Daubentonia*, 1; *Nycticebus*, 9; *Perodicticus*, 1; *Galago*, 10; *Tarsius*, 8

NEW WORLD MONKEYS

Leontocebus, 28; *Aotus*, 9; *Cacajao*, 2; *Pithecia*, 1; *Cebus*, 25; *Saimiri*, 49; *Alouatta*, 4; *Ateles*, 74; *Lagothrix*, 2

OLD WORLD MONKEYS

Macaca, 27; *Cercocebus*, 3; *Papio*, 5; *Cercopithecus*, 3; *Erythrocebus*, 2; *Presbytis*, 14; *Nasalis*, 26; *Colobus*, 2

HOMINOIDS

Hylobates, 78; *Pongo*, 13; *Pan*, 26; *Gorilla*, 6; *Homo*, 25

Genera and numbers of specimens studied: Jouffroy and Lessertisseur's data

Hapalemur, 7; *Lemur*, 46; *Lepilemur*, 9; *Cheirogaleus major*, 6; *Cheirogaleus medius*, 3; *Microcebus*, 15; *Avahi*, 6; *Indri*, 6; *Propithecus*, 7; *Galago*, 39; *Euoticus*, 9; *Tarsius*, 8

Frame 1

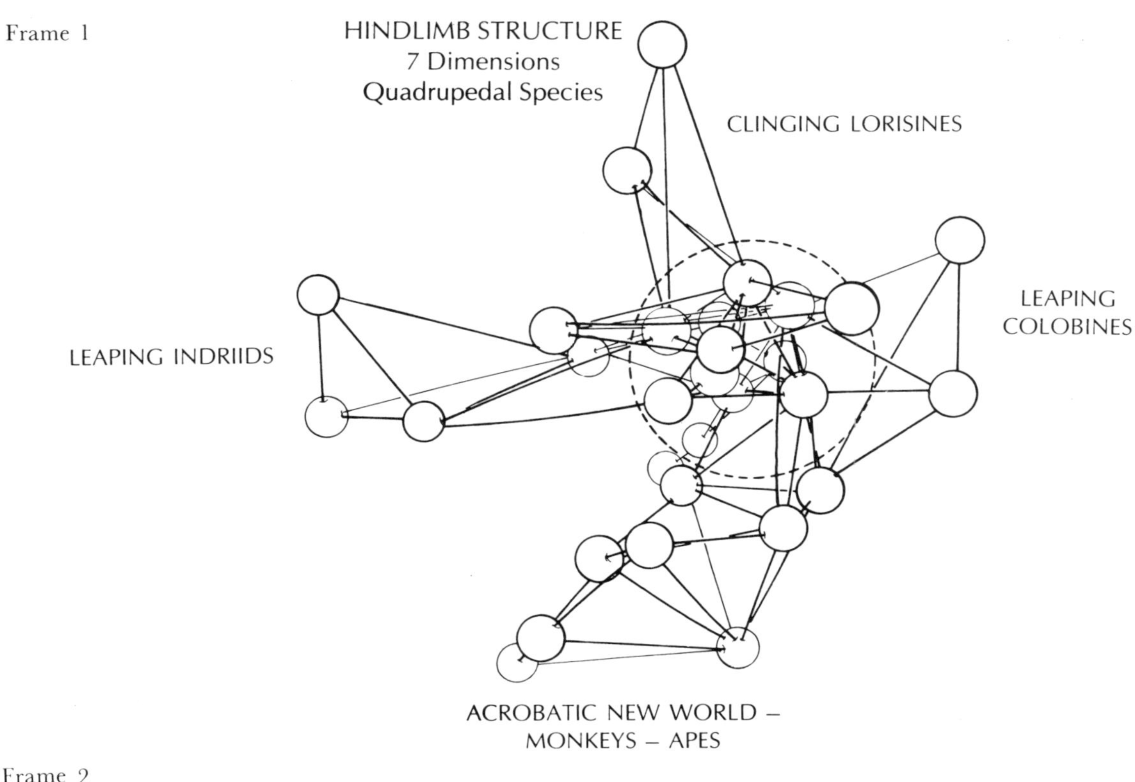

Frame 2

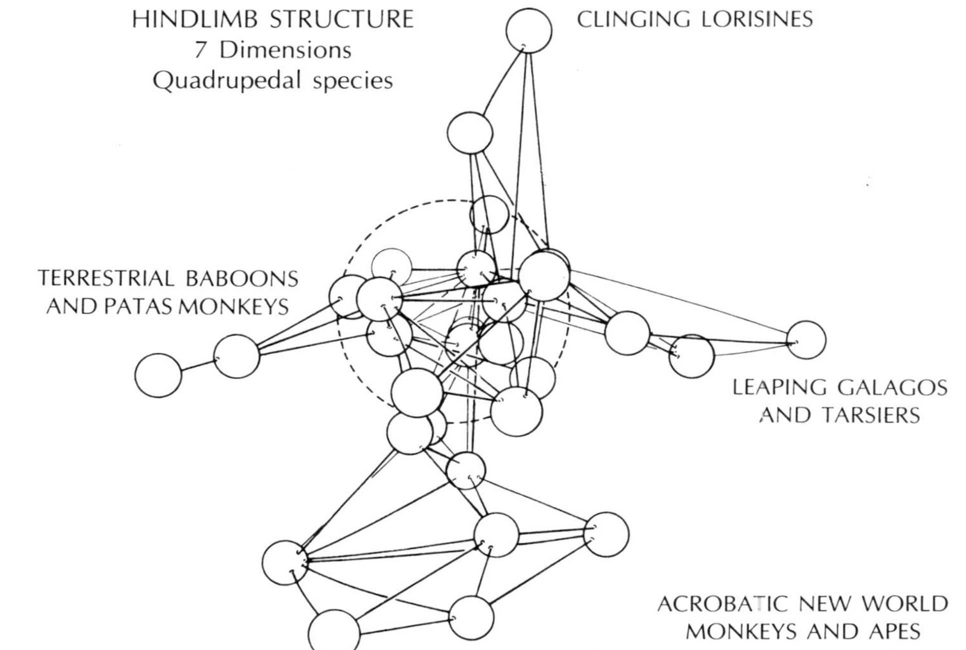

Fig. 6.39. The arrangement of various primate genera by multivariate study of overall proportions of the lower limbs. A three-dimensional model has been constructed from the minimum and near minimum generalized distance connections between genera. The overall scale of the diagram is some forty standard deviation units. Because of the manner of display, only those distances represented by connected lines are actually correct.

The basic form in a star: it is fascinating that all of the species at the centre of the star (located within the dotted circle) are arboreal quadrupeds. Species in various rays of the star are linked only by parallels in locomotor activity of the lower limb.

mangabeys – vervets and patas monkeys to other cercopitheques, as examples). But vervets, patas monkeys and baboons do have in common that they are more adapted for a degree of cursorial terrestrial locomotion than other Old World monkeys (though vervets are, of course, least so adapted, patas monkeys most).

Another ray includes tarsiers and many bushbabies, taxonomically distinct forms that share the functional association of being superb at leaping (this is discussed in more detail earlier in this chapter).

Yet another ray includes the lorisines. In this case all species belong to the same taxonomic group; yet the main difference between these species and those in the centre of the star to which they are most closely linked is that they are capable of curious slow acrobatic types of movement unique among the primates.

Even a further ray includes both the gibbon and siamang together with spider, woolly spider and woolly monkeys. Again, two totally different taxonomic groups are represented here. Their association surely lies in their possession of somewhat similar lower limb functions during the various acrobatic and brachiating movements of which all are capable.

Functional associations like these are found in whatever part of the star we look. Sometimes, to be true, the functional association is not clearly different from the taxonomic one, but this mild ambiguity is only found in those special cases (such as lorisines, indriids and pithecines) where the particular taxonomic groups happen to be the only ones among the primates that contain members moving in specialized ways (for lorisines, peculiar slow acrobatic climbing; for indriids, specialized leaping; for pithecines, curious acrobatic climbing with some hind limb hanging). It would seem, from the overall picture, that the functional association, rather than the occasionally coincidental taxonomic one, produces the star-shaped model that we see.

The analogy that is being offered here can easily be seen if we draw on the model of lower limb function for the various primates discussed in Chapter 4. This is repeated in Fig. 6.40; it was suggested from study of lower limb function long before the three sets of results shown here were discovered.

It is worth looking for a moment or two at the position of *Homo* within these various studies. This is best given by the further view (Fig. 6.41) of the model of Fig. 6.39. Humans lie in such a totally offset position that they cannot be considered to be within the star at all; their morphology must therefore be unique; this, in turn, implies uniqueness for the average biomechanical situation in their lower limbs. And this last we know that they have: the functions of human lower limbs are, indeed, uniquely different from those of any other primate.

It is worth also thinking for a moment or two about fossils. For this study makes it clear that the structure of a fossil, if analysed in this way, might allow us to say a great deal about the lower limb of that fossil. Depending upon where the fossil fell within the star-shaped spectrum, we might be justified in being quite detailed about (a) the degree to which the fossil lower limb was used in a general quadrupedal mode (i.e. a fossil lying centrally within the star) or (b) the degree to which the fossil lower limb was adapted for bearing forces associated within one or other of the specialized activities represented by one or other of the outlying rays of the star (i.e. a fossil lying within an appropriate ray of the star) or even (c) the degree to which the fossil lower limb was adapted for some specialized activity not represented by that of any extant form (i.e. a fossil lying in such a position that it formed a new ray to the star).

Such an analysis would not necessarily give any information about the taxonomic group to which

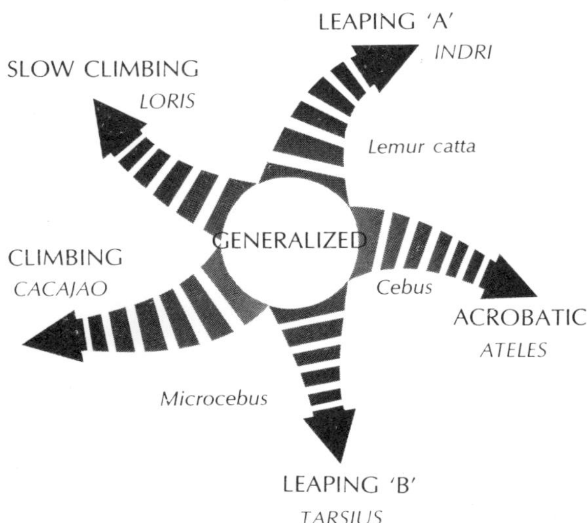

Fig. 6.40. A theoretical spectrum of lower limb activities in various primates. A star-shaped model is similar to the actual model of the last figure for lower limb structures.

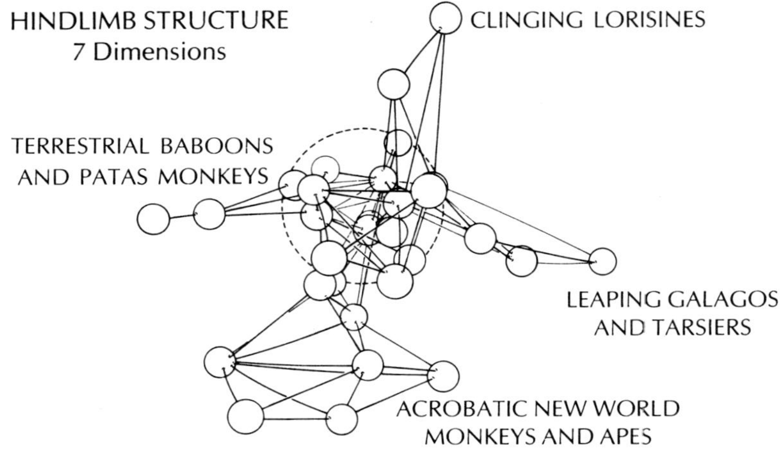

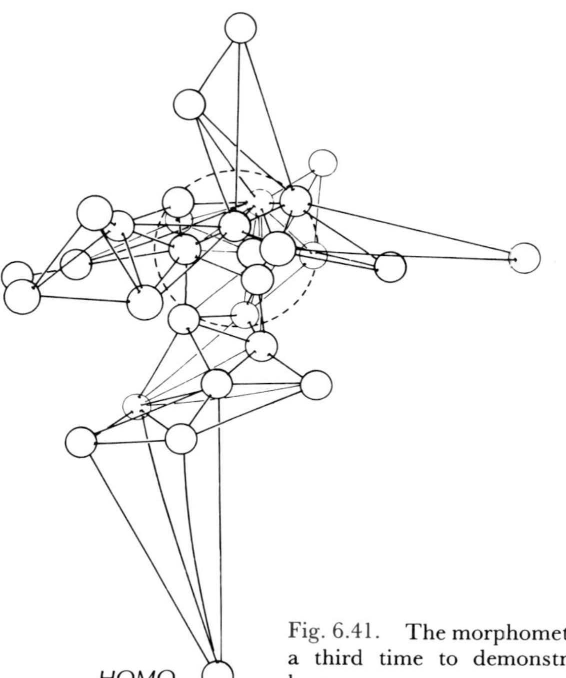

Fig. 6.41. The morphometric model of Fig. 6.39 rotated a third time to demonstrate the unique position of humans.

the fossil lower limb belonged. If, for instance, it fell within the centre of the star, though clearly adapted for bearing the forces of regular quadrupedalism, we could not tell whether it was prosimian or New or Old World monkey; all occupy the centre of the star. We could not even deny that it was an ape for, though no extant ape occupies that position, who knows whether or not regular quadrupedal apes may have lived at one time? In the same way if it fell within outlying rays of the star, whether already existing or new, this would not necessarily provide information about taxonomic position. Many of the outlying rays contain species quite widely separated taxonomically.

Because, however, humans as a group are so utterly separated from all other primates in the morphology of their hind limbs, we could well expect human precursors to be intermediate between the positions of humans and the entire star. So far, no such fossils have been discovered. Certainly, such information as we have about the australopithecines does not place them in this intermediate position (see Chapter 10).

Lower limbs: taxonomy

An initial reaction to these results might be to question them because they are so different from the taxonomic arrangements of the primates outlined in Chapter 2. Yet there is so much confirming evidence from the different studies that, as with the upper limbs, we are forced to accept the functional view. As with the upper limbs, however, it is worth asking if any aspect of the form of the lower limb seems related to the overall relationships of the primates. For it is well known that individual meristic features are so related; a pre-pubic bone is, for instance, especially diagnostic of primitive mammals. Such features have long been used to classify fossils.

Equivalent characters are indeed found among the various quantitative features described above. Thus, though the main aim of the study of the pelvis is to define features of distinct locomotor import, it also proved possible to define other features whose locomotor association is indeterminate or even nil. These 'residual' features (as they are styled in the original publications and summarized in table 6.8) especially demonstrate, on both univariate and multivariate examination, arrangements of primate genera more similar to taxonomy than to function. But the analyses show that these features also contain considerable functional information, and, when they are allied to the suite of locomotor features in the overall analysis, it is the functional associations that stand most clearly revealed. Nevertheless, a measure of classificatory information is contained within these dimensions, and this is a facet of the results to which we shall return in Chapter 8.

Summary. This review of structural-functional associations within primate lower limbs starts first with a more restricted problem: structural-functional associations within the lower limbs of those prosimians that are the most specialized leapers. The idea that such leaping within prosimians is a single biomechanical entity was useful at earlier times. But new information and ideas about prosimian locomotion and structure now reveal that lower limb function in leaping is better described in other ways.

We thus recognize the underlying difference between the many different forms of prosimian leaping that exist in living species: e.g. leaping with the trunk extended, leaping with the body curled up, leaping with the limbs hanging down. And we also note their relationships to the more quadrupedal habits of other less specialized prosimians. The average biomechanical situation for the lower limbs of these creatures is best seen as a star-shaped spectrum with the generalized species lying centrally within the nucleus of the star and the different more specialized leaping species lying peripherally, each within its own separate ray of the star.

Study by the multivariate statistical analysis of the third chapter of the hip, thigh and overall proportions of the lower limb, indicates structural arrangements of these prosimians that are also in the form of star-shaped spectra.

Again, the degree of concordance between these functional and structural spectra is very big and suggests that this quantitative view of lower limb structure truly mirrors the essential nub of lower limb function in these species.

A similar study of the entire order of primates indicates that lower limb function, when widened to include that of other types that leap much less but that do other special things with their lower limbs (e.g. hang by their feet, run cursorially on the ground), also forms a functional spectrum that is star-shaped. And again, when we view the multi-variate statistical result of the arrangements of the non-human genera in studies of the hip, thigh, foot and overall proportions of the lower limbs in the entire Order, a star-shaped structural spectrum is found. Humans, in contrast, do not lie within the star; as befits their totally unique form of locomotion and therefore function of the lower limb within it, they lie completely outside the star.

As with the study of the upper limb of the fifth chapter, the degree of concordance between the structural and functional spectra is very high. Again, therefore, we must conclude that the morphometrics of lower limbs and lower limb parts speak most strongly to function. And this implies that the examination of these parts in fossils is most likely to give information about function in fossils.

CHAPTER 7

Four Limbs and Quadrupeds

Abstract. In this chapter we discuss form and function in those primates which are usually described as quadrupeds. At this point we look at both upper and lower limbs, for both partake in quadrupedal posture and movements. Again, the early simpler view tends to separate each of the more extreme forms of locomotion among the primates and to place, in a single group and somewhat as a miscellany, all the less extreme primates as generalized quadrupeds. But more detailed and new natural history observations suggest that differences can be observed among the various generalized quadrupedal species. These differences, though small, are large enough to imply their own various average biomechanical situations with quadrupedalism (as discussed generally in the fourth chapter). Thus, we can identify separate functional spectra within the generalized quadrupeds of both New and Old World monkeys.

In the same way, at first sight, structural descriptions of various parts of both upper and lower limbs appear to confirm the notion of a single group of quadrupedal forms. But when more detailed attention is focused upon the quadrupedal species alone, and when methods of display such as the high-dimensional technique of Andrews (Chapter 3) are brought to bear upon the results, then a series of structural spectra can be discerned.

Once again, there is a marked concordance between the functional and structural spectra. Again, we see that quite clear differences in overall structure can be associated with differences in function, even though the separations described here are considerably finer than those in prior chapters.

In this case there are a number of very special caveats relating to fossils, because primate quadrupedalism of one kind or another may have been continually recreated during the long evolution of the entire Order.

Introduction

Although the subject of our last chapter, leaping, is a conspicuous activity of many primates, the most general form of locomotion among primates is simple four-footed movement. And this, of course, though it may depend somewhat more upon one pair of limbs than upon the other, enjoins for the most part relatively equal participation of all four limbs. It is an activity of which all primates, save humans, are capable and which characterizes many of them, including representatives of all major taxonomic groups. Paradoxically, perhaps, primate quadrupedalism has been much less studied than the more exotic forms of primate movement, such as leaping and brachiating, considerably less even than the quadrupedalism of many other animals.

What is animal quadrupedalism: essays into biomechanics?

Biomechanical and anatomical studies of such interesting modes of locomotion as swimming, flying and leaping have been pursued by many investigators over the years (e.g. summarized in Gray, 1968). Nevertheless, although less remarkable in its manifestations, that simple form of movement exhibited by many land vertebrates, quadrupedalism, has also received much general attention.

As with many other locomotor studies, the basic principles inherent in Newton's laws of motion are one major starting point. They remind us that if an animal is to move its own body by its own efforts, then it must elicit a force from its environment; this is, of course, especially so if it is radically to change its speed or direction of motion. They remind us that, as with engines so with animals – relationships must exist between forces exerted by motors (muscles), times over which such forces act, speeds and accelerations, and masses being moved. And finally they remind us that muscles with their tendons can be considered both as power-producing engines and as flexible tension-bearing links; that bones and their ligaments can be thought of both as mechanisms such as levers and pulleys of various sorts and as pressure and tension-bearing struts and ties. Such concepts apply to all animal forms and are summarized in such writings as those of Gray (1968), Alexander (1968) and Hill (1970), who themselves have been among the foremost contributors to this field for many years.

Gray, for instance, has provided the analogy of the quadruped as a loaded, flexible beam (the body) both supported and propelled by four flexible and extensible pillars (the limbs). This model has shown us how the entire skeletal musculature of the trunk and limbs must be regarded as a single very highly co-ordinated functional unit and that tensile force developed in any one muscle is very precisely, if complexly, related to those developed in all of the others.

This model has provided us with insight into the notion that the limbs, in effecting animal propulsion in quadrupedalism, function not only as propulsive levers but also as propulsive struts. The propulsive lever is readily recognized and an excellent description is that provided by Gray himself, i.e. as equivalent to an oar in propelling a skiff. The propulsive strut is less intuitively obvious, but is again readily recognized through Gray's analogy with a pole propelling a punt.

Finally, this model has also allowed us to recognize how the diagonal pattern of co-ordination between the four limbs in many quadrupeds allows better control of the movements of the centre of gravity of the body, in particular a minimization of less stable and less efficient pitching and rolling movements, and a preparation for the stable postures of animals when coming to rest.

The quadrupedal theme in vertebrates

Within the limits imposed by these reflections on the implications of Newton's laws for moving animals, other investigations have been carried out to study differences between different kinds of quadrupedal locomotion. A first, very obvious and, as it turns out, somewhat misleading differentiation has long been thought to exist between quadrupedalism in amphibians and reptiles on the one hand and in mammals on the other.

While cranio-caudal swinging movements of the limbs are obvious in any tetrapod, the movements displayed by most amphibians and reptiles are additionally characterized by relatively horizontally aligned limbs supporting the body in splayed or sprawling positions and by alternate horizontal bending of the trunk increasing the reach of the stride on the convex side.

Anatomical arrangements in animals that move in these ways are clearly appropriately suited. The feet of such amphibians and reptiles comprise a large number of bones that allow them to accommodate themselves to a large and irregular contact with the supporting surface. The distal levers of the limbs (tibia and fibula, radius and ulna, respectively) are provided with rotatory capacities allowing these bones to cross over one another so that the feet can be maintained in forward pointing positions all the time that they are on the ground, however the positions of the more proximal limb elements may change. The proximal levers of the limbs (femur and pelvic girdle, humerus and shoulder girdle) are so structured that the extra strength and stability required to maintain sprawling positions are provided through large and stable skeletal and muscular struts and ties with the trunk. And, finally, the structure of the trunk itself is such as to allow and promote the lateral flexion that is so important an element in helping the swinging movements of the limbs to provide increased reach and greater propulsive force against the ground.

Mammals, in contrast, though displaying cranio-caudal swinging movements of the limbs just as in other forms, possess locomotor patterns additionally characterized by limbs tucked underneath the body with relatively vertical dispositions and by vertical bendings (flexions and extensions) of the trunk which increase stride length during extension of the vertebral column.

Anatomical architectures in mammals are equivalently adapted. The feet are arranged so that the points of contact are very much smaller, frequently only the fore foot, often only the free toes, sometimes even only the tip of a single digit. The distal levers of the limbs (tibia and fibula, radius and ulna, respectively) are usually much less capable of rotary movements; indeed, in some cases they are actually bound by ligaments or fused by bone so as to prevent rotation. The proximal levers of the limbs (femur and pelvic girdle, humerus and shoulder girdle) are so structured that, although strength and stability are present, the extra stabilities imposed by sprawling postures do not exist. Finally, the structure of the trunk itself is such as to allow and in some forms promote the vertical (dorso-ventral) bending movements which are so clearly seen in such extreme running forms as the cheetah and the greyhound.

These conventional descriptions of reptilian and mammalian quadrupedalism have, in more recent years, been shown to be not as clear-cut as was earlier believed (Jenkins, 1971; Crompton and Jenkins, 1973). It is rather likely, for instance, that the reptilian sprawling posture is not typical of all reptiles. Recent investigations suggest that perhaps a number of the larger fossil reptiles were, indeed, able to approximate somewhat more closely to modes of movement resembling, although probably only in parallel ways, what occurs in some mammals. Thus, although in none of the extinct reptilian groups were the limbs carried totally underneath the body with vertical (dorso-ventral) body flexion as in mammals, there is some evidence that in pelycosaurs, for example, increased size, increased power and speed and perhaps other factors may have resulted in adaptations paralleling to a degree what is found in some mammals today.

And it is also the case that the picture described as typical for all mammals is, indeed, found only in those mammals that are specialized for fast terrestrial locomotion. Thus, although such species as the cat and the dog, antelope and zebra, conform rather well to the description postulated as 'typically mammalian', recent studies of other species (Jenkins, 1970, 1971), such as some monotremes, marsupials, rodents and shrews, demonstrate that the classical concept is quite inaccurate as a description of what transpires in many mammalian quadrupeds. In these latter species the feet may maintain far greater areas of contact with the ground, considerable rotatory movements of the more distal portions of the limbs may occur and the more proximal portions of the limbs may be placed very far from vertically.

Forming yet another linking feature between mammalian and reptilian locomotor patterns, studies of the monotremes of Australia, animals such as echidna, traditionally described as having a reptilian mode of locomotion with sprawling limbs, have shown that they are far more mammalian in character than previously believed. In echidna, the limbs support the trunk well away from the ground, even when walking slowly or at rest; and the movements and positions of the limbs are little different from those found in some of the other non-cursorial mammals described above (Jenkins, 1974).

Even vertebral column function may be less decisively different in the so-called reptile-mammalian locomotor dichotomy than previously believed. Radiographic and cine-radiographic studies of trunk movements in living mammals (and even casual observation of some species) indicate that horizontal bending movements of the trunk do exist in quadrupedal movement in at least some mammals, especially the least cursorial ones. They are not, of course, anywhere near as great as the horizontal movements found in reptiles and amphibians, and they are of so small a degree that they are unlikely to have the same enhancing effect on stride length; but they do indeed exist, and their significance remains open to investigation.

Finally, certain recent studies are showing that among those reptilian species that may be telling us something of the evolution of mammals, there is a lack of a simple dividing line between reptilian and mammalian locomotor quadrupedalism. In many features of their post-cranial anatomy, certain fossil reptiles, cynodonts, show in spite of the very non-mammalian appearances of some parts (e.g. the humerus, the femur) features of other parts (e.g. the shape of the limb girdles and the articular relationships of the proximal long bones) that suggest some elements of locomotion and posture not overly far removed from what can be seen in mammals (Crompton and Jenkins, 1973).

The concept that quadrupedalism in mammals is no single type of movement but consists of a variety of different forms has also been appreciated through rather different studies of gait patterns. Some of these investigations are longstanding; Muybridge (1899) was interested in such matters in the last century and others are doing work at the present day (e.g. Hildebrand, 1967, 1976). Hildebrand places on a scientific footing much information that has been known to animal husbandry for a long time. And, of course, he takes such studies very much further.

Hildebrand describes most succinctly some of the different forms of gait as defined through the pattern of relative placement and time of limb contact during the locomotor cycle. He is able to give precise descriptors for such symmetrical gaits of mammals as walking, trotting and running, and such asymmetrical gaits as the half bound, moderate gallop and fast gallop. These investigations have not yet been conspicuously used to find morphological correlates among the animals concerned. The horse, for instance, is capable of every one of the gaits named above. But Hildebrand (1976, 1977) has shown how certain parts of the actual gait universe are confined to some animals alone, and how other parts of the possible gait universe are not occupied by any known species. The morphological implications of such studies remain open questions, although Hildebrand (1980) has started to supply some biomechanical answers.

In summary, however, it is clear that posture and locomotion in the Class Mammalia should no longer be defined in terms of the adaptations of those cursorial species, such as horse and cheetah, which have traditionally been described. The diverse limb postures and movements of living mammals do not fit a simple monotypic scheme. And, if there exist living mammals that retain the postural and locomotor patterns of their Cretaceous or early tertiary ancestors, they are to be found among the non-cursorial forms, such as the platypus, the mole and the shrew.

Arboreal quadrupedal behaviour of primates

This, then, is the background of quadrupedalism in land vertebrates against which we must set our ideas about quadrupedalism in primates. When we come to study what is known about primate quadrupedalism, we discover that rather little is available. Most studies, both of the behaviour of primates and of their morphology, have drawn attention to the more extreme forms of locomotor adaptation (e.g. leaping, arm-swinging, leg-hanging, knuckle-walking) that exist within the Order. Some of these are outlined in other chapters. In considering functional and evolutionary questions relating to the structural concomitants of the more bizarre locomotor behaviours, investigators have often compared them with 'generalized arboreal quadrupeds' (e.g. Napier and Napier, 1967). Sometimes the term means those primates with locomotor patterns describable as general four-footed manoeuvrability within many different parts of the arboreal environment; sometimes it refers to presumed appropriate morphological adaptations to such a locomotor mode; sometimes it implies presumed similarity to generalized phylogenetic precursors.

Many studies have, either directly or indirectly, depended upon the supposition that, indeed, the generalized arboreal quadruped (among primates) truly exists.

In one sense, this is a fair assumption; if we are interested in the nature of the differences of extreme species, it may, indeed, be useful to compare them with a bulk of non-extreme, apparently relatively general forms. And useful information has arisen from such comparisons (e.g. Napier and Napier, 1967). Some of this author's own findings, as in the locomotor 'star' of Chapter 6 and the locomotor 'band' of Chapter 5, owe much to such comparisons. But it is entirely possible that in the original assumption of the existence of a single group of non-extreme average forms — the 'quadrupeds' of so many authors — a great deal of important information may have been overlooked.

First, comparison of the extreme forms with the arboreal quadruped may produce information that is specialized, that is restricted to the extreme form under investigation and that does not, therefore, speak towards the more generalized adaptation of arboreal quadrupedalism itself. Such studies may, in fact, actually hide those more generalized adaptations of arboreal quadrupedalism that led towards the extreme forms.

Second, the very term 'arboreal quadruped' may be a fiction; arboreal quadrupedalism may exist in many different forms, having a rich internal structure of its own.

The inter-relationship of these two points depends upon whether or not the minor differences among arboreal quadrupeds resemble or differ from the major differences characterizing the extreme species; it is entirely possible that both resemblances and differences, both parallel and distinct trends exist among the different arboreal quadrupeds.

Many previous studies have recognized differences among the structures of animals that are generally grouped (e.g. Napier and Napier, 1967) as arboreal quadrupeds. And it is probably also a matter of common knowledge among those interested in the living primates that even simple locomotor profiles vary greatly for different primate quadrupeds. Few investigators would confuse the three quadrupedal behaviours exhibited by, say, marmosets, macaques and lemurs even if a momentary glimpse of the running animal was all that was available (and even if the real interests of these investigators were the sexual activities or communicative behaviours of these respective genera). Thus, the recognition and study of somewhat more subtle differences among various primate quadrupeds have, indeed, already been made by some investigators.

A few earlier workers (e.g. Mollison, 1910; Priemel, 1937; Erikson, 1963) note behavioural differences among quadrupeds and discuss individual structural differences that they think are adaptively related to the behaviours that they distinguish. This results in the recognition of behavioural and structural features capable of distinguishing (although not, of course, completely, because there is much overlap) between, for instance, regular branch-running quadrupeds, springing quadrupeds and climbing or acrobatic quadrupeds. These might be represented by genera of New World monkeys for example, such as marmosets, squirrel monkeys and uakaris, respectively.

More recently, variations and differences of even finer grade have become recognized. Thus, even within individual genera various investigators have summarized enough behavioural information that it is possible to note differences in quadrupedal behaviour (e.g. Tappen, 1960). As long ago as two decades, a review of the literature (Ashton and Oxnard, 1964a) notes that among different cercopitheques, for instance, some forms prefer simpler walking and running quadrupedal movement on larger boughs, while others more often frequent the smaller-branch milieu in which quadrupedalism includes more acrobatic locomotor activities that are correspondingly more complex. Ashton and Oxnard (1964a) further note that among the various langurs there are forms practising different kinds of quadrupedalism, ranging from those that spend a great deal of time on the ground in simpler quadrupedal activities, through to those most usually observed within the high canopy permitting more complex acrobatic movements within their basic quadrupedalism. Similar descriptions are presented for yet other Old World primate genera (e.g., mangabeys and macaques, Ashton and Oxnard, 1964a). And from these summaries, it even seems likely that equivalent differences exist in other primate groups; examples include such genera as howler and titi monkeys, but in these cases less is surely known about the animals so that detailed diagnoses can be less readily assumed.

Since that study of almost two decades ago, itself essentially a review of a much older literature, many newer investigations have provided information from field studies. And, although in a number of cases the new data have modified, indeed even radically changed our views as to what some of the animals actually do (who would have thought, for instance, that the male orang-utan was a competent ground-moving creature capable of outrunning a man among the forest trees: MacKinnon, 1974; or that Allen's bush-baby leaps primarily from a single foot (Jouffroy and Gasc, 1974), those data have also often confirmed and amplified many of the notions presented earlier. They have confirmed in detail the idea that not all species of genera, for instance langurs (Fleagle, 1976), colobs (Morbeck, 1976, 1979) and sakis (Fleagle, 1976, 1978), move in the same way.

It is rather likely, moreover, that the existence of such phenomena within different species accounts for the sometimes apparently contradictory views held by different authors. Thus, Erikson (1963) clearly believes that howler monkeys are capable of a degree of acrobatic activity in addition to their well-known quadrupedalism. Schon (1968) thinks that this is not so and takes some issue with Erikson. Such a contradiction probably does not rest upon mistaken observation (as this literature controversy would have one believe), but rather upon the likelihood that different howler monkeys do truly move in somewhat different ways, and even upon the possibility that the same howler monkey moves in different ways when in a different environment or under other different conditions of some kind or another (see Chapter 5). We are talking here of variation over and above that usual in most species.

The same sort of explanation may be true for other apparent differences of opinion in the literature. Thus, in her earlier studies of Hanuman langurs, Ripley (1967) took issue with the notions that Ashton and Oxnard (1964a) obtained from the literature, that langurs are capable of degrees of acrobatic arboreal activity allied to their quadrupedalism such that the fore limbs may be subject to tensile forces to a somewhat greater degree than in other quadrupedal species. However, Ripley's own later studies of grey langurs (1976a, b), now well-confirmed by other investigators (e.g. Fleagle, 1976, 1978) have shown that her disagreement is spurious; many langurs do actually perform activities resulting in tensile forces in the upper limb (see also Chapter 5).

Finally, an example can be proffered from

among the Prosimii. For it is likely that one known population of *Hapalemur simus* living in a particular milieu in Madagascar practises a rather different form of quadrupedal locomotion to the other known populations living in a different ecological habitat (personal communication, J. E. McArdle).

Structural correlates of quadrupedalism

More recent studies at levels below the genus have not only noted behavioural differences such as those above but have started to identify some of the associated morphological adaptations. For instance, one of the earlier investigations of the shoulder (Oxnard, 1967), although mainly aimed at describing major differences within the entire Order, notes as an aside that within the genera *Cercopithecus*, *Presbytis*, *Macaca* and *Cercocebus* (the only genera sufficiently largely represented in that study to allow this kind of breakdown), morphological differences exist between the more terrestrial and the more arboreal forms, the more four-footed and the more acrobatic species, those living more on larger supports as compared with those on finer supports, those living within the lower parts of the forest contrasted with those among the higher branches.

This could be observed, although with difficulty, in the examination of, for example, individual features of the primate shoulder and arm. It is especially evident in a study of the relative twist of the shoulder blade (Oxnard, 1967) or the relative length of the elbow part of the ulna (ulnar olecranon, Oxnard, 1963).

But it is considerably more evident in pilot multivariate analyses of those same data of Oxnard carried out by Manaster (1975). In each case, differences among species can be discerned that not only are such as to be associated with the behavioural differences just noted, but are also interpretable in ways making biomechanical sense, given the particular behaviours.

Since the tentative indications in those pilot studies, several investigations have been aimed directly at defining locomotor and structural differences at similar finer taxonomic levels. The first of these is the further work of Manaster (1975, 1979) who carried out detailed parallel studies on the genera *Cercopithecus*, *Cercocebus* and *Presbytis*. The results confirm and greatly extend the above suggestions.

Thus, Manaster finds structural differences that correspond rather well with the spectrum of known locomotor activities exhibited by species within these genera. Some of the findings suggest that there are anatomical trends that change in a gradual fashion from one end of the functional spectrum to the other. The changes make considerable biomechanical sense when viewed in the light of the functional differences. An example of this is an anatomical feature, such as the width of epicondyles of the humerus which may be thought of as being related to the general amount of forearm musculature. The bony structure is smallest at one end of the continuum, in the more terrestrial *Cercopithecus aethiops* (that also has less massive forearm musculature); it is intermediate in the semi-arboreal, semi-terrestrial cercopitheques, it is largest at the other end of the spectrum in the middle and upper canopy cercopitheques; such as *C. diana* (that have the most massive forearm musculature). The increased size of the forearm musculature may be associated with increased grasping in acrobatic activities throughout this spectrum.

On the other hand, others of Manaster's anatomical features vary in a different way, increasing to a peak and then decreasing from one end of the functional spectrum to the other. This suggests that the species that live in the middle regions are relatively more generalized and similar to one another regardless of taxonomic group and that, towards each end of the behavioural and environmental spectra, there are more specialized and more widely divergent anatomical features. An example of this type of feature is the form of the radius. This bone is shorter and more robust in the semi-arboreal, semi-terrestrial species (e.g. *Cercopithecus mitis*) living in the middle of the behavioural environmental continuum. It is longer and less robust at both of ends of this continuum (i.e. in both the more terrestrial forms, such as *C. aethiops*, and the more arboreal forms, such as *C. diana*). Robusticity may be less in both the extreme forms because each in its different way is involved with smaller forces in various activities. The more terrestrial forms bear smaller forces because they leap less than those species living among the large branches. The more arboreal forms bear smaller forces because, although they do, indeed, make considerable leaps within their more three-dimensional environment, some of the load of leaping is taken up by the increased elasticity of the smaller branches among which they operate. Thus, the similarity in anatomical features between more extreme terrestrial and arboreal forms, as compared with the intermediate species, may actually be spurious.

Subsequent to the work of Manaster, and using somewhat different methods, Fleagle (1976, 1978) has shown equivalent variations in other species. Thus, in one study he demonstrates that two species of *Presbytis* have different, biomechanically relevant adaptations which are most explicable on the grounds of equivalent locomotor differences that he himself has elaborated. For instance, he shows that the long bones of the hand of *Presbytis obscura* are more robust than those of *P. melalophos*, and that this is probably associated with the increased bending forces presumed to exist in *P. obscura* because of its increased propensities for quadrupedal progression. Likewise, he shows that the humeral head of *P. obscura* is less well-rounded than that of *P. melalophos* and this associates well with the increased shoulder mobility of *P. melalophos* in its more acrobatic climbing mode, as compared with the more fore and aft mobility of the humerus of *P. obscura* associated with its more quadrupedal mode.

Such features, anatomically much more marked, have been noted in many creatures such as spider monkeys and gibbons that are extremely acrobatic in the trees as compared with others, such as macaques and mangabeys, that are not. And Fleagle's findings associate very well with the similar findings of Ashton and Oxnard in studies over many years, in which distinctions are noted both between Hanuman langurs (more terrestrial and therefore less acrobatic) and other langurs (more arboreal and therefore more acrobatic), and among different grades of arboreal langurs (between middle and high canopy species) in various parameters of the shoulder, arm, forearm and pelvis (Oxnard, 1973a, 1975a).

Fleagle believes that this may also be the case in yet another group, the New World monkeys. For instance, he suggests that small differences in habitat utilization and locomotor behaviour between the species of the sub-family Pithecinae (*Pithecia* and *Chiropotes*) may well show somewhat similar patterns. *Pithecia* seems to be documented as a somewhat leaping arboreal form; *Chiropotes* is much less well known but may, indeed, be more of an arboreal climbing quadruped. Associated anatomical specializations may exist between these two species and may parallel to some degree those between the more acrobatic *Presbytis melalophos* and the more quadrupedal *Presbytis obscura*.

Yet other workers have started to study locomotor adaptation at this level of detail. Kinzey (1976) noted in passing a few differences between two types of titi monkey. Mittermeier (1978) in a more formal way noted considerable differences between two species of spider monkeys. Rodman (1979) quite overtly compares two forms of macaque.

Even among prosimians, similar interspecific, even intraspecific, differences may be found. Thus, although many authors, including myself at prior times, have recognized bush-babies as generally leaping forms, it is in fact fairly well known that there is a spectrum of leaping in this group.

McArdle (1978, 1981) has shown that many individual anatomical features of the lower limbs (e.g. relative limb proportions, extent of many muscle insertions and apparently related osteological dimensions around the hip and thigh) are arranged in a similar spectrum. In each case, the form of the particular anatomical feature seems to make considerable biomechanical sense when viewed in the light of the position of individual species in a behavioural spectrum that ranges from less able leapers (*Galago crassicaudatus*) to those that are among the best leapers of all (e.g. *G. alleni*, *G. senegalensis*). This spectrum is very much more obvious in multivariate statistical studies of these same genera (Oxnard, German and McArdle, 1981; Oxnard, German, Jouffroy and Lessertisseur, 1981).

A similar pattern exists within the cheirogaleines where, although often examined as a single genus because of shortage of materials, the anatomical study of the three separate species easily identifies anatomical features of the hind limb that seem to be associated with their differing quadrupedal, running, leaping abilities (Oxnard, German and McArdle, 1981; Oxnard, German, Jouffroy and Lessertisseur, 1981).

And finally, among various hominoids, there are equivalent findings. Differences between small and large-bodied gibbons (*Hylobates* and *Symphalangus*) and between slightly smaller and slightly larger-bodied chimpanzees (e.g. *Pan troglodytes* and *P. paniscus*), may be far more than simply size and may accompany differences in degree of arboreal locomotion (e.g. in the structure of the cancellous networks within the lumbar vertebrae: Yang and Oxnard, 1979; Oxnard and Yang, 1981).

The structure of quadrupeds: a deeper view

The evidence of previous sections demonstrates that function in different anatomical regions is reflected in morphological features (Chapter 4) and that the sensitivity of such features in mirroring functional differences is very much greater when they are summarized using multivariate statistical descriptors (Chapter 3).

Thus, single features, such as the twist of the scapula or the length of the olecranon process in the upper limb, easily differentiate animals which differ markedly in their locomotor activities; a highly twisted scapula and a short olecranon are found in creatures in which the extended elbow and shoulder participate in bearing tensile forces as a result of hanging activities (e.g. gibbons); these can be compared with an untwisted scapula and a longer olecranon in animals that perform these sorts of activities only rarely and inexpertly and which more usually move in quadrupedal ways (e.g. baboons). But such single features do not clearly separate the many intermediate primate genera which differ functionally to much lesser degrees.

The twist of the scapula, perhaps the very best of these features, separates *all* primate genera by only approximately six standard deviation units; differences between genera with small locomotor differences are correspondingly less, and it is scarcely possible at all, with knowledge of such a single feature, to characterize any single unknown specimen unless it falls in some fairly extreme position. The length of the olecranon process likewise, when taken by itself, differentiates the extreme forms by little more than five standard deviation units. But it scarcely allows for any finer morphological gradations among the less extreme species.

When, however, suites of such features are taken together, utilizing the multivariate statistical approach, the separations of genera are so much increased that the spread between extreme forms may be as much as 20 or 30 standard deviation units (as occurs both in the examination of the shoulder including the character defining the scapula twistedness and in the study of the arm and forearm including the character defining the relative length of the olecranon process).

Such increased sensitivity not only reveals very clearly the known differences between extreme forms, but it also allows us to seek differences among rather similar creatures that we may not have been able to separate clearly on the basis of visual assessment of univariate examination of individual characters alone.

Fine differences in quadrupedal behaviour. A series of finer studies have thus been carried out upon different regions of both the upper and lower limbs. And even here, the fineness of the distinctions is such that it is not particularly easy to see the contained information. Though the univariate studies reveal only a single morphological group of 'quadrupeds', even the multivariate studies reveal *at first glance* only a single grouping of 'quadrupedal forms'. Thus, the separate studies of the upper and lower limbs examined in previous chapters for the positions of extreme primates show non-extreme quadrupedal forms grouped into a single cluster contained within the dotted circles in Figs. 7.1 and 7.2.

However, detailed study of the matrix of generalized distances and of the total of significant canonical axes into which these distances may be partitioned, and of which these figures form merely a convenient summary, indicates that considerably more information exists. Accordingly, then, the data are re-examined here, *first*, in such a way as to look for differences among those species usually characterized as 'simply quadrupedal', and *second*, involving methods of display (high-dimensional analyses, see Chapter 3) that are capable of reflecting the more detailed and complex arrangements.

It turns out that analysis of individual quadrupedal species alone produces some marked groupings. One such group of the quadrupedal Old World genera includes those species that are more terrestrial and less acrobatic, more confined to larger (relative to the size of the animal) branches when in the trees and less at home in the small-branch milieu. These comprise some cercopitheques such as *Cercopithecus aethiops*, some mangabeys, such as *Cercocebus torquatus*, *C. atys* and *C. galeritus*, some macaques such, as *Macaca irus*, *M. mulatta* and *M. arctoides* and some langurs such as *Presbytis entellus*. They are labelled 'A' in the series of diagrams that follow.

A second cluster of Old World monkeys comprises those species that are least terrestrial (in fact, they rarely, if ever, go upon the ground) and most acrobatic, least frequently found in the large-branch milieu and most confined to smaller branches when in the trees. These species comprise forms such as *Cercopithecus diana* among the cercopitheques, *Presbytis obscura* and *P. kasi* among the langurs, together with other colobines such as the proboscis monkey (*Nasalis*), *retroussé*-nosed langur (*Rhinopithecus*) and various *Colobus* species. These forms are all labelled 'C' in the series of diagrams that follow.

Intermediate between these two sets in the locomotor, environmental and ecological definitions that have just been outlined are a series of other Old World genera. Among the cercopitheques these include various main-canopy species such as *Cercopithecus cephus*, *C. mona* and *C. nictitans*; among macaques they include the more arboreal forms (but not so extreme as *Cercopithecus diana* and *Presbytis obscurus*) such as *Macaca silenus*, *M. radiata* and *M. nemestrina*; among the mangabeys they include likewise the somewhat more arboreal forms, e.g. *Cercocebus albigena* and *C. aterrimus*. These various species are grouped together and

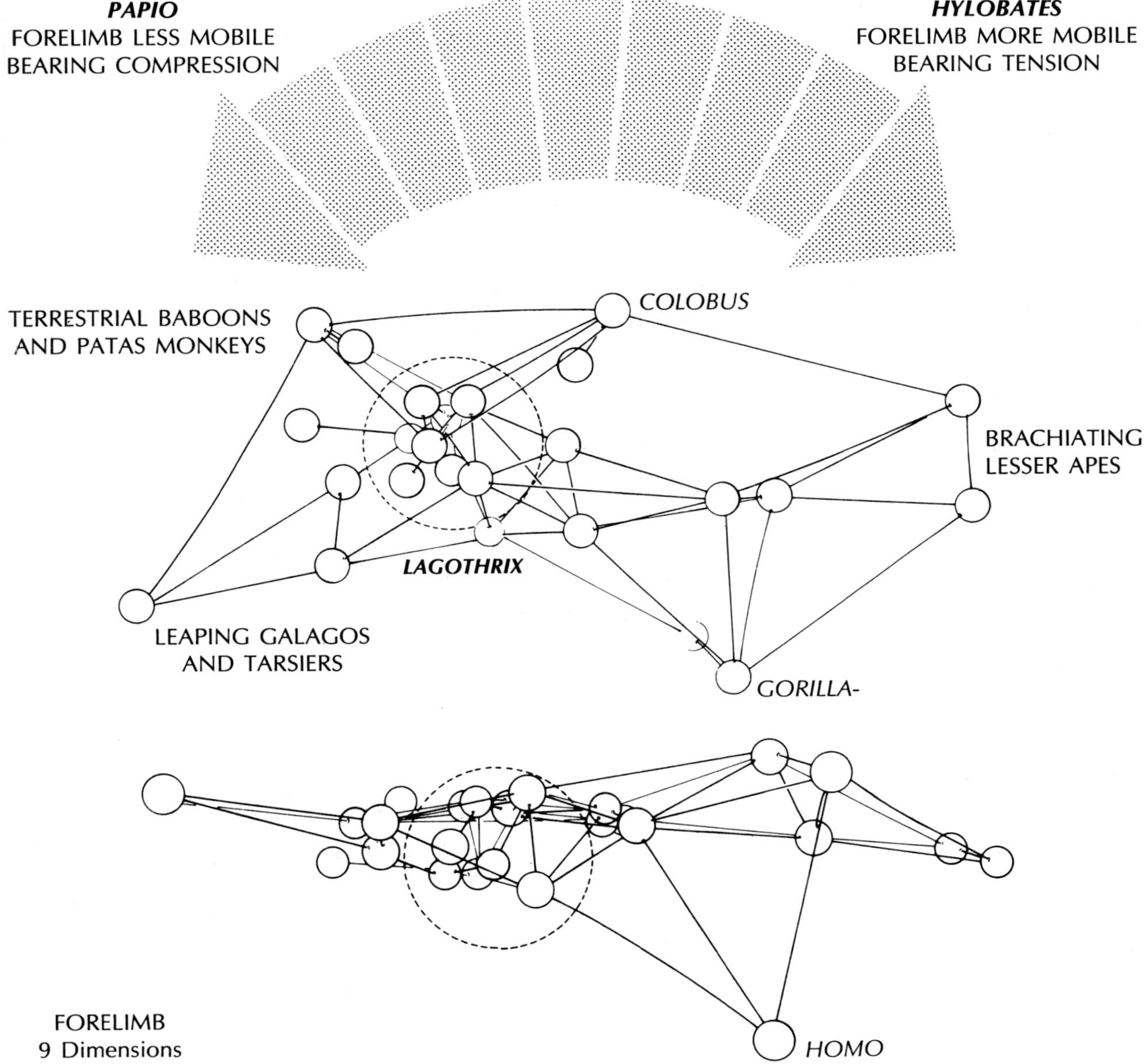

Fig. 7.1. This summarizes the result obtained from the multivariate statistical study of the overall proportions of the upper limb in primates (see Chapter 5). The large group of genera which are usually characterized as regular arboreal quadrupeds all lie close together within the dotted circle. The model is constructed from the generalized distances between the genera and its scale is some thirty standard deviation units.

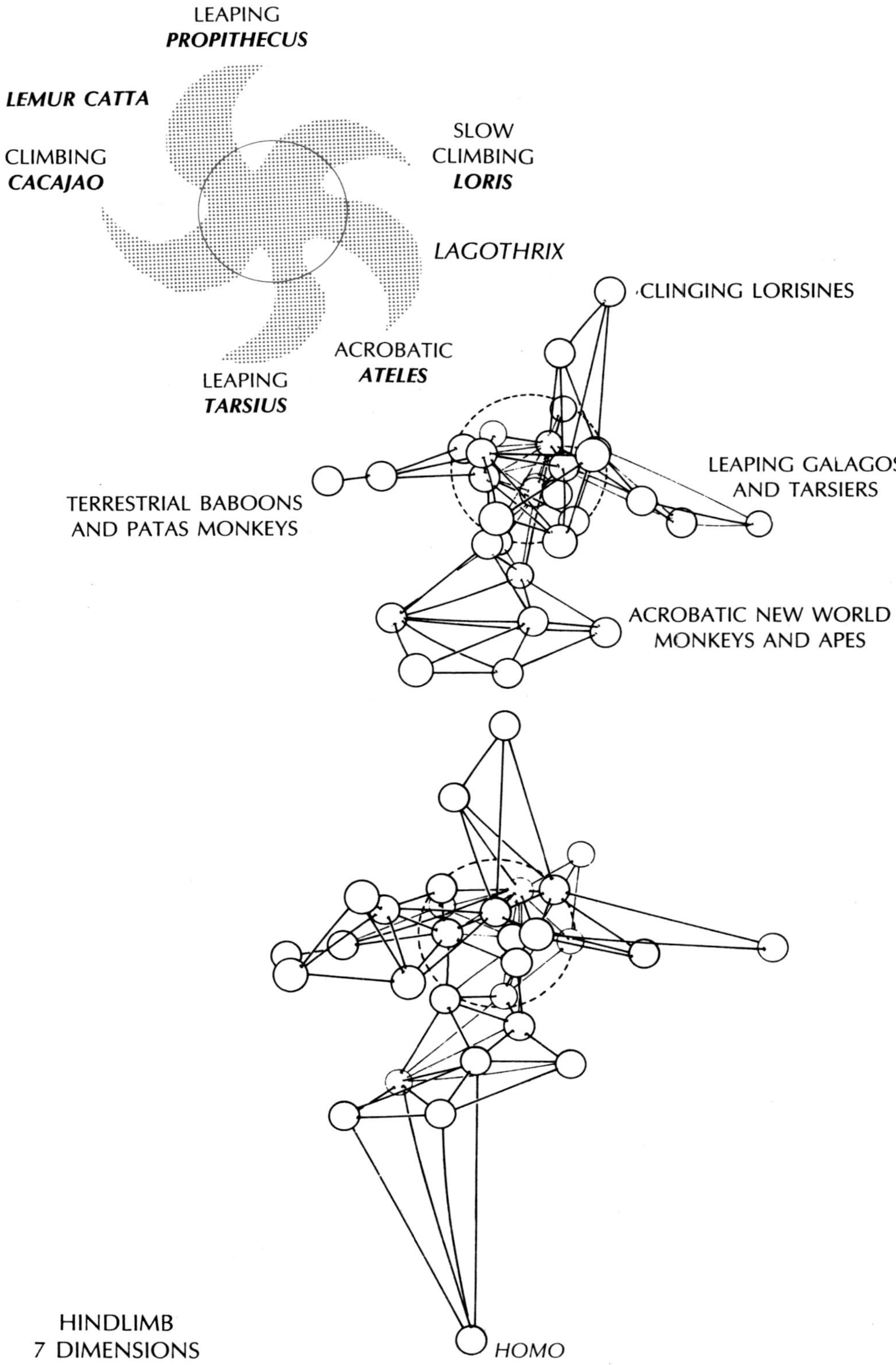

Fig. 7.2. This summarizes the result obtained from the multivariate statistical study of the overall proportions of the lower limb in primates (see Chapter 6). The large group of genera that are usually characterized as regular arboreal quadrupeds all lie close together within the dotted circle. The model is constructed from the generalized distances between the genera and its scale is some thirty standard deviation units.

labelled 'B' in the series of diagrams that follow.

In the same way, but carried out separately because of taxonomic differences between the major groups, a number of clusters of New World monkeys have been found. One of these has been labelled 'A' and comprises those forms that have more regularly quadrupedal, squirrel-like gaits, that seem to be confined to the larger boughs (relative to animal size) and that less frequently carry out non-quadrupedal acrobatic types of activities; included here are marmosets, tamarins and owl monkeys (the genera *Callithrix*, *Leontocebus*, *Callimico* and *Aotus*).

There is a second cluster of New World genera. These species are known to be fairly highly acrobatic in their different ways and to occupy, compared to body size, a smaller-branch milieu. They have been labelled 'C' in the diagrams that follow and comprise woolly and howler monkeys (the genera *Lagothrix* and *Alouatta*). The spider and woolly-spider monkeys (*Ateles* and *Brachyteles*) are exempted from this study on the grounds that they probably really are more comparable with the extreme brachiating primates, gibbons and siamangs. There could even be good reason for eliminating the woolly monkey on this basis; however, it is considerably more quadrupedal than *Ateles* and *Brachyteles* and approximates more to the quadrupedality of the howler monkey.

Finally, between these two are clustered a series of New World genera that have been labelled 'B' in the following diagrams: these include sakis, uakaris and squirrel monkeys (the genera *Cacajao*, *Pithecia*, *Chiropotes* and *Saimiri*) and they are considerably more acrobatic and versatile than the species in 'A' but also considerably less acrobatic than those in 'C'.

The single genus *Cebus* is, in an analysis like this, very difficult to place. It is not clear whether it is closer to 'B' or 'C' and it is, accordingly, kept distinct in the diagrams that follow.

In summary, then, those Old and New World monkeys that are generally regarded as quadrupedal are located in three groups each; and each of the three groups comprises genera that can be assigned as least acrobatic: 'A'; more acrobatic: 'B'; and most acrobatic: 'C' — even though the 'A's, 'B's and 'C's are different in the Old World monkeys than in the New. This display is, of course, somewhat artificial because the various species, even those within the same cluster, *can* be statistically differentiated from each other. But there is an overall reality to the descriptions. And in the display of the morphological similarities within the clusters, taxonomic effects have been overcome because each cluster combines species from several taxonomic groups. (This is an important part of the design of observations as mentioned previously in Chapter 4).

What then are the results of studying the morphologies of these creatures examined alone? They are the following and are best described by reference, separately, to the studies of the upper and lower limbs. The materials, measurements and methods are those already outlined in Chapter 3 (table 3.4) and in the examination of the upper limb (Chapter 5) and lower limb (Chapter 6) separately.

Fine differences in quadrupedal structures. Study of the shoulder (Fig. 7.3), arm and forearm (Fig. 7.4), shoulder, upper arm and forearm (Fig. 7.5) and overall proportions of the upper limb (Fig. 7.6) are based upon the materials and methods outlined in Chapter 5. Each study results in arrangements of the individual separate species (described using Andrews' high-dimensional display) so that, in both the Old and New World species, groups 'A' lie always at one extreme, groups 'C' lie always at the other and groups 'B' lie always intermediately. Among the New World monkeys the genus *Cebus*, which was not defined as belonging to any particular group, usually lies (in those studies in which it can be separated) intermediately between New World groups 'B' and 'C'.

In the Old World scapular studies (Fig. 7.3) and the New World overall upper limb studies (Fig. 7.6) the groups 'A', 'B' and 'C' are linearly separated with reference to the vertical scale of the diagrams. This means that in those particular studies a single discriminant axis (the first) has been mainly responsible for that particular separation.

But the linear separations of groups are not necessarily linear within a single dimension of the discriminant space of the various analyses. Thus, in most of the other studies, although there is always a separation so that 'B' is intermediate between 'A' and 'C', the waviness of the plots results in the separations being A-B-C in the vertical scale at some locations along the plots and C-B-A in the vertical scale at other locations. This means, in contrast to the above, that a discriminant axis other than the first, and usually more than one such axis, is responsible for the separations that exist (Figs. 7.3 through 7.6).

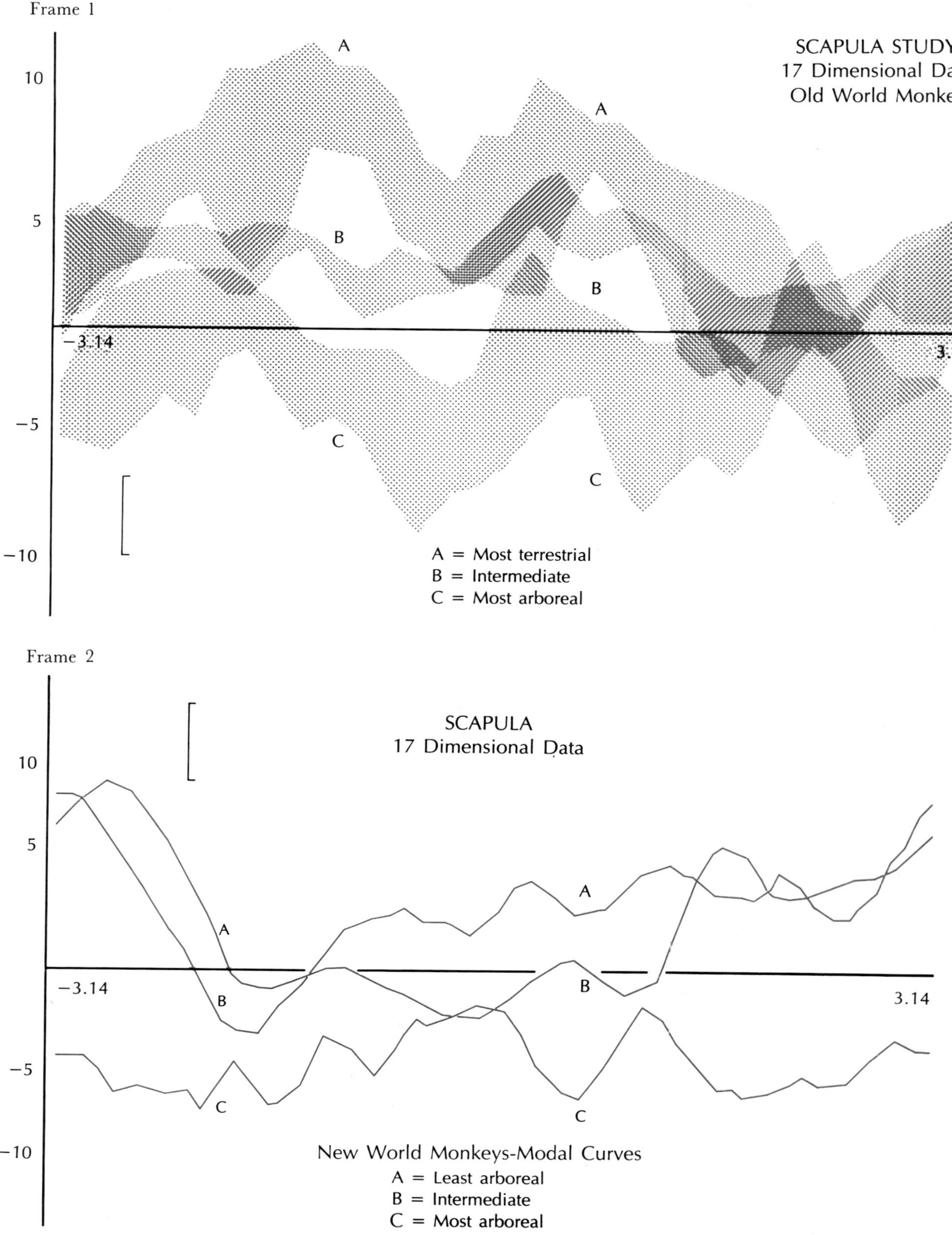

Fig. 7.3. Study, using Andrews' high-dimensional technique, of the regular arboreal quadrupeds shown within the dotted circles of Fig. 7.1. In this case data from the shoulder are examined. The Old World monkeys are separated as indicated by the shaded envelopes for the groups of genera; the separation between the New World genera are rather smaller and more easily seen by studying the modal (average) curves for the envelopes of the groups of animals. As explained in the next, the structure of the shoulder separates these arboreal quadrupeds according to the degree to which they are acrobatic in their usage of the upper limb in locomotion.

The fact that the order A-B-C is maintained irrespective of which or how many axes are involved demonstrates that the overall linear relationship of the morphology persists. It is this persistence that is the principal evidence suggesting that the morphological relationships exhibited are, indeed, associated with the similarly linearly related functional groupings posited before the analysis was made. Had some other feature such as taxonomy or size been the associated factor, then the arrangement of the groups would not have consistently followed this pattern.

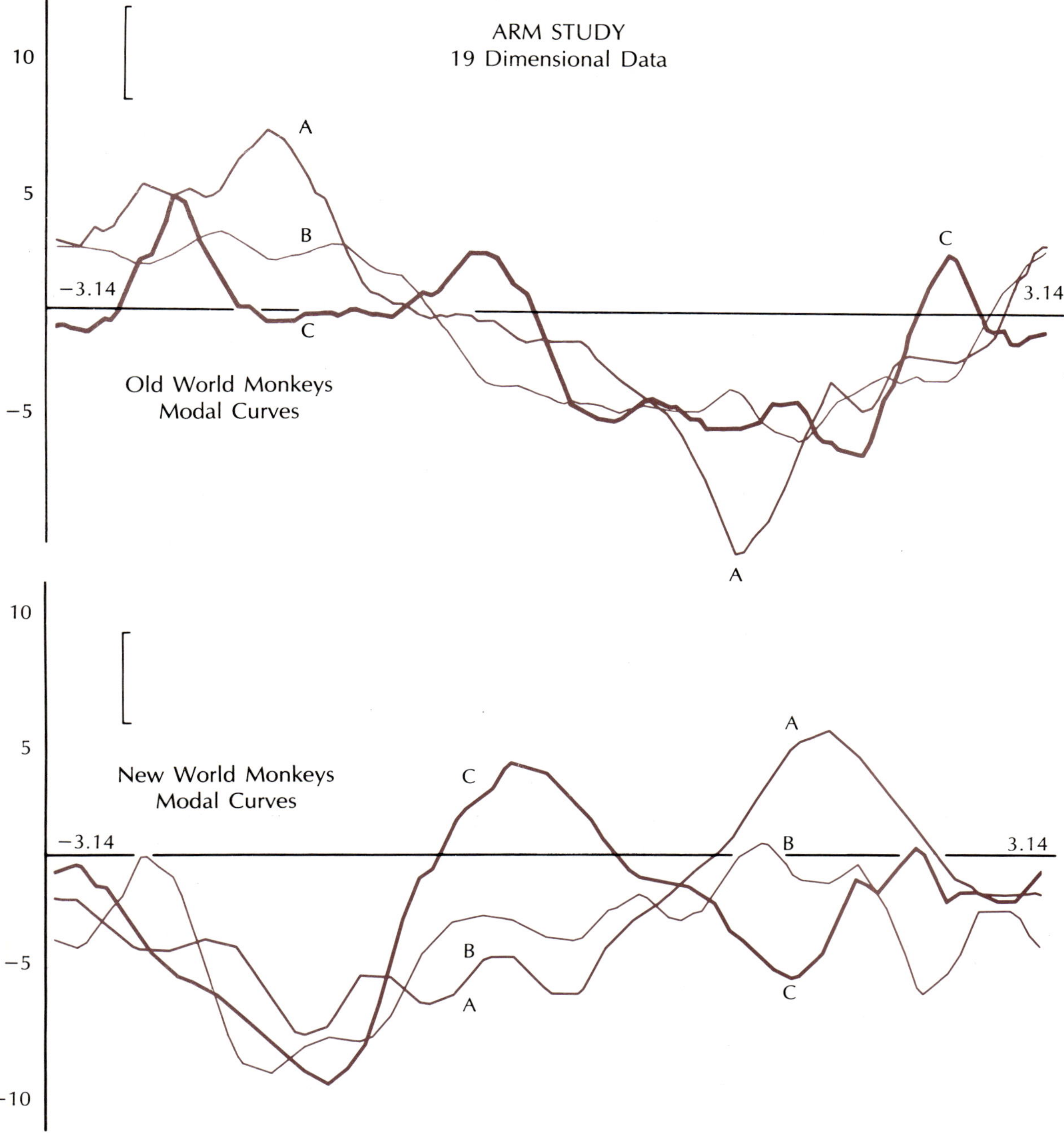

Fig. 7.4. Study, using Andrews' high-dimensional technique, of the regular arboreal quadrupeds shown within the dotted circles of Fig. 7.1. In this case data from the upper arm and forearm are examined through modal curves for each group of animals. As explained in the text, the structure of the upper arm and forearm separates these arboreal quadrupeds according to the degree to which they are acrobatic in their usage of the upper limb in locomotion.

And the fact that this pattern is replicated in each of the different anatomical regions of the upper limb likewise suggests that different internal functions of different anatomical areas (such as shoulders, elbows, wrists) are not the factors resulting in these separations. The factor or factors must, then, be acting equivalently in every area of the upper limb; and although it is possible that there are others, the most likely are the average biomechanical situations stemming from the locomotor functions of the upper limb as outlined above in the band-shaped spectrum.

In exactly the same way, studies of various anatomical regions of the lower limb, pelvis (Fig. 7.7) and lower limb as a whole (Fig. 7.8), have been carried out with the materials and methods given

Fig. 7.5. Study, using Andrews' high-dimensional technique, of the regular arboreal quadrupeds shown within the dotted circles of Fig. 7.1. In this case data from the shoulder, upper arm and forearm all taken together are examined. The Old World monkeys are separated as indicated by the shaded envelopes for the groups of genera; the separations between the New World genera are rather smaller and more easily seen by studying the modal (average) curves for the envelopes of the groups of animals. As explained in the text the structure of the shoulder, upper arm and forearm separates these arboreal quadrupeds according to the degree to which they are acrobatic in their usage of the upper limb in locomotion.

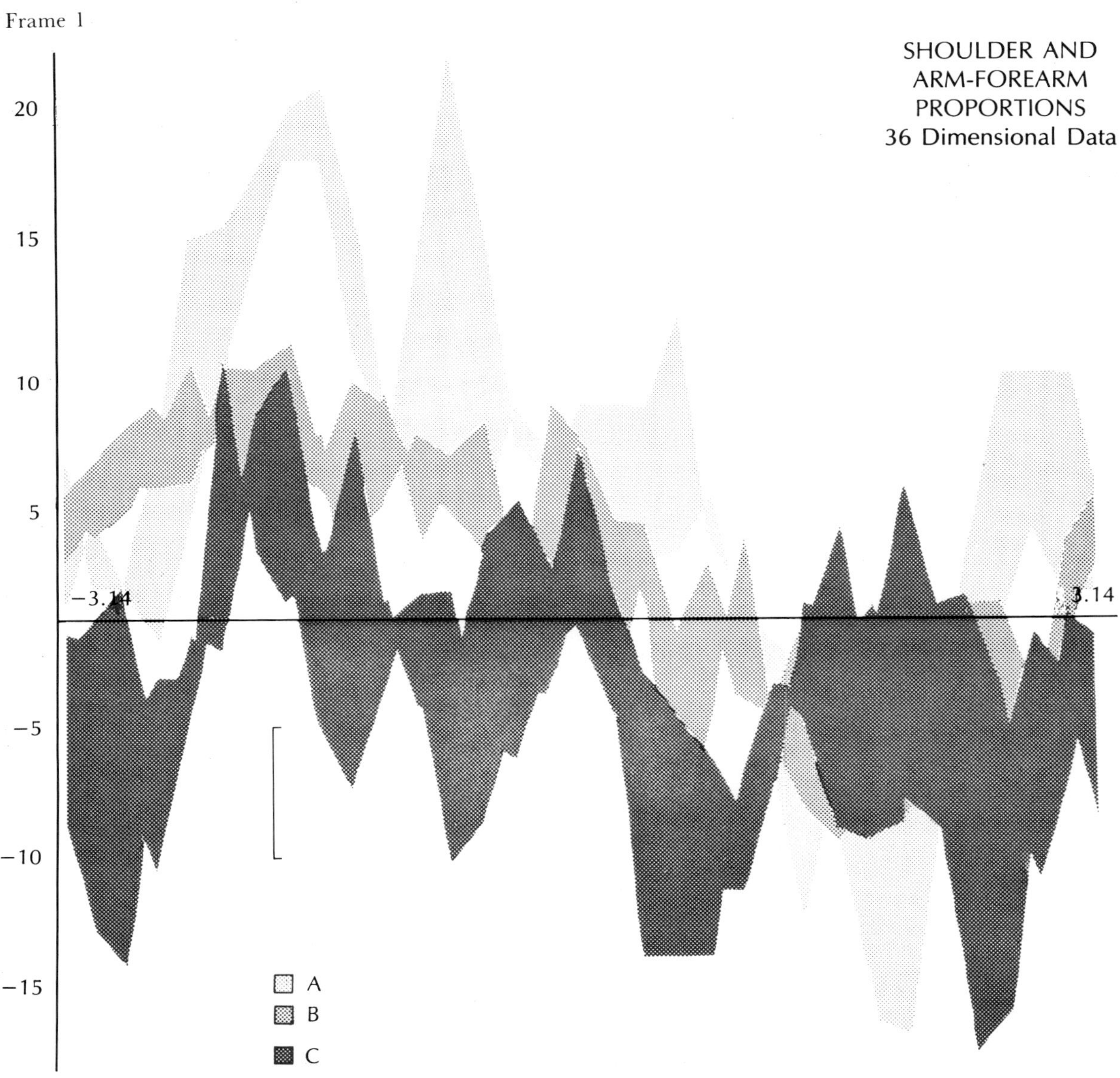

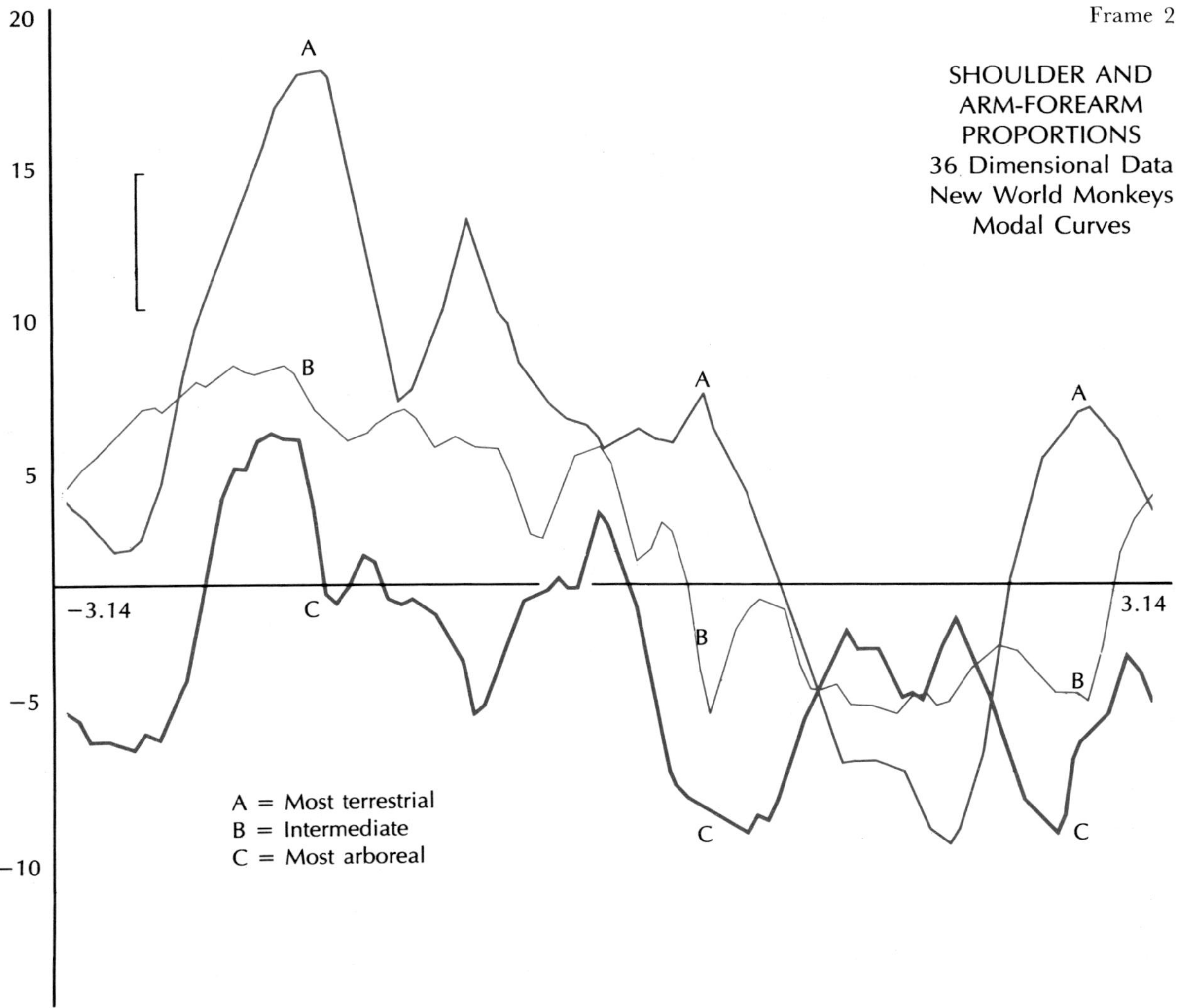

in Chapter 6. The studies proffer similar results. The separations of the genera, when displayed by Andrews' (1972, 1973) high-dimensional method, are such that they fall into the same clusters: 'A', 'B' and 'C'. And coincident with this, the clusters are again always in the linear series A-B-C and support the hypothesis that the arrangement relates to the locomotor, environmental and ecological associations that were the basis of our original descriptions of the species. For the same reasons as in the upper limb, it is unlikely that other biological factors such as size, taxonomy or special internal functions of individual anatomical subunits have resulted in this picture.

What may be thought to be a curiosity in these results is the apparent similarity between the picture for the upper limb and its various parts on the one hand, and that for the lower limb and its fractions on the other. Certainly, in the study of all primates, arrangements for upper and lower limbs are quite different; Figs. 7.1 and 7.2 show that while the upper limb arrangements are a broad, bi-polar band or swath, the lower limb arrangements are a multi-polar star. However, it is not necessarily the case that these two A-B-C arrangements indicate direct similarities between upper and lower limbs. For in an example like this, where we are restricted to only three groups, the only possibilities are either randomly arranged groups, such as A-C-B and C-A-B for the various studies, or a linear A-B-C relationship in each case. And with a linear arrangement of only three groups, we cannot differentiate between three groups in a simple bi-polar formation or three groups forming part of a larger multi-polar structure.

This difference between upper and lower limbs is

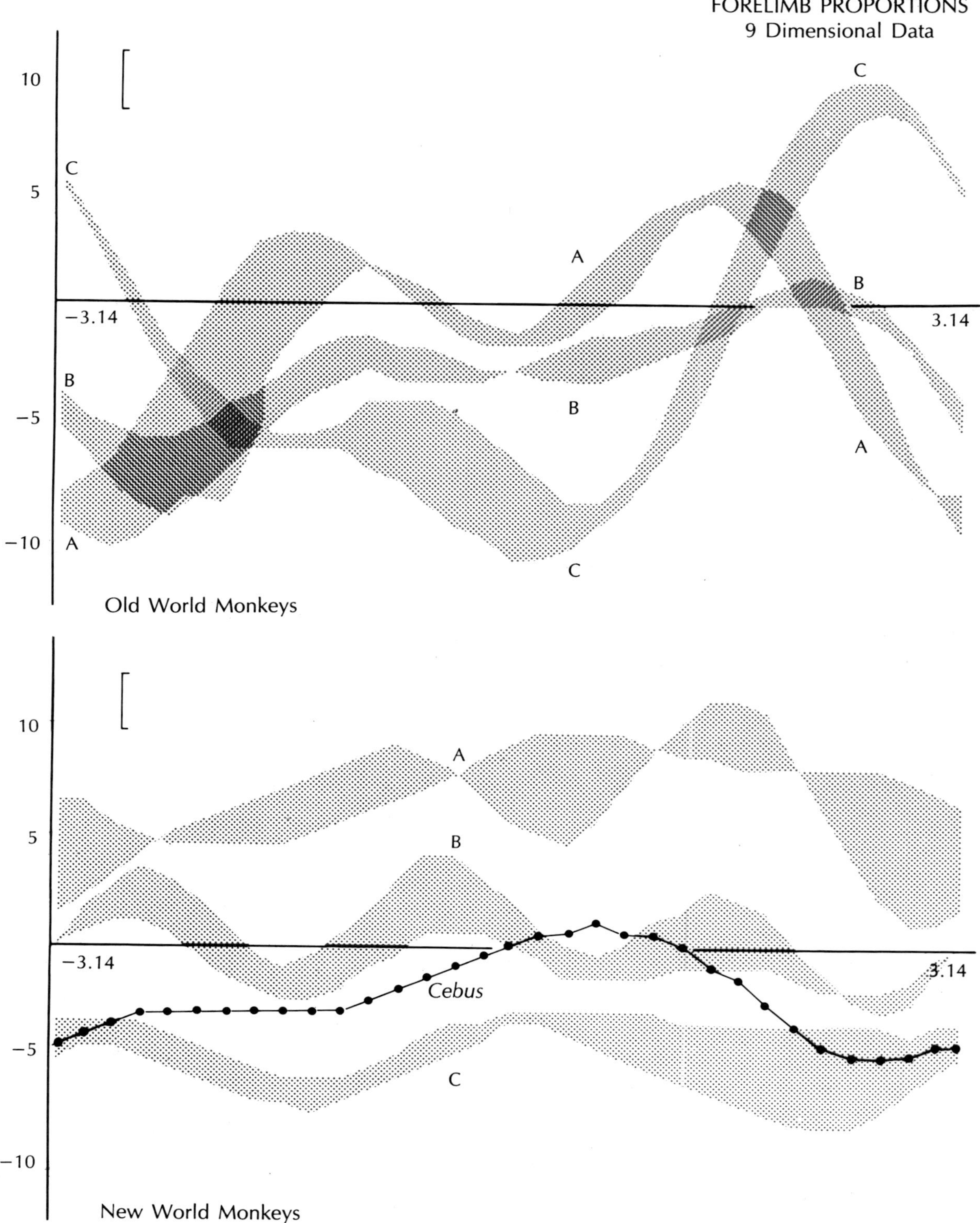

Fig. 7.6. Study, as in Figs. 7.3 through 7.5, of data from the entire upper limb. Again, the regular quadrupedal genera of Fig. 7.1 are separated according to the degree to which they display acrobatic usages of the upper limb within their regular quadrupedalism.

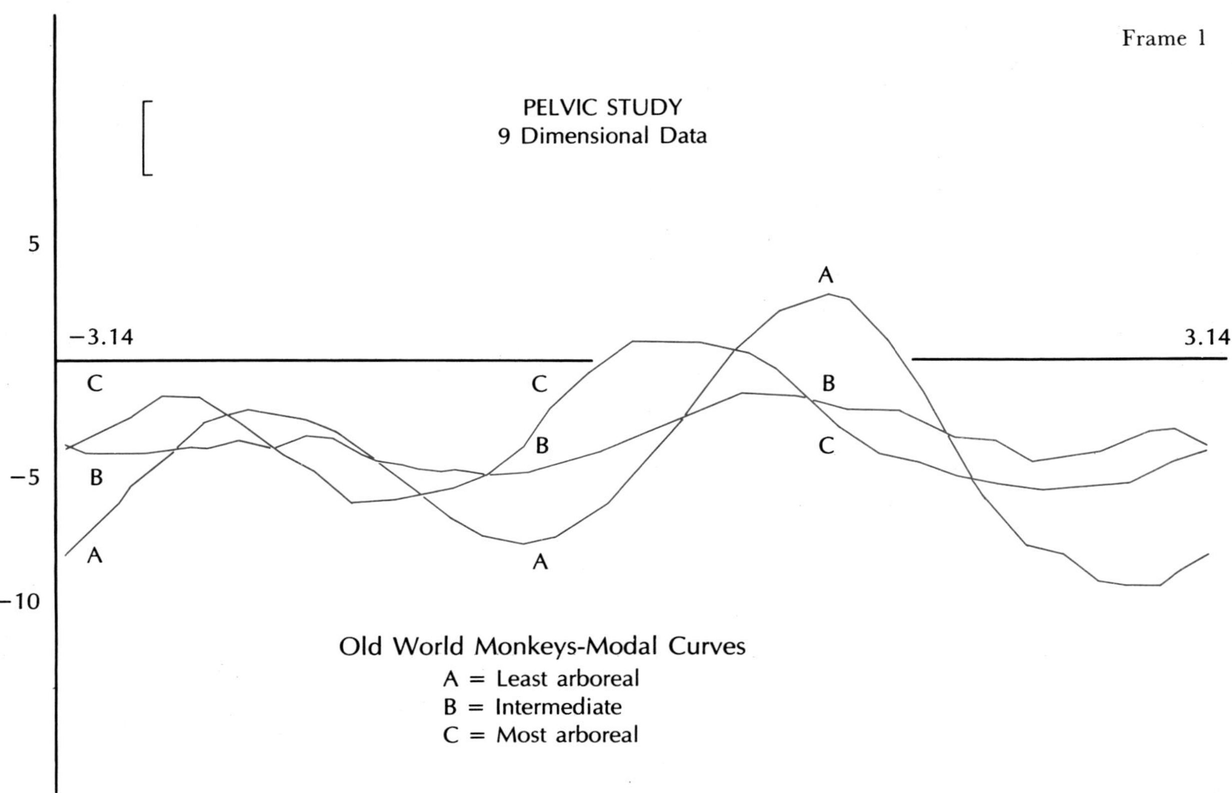

Fig. 7.7. Study, using Andrews' high-dimensional technique, of the regular arboreal quadrupeds shown in the dotted circle in Fig. 7.2. In this case data from the pelvis are examined through modal curves for the generic groups. As explained in the text, the various species are separated according to degrees of acrobatic activities of the lower limbs within their regular arboreal quadrupedalism.

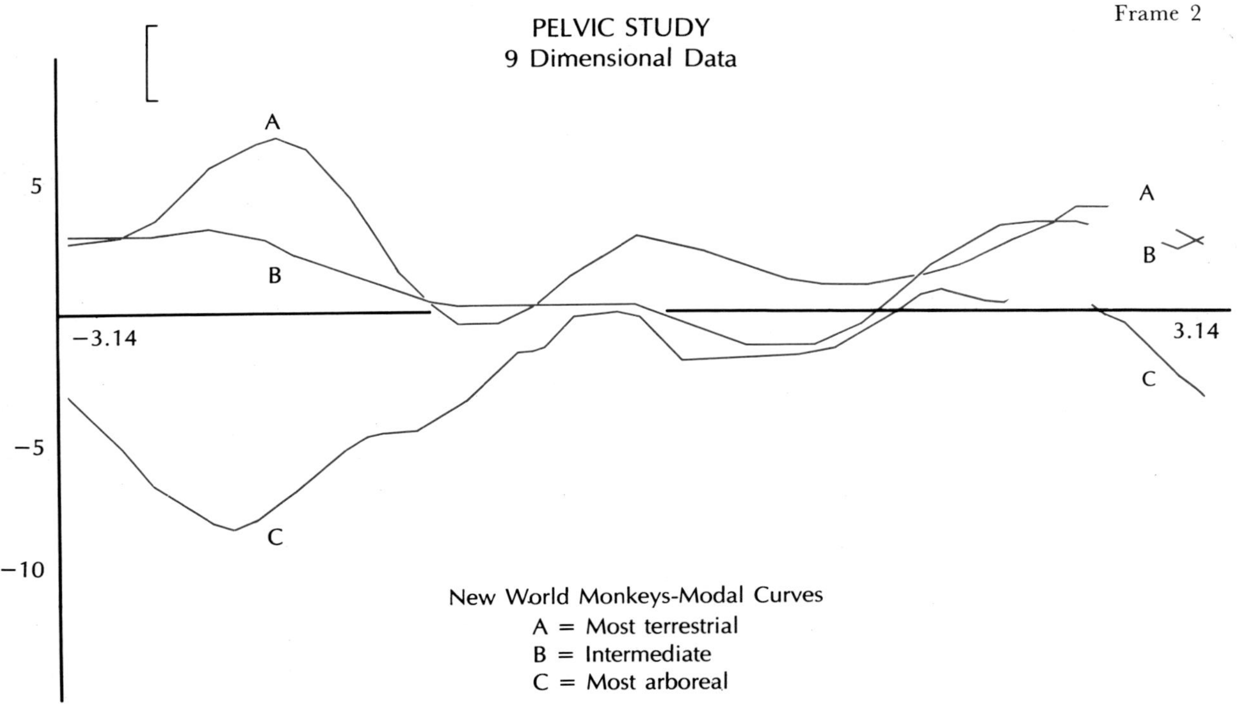

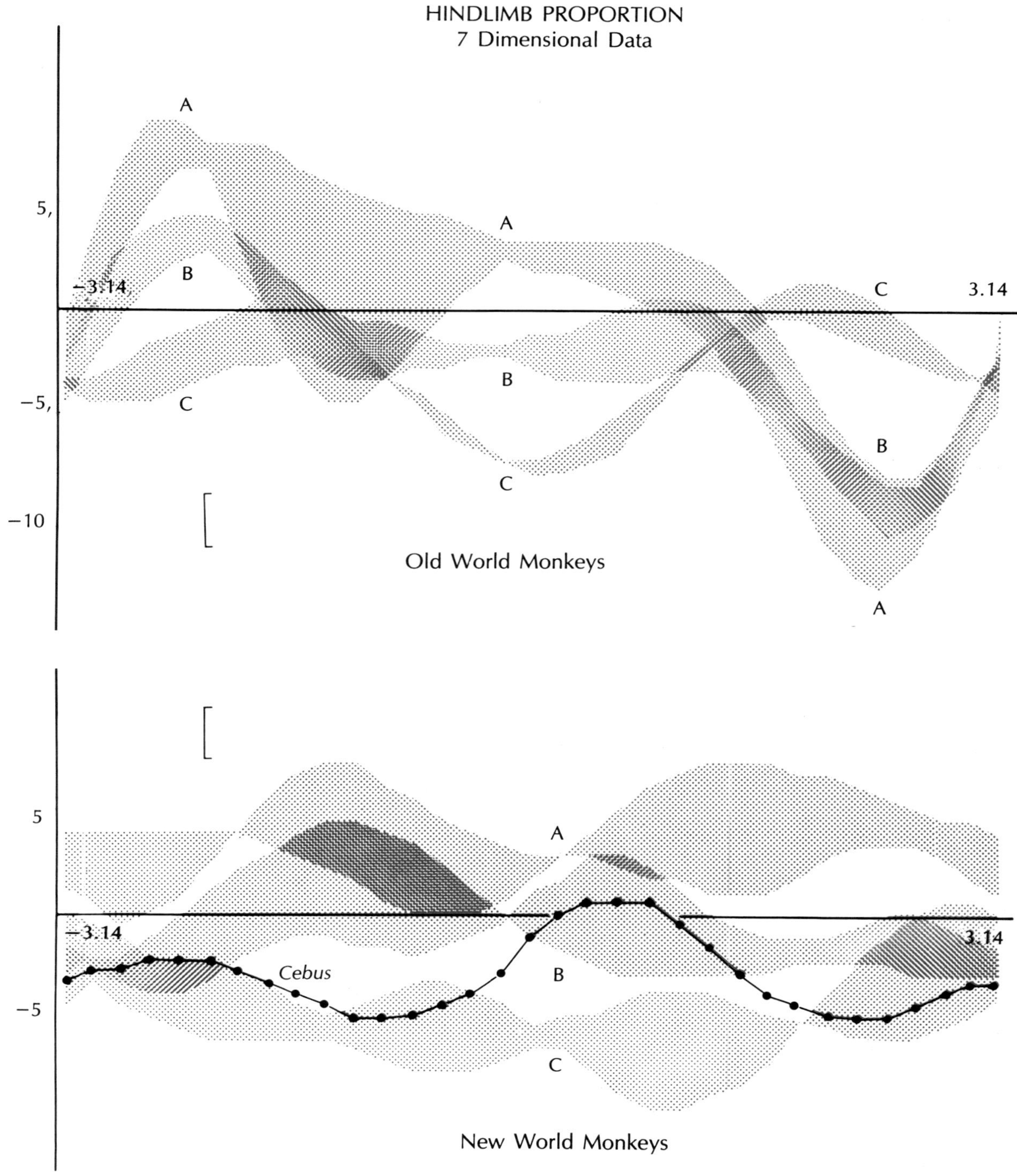

Fig. 7.8. Study, as in Fig. 7.6, of data from the entire lower limb results in clear-cut envelopes of the different animal groups. Again, regular arboreal quadrupeds are separated according to the differences in function of the lower limb within their arborealism as explained in the text.

highlighted when we examine correlations within and between them. Thus, within each limb the correlations are generally high rank orders of genera are generally similar (Fig. 7.9 and 7.10). But the apparent similarity between upper and lower limb patterns is spurious. This is shown by

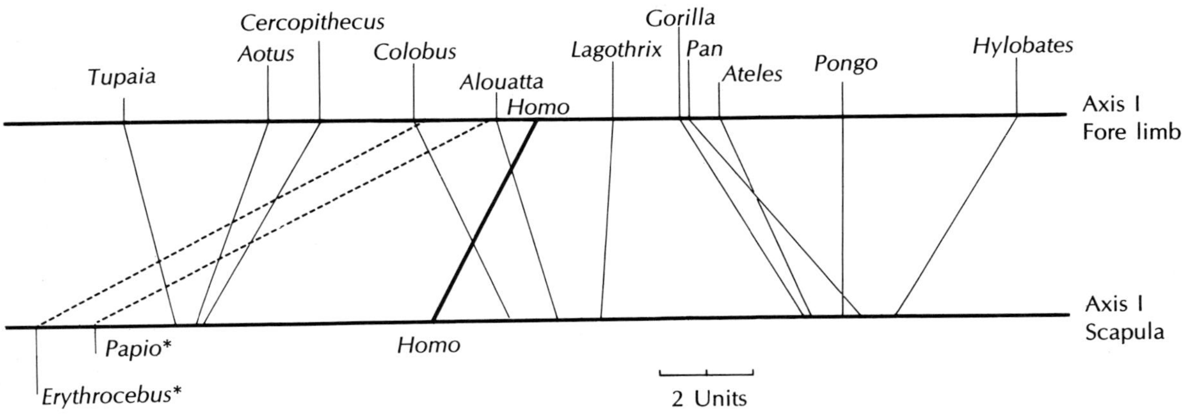

Fig. 7.9. The studies of Figs. 7.3 to 7.6 suggest similar pictures from various parts of the upper limb. This diagram shows that, indeed, the correlation between the different parts of the upper limb is high; the rank orders of the various genera in the first axes of these studies are very similar; that is, lines connecting the same genus in each study do not cross one another over much.

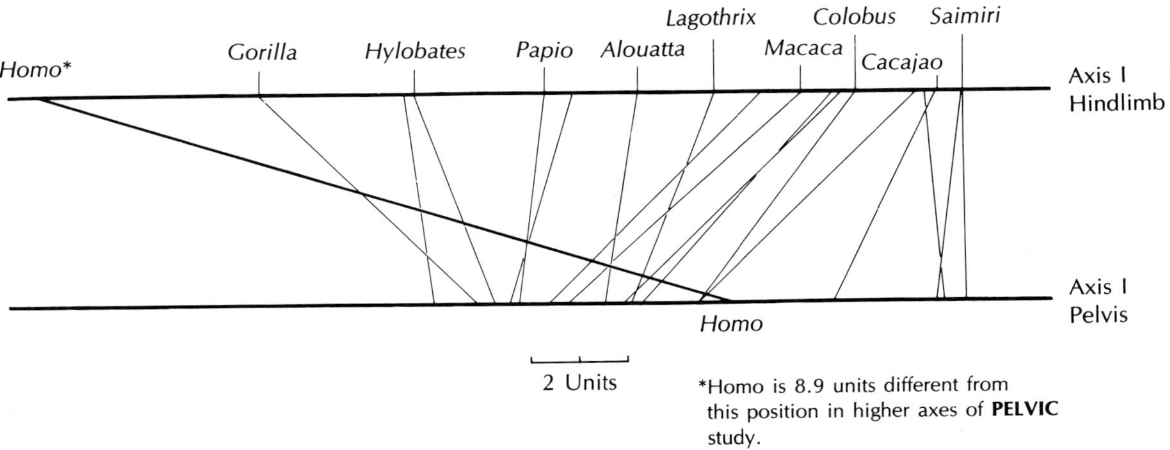

Fig. 7.10. In the same manner as Fig. 7.9, the studies of Figs. 7.7 and 7.8 suggest similar pictures from two parts of the lower limb. This diagram shows that the information presented by these two studies is indeed consonant; the rank orders of the various non-human genera in the first axes of these studies are very similar; that is, lines connecting the same genus in each study do not cross one another over much.

the low correlations that exist between equivalent upper and lower limb analyses (Fig. 7.11; see also Fig. 7.12 for explanation of these diagrams). This indicates that the apparently similar patterns for the studies of each limb as a separate unit are actually produced by different relationships between the variables within each limb respectively. And it is this that presumably results in the overall different patterns (band-like and star-like, respectively) when all primate species are included in the analyses.

Another feature of these results that speaks to

this matter can be seen by examining the results of study of the different anatomical areas. It is provided by the fact that the pattern, although linear in each study, is not identical in each study. Thus, in the upper limb and lower limb studies, the groups 'A', 'B' and 'C' of the New World monkeys are displaced vertically in the plots. It is rather likely that what is being replicated is a minor reflection of the similar first discriminant axis separation of the study of all primates: that is, for the New World monkeys, the separations are perhaps those equivalent to the same major separations found within the primates as a whole. And for the primates as a whole, as we have seen, a large

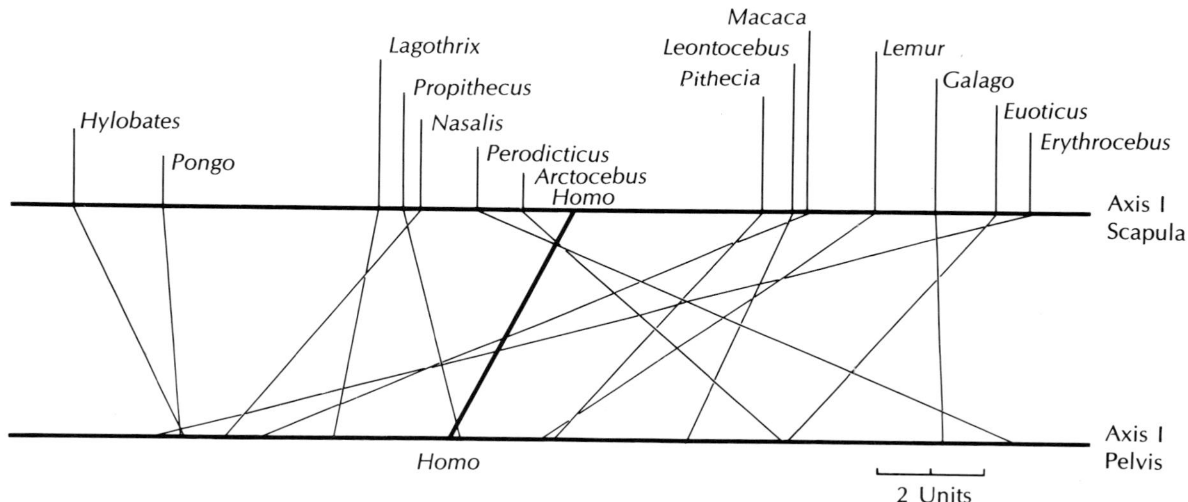

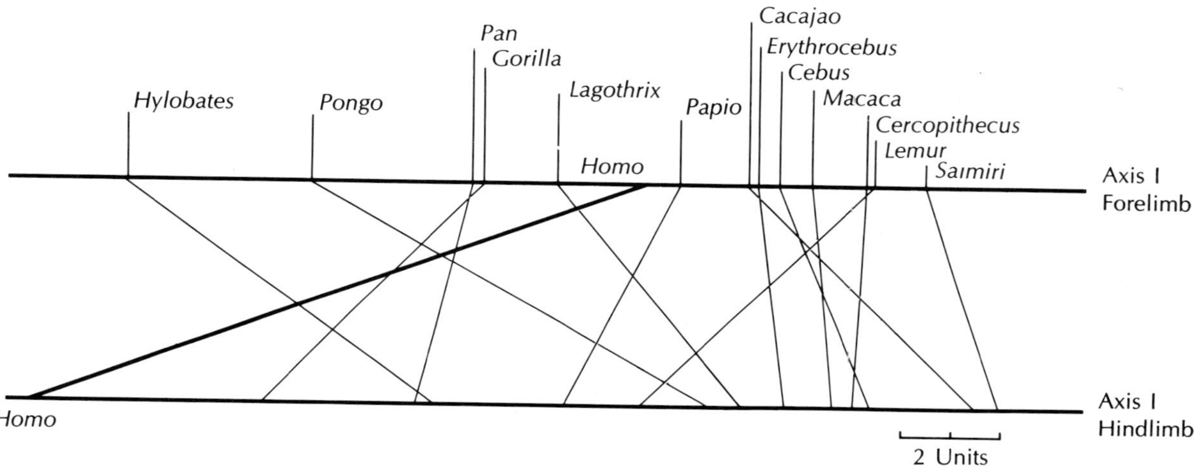

Fig. 7.11. Although there are general similarities between the pictures presented for both upper and lower limbs (i.e. genera separated according to degrees of arboreal acrobatics within regular arboreal quadrupedalism), and although therefore we might think that the information presented by these two anatomical areas is similar (as is the case within each limb), the two frames of this figure demonstrate otherwise.

There is little correlation between the rank orders of genera within the first axes of either scapular and pelvic studies (upper frame) or upper and lower limb studies (lower frame); that is, lines joining the same genera in each study cross almost randomly. For further explanation see text.

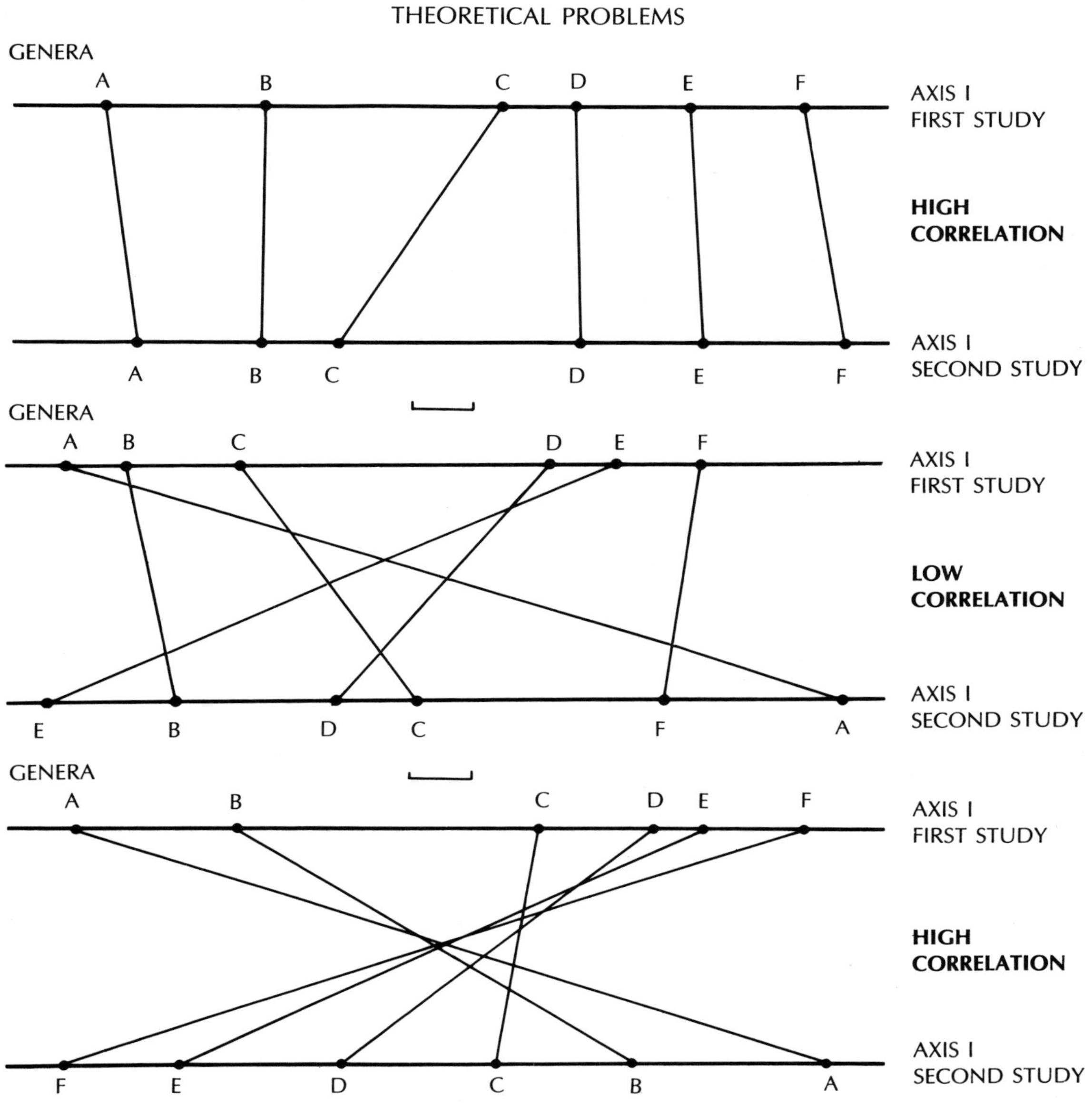

Fig. 7.12. This figure reminds us that completely parallel links *and* completely crossed links both imply high correlation. It is a mixture of crossed, parallel and every other angle of link that is observed when correlations are near zero. Backed by statistical tests.

body of data suggests strongly that this is associated with tensile-compressive forces in the upper limb and multiple activities in the lower limb (Oxnard, 1973a, 1975a).

But for the Old World monkeys the separations in both upper and lower limbs are more complex, involving several discriminant axes, to wit, the increased waviness of the plots. They do not, therefore, correlate so well with the broader associations for the entire primates. This may be because, within the comparisons afforded by the more restricted groups of Old World monkeys, a new mix of locomotor activities is evident. Thus, all the animals can leap, climb and run in ways that are important in their various life styles, even though there are differences in the overall locomotor picture for each species. And it is also possible that the particular adaptations that we see are modified by changes backwards and forwards along a terrestrio-arboreal axis in times past. Such possibilities may result in the more complex picture that we see in the Old World monkeys.

We can be certain that there exists a separate series of lesser behavioural and structural associations found among those species normally regarded as 'only' quadrupeds.

Some of the minor associations among the quadrupeds (especially the New World forms) are similar in direction to those more extreme ones among the entire group of primates. Others (especially among the Old World species) are not, being rather adaptations reserved for special differences among the Old World quadrupeds themselves.

These ideas can be fully documented by more extensive investigations in which larger samples and better sets of dimensions are studied and with clearer information about the functional differences between the different animals. To carry out these better studies for the entire primate Order is a large task. But a start has already been made on this in our own laboratories in the work of Manaster (1975, 1979) aimed, initially, at the three genera of Old World monkeys, and in the work stemming from extensive collaborations and pointed, in the first instance, at the various prosimii (Oxnard, German and McArdle, 1981; Oxnard, German, Jouffroy and Lessertisseur, 1981).

Some conclusions on 'generalized' primate quadrupedalism

There are several values in having the results of the above studies. *First*, it is possible that the rather more subtle structural differences associated with minor functional differences may provide clues as to how the more extreme differences of the extreme locomotor forms among the primates may have arisen. Such structural variations may allow for rather direct adaptation eventually leading to the more spectacular end products that are easily recognized. This may be more the case among the New World monkeys.

Secondly, however, others of the features that may be recognized through such investigations may not be related at all to the more extreme locomotor modes but to specializations within more general quadrupedal locomotor modes themselves. Such features may, if identifiable in fossils, provide some information about the overall evolution of arboreal quadrupedalism. And although the usual tendency is, in the first instance, to look for evidences of a single origination of locomotor modes, our prior experience prompts us to look for more than one, perhaps even many different essays towards arboreal quadrupedalism in the earlier days of primate evolution. This may especially be the case among Old World monkeys.

Thirdly, such data may provide information about local reversals during the evolution of individual groups. For although it is possible that a particular extreme locomotor mode arose through a continuous trend from some form of general quadrupedalism, through intermediate adaptation, to the eventual extreme form, it is rather more likely that such long term trends are actually made up of a series of smaller movements both towards *and away from* the eventual specialization.

We can again remind ourselves, for instance, of the terrestrial quadrupedal abilities of some langurs that seem to be fairly recent, consequent, perhaps, upon concomitant evolution of human social groups, together with the opportunities for ground-foraging for langurs that the garbage of such groups might provide. We can remember, in contrast, the terrestrial abilities of those rather more extreme species patas monkeys (genus *Erythrocebus* but rather obviously really close relatives to, or even part of, the genus *Cercopithecus*) that presumably arose a rather longer time ago under different conditions. And we are mindful of yet other cercopitheques (e.g. *C. aethiops*) that may have become terrestrial at a time intermediate between these two examples. A basically arboreal quadrupedal stock such as the Old World monkeys may thus have given rise to separate terrestrial variants several times and rather readily.

In its turn such a notion must be linked with the further idea that, if such terrestrial developments occur so easily, so must the reverse also be a common feature of the evolution of locomotor patterns among such groups. In other words, over many millions of years such groups may exhibit locomotor evolution that includes changes from more arboreal to more terrestrial modes *and back again*. Such reversals, if they were complete, could of course, never be detected. But the likelihood is that they would not be complete and might, given adequate materials, be detectable. It is possible that the difference between the quadrupedalism of many Old World monkeys from the quadrupedalism of the monkeys of the New World may stem from the existence of more terrestrial ancestors at a number of intermediate stages in the Old World groups.

The existence of this idea, of this possibility, means that our functional assessments and estimations must be carried out with caution. It suggests that the notion of defining primitive and derived features in such systems is far more complicated than generally thought (Chapter 4). It may be impossible to decide that a given feature is primitive or derived because a given feature may be associated with a series of different equivalent positions in the evolution of a lineage. In such a situation, the detection of primitive or derived conditions would only occur through study of mixes of characters; only in combination would it be possible to determine that particular features were primitive or derived, and then that determination would itself be the end result of the study and could not be used for further speculation about the evolution of that group.

Several investigators are now involved in the study of evolutionary adaptation at these very fine levels. And as results of these labours outline for us, in greater and greater degree, the rich and subtle meshwork of functional morphological variations that almost certainly exist throughout most of the primates, then such considerations will loom even larger.

It will also become more and more difficult to disentangle those structural-functional associations that do rest upon an evolutionary (genetic) basis from those that are merely inherent in the functional-structural relationships that stem from causal factors within individual ontogenies. That is, it may well be most difficult indeed to differentiate between a terrestrial and an arboreal variant of a given species due to micro-evolution and those other ontogenetic differences between individual natural specimens of arboreal species and particular specimens that have been forced, for whatever reason, into a terrestrial environment during their entire individual lives.

But the principal implication of this chapter remains the surprising degree of detail in the morphological-behavioural interface that seems truly to exist when methods such as the multivariate morphometric are used. Certainly this detail suggests that further investigation of every primate genus should be undertaken. This detail is, indeed, so fine that, given studies of extensive fossil remains using these methods, it should one day be possible not only to suggest broadly what given fossils may have been doing but also to make much more subtle estimates of resultant or average biomechanical situations. And, as we shall see, there is yet further information about the functions of upper and lower limbs taken together; but this must await the more complex *studies of the whole* in the next chapter.

Summary. This review of structural-functional associations within both upper and lower limbs of essentially quadrupedal primates started from the old notion that, indeed, a unified group of quadrupeds existed. But new more detailed information and new views of locomotion and other activities in many monkeys now reveal that gradations exist within quadrupedalism.

We thus recognize that among the less specialized Old World monkeys there is a spread of habitats from those that are partly terrestrial, through those that live mainly on the larger branches, to those that live more often in the small-branch milieu. These habitat differences imply differences in the habits of, and abilities for, more acrobatic types of movements grafted onto a basic quadrupedalism.

Similarly, though New World monkeys rarely go on the ground, there are ranges of acrobatic activities that are grafted onto their basic quadrupedalism among the branches. These extend from species which mainly run and climb on branches that are (relative to their own body sizes) fairly large and therefore somewhat like main roads, to those that increasingly inhabit the finer-branch milieu and partake to a greater extent in climbing and acrobatics as compared with running. There thus exist functional spectra within each of these groups of animals, although considerations mentioned above make it unlikely that these spectra will be the same in both the New and Old Worlds.

Multivariate statistical studies of structural parts from both upper and lower limbs indicate that equivalent structural spectra also exist. Considerations of the precise form of those spectra indicate that they differ in the two geographic areas.

Comparisons of the functional with the structural spectra, as was also the case for the upper and lower limbs separately in the fifth and sixth chapters, show marked concordances with one another.

It would seem evident that, once again, structures, when viewed in these quantitative ways, suggest a great deal of information about special elements of behaviour – to wit, the average biomechanical situations resulting primarily from locomotion. Because arboreal quadrupedalism must have been so central an adaptation during the whole span of primate evolution, we may need to be especially careful how we interpret such findings for fossils. Quadrupedalism must have given rise to other locomotor modes many times separately; conversely, it is not at all unlikely that in times past, other more extreme locomotor modes may have given rise, in their turn, to forms of quadrupedalism. Separating out these cycles may not be easy, but it may be a fascinating task for the future.

CHAPTER 8

Whole Primates – Their Arrangement by Anatomies

Abstract. In the previous chapters we have seen, unequivocally, that the structural arrangements of the primates (based upon study of individual post-cranial anatomical regions) concord most strongly with functional arrangements of the primates (based upon average biomechanical situations that result from function, mainly locomotion, in those same anatomical parts).

It would have been of great interest, therefore, to have known what arrangements of the primates would come from studying the details of all anatomical regions at once. But this larger study is not yet logistically possible and awaits at least a decade's more work. However, an appreciation of the answer can be obtained by doing the next best thing: that is, by examining combinations of all major bodily parts.

In this chapter we examine the results of doing this. First, the addition of all dimensions of both limbs gives an arrangement of the primates that speaks to function, but in a way rather different from what we have previously encountered. Second, the addition of all transverse dimensions also provides a somewhat surprising result, in this case about sexual dimorphism.

But the third key investigation stems from analyses in which dimensions from all anatomical parts are added together. This is done in two ways. *First*, the overall proportions of lower limb, upper limb, thorax, abdomen, and head and neck are summated through a total morphometric analysis of the entire data. *Second*, separate morphometric analyses of more detailed dimensions of individual smaller anatomical parts — shoulder, arm, forearm, upper limb as a whole, pelvis, foot, lower limb as a whole, trunk and head and neck — are summated using the high-dimensional technique. In both cases similar results are found. These results do not reflect function, they do mirror taxonomy. However, though the concordance of the new morphometric view with the current morphological taxonomy of the primates is, in general, close, it differs from current taxonomy in important details. The general concordance and local discordances are examined in the light of other information about the primates, for example data from biochemical and biomolecular investigations.

Two particular discordances with conventional taxonomy are somewhat larger than the rest. These are the specific cases of the relationships, separately, of tarsiers and aye-ayes. Accordingly, these two cases are examined in more detail and the results read both in the light of current taxonomy and against the broad background of the histories of the taxonomic placement of these forms.

We finally pass to a discussion of how these particular concordances and discordances may have come about, and why the total data, though functional in its parts, is taxonomic when viewed as a whole.

Introduction

The studies of prior chapters have shown us several views of primate structure. Chapter 2 demonstrates that, if we look at a wide variety of anatomical features, such as external characteristics of faces, genitals, hands and feet, internal features of teeth, jaws and skulls, and varieties of pelage colours and types, we find the currently accepted arrangements of the primates. This shows the generally linear array from prosimians at one extreme, through New World monkeys, Old World monkeys, lesser apes and great apes in sequence, to man at the opposite extreme.

Chapters 5 through 7 demonstrate, in contrast, that if we study suites of characters of various anatomical parts in such a way that anatomical features (a) are chosen on the basis of the functional anatomy of muscles, bones and joints and (b) are defined quantitatively and studied using methods that allow for their intercorrelations, then the primates are arranged in ways reflecting the main functions that each anatomical part performs in the life styles of the different species. Chapters 5 through 7 also show that when other quantitative measures are studied that have not been chosen on the basis of prior functional ideas about the bone-joint-muscle unit (e.g. the overall limb proportions taken by Schultz using carefully defined methods: Schultz, 1929) then the same statistical methods again reveal the functional picture for each anatomical region.

That these two pictures (of Chapter 2 and Chapters 5, 6 and 7, respectively) differ is not unreasonable, for most of the features used in traditional assessments of primate relationships, save perhaps those of hands and feet, are not as obviously functionally adaptive in locomotion as those studied here.

A most important question obviously is: what happens if we examine the quantitative features of the different anatomical regions taken together using these statistical methods? Unfortunately, it is rather difficult to do this in the best way because it requires that the same individual specimens be used in obtaining all the data.

The amount of work that is required in such an undertaking is such that it may well be a decade or more before our laboratories will be able to carry it out. Yet this will eventually be done, and we should record that we have made a start by adding together the data from the shoulder, arm and forearm (Ashton, Flinn, Oxnard and Spence, 1976). A next step will certainly be the addition to these data of the information from the investigations of the pelvis (Zuckerman, Ashton, Flinn, Oxnard and Spence, 1973; and Ashton, Flinn, Moore, Oxnard and Spence, 1981). The hope must be that this will eventually be possible for all anatomical regions.

In the meantime, however, it is possible to add together many aspects of these data from all bodily regions by methods that, though less than perfect, can still provide us with an overall picture.

First, the overall bodily proportions taken by Professor A.H. Schultz are available on the same specimens and do cover every region of the body, if not in the same detailed manner as in the regional studies. And although Professor Schultz's measurements were not chosen on the same functional basis as the dimensions of the more restricted anatomical parts, studies of his data show that the proportions of the limbs provide results similar to those derived from the more functionally determined measures of sub-units of those same limbs (Oxnard, 1978c). Accordingly, it seems that they contain roughly comparable information.

This approach possesses an important advantage: that it adds together all the data in a single morphometric analysis to give a single view of overall body structure. The approach suffers from the disadvantage that it only represents the structure of the body rather crudely through its overall proportions.

Second, though the more detailed dimensions on the smaller bodily regions cannot be analysed by a single morphometric study, it is possible to add together the results of several morphometric studies taken on each individual region using the high-dimensional method. The advantages and disadvantages of this approach are the mirror image of those for the study of overall proportions. The advantage is that, indeed, the structures are represented by all of the detailed measures rather than the cruder overall proportions. The disadvantage is that the approach cannot represent the statistical structure fully because interactions between the separate regions are missing.

However, before carrying out these very large undertakings, let us look at a number of simpler additions that move us towards the total picture, if not all the way.

First is an analysis of the data from upper limbs added to that from lower limbs. *Second* is a study of the addition of variables arranged so as to reflect aspects of the form of the head, neck and trunk, anatomical regions not so far considered in any detail. *Third* is an analysis of dimensions grouped so that all longitudinal measures are combined and, separately, all transverse measures.

The combination of upper and lower limbs

Let us look first, then, at the addition of upper limbs to lower limbs. As indicated in Chapter 7, the results of studying various parts of the same limb (whether upper or lower) are generally rather similar. When, however, we come to add together information in upper and lower limbs, it is useful to remember that intercorrelations between limbs are very much lower than within limbs (again, Chapter 7). This implies that a great deal of relatively independent information is enclosed within each anatomical region. We therefore expect that, in the addition to these data, new arrangements of the primates should be obtained. The proportions that are being added are shown in diagrammatic form in Fig. 8.1. The measurements, materials and methods are already outlined in Chapters 3, 5 and 6.

When the combination of upper and lower limbs was first carried out, my mind was already channelled by the results of the investigations on each limb separately. Because of this, I already had the following expectation: given that the upper limb arranges animals in a manner associated with upper limb function, and given that the lower limb arranges animals in accordance with lower limb function, it seemed to me that I would find that upper and lower limbs taken together would

Fig. 8.1. A skeletal diagram of the suite of elements defined by the measures in the study of the upper and lower limbs of the primates combined.

Fig. 8.2. A repeat of the results of Chapters 5 and 6: the morphometric results of studying overall proportions of upper limbs and lower limbs separately and the functional analogies to which they are related.

Frame 1

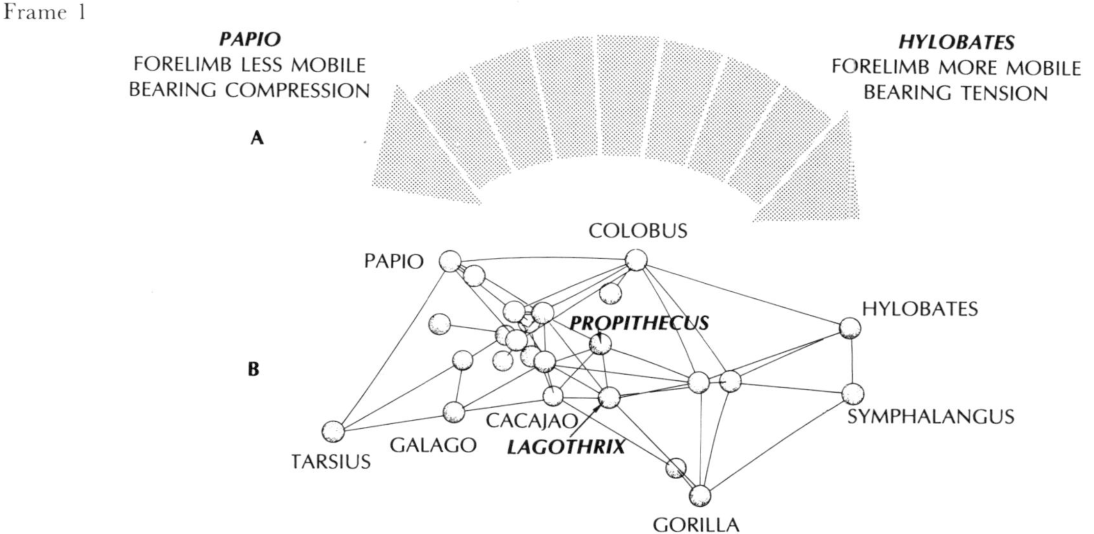

Frame 2

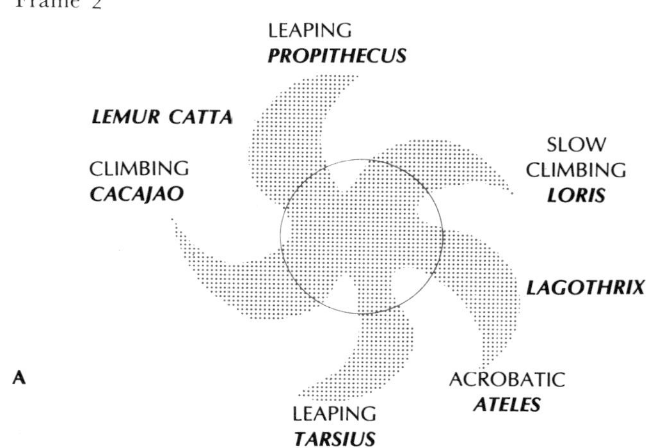

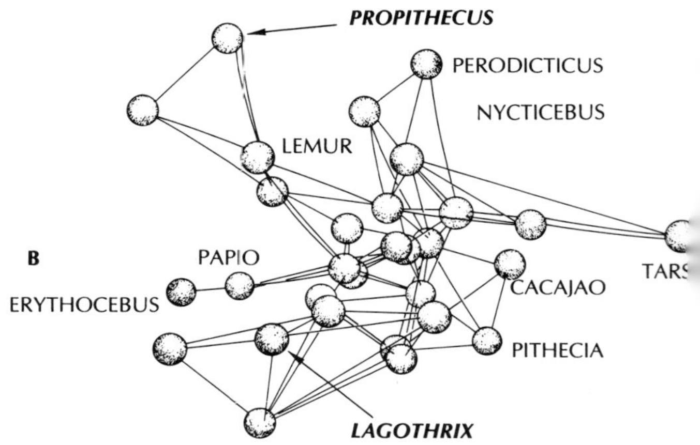

arrange animals in accordance with some overall function of both limbs, perhaps similar to a locomotor classification of the Napier type. But this I did not find; and accordingly my thoughts were in some confusion.

As discussion and thinking about locomotor patterns continued, and especially as behaviourists demonstrated that classifications of overall locomotor patterns of animals are not of much value (e.g. Ripley, 1976a, b), and as morphologists determined that regional functional classifications are more appropriate (Oxnard, 1975c), I returned once again to the confusing result. What I might have more realistically expected was some arrangement of the primates based upon a restricted functional concept for upper and lower limbs: functions within average usages in the animals – in effect, our average or resultant biomechanical situation.

Now that the clouding arguments have been removed, the biological meaning of the result of combining the data from the two limbs is immediately obvious (so obvious, in fact, that I scarcely see how I missed it before). Let us remind ourselves that the upper limb morphological arrangement is in the form of a spectrum shaped like a broad band paralleling a band-shaped spectrum of compression and tension-bearing. Let us remember also that the lower limb morphological arrangement is in the form of a spectrum shaped like a star paralleling the star-shaped spectrum of lower limb functions (Fig. 8.2). *In stark contrast, the morphological arrangement of both limbs combined resembles, crudely, a doughnut or, better, a signet ring.* This is shown in Fig. 8.3.

Whether or not a signet ring is what we 'ought' to obtain upon the 'addition' of a band and a star. I do not know. But whatever 'ought' to result from such a combination, we have to remember that we are not merely adding these two spectra. In addition to the information contained within each, we are also including new information, that resident within the interactions between each.

Let us now examine the positions within the signet ring of the various primate genera. A thickened part of the ring, the 'seal' or 'bezel', encloses a large number of very different genera. Thus, baboons and some cercopitheques lie close to slender lorises and pottos. With them lie various New World (sakis, douroucoulis) and other Old World (macaques, mangabeys, colobus) monkeys. All these seem to be fairly closely intermixed with some prosimians (a number of different lemurs and several tree-shrews, to the extent that we believe that tree-shrews are prosimians). This agglomeration of such enormously different genera is one of the factors that produced my prior blindness to the significance of the result.

Stemming from the seal along one 'haft' of the ring are a variety of other creatures: indriids, bush-babies, tarsiers, cheirogaleines and some lemurs. These forms have in common only one locomotor feature: *a heavy locomotor dependence upon the lower limb*. And stemming from the seal along the other haft of the ring are another series of creatures: woolly monkeys, spider monkeys, gibbons, siamangs and orang-utans. These have in common *a very heavy dependence upon the upper limb*.

With the hindsight of this arrangement, the meaning of the curious mix of creatures in the seal of the ring becomes evident. Thus, we realize that, (a) whether of different taxonomic groups (prosimians versus anthropoids), (b) whether terrestrial or arboreal (baboons versus cercopitheques), (c) whether primarily runners alone or runners together with some leaping (some cercopitheques and baboons on the one hand, and mangabeys, langurs and colobs on the other), (d) whether less

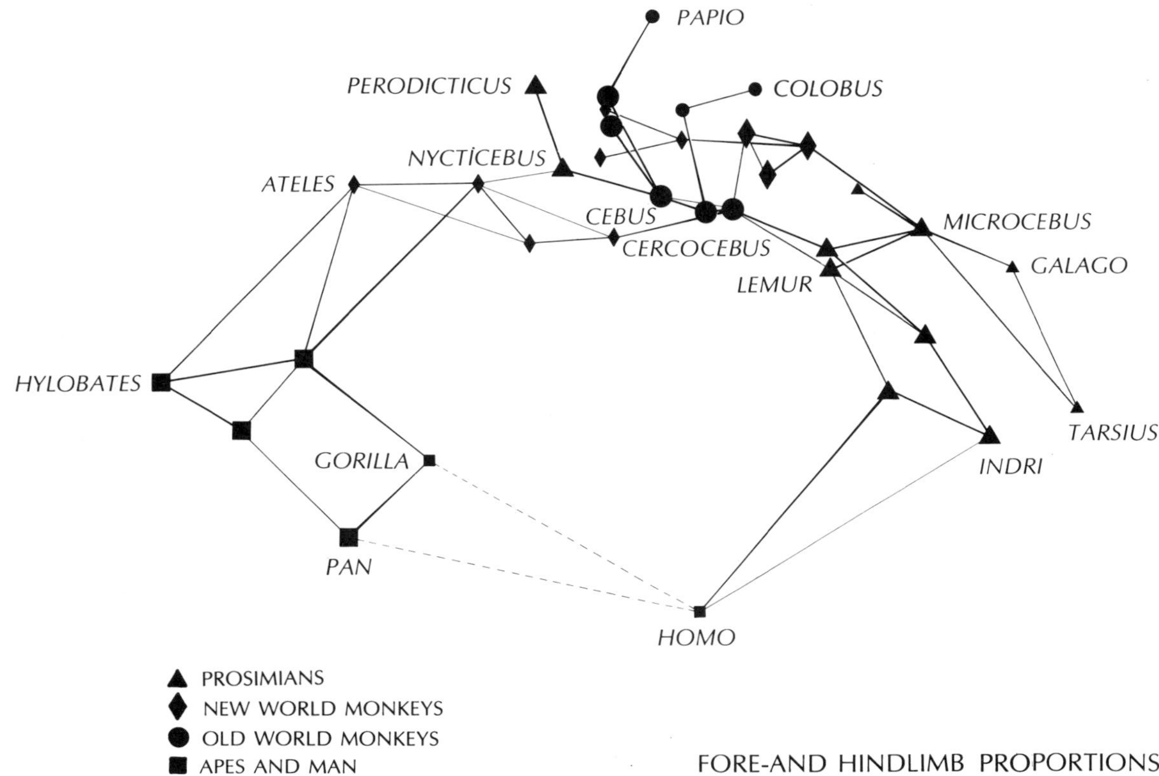

Fig. 8.3. The multivariate morphometric results of studying the data of Fig. 8.1 displayed as a perspective diagram of minimum and near minimum connections among the various genera. The genera are arranged in the form of a signet ring. The overall scale of the diagram is more than 30 standard deviation units.

acrobatic or with a more varied acrobatic repertoire among the branches (douroucoulis as compared with uakaris), or even (e) whether plying their locomotor trade mainly above the branches or not infrequently below (cercopitheques as contrasted with pottos), the many different species lying within the seal of the signet ring do have in common that their locomotor patterns are neither strongly upper limb nor strongly lower limb dominant. All these locomotor patterns, different as they are, ***depend upon generally equal usages of the two limbs.***

We may also look at some peculiar genera that fall on the opposite side of the signet ring. Fairly closely associated with the upper limb dominant forms (gibbons and siamangs), but nevertheless placed away from the upper limb dominant pole of the ring, lie gorillas and chimpanzees. These are two species whose anatomy proclaims them as having been far more upper limb dominant at one time but who, in recent times, have come to possess a more equal usage of the limbs within their own curious knuckle-walking locomotor mode. Yet they are also still able to perform in the trees in a manner that indicates upper limb dominant abilities. Their position in the ring thus correlates rather well with this combination of limb activities and abilities.

Also lying on the opposite side of the ring, but most closely associated with the lower limb dominant indriids, is *Homo*. Humans are, of course, the most highly lower limb dominant of all the primates and so this position, on the lower limb dominant side of the ring, is not incompatible with that functional concept. And yet, whatever guesses we may wish to make about human evolution, we must postulate humans as having arisen originally from forms something like apes and, therefore, at some earlier time, heavily upper limb dominant, as are all known apes. The curious position of *Homo* within the ring accords well with that idea.

Fig. 8.4 indicates the functional analogy that the morphology seems to suggest.

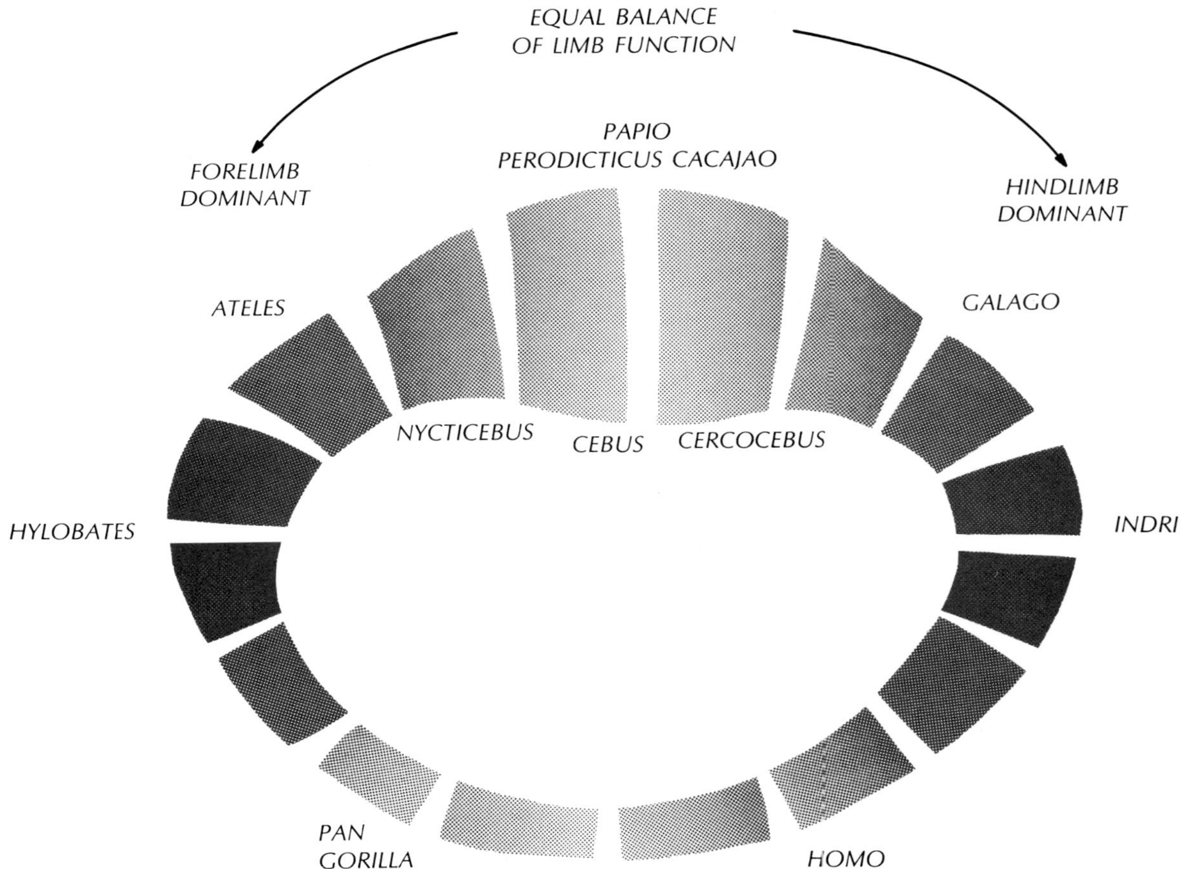

Fig. 8.4. The functional analogy suggested by the generic arrangements found in Fig. 8.3.

It should be emphasized that these various arrangements must be used very carefully in any speculations that we may make about evolution. There is no way, for instance, that I believe this result suggests that *Homo* has arisen from indriids merely because they lie close together; their somewhat close relationship stems only because, within their very different locomotor patterns, they share a degree of dependence upon lower limb activities; the location of humans near to indriids presumably results from further progression from species that were more upper limb dependent at earlier times.

There are other features of these arrangements that are of extreme interest. Why are there no species within the centre of the ring resulting from the addition of upper limbs and lower limbs? We do not know. And we can also ask: is it possible that some species in times past may have been of such and such a limb morphology that they *did* occupy the centre of the ring? Again, we do not know. But answers to these questions may be most exciting.

One answer could involve our discussion of prior human ancestors. For although the form of the ring-shaped spectrum makes it appear as though prior human ancestors may have lain *on* the ring, and been, therefore, similar to gorillas and chimpanzees, it is certainly not impossible that the pathway to modern humans may have *crossed* the ring, by-passing the knuckle-walking quadrupedal adaptations found in the African great apes today. Perhaps that pathway proceeded more directly from creatures whose mix of limb functions was reminiscent of those in climbing arboreal apes that were functionally somewhat similar to today's brachiating lesser apes, arm-swinging acrobatic New World monkeys or even cautious acrobatic climbing orang-utans (see Stern, 1975, for an interesting discussion of 'before bipedality'). Perhaps this speculation may more readily accord with new information resulting from studies of australopithecines (see Chapter 10) than do conventional ideas about those fossils. Certainly, in the few fossil remnants that we do possess, there is no indication of knuckle-walking in human ancestors.

Other examples of forms that may have lain within the now empty centre of the ring are the unique Malagasy prosimian fossils of giant size and the unique groups of hominoid forms known as *Gigantopithecus*. We know little about the locomotor patterns of these species. But presumably each in its turn is markedly different from anything that we know today. And for the Malagasy fossils, materials are actually available that would allow interpolation into these analyses. These additional studies have not yet been performed (though presumably the data of Jungers, 1976, might be so applied).

A final example that might be interesting is the aberrant genus *Daubentonia*, the locomotion of which is different from that of most other primates. However, that case is more fully covered later in this chapter.

I have explained the difficulties that were encountered in arriving at the above interpretations. It must be borne in mind that the models of upper and lower limb function (the band and star, respectively) were arrived at by thinking about locomotor function prior to doing the morphometric studies. But the new ring-shaped model has been derived only from seeing the morphometric result. I do believe that we could have postulated it before seeing the results had various controversies about locomotor classifications not obtruded. However, because of the temporal inversion of the logic, we have to be especially careful in describing and using the ring-shaped analogy.

The head, neck and trunk

Let us now look at the combination of the head, neck and trunk. Studies of Schultz's dimensions (table 8.1) of the head and trunk, either separately or taken in combination, provide information that is complex. Univariate examination describes a confusion of overlapping groups. But multivariate (canonical variate) analysis of the data is more helpful. First and second discriminant axes separate to some degree the three major subdivisions of the primates: hominoids, monkeys and prosimians. Yet in spite of the quite large scale of the separations (20 standard deviation units in a first axis, 15 in a second) there is still enormous overlap between component genera in these groups; little can be determined from the plots with certainty.

Study of higher axes from the investigations of the head, neck and trunk makes it clear that no further information relating to overall groupings of genera resides there. But the higher axes do show that the data for the head, neck and trunk differ from those for upper and lower limbs. Thus, the higher axes provide many individual separations of particular primate species. The particular species so distinguished are those that are usually regarded as being somewhat more different from their fellows than most other species.

Table 8.1

DIMENSIONS AND ANIMALS IN MORPHOMETRICS OF THE HEAD, NECK AND TRUNK

Anatomical features: Schultz's data on head, neck and trunck

Relative chest circumference: as per cent of trunk height
Relative shoulder breadth: as per cent of trunk height
Relative hip breadth: as per cent of trunk height
Chest index: chest breadth as per cent of saggital chest diameter
Relative head size: average head diameter as per cent of trunk height
Relative face height: face height as per cent of trunk height
Relative upper face height: as per cent of average head diameter
Cephalic index: head breadth as per cent of head height
Interocular index: inner eye breadth as per cent of face breadth
Relative ear size: ear height × ear breadth as per cent of head length × total head height

Genera and numbers of specimens studied: Schultz's data

PROSIMIANS

Tupaia, 5; *Lemur*, 3; *Microcebus*, 3; *Lichanotus*, 1; *Propithecus*, 1; *Indri*, 1; *Daubentonia*, 1; *Nycticebus*, 9; *Perodicticus*, 1; *Galago*, 10; *Tarsius*, 8

NEW WORLD MONKEYS

Leontocebus, 28; *Aotus*, 9; *Cacajao*, 2; *Pithecia*, 1; *Cebus*, 25; *Saimiri*, 49; *Alouatta*, 4; *Ateles*, 74; *Lagothrix*, 2

OLD WORLD MONKEYS

Macaca, 27; *Cercocebus*, 3; *Papio*, 5; *Cercopithecus*, 3; *Erythrocebus*, 2; *Presbytis*, 14; *Nasalis*, 26; *Colobus*, 2

HOMINOIDS

Hylobates, 78; *Pongo*, 13; *Pan*, 26; *Gorilla*, 6; *Homo*, 25

Thus, among the prosimians such analyses tend to isolate, in different ways, forms such as aye-ayes, tree-shrews, pottos and spectral tarsiers. Among the anthropoids, genera such as patas monkeys and baboons separately from the cercopithecine monkeys, proboscis monkeys from the colobines, uakaris and howler monkeys separately from the cebid monkeys and, of course, humans from among the hominoids tend to be most distinct. These individual characterizations involve large deviations for these special genera in individual statistically significant discriminant axes as high as the third to the ninth in an analysis involving 11 original variables (and, therefore, given a larger number of species, 11 discriminant axes).

In general it is rather difficult to place any detailed biological interpretation upon such results other than to note that these genera tend to be extreme in various ways; but the lack of any biological meaning save this is not to say that the separations are not real nor of very distinct biological (probably systematic) importance; certainly the scale of these separations is such that not only are they statistically significant but also so big that they must be taken into account in any assessment of these genera (table 8.2). Thus, the largest of these separations is +9 standard deviation units in the fourth axis for the baboon, but even in the seventh axis a distance of −7.6 standard deviation units is evident for the aye-aye.

Table 8.2

STUDY OF THE HEAD, NECK AND TRUNK:
RESULTS OF EXAMINATION OF HIGHER AXES

Axis 5	Axis 6	Axis 7	Axis 8
Daubentonia: −5.0	*Tupaia:* −3.9	*Daubentonia:* −7.6	*Perodicticus:* −3.7
Nycticebus: −4.1	*Cacajao:* −3.8	*Alouatta:* −4.7	*Cercopithecus:* +3.9
Erythrocebus: −3.4	*Presbytis:* +3.5	*Erythrocebus:* +3.2	*Pongo:* −3.8

Other individual genera are separated by yet higher axes. Thus, axis 9 separates *Arctocebus* and axis 10 separates *Papio, Hylobates* and *Cacajao*. Each of these axes is statistically significant but contributes only a few per cent to the separations of the entire range of primates. Because, however, what little separation exists is aimed at only one or two genera, their separations are not only statistically significant, but also big enough to have biological significance for those genera.

Other anatomical combinations

Let us move along to other anatomical combinations. These include analyses of (a) all longitudinal measures and (b) all transverse dimensions.

All longitudinal measures. Examination of a series of measures of overall bodily lengths provides information that is fairly closely similar to that from the limbs that we have just examined. This is perfectly reasonable because, of course, the majority of measurements of the proportions of the limbs are, indeed, measures of longitudinal elements of the body.

All transverse measures. Discriminant analysis of the suite of transverse dimensions does not provide any very meaningful overall arrangement of primate genera; but it does separate, in a very interesting manner, the sexes of the individual genera. Difference in size between the sexes occur in almost all primate groups although, of course, it differs in degree. Males may be as much as twice the size of females in baboons, proboscis monkeys, orang-utans and gorillas. In contrast, in squirrel-monkeys, some langurs, colobus-monkeys and gibbons, differences in size between the sexes are quite small. In some lemurs, some marmosets and in spider-monkeys these differences may even be reversed with larger females than males (e.g. Schultz, 1969).

However, in addition to size differences, it is also well documented that there are differences in shape between the sexes of various species. For instance, chest circumference and shoulder breadth are relatively larger in the males of many primate species than in the females, and the difference is most pronounced in those species with the greatest size differentials. Thus in the orang-utan, chest girth is over 200 per cent of trunk length in males but only about 170 per cent in females. In howler-monkeys the proportional differences are even greater, 150 per cent in males as compared with 90 per cent in females.

Although these differences in shape and size differ among the various primates, the nature of the difference has generally been assumed to be similar. Thus, structural sexual dimorphism is often thought of as a single descriptive idea — larger size differences go with larger proportional differences, and both are related to larger differences in robusticity. Certainly, within the primates, this notion of a single phenomenon of sexual dimorphism with differential expression seems to be confirmed by the studies of A. H. Schultz (1969) as he examines, one by one, his proportional measurements of the various segments of the body. And this is also the specific finding of a more detailed study using the methods of correlation, regression and Penrose's size and shape factors for studying finer measures of five particular primate species. Indeed, in that study, Bernard Wood (1975) reports that the idea that considerable differences in shape exist between males and females must be rejected; thus, he suggests that even those differences in shape that we see are really the result of differential size.

Yet a glance at the results (Figs. 8.5, 8.6 and 8.7) indicates that the idea of sexual dimorphism within

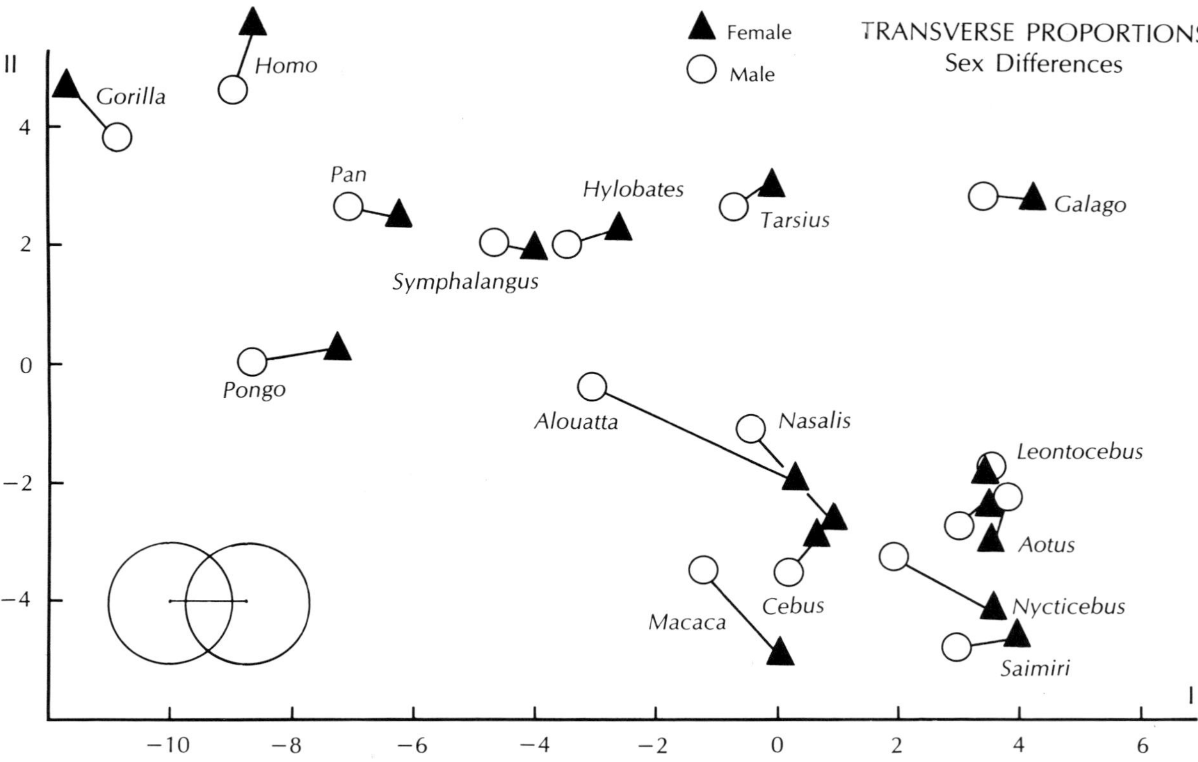

Fig. 8.5. The multivariate-morphometric result of studying the suite of transverse dimensions of the primate displayed as the bivariate plot of the first two axes. The scale of the diagram is presented in standard deviation units both on the axes and in the circles around a point. The sexes are separated in the directions shown.

Fig. 8.6. The first and third axes of the study of figure 8.5. Further differences between the sexes are evident.

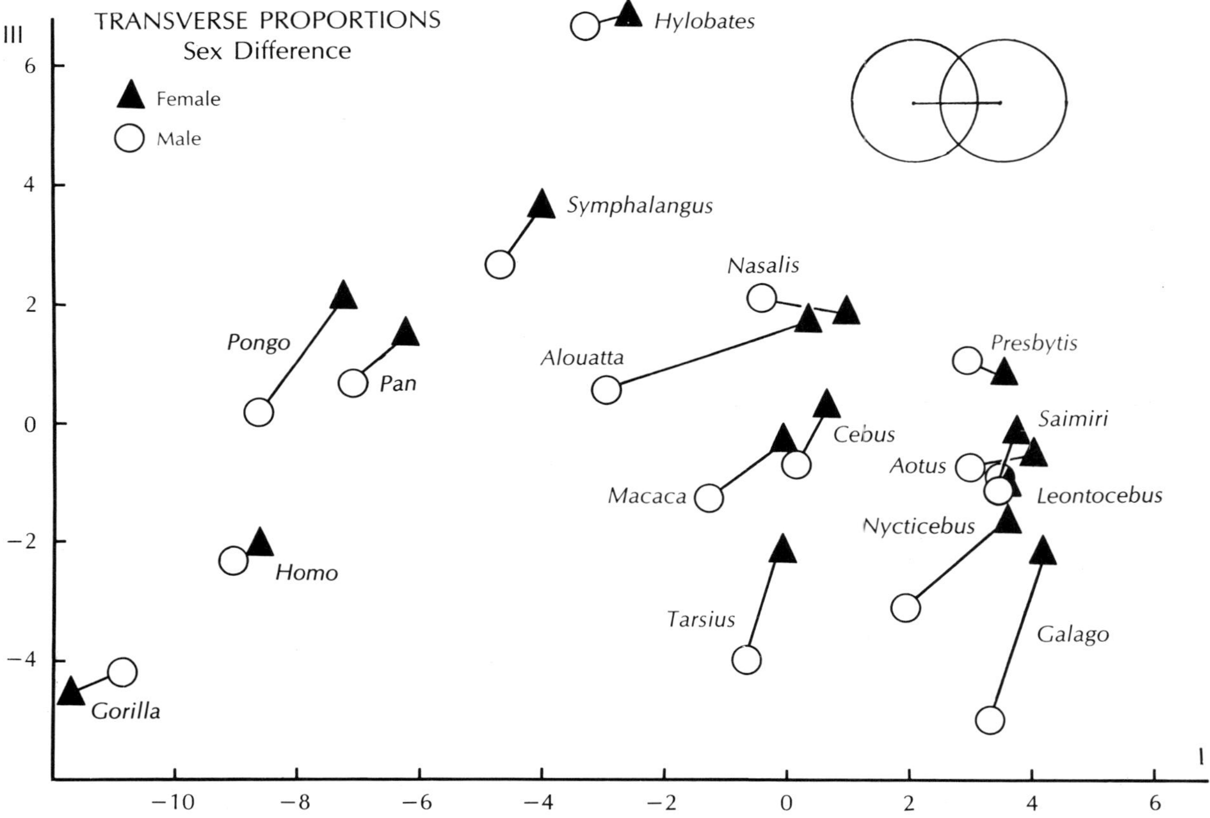

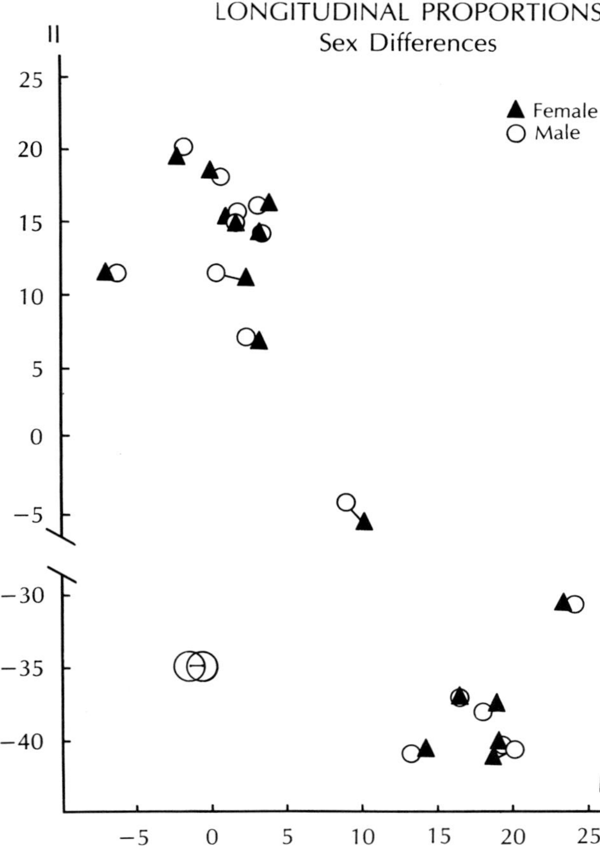

Fig. 8.7. In contrast to the results shown in Figs. 8.5 and 8.6, the separations of the sexes in a study of longitudinal-dimensions are small.

the primates is rather complex. It is true that most of the sexual differences between the different species do, indeed, parallel one another (the difference between male and female is similar in direction and scale for most genera; see the directions and lengths of the arrows in Figs. 8.5 and 8.6). But for a considerable number of genera, the scales and directions of the differences between the sexes are quite different.

In the cases of langurs, capuchins and humans, for example, the differences are located approximately at 90 degrees to those generally found among the primates. In gorillas and night monkeys, the differences, at 180 degrees, are actually reversed compared with those between the sexes of most genera. And there are yet other individual genera that depart from the general pattern; for example, orang-utans, macaques and proboscis monkeys are angulated, each differently, at approximately 45 degrees away from the regular primate pattern.

This result has at least two implications. *First* is that structural differences between the sexes of primate genera tend *generally* to be similar from one genus to another; this may be related to the well-known differences in overall size that exist between the sexes of many groups (this does not mean, of course, that size is the sole determining factor; see Albrecht, 1980).

The *second* is that differences between the sexes are sometimes quite peculiarly arranged; in specific genera, therefore, they may be due to mechanisms that we may not have realized exist. Given the general tendency to dismiss sexual dimorphism as being much the same thing (save for degree) amongst most species, further investigation of this point would appear to be most important.

For it is usual to point to sexual dimorphism as being most strong in species where there is, indeed, a large difference in body weight between the sexes. But the present results suggest that structural sexual differences, not necessarily related to size alone and hidden internally in the intercorrelations among dimensions, may exist as forms of sexual dimorphism every bit as marked as that related to simple size and robusticity. Thus, sexual dimorphism in bush-babies and tarsiers is, by the relative measures here (the lengths of the arrows in Figs. 8.5 and 8.6), considerably greater than in gorillas and orang-utans which are visually recognized as the most obviously sexually dimorphic primates because of great size and robusticity differences between their sexes. Sexual dimorphism, by these measures, is greatest of all in howler monkeys, but that is probably not a big surprise. Undoubtedly these results are because the use of indices has removed simple size differences (though not, of course, allometrically related size differences) and thus allows us to see other, perhaps many other, more subtle shape differences.

This result is so contrary to the conventional wisdom on this score that it forces us to study the differences between the sexes in the various primate genera in more detail. This involves viewing not only the differences between the sexes in the first three canonical variates (as in Figs. 8.5 and 8.6) but the entire pattern of differences portrayed in the full suite of significant canonical axes. It requires, therefore, the use of Andrews' high-dimensional display.

In order to make sure that sexual differences are primarily confined to the suite of transverse dimensions, Andrews' curves were plotted for the canonical values for each sex derived from the analysis of longitudinal measures. The result is summarized in Fig. 8.8 demonstrating the nature and size of differences between the sexes in selected genera. The picture, that there is very little difference between the sexes, is found throughout the entire primates even though only a selection of genera can be shown in this diagram.

Andrews' curves for the suite of transverse dimensions for each genus are quite different. In addition to much larger separations between the sexes not one but several separate patterns exist. Of the 20 available genera, 6 demonstrate the common pattern shown in Fig. 8.9, and another 6

Frame 3
Presbytis: Lengths
−1 PI
−.5 PI
0 PI
.5 PI
1 PI

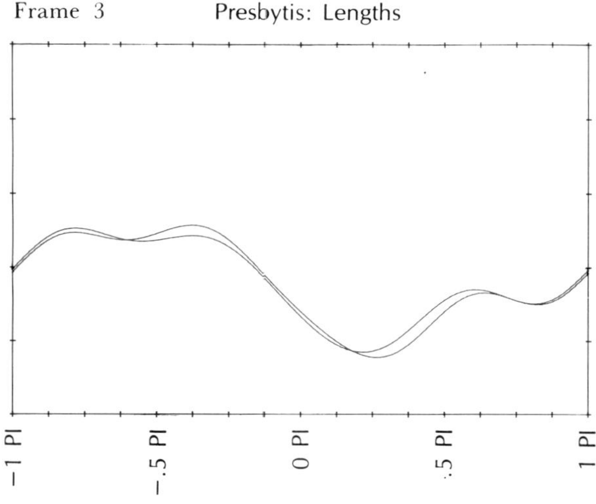

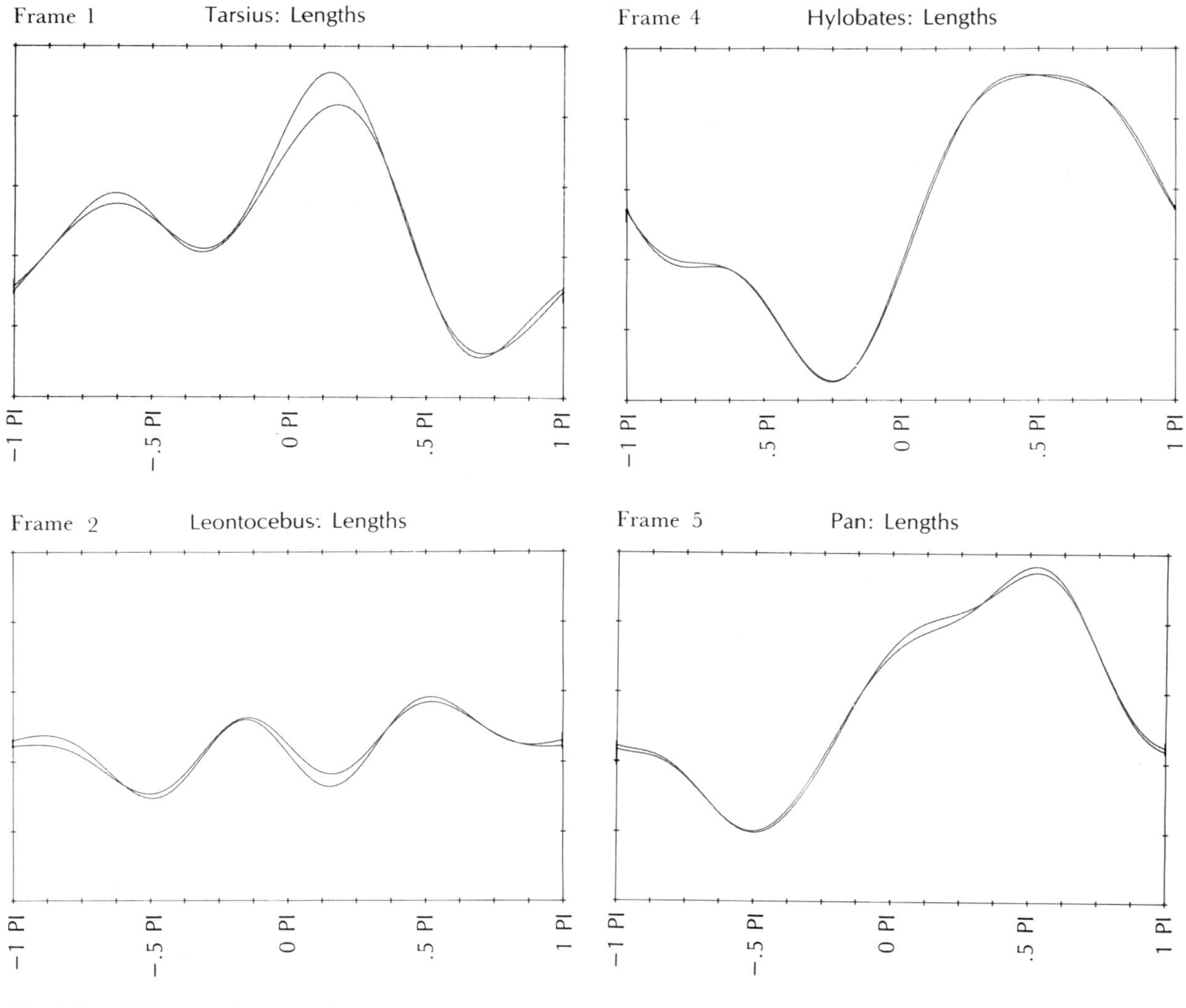

Frame 1
Tarsius: Lengths
−1 PI
−.5 PI
0 PI
.5 PI
1 PI
Frame 4
Hylobates: Lengths
−1 PI
.5 PI
0 PI
.5 PI
1 PI
Frame 2
Leontocebus: Lengths
−1 PI
−.5 PI
0 PI
.5 PI
1 PI
Frame 5
Pan: Lengths
−1 PI
−.5 PI
0 PI
.5 PI
1 PI

Fig. 8.8. Differences between the sexes in five selected primate genera as reflected in the Andrews' high-dimensional plots for all significant canonical axes of the study of longitudinal dimensions. The curvature of the plot for each genus is different because the genera are different one from another. But the difference in curvature between the sexes is very small, as is evident from the small area between the curves for each sex. Although not illustrated here, this is the case for every primate genus examined.

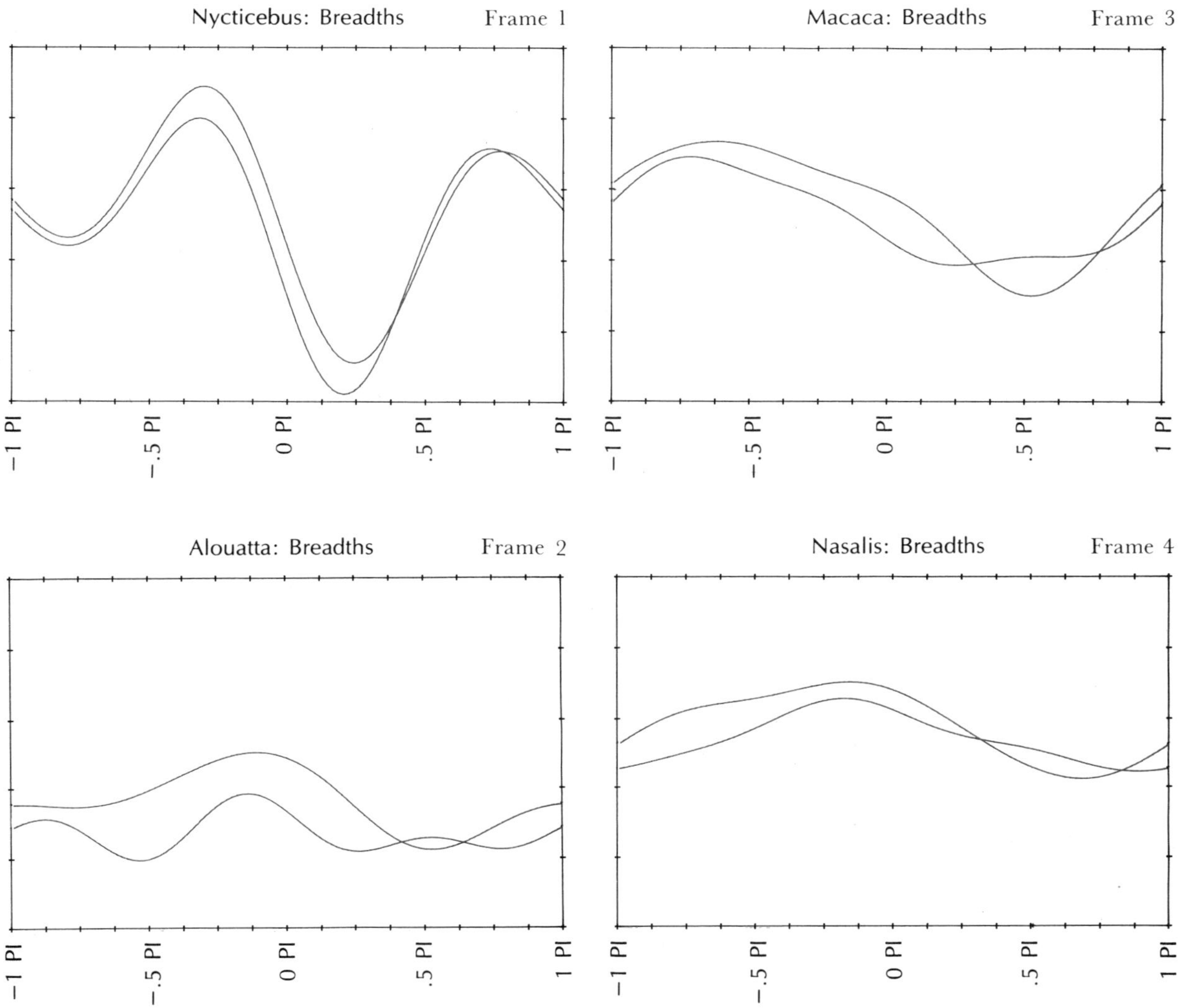

Fig. 8.9. In contrast to Fig. 8.8, differences between the sexes are rather large when the results that are displayed are those from analysis of transverse dimensions. In addition, as demonstrated here, there is a pattern of differences that is shared among many genera – a pattern in which the curves are separate for most of the plot but cross twice at its positive end.

show a second consistent pattern as in Fig. 8.10. The remaining genera each show individual patterns unique unto themselves. Four of these include orang-utans, chimpanzees, gorillas and humans (Fig. 8.11).

Several ideas flow from these findings. First, we can no longer adhere to the view that sexual dimorphism is a single phenomenon with differential expression (as in Schultz, 1969), nor to the view that it is related only to size differences between the sexes (Wood, 1975). The reality is that several different sexual dimorphisms exist and these must be associated with a complex of different factors, of which size can be only one possible factor among many (Oxnard, 1983b).

We can no longer lean, furthermore, upon a single evolutionary mechanism as causally related to sexual dimorphism. Early studies (reviewed by Schultz, 1969 and Wood, 1975) emphasized simple notions such as troop defence being the responsibility of males. Later studies suggest more complex ideas, such as inter-male competition for females within the specific social organization of the species (e.g. Gautier-Hion, 1975; Clutton-Brock, Harvey

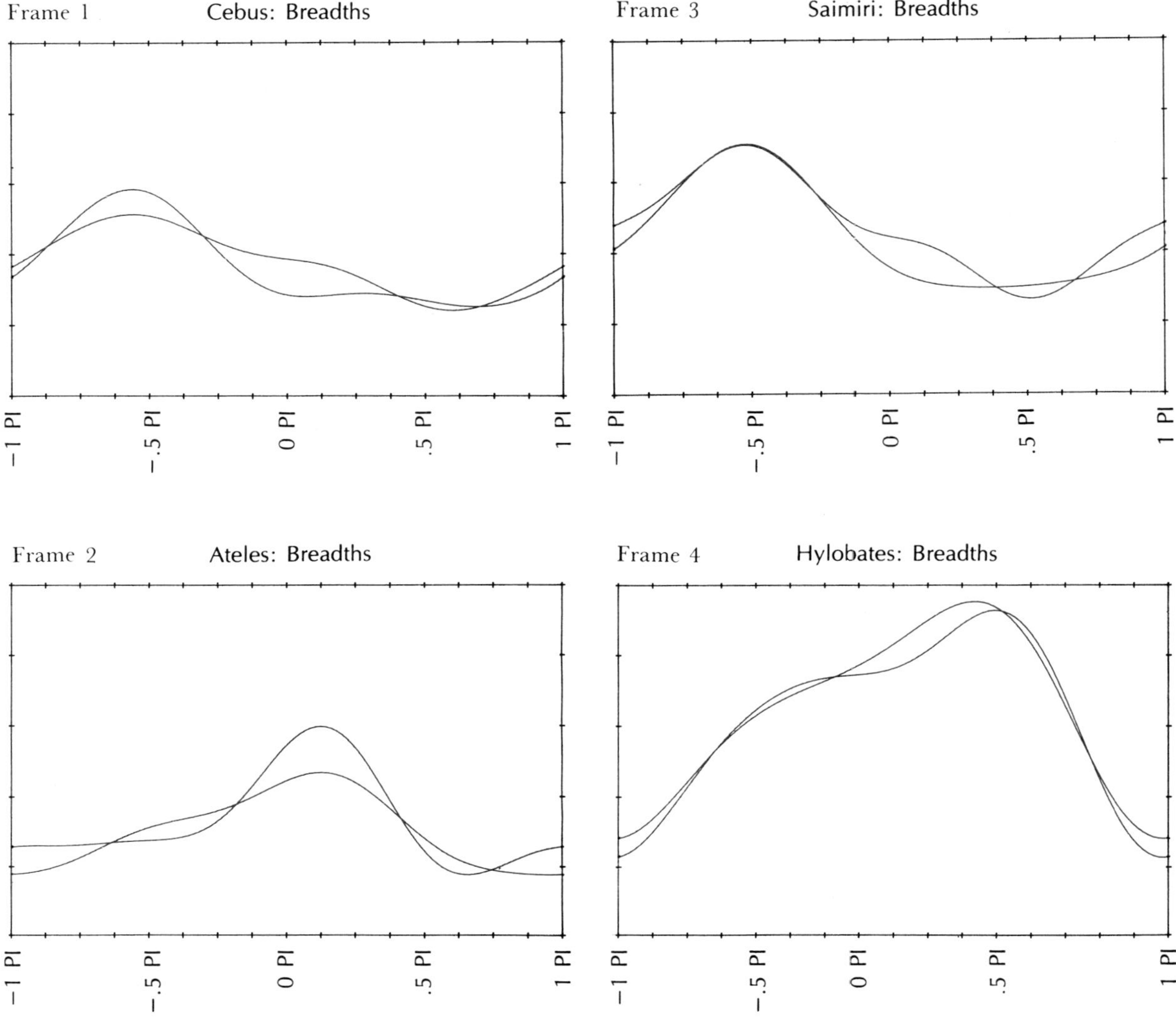

Fig. 8.10. Further studies similar to those of Fig. 8.9 demonstrate a second pattern of sexual differences that is common among the primates. In this case the pattern of results among the canonical variates is such as to cause the curves for each sex to cross four times, as seen here. Other conventions as in Fig. 8.9.

and Rudder, 1977; Leutenneger and Kelly, 1977 Harvey, Kavanagh and Clutton-Brock, 1978). If, sexual dimorphism is really several different phenomena, then it is likely that causation involves a mix of several factors and different mixes among the different animals. These factors may include phenomena such as social organization, ecological niche, feeding pattern, reproductive efficiency, developmental pattern and even other, hitherto as yet undiscovered possibilities (Oxnard, 1983b).

The findings also lead into the idea that some structural sexual differences within the primates (additional, presumably, to those shared with prior mammalian ancestors) must have evolved more than once within the group. For even the two major patterns found here, because of their widespread representation among different taxonomic groups, are likely to have evolved in parallel a number of times. And, among the hominoids, whatever may have been the original sexual dimorphism existing prior to their separate evolution, there must have been much evolutionary change since their divergence.

These findings mean that when we attempt to view differences between the sexes in pre-human, pre-hominid, or even pre-hominoid fossils, we can

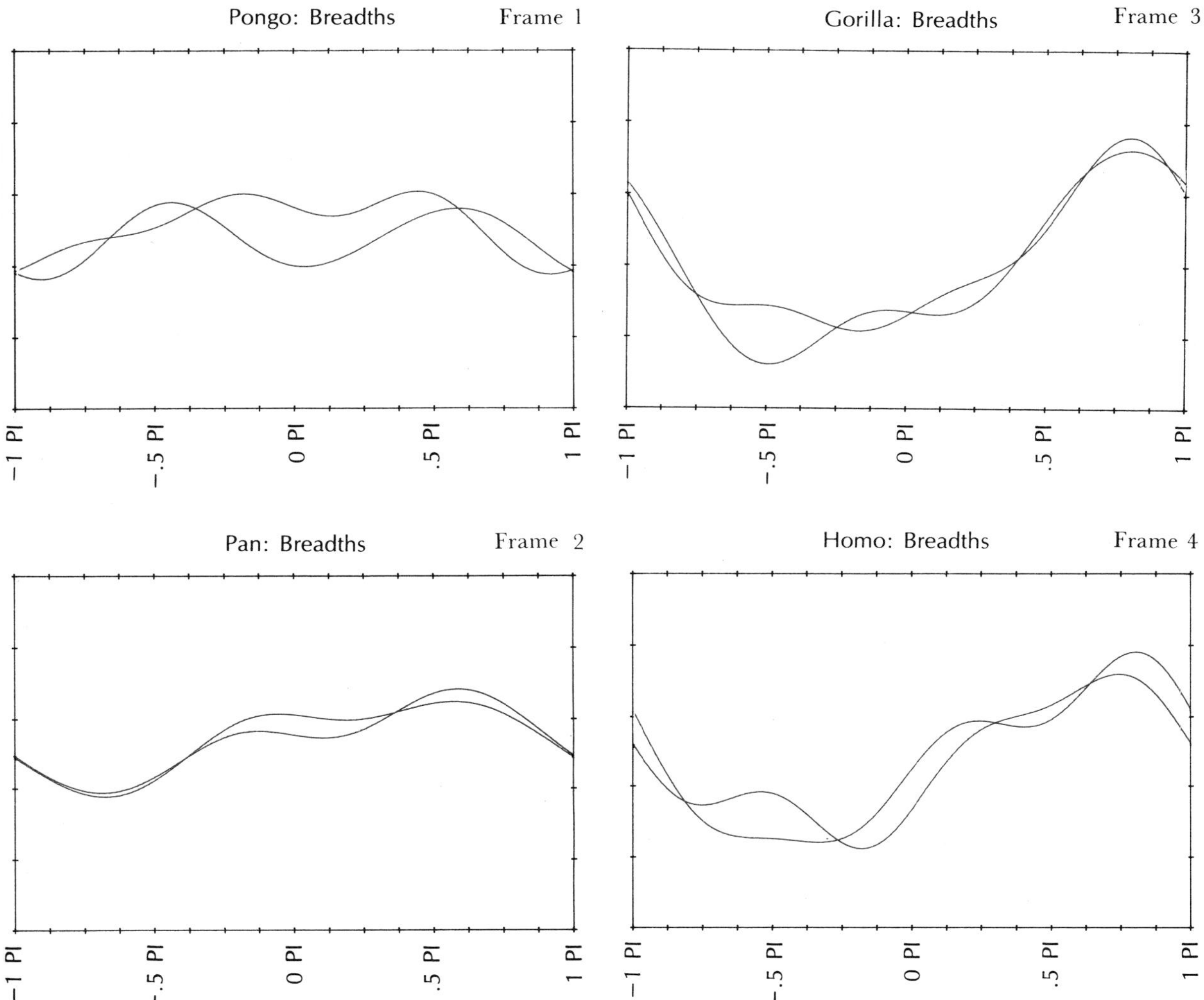

Fig. 8.11. A number of genera do not conform to either of the common patterns found among the primates. These include the hominoids: the examples for humans, gorillas, chimpanzees and orang-utans are shown here. Each curve differs from those of Figs. 8.9 and 8.10. In addition, each curve differs from each of the others. Thus, at least some elements of sexual dimorphism in these genera may have evolved since they split from one another – presumably only a short time ago, geologically speaking.

no longer assume (as do Johanson and Edey, 1980: Fleagle, Kay and Simons, 1980: and Gingerich, 1981) that a single spectrum of sexual dimorphism with a single explanatory association (social organization) covers all genera. Now we must be prepared to ask, 'is the sexual dimorphism of a particular fossil population like that of humans or gorillas or chimpanzees or even orang-utans (species very far removed from human ancestry) or even one of the two general patterns that exist more widely among the primates (including the lesser apes)? For, whatever sexual dimorphism exists in our fossil group it cannot be similar to all these at once. Indeed, we might even wish to ask, 'could the sexual dimorphism of our fossil group be yet another pattern unique unto the fossils themselves? And, in asking such questions we cannot automatically assume that the explanation is a single difference in social organization, that is, the difference between multi-male multi-female structure and nuclear family structure. Other factors will have to be evaluated.

Finally, the findings mean that we have to re-evaluate aspects of non-structural differences between the sexes in humans. As long as human structural differences were thought to be part of an overall primate pattern that might be as old as 70 million years or even more, then we could be pardoned for thinking that perhaps the genetic basis for that dimorphism must be equally old. This might provide a most ancient and perhaps very rigid set of biological constraints, within which other sexually dimorphic features (for instance, social cultural, psychological, intellectual) might have to be set.

Once, however, it is apparent that even the structural differences may have undergone separate differentiation as recently as the time of a human-ape progenitor (which, however long ago that may have been was surely much less than 70 million years – perhaps only 5 or 10 million) then we must recognize that the genetic basis of that structural dimorphism may be very much shorter than we may have previously thought. We can, therefore, far more reasonably expect non-structural human sexual differences to be much less constrained by an ancient and rigid genetic base, and much more able to be changed by far more flexible, non-biological phenomena such as individual decision-making, social change, education and legislation.

Let us now, however, depart from this digression into what I now believe are the sexual dimorphisms of the primates, and return to the summation of data for as much of the entire body as possible.

The appreciation of the whole

The first way of attempting to appreciate the whole animal uses all of Professor Schultz's overall proportions analysed in a single morphometric analysis (Oxnard, 1981b). This study thus includes dimensions of the upper and lower limbs, of the trunk and head, whether of transverse or longitudinal type, a total of 23 overall bodily proportions. The dimensions are outlined in the exploded diagram of Fig. 8.12 and in table 8.3. The precautions of table 3.4 have been adopted in the analyses.

Fig. 8.12. A skeletal diagram of the full suite of elements defined by the measures in a study of the overall proportions of the entire body in the primates.

Table 8.3

DIMENSIONS USED IN MORPHOMETRICS OF OVERALL BODILY PROPORTIONS

Anatomical features

Schultz's data on upper limbs	*Schultz's data on lower limbs*
Relative chest circumference	Relative hip breadth
Relative shoulder breadth	Relative lower limb length
Relative upper limb length	Intermembral index
Intermembral index	Crural index
Brachial index	Length of foot relative to limb
Relative hand length	Length of foot relative to trunk
Relative hand breadth	Relative foot breadth
Relative thumb length	Relative trunk height
Relative trunk height	

Schultz's data on head, neck and trunk

Relative chest circumference: as per cent of trunk height	Relative face height: face height as per cent of trunk height
Relative shoulder breadth: as per cent of trunk height	Relative upper face height: as per cent of average head diameter
Relative hip breadth: as per cent of trunk height	Cephalic index: head breadth as per cent of head height
Chest index: chest breadth as per cent of saggital chest diameter	Interocular index: inner eye breadth as per cent of face breadth
Relative head size: average head diameter as per cent of trunk height	Relative ear size: ear height × ear breadth as per cent of head length × total head height

Analyses of total proportions performed without duplication of variables

The second way of attempting to appreciate the whole animal employs several different morphometric analyses but draws them together using the high-dimensional display of Andrews (Oxnard, 1983c). This study includes a summation of the more detailed measures of shoulder, arm, forearm, forelimb as a whole, hip, thigh, foot, hindlimb as a whole and, when available, overall proportions of the trunk, head and neck. The dimensions are those outlined in previous chapters.

Based upon prior discussion, we may well expect the result to be an arrangement of primate genera, reflecting in some global way total bodily function. Of course, total bodily function is a most complicated concept: with the information that we currently have available, we could expect it to reflect the major features of primate locomotion.

In fact, however, that is not the picture that we find. *First*, Figs. 8.13 and 8.14 provide summary results of combined multivariate statistical studies of overall bodily proportions. What is immediately obvious is that they do not add up to any recognizable functional answer. But they do resemble most closely the generally accepted relationships of the primates. Not only are the various major groups of prosimians clearly evident (e.g. indriids, galagines, lorisines) but so also are many of the subgroups of the anthropoids obvious (Old World monkeys, and within them, cercopithecines and colobines; New World monkeys; hominoids).

Second, figures 8.15 and 8.16 provide the equivalent summary of the high-dimensional analysis of the measures of the detailed anatomical regions. Again, a functional answer is not evident. The answer displays the generally accepted relationships of the primates: prosimians and anthropoids, and within the latter, New and Old World monkeys and hominoids.

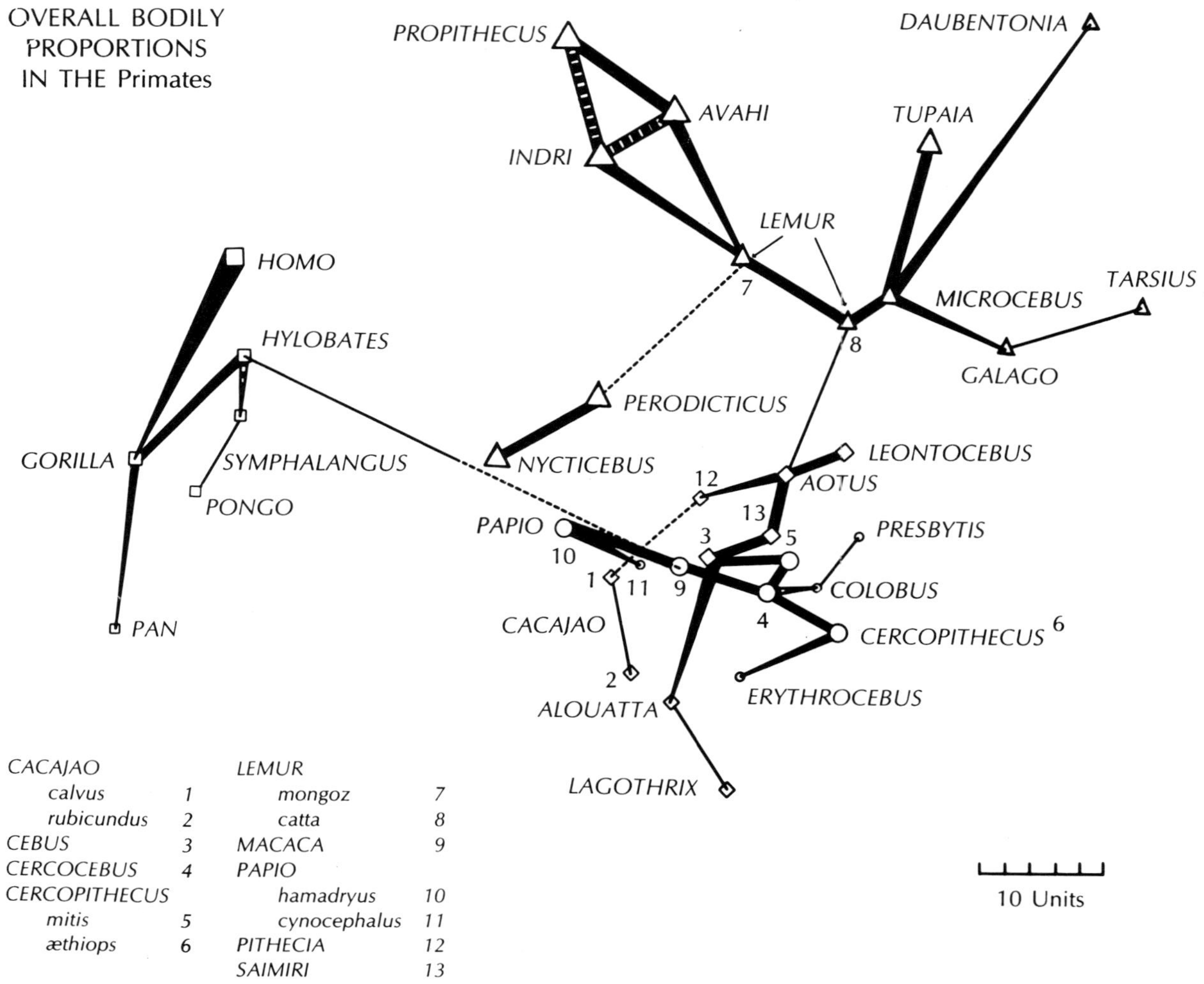

OVERALL BODILY PROPORTIONS IN THE Primates

PROPITHECUS
DAUBENTONIA
AVAHI
TUPAIA
INDRI
LEMUR
HOMO
7
TARSIUS
MICROCEBUS
8
HYLOBATES
GALAGO
PERODICTICUS
GORILLA
SYMPHALANGUS
NYCTICEBUS
LEONTOCEBUS
12
AOTUS
PONGO
13
PAPIO
3
5
PRESBYTIS
10
1
11
9
COLOBUS
CACAJAO
4
PAN
CERCOPITHECUS 6
2
ALOUATTA
ERYTHROCEBUS
LAGOTHRIX

CACAJAO		LEMUR	
calvus	1	mongoz	7
rubicundus	2	catta	8
CEBUS	3	MACACA	9
CERCOCEBUS	4	PAPIO	
CERCOPITHECUS		hamadryus	10
mitis	5	cynocephalus	11
æthiops	6	PITHECIA	12
		SAIMIRI	13

10 Units

Fig. 8.13. The multivariate morphometric study of the full suite of dimensions of Fig. 8.12 displayed as a minimum spanning tree drawn in a degree of perspective. The genera are arranged as a Fig. '6', throughout which hominoids are linked with hominoids (squares), Old World monkeys with Old World monkeys (circles), New World monkeys with New World monkeys (diamonds) and prosimians with prosimians (triangles). The general scale of the diagram is more than thirty-five standard deviation units.

Fig. 8.14. The data of the minimum spanning tree of Fig. 8.13 are here simplified by being straightened out in a diagrammatic form. This renders considerably clearer the taxonomic analogy with various data of chapter 2 (conventions are the same as in Fig. 8.13).

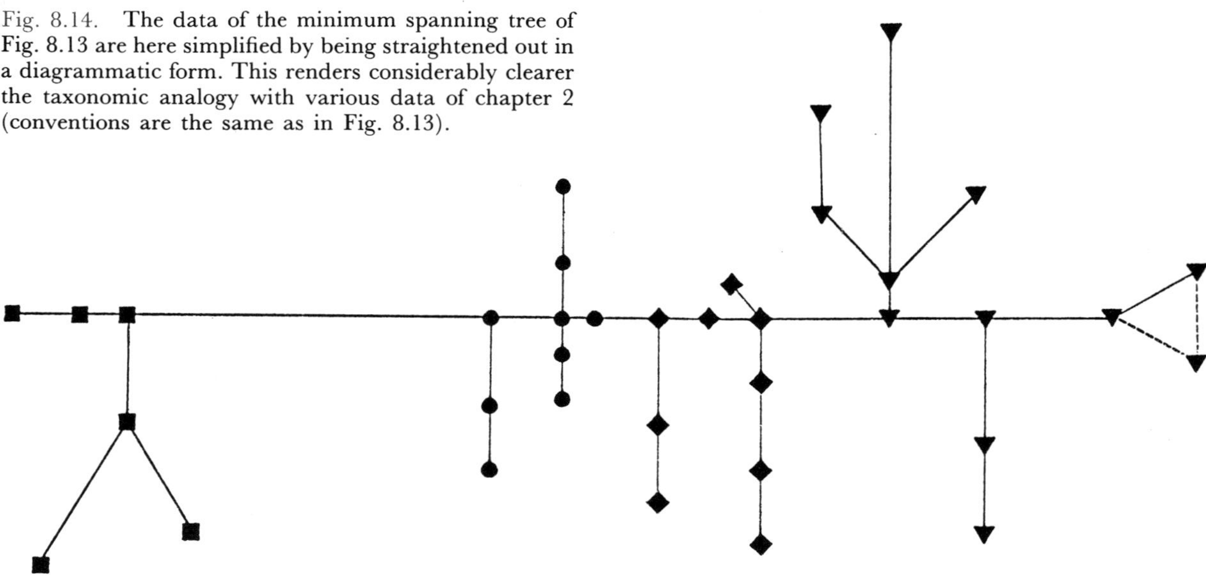

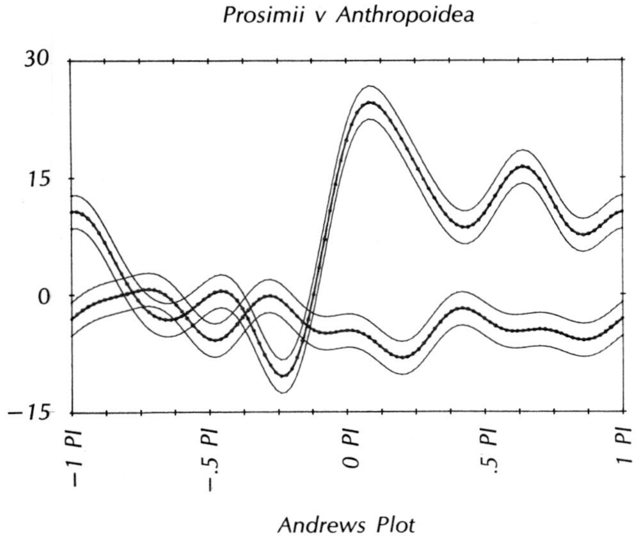

Prosimii v Anthropoidea

30
15
0
−15

−1 PI
−.5 PI
0 PI
.5 PI
1 PI

Andrews Plot

Fig. 8.15. The high-dimensional display resulting from the combination of detailed anatomical parts as described in the text when applied to the primates as a whole. The separation between prosimians (upper curve) and anthropoids (lower curve) is remarkably clear. The thinner curves parallel to the main curves are three standard deviation units distant in each direction from the main curves. This helps to provide us with a feel for the very large difference between the two main curves – between, therefore, prosimians and anthropoids.

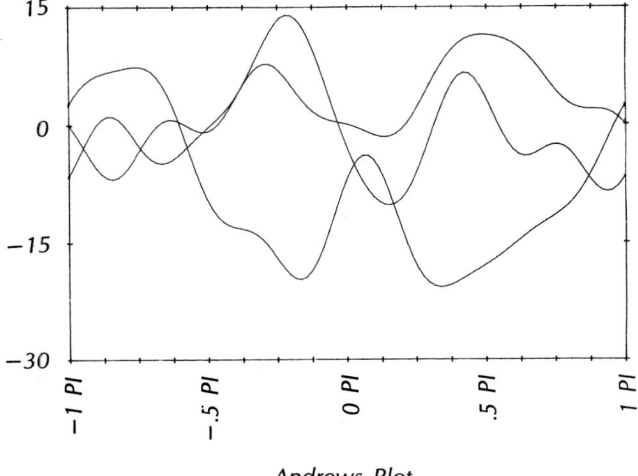

Fig. 8.16. The high-dimensional difference between the three superfamilies of the anthropoids displayed using the same method as Fig. 8.15. The measure of variability is the same (although not shown in this graph). Again, a very clear difference between these taxonomic groups is evident: upper curve – New World monkeys, middle curve – Old World monkeys, lower curve – hominoids.

We now have three types of information resulting in the definition of the major groups of the Primates: the structural information classically examined in organismic studies (see the early part of Chapter 2), the molecular results stemming from a study of biochemical parameters (see the later part of Chapter 2) and now, in this chapter, two different biometric appreciations of morphometric data summarized through multivariate statistical analysis. These last are, of course, data that are known, through investigation of individual anatomical regions, to contain information of very strong functional content.

These three methods, the classical morphologic, the biomolecular and now the multivariate morphometric, provide the same overall view of the primates — prosimians, New World monkeys, Old World monkeys, apes and humans — that we summarized in Chapter 2. And whenever we look closely at individual primate groups, e.g. localized variants of a single species group, such as the celebes macaques, we also find close associations between the results of study by the same three methods (Albrecht, 1978). It is in the study of intermediate levels that controversy arises.

Scarcely any two classical morphologists actually agree about the view of intermediate levels derived from classical data. Every shade of difference is found between the conservative 'lumping' attitudes of Simpson as he reviews the work of primate anatomists before 1945 and the extreme 'splitting' characterizing the works of others such as Osman Hill (1953–66) and Hershkovitz (1977). These differences are not, in general, differences of clustering patterns, rather are they differences of evaluation of the level of clustering. What one worker would regard as a species, another believes to be a genus; what one worker dubs a genus, another desires to call a sub-family. An example of this type of discordance is shown in Fig. 8.17.

Equally, by no means do the various molecular evolutionists agree totally with one another. In this case, however, the differences among them do not generally affect patterns of clustering; they concern, rather, matters of interpretation: whether or not biochemical evolution has proceeded linearly, curvilinearly or in some other manner; whether or not the methods can be interpreted to give estimates of absolute time of evolutionary divergence and so on (Fig. 8.18). Their discordances are less about levels of clustering and more about distances of clustering.

Notwithstanding the general agreement about

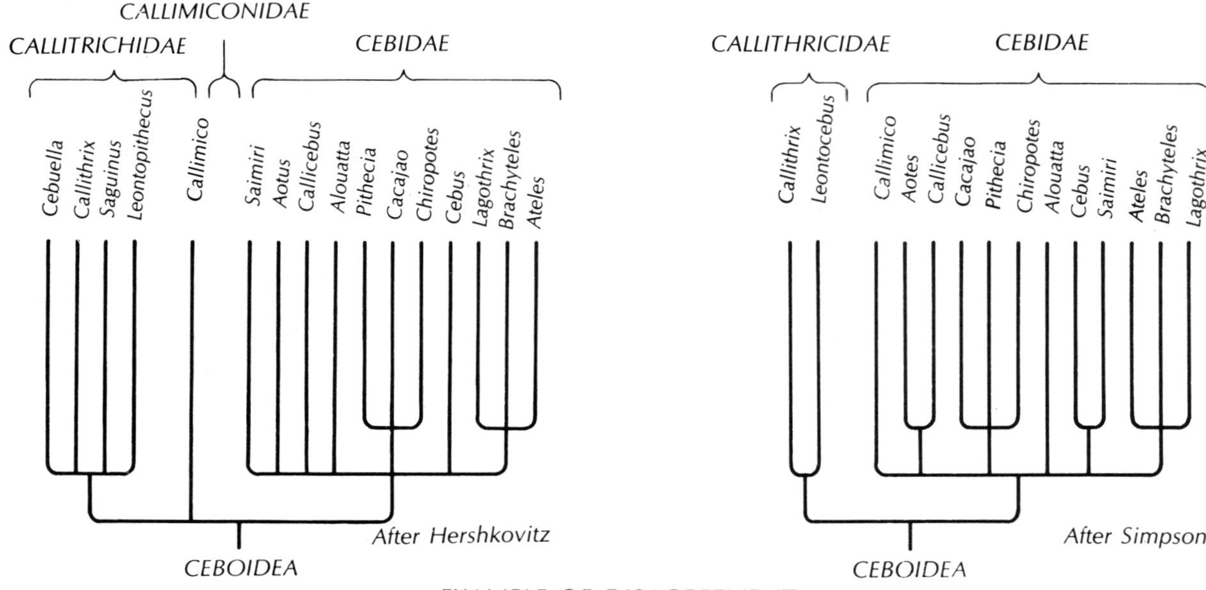

Fig. 8.17. Discordances among classical morphologists generally relate to differences in 'patterns' between 'lumpers' and 'splitters'.

Fig. 8.18. Discordances among biomolecular evolutionists generally relate to differences between distances of groups (some controversial distances are given by the numbers on special links in the diagram).

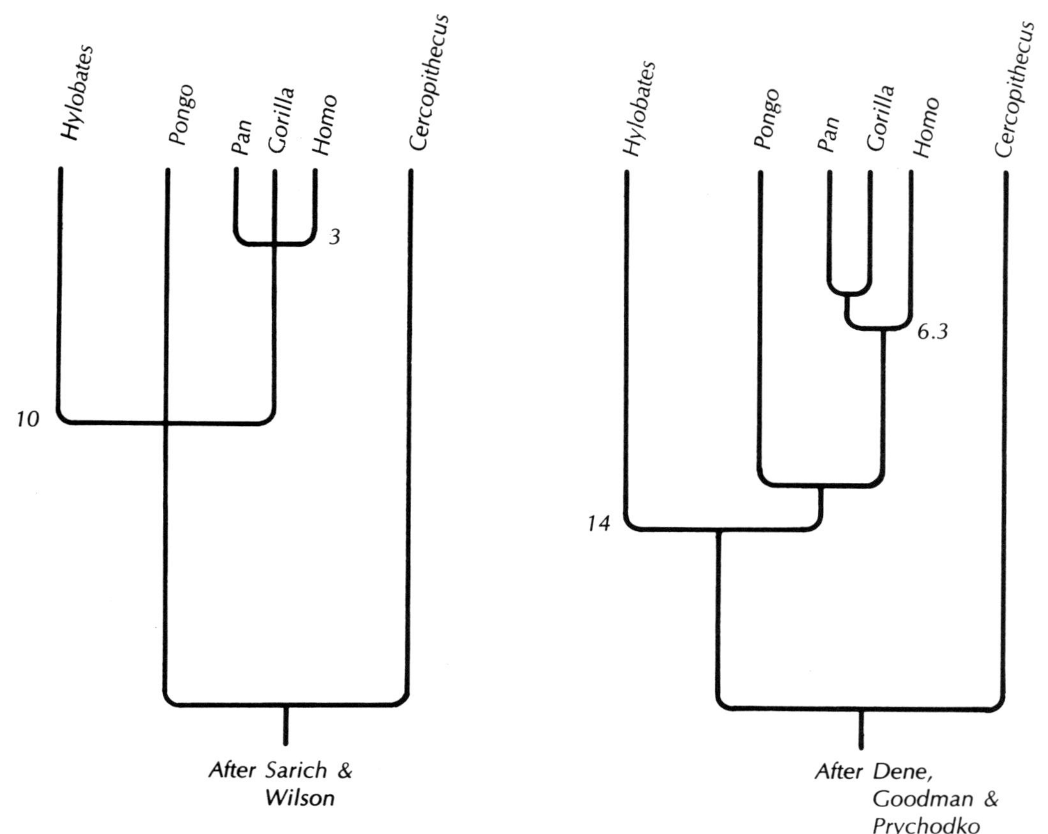

the overall arrangements of the primates (prosimians, New World monkeys, Old World monkeys, apes and humans; see Chapter 2), and notwithstanding the internal disagreements *within* each of these categories of investigator, it is apparent that there is a different level of disagreement *between* most molecular investigators on the one hand and most of the more classical morphologists on the other. These discordances can be seen by studying more closely the intermediate levels of clustering (e.g. clustering of genera into sub-families, of subfamilies into families) within the entire order.

Humans and apes

Within the hominoids, for instance, almost all classical studies of the morphology of the extant species place *Homo* as separate from the great apes, which later are seen as forming a coherent group. This results in the two families, the Hominidae, containing only humans, and the Pongidae, including gorillas, chimpanzees and orang-utans.

This is the case whether we look at the work of someone like Simpson (1945) who is generally rather conservative about the creation of new

Groupings Inherent in Classical Morphology

HYLOBATIDAE PONGIDAE HOMINIDAE

Hylobates *Symphalangus* *Pongo** *Gorilla* *Pan* *Homo*

HOMINOIDS

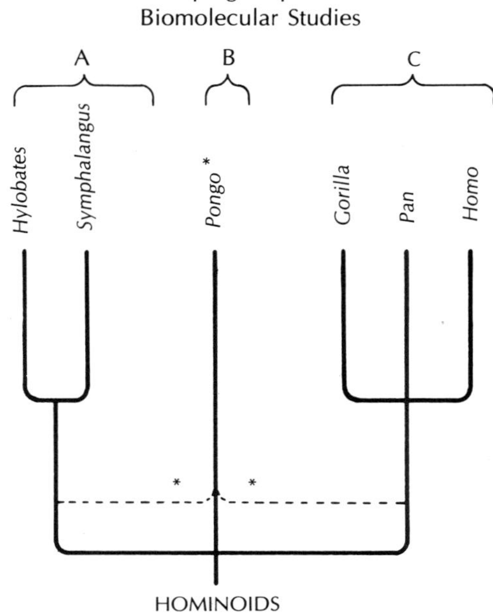

DISCORDANCES:
DETAILED DIFFERENCES
IN HOMINOID
GROUPING PATTERNS

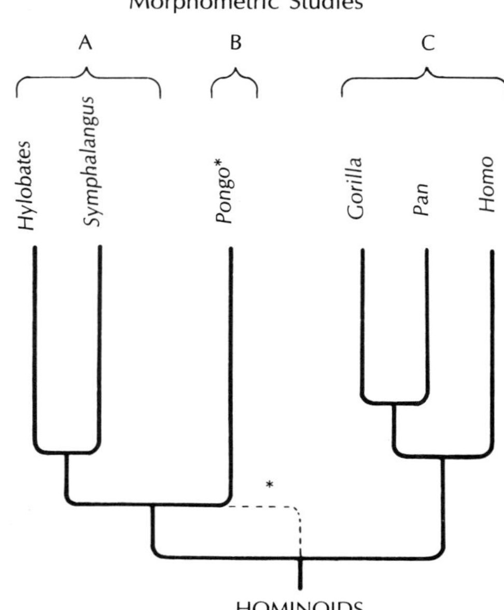

Fig. 8.19. Discordances between classical morphologists and molecular evolutionists in the hominoids. The new morphometric result mirrors most closely the biomolecular view. Asterisks indicate alternative interpretations.

groups, or whether we look at the investigations of Osman Hill (1953–66) who freely recognizes more species, genera and sub-families. Even the studies of a primatologist such as Tuttle (1977), who finds the relationship between the gorilla and the chimpanzee so close that he suggests they should be grouped in the same genus, *Pan*, provide a picture of hominoid clusterings that allies all the great apes together in contrast to *Homo*.

In contra-distinction, almost all the molecular investigators are agreed that the relationships are more likely a division into two other groups: one contains *humans together with gorillas and chimpanzees*; the other contains *orang-utans alone*, or includes, sometimes, a distant linkage *of orang-utans with the Asiatic lesser apes, gibbons and siamangs*. In this case, the evidence depends upon a much smaller bulk of work carried out over a relatively short period of time. Yet it is impressive.

Thus Goodman's laboratory has studied antigenic distances (Dene, Goodman, Prychodko and Moore, 1976) and beta-haemoglobin sequences (Goodman, 1976); Sarich and Cronin (1976) have investigated albumins, transferrins and DNA phylogenies; Kohne (1975) and Benveniste and Todaro (1976) have studied DNA nucleotide sequencing data; Romero-Herrera and co-workers have investigated myoglobin amino acid sequences (Romero-Herrera, Lehman, Joysey and Friday, 1976). All these studies (and others that I have not referenced) provide assessments of the Hominoidea that place humans together with gorillas and chimpanzees (e.g. King and Wilson, 1975). They locate as far distant the lesser apes, gibbons and siamangs; and when orang-utans are included in different individual studies they are placed either closer to the lesser apes or intermediate between them and the human-African ape complex. Rarely are they situated with the great apes in a group that might be named the Pongidae as distinct from the Hominidae (but see Bruce and Ayala, 1978, for a single dissenting view).

Fig. 8.19 demonstrates the difference between this summary of the classical morphological and the molecular biological pictures. Notwithstanding the controversies within each group (as demonstrated in Figs. 8.17 and 8.18), it is clear that there really is a major difference between the clusterings of these two types of study.

It is therefore fascinating to return to the morphometric studies for a view of the relationships of the hominoids. This demonstrates *a network of relationships of the hominoids that mirrors the biomolecular*

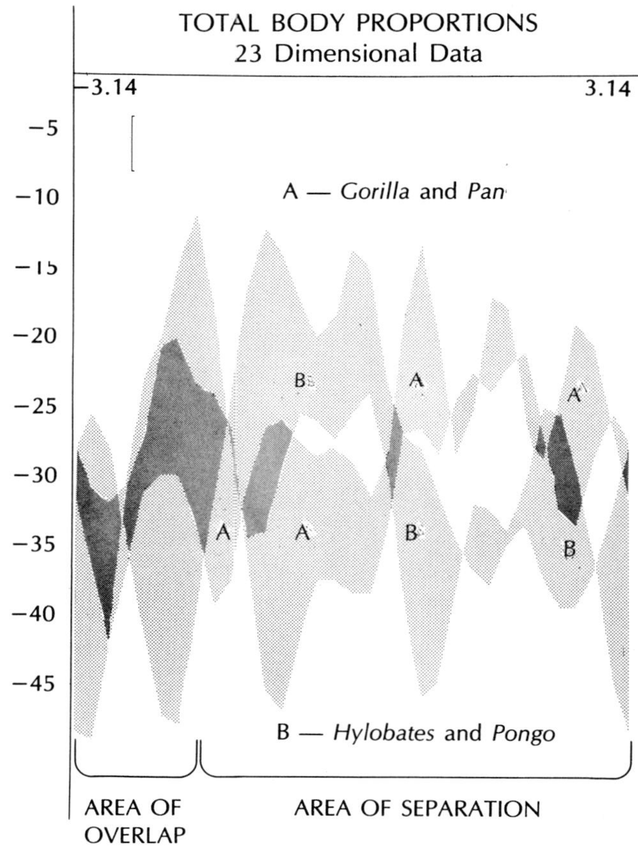

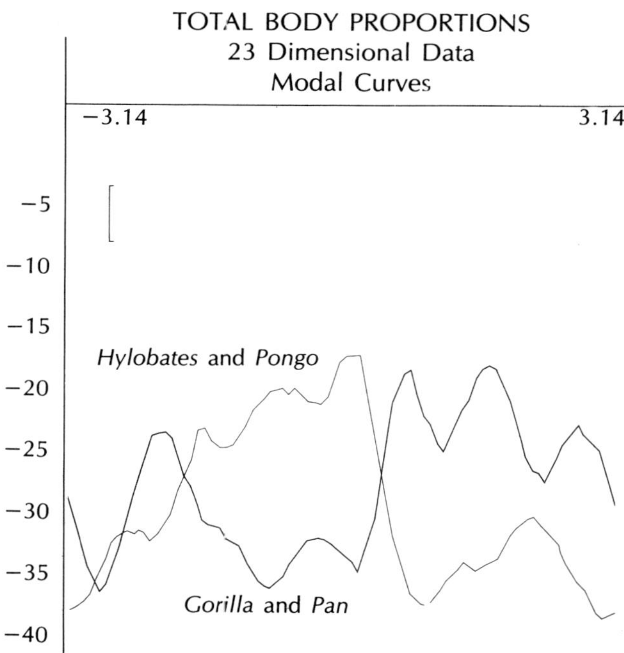

Fig. 8.20. The separation between the Asiatic apes on the one hand and the African apes on the other that stem from the morphometric study of the entire suite of overall proportions. Demonstrated by high-dimensional display.

rather than the classical morphological picture. Thus, the study of overall proportions as in figure 8.20 (see also Oxnard, 1981b) shows the marked difference between orang-utans and gibbons on the one hand, and African apes on the other. The summated study of the smaller anatomical regions as in figure 8.21 provides a confirming picture (see also Oxnard, 1983c).

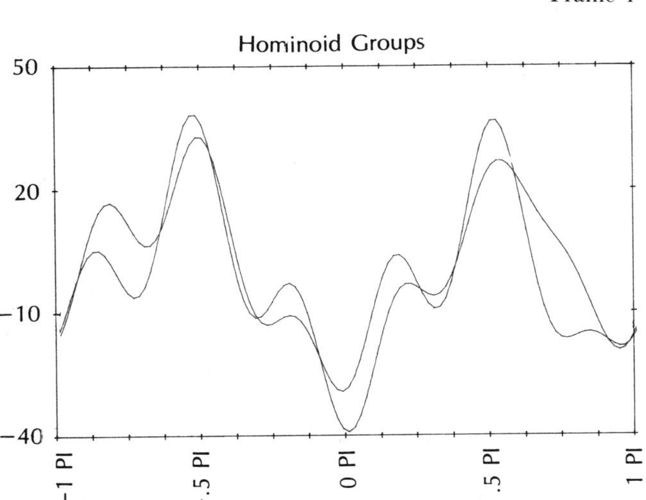

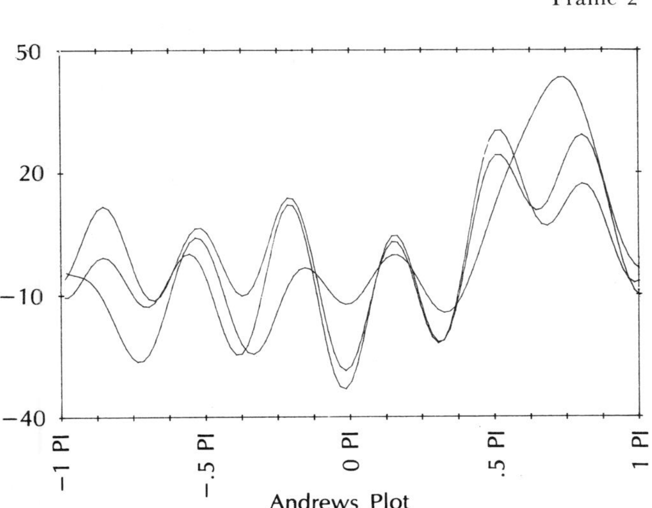

Fig. 8.21 The separation between the Asiatic apes on the one hand (first frame) and the African apes and humans on the other (second frame) that stem from the morphometric study of the detailed anatomical regions. Demonstrated by high-dimensional display. Humans have differences, but are more closely allied with African apes.

Old World Monkeys

When we move to the Old World Monkeys we again observe a difference between the classical morphological picture and that of the molecular biologists. Classical morphology groups large numbers of species and genera in only two divisions, the cercopithecine monkeys and the colobine monkeys (e.g. Simpson, 1945; Osman Hill, 1953–66; Hershkovitz, 1977). Even this difference is not great, Schultz noting as recently as 1970 that there is very little variation among Old World monkeys.

Biomolecular studies of various kinds almost always suggest a different picture. Three more or less equal groups are defined by such studies rather than two. This is so in Goodman's work based upon antigenic distances (Dene, Goodman, Prychodko and Moore, 1976), the studies of Sarich and Wilson (1976) on albumins and transferrins, Benveniste and Todaro (1976) on cellular DNA, Hewett-Emmett, Cook and Barnicot (1976) on haemoglobins, and many other studies since. In almost every case, although the precise species included in each given study varies, tripartite clustering of Old World genera is found as indicated in Fig. 8.22.

Other types of study also support these ideas. For instance, the results of a number of breeding experiments, both natural and experimental, are now becoming available. Using a 'biological species' concept, they suggest that there are truly far fewer 'real' biological groups than classically believed. The number of some species, indeed in some cases genera, between which it is possible to produce viable, often fertile offspring is quite remarkable. This has been demonstrated, for instance, among seven species of *Papio*, nine of *Macaca*, and one each of *Cynopithecus* and *Cercocebus* (reviewed in Chiarelli, 1973).

It is therefore of great interest to discover that *the separations achieved by the new morphometric studies of Old World Monkeys here reported is into a small number of apparently separate groups (Fig. 8.22) that mirror the biochemical rather than the classical picture*. The result of looking at overall proportions of the Old World Monkeys that is so evident in studying the upper and lower limbs (divisions into more terrestrial and more arboreal species) is not present. Instead the major clusterings are in only four groups: (1) the baboons and mandrills, (2) the macaques and mangabeys, (3) the cercopitheques, including patas monkeys and Allen's guenon and (4) the various colobines (Fig. 8.23). Of these four groups

Groupings Inherent in Classical Morphology

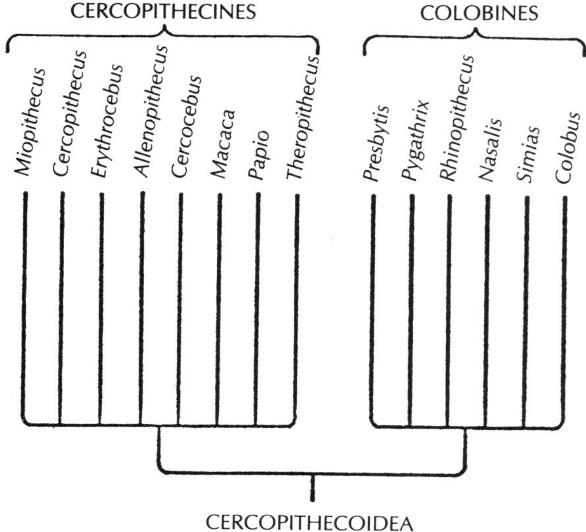

Groupings Explicit in Biomolecular Studies

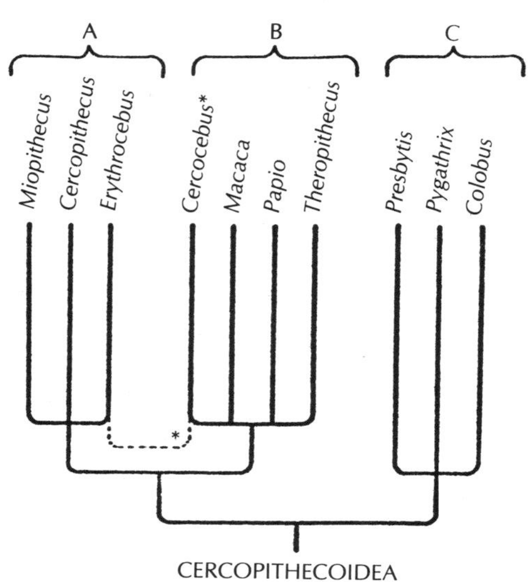

DISCORDANCES: DETAILED DIFFERENCES IN CERCOPITHECOID GROUPING PATTERNS

Groupings Explicit in Morphometric Studies

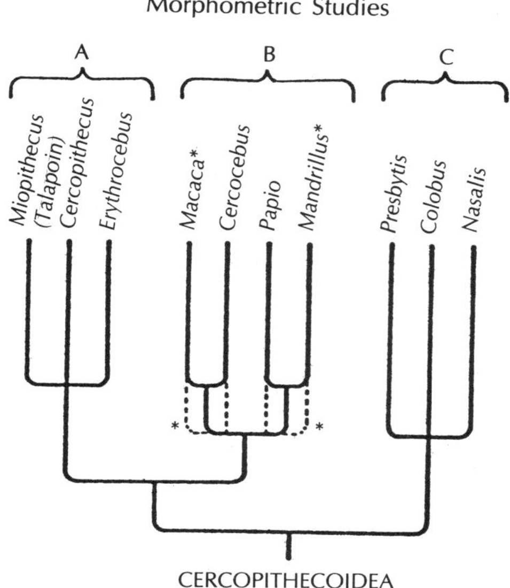

Fig. 8.22. Discordances between classical morphologists and molecular evolutionists in the Old World monkeys. The new morphometric result mirrors most closely that summarized in the molecules. Asterisks indicate alternative interpretations.

there is some closer relationship between the first two, and an almost equally probable pattern of grouping into three is possible. With this arrangement, a combined group of mandrills, baboons, macaques and mangabeys is a little more variable than the other groups but is not impossibly broad.

The tripartite arrangement of the Old World monkeys is even more clearly revealed in the summated high-dimensional studies of detailed anatomical regions (Fig. 8.24; and summarized in Oxnard, 1983c).

Thus, though this result is somewhat discordant with that obtained from classical studies (though not entirely so — a few workers using mainly dental information have identified lower level groups, 'tribes' that are somewhat similar but not identical to two of the above groups: e.g. Delson and Andrews, 1975; Simons and Delson, 1978), it accords very well with that from most biomolecular investigations.

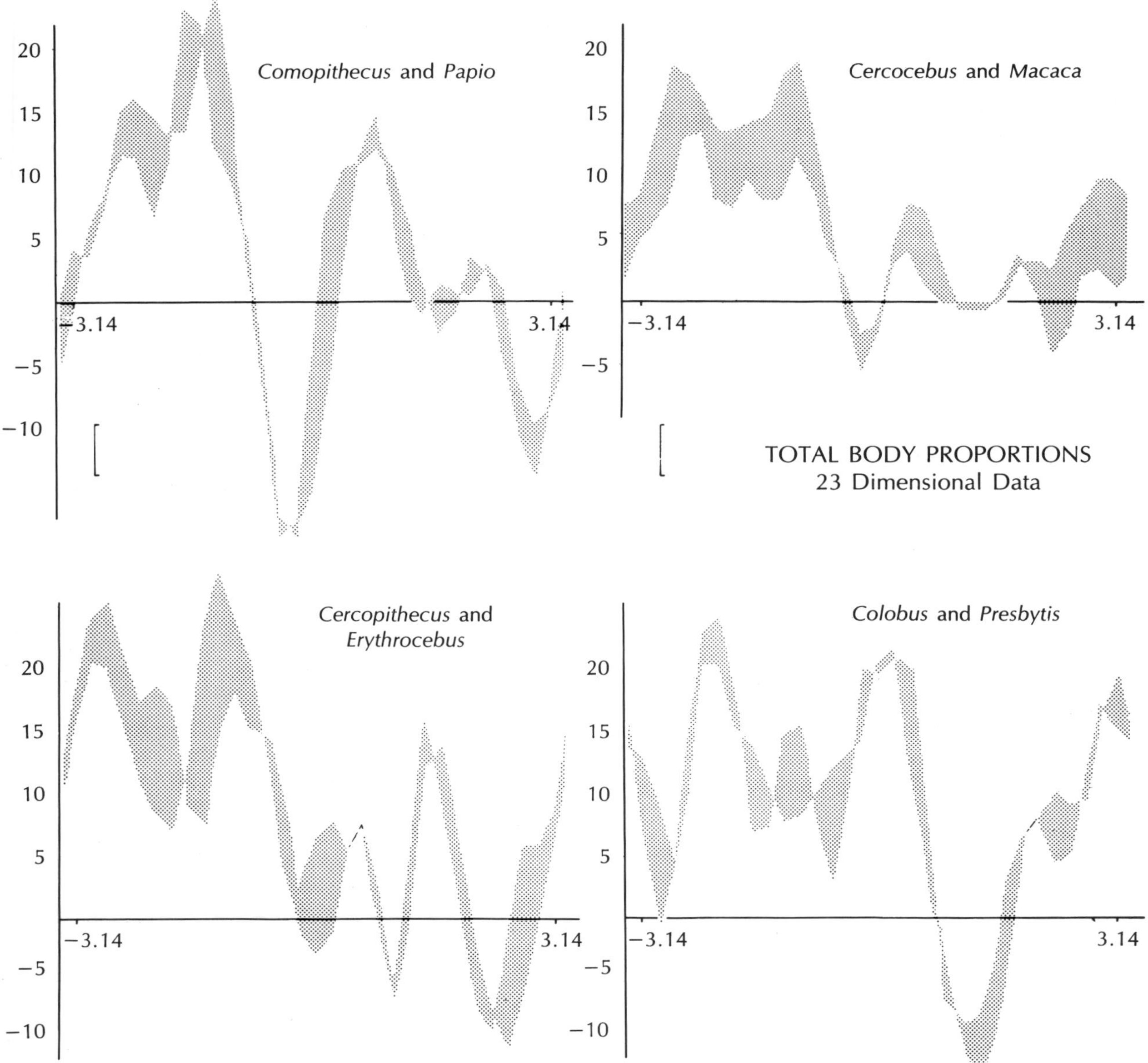

Fig. 8.23. The separations among the Old World monkeys that stem from the morphometric study of the suite of overall proportions. Demonstrated by high-dimensional display. The four named groups are easily evident. And there are considerable similarities between the top two envelopes so that an almost equally probable arrangement is into only three groups: baboons, mandrills, macaques and mangabeys being combined as a single group.

New World monkeys

As with the Old World monkeys, so with the New: there is a distinct difference among the details of the intermediate levels of grouping resulting from the different approaches. The differences between various morphologists is best seen, perhaps, by comparing Simpson's (1945) classification of New World monkeys with that of Hershkovitz (1977). Simpson lists 14 genera spread among 7 sub-families (Aotinae, Pitheciinae, Alouattinae, Cebinae, Atelinae, Callimiconinae and the single sub-family that may be presumed to be associated with the family Callithricidae). Of these 14 genera only two, *Callimico* and *Alouatta*, represent sub-families containing no other genera. In contrast, Hershkovitz lists a total of 16 genera spread around 9 sub-families (Saimirinae, Aotinae, Callicebinae, Alouattinae, Pitheciinae, Cebinae, Atelinae and the two subfamilies that may be presumed to be

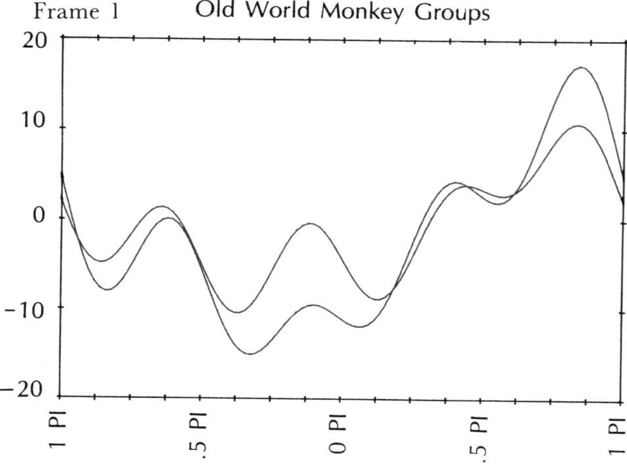

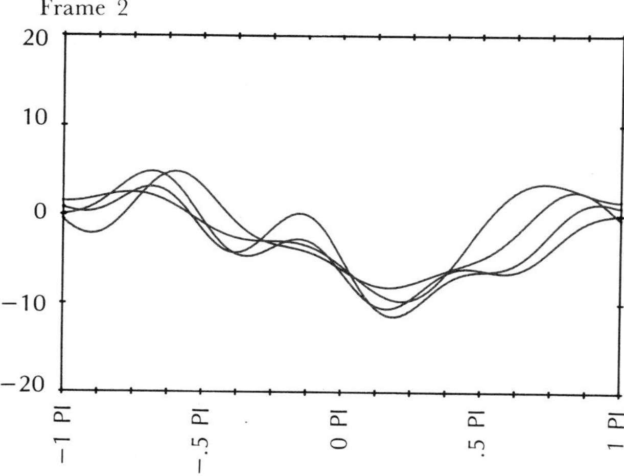

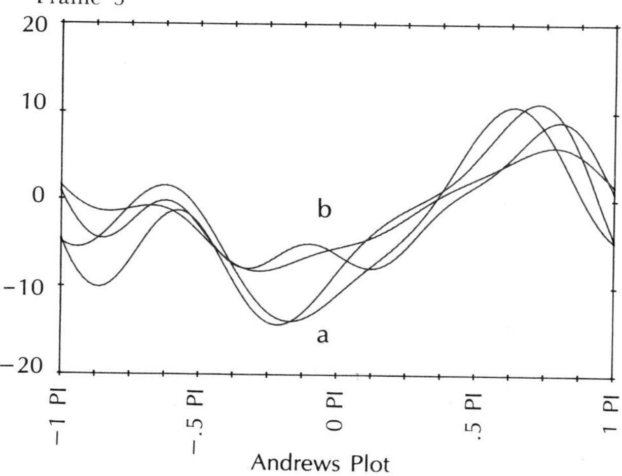

Fig. 8.24. The separations among the Old World monkeys that stem from the morphometric study of the detailed anatomical regions. Demonstrated by high-dimensional display. The division into three groups is clear, upper group: vervets and patas monkeys, middle group: the various colobines, lower group: baboons, mandrills, macaques and mangabeys. The lowermost group does show evidence within itself of two subgroups: a) baboons and mandrills and b) macaques and mangabeys.

associated with the families Callitrichidae [sic] and Callimiconidae). Of these sub-families, as many as 6 contain only a single genus. Yet, notwithstanding these differences, there is a basic agreement in the clustering picture provided by even these divergent authors. For instance, both recognize the primary sub-division into Callithricidae and Cebidae (Fig. 8.25).

Of course, in comparison with hominoids, in comparison even with Old World Monkeys, very much less is known about biomolecular parameters of New World monkeys. However, some of what is known differs considerably from the classical picture in that differences among many of the genera are small or undetectable except at the highest, familial levels. Thus, urinary amino acids (Fooden, 1961), though separating cebids from callithricids, show very few differences within each of these groups. Likewise, such information as is available from karyology and from reproductive biology (e.g. Chu and Bender, 1961; Luckett, 1974) provides no basis for separating groups of New World monkeys.

But investigations of albumins and transferrins (Sarich and Cronin, 1976), of antigenic distances (Dene, Goodman, Prychodko and Moore, 1976) and of alpha haemoglobin sequences (Goodman, 1976) together with other even more recent studies (e.g. Baba, Darga and Goodman, 1979) all conform in separating three groups of New World monkeys. One group consists of *Leontideus*, *Callimico*, *Callithrix*, *Cebuella*, *Saguinus*, *Saimiri* and *Aotus*; the second includes *Ateles*, *Lagothrix*, *Alouatta* and *Cebus*; the third includes, to the degree that these genera have been studied, *Pithecia* and *Cacajao*. The numbers of genera, species and individual specimens included in any single one of these studies is small enough that a degree of tentativeness must attach to these results. At present, however, this is the overall picture (Fig. 8.25).

It is again, therefore, of interest to note the strength of the clusterings produced by the new morphometric analyses of the overall proportions of these animals. As with the Old World monkeys, the sensitivity of the separations is large because it depends upon the addition of so much relatively independent information from so many features. And again, as with the Old World monkeys, the locomotor or functional results that seem to relate to locomotor abilities within the trees obtained in studying restricted anatomical regions alone (Chapter 7) are not seen.

These major clusterings are shown both in the

Groupings Inherent in
Classical Morphology

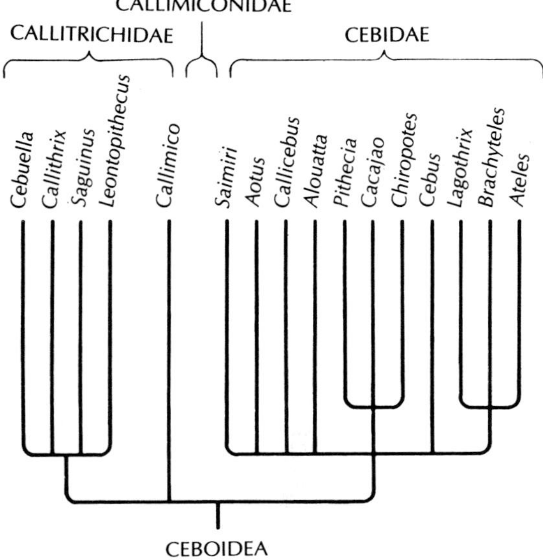

Groupings Explicit in
Biomolecular Studies

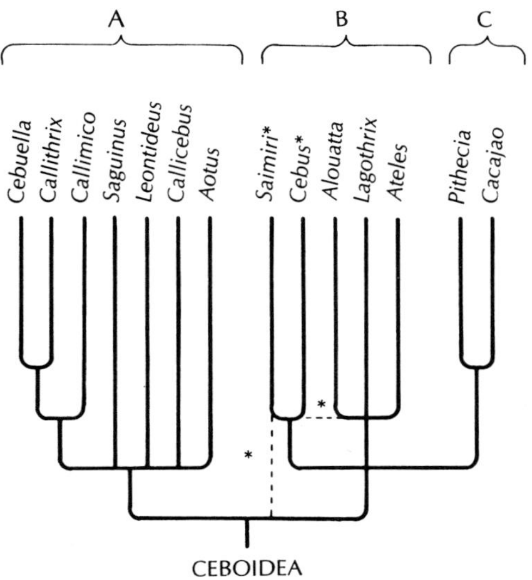

Groupings Explicit in
Morphometric Studies

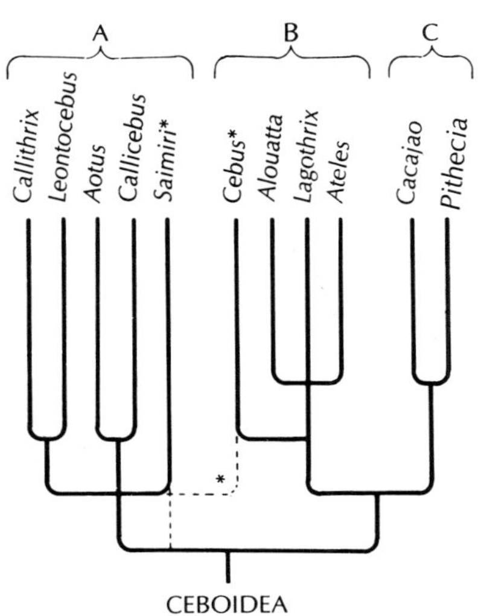

DISCORDANCES:
DETAILED DIFFERENCES
IN CEBOID
GROUPING PATTERNS

Fig. 8.25. Discordances between classical morphologists and molecular evolutionists in the New World monkeys. The new morphometric result mirrors most closely that obtained from the study of molecules. Asterisks indicate alternative interpretations.

studies of overall proportions (Fig. 8.26 and Oxnard, 1981b) and in the studies of more detailed anatomical regions (Fig. 8.27 and Oxnard, 1983c). They demonstrate that at most the New World monkeys form three groups. *One* consists of the Atelinae, Alouattinae and *Cebus* of Simpson taken together; a *second* contains the three pithecine genera, *Cacajao*, *Pithecia* and *Chiropoles*; and a *third* includes the Aotinae, Callithricidae and *Saimiri* of Simpson.

Knowing what we do of these various genera it is easy to see them as (a) a markedly generalized group (with callithricid and non-callithricid parts), (b) an intermediately specialized group (Pithecinae) and (c) a highly specialized group with prehensile or semi-prehensile tails. This clustering makes every bit as much biological sense as any of the more 'split' classifications, although it is true that it places less weight upon certain peculiar anatomical features such as the presence of 'claws'

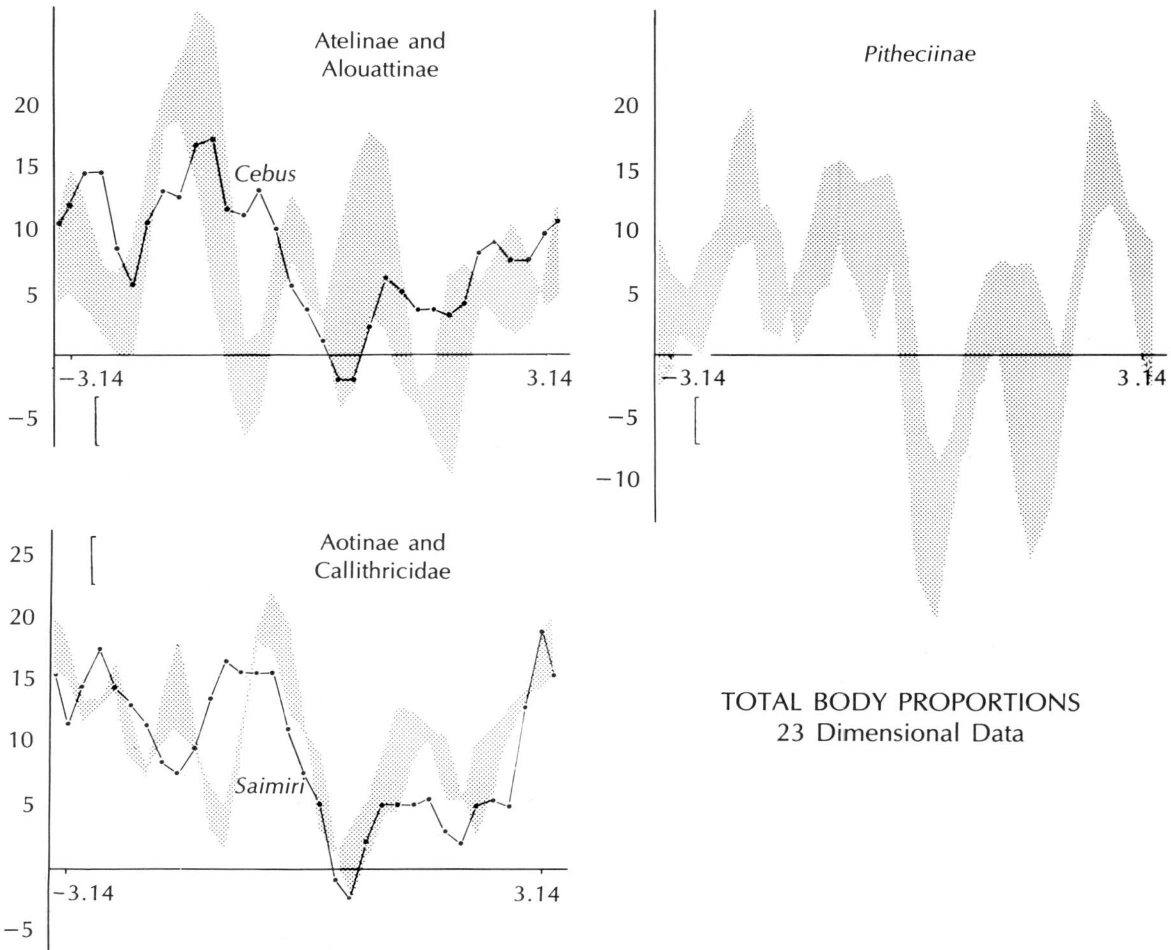

Fig. 8.26. The morphometric separations of the New World monkeys as displayed by the high-dimensional method applied to the study of overall bodily proportions; the three named groups are evident.

in marmosets and tamarins. *The general multivariate morphometric picture is again discordant with classical morphology and fairly well concordant, although not completely so, of course, with the average view obtained from the study of biomolecules (Fig. 8.25).*

It may be thought that these results for the Anthropoidea are merely due either to a policy of lumping or to lack of sensitivity in the analyses. This is not so.

The envelopes of Figs. 8.20, 8.21, 8.23, 8.24, 8.26 and 8.27 are very narrow; intra-group distances are small; clusters are real. The overall intergroup distances are far greater than those normally achieved by data of a classical sort. The results are not due to artificial lumping of groups.

The sensitivity of individual morphological features in separating groups is not high; for instance, in those morphological features which are quantitative, the entire order of primates may be enclosed within little more than four or five standard deviation units. In contrast, sensitivity is so greatly increased by the use of the multivariate statistical tool that as many as 30 or 40 standard deviation units may enclose the entire order.

The statistical significances of the groups just outlined for the Hominoidea and the Old and New World monkeys are not in doubt. Tests of the type outlined in table 3.4 show that the groups found above are vastly different from one another and the differences between their component species are very small. The results are real and due neither to lumping nor lack of sensitivity.

Such statements cannot be applied to the

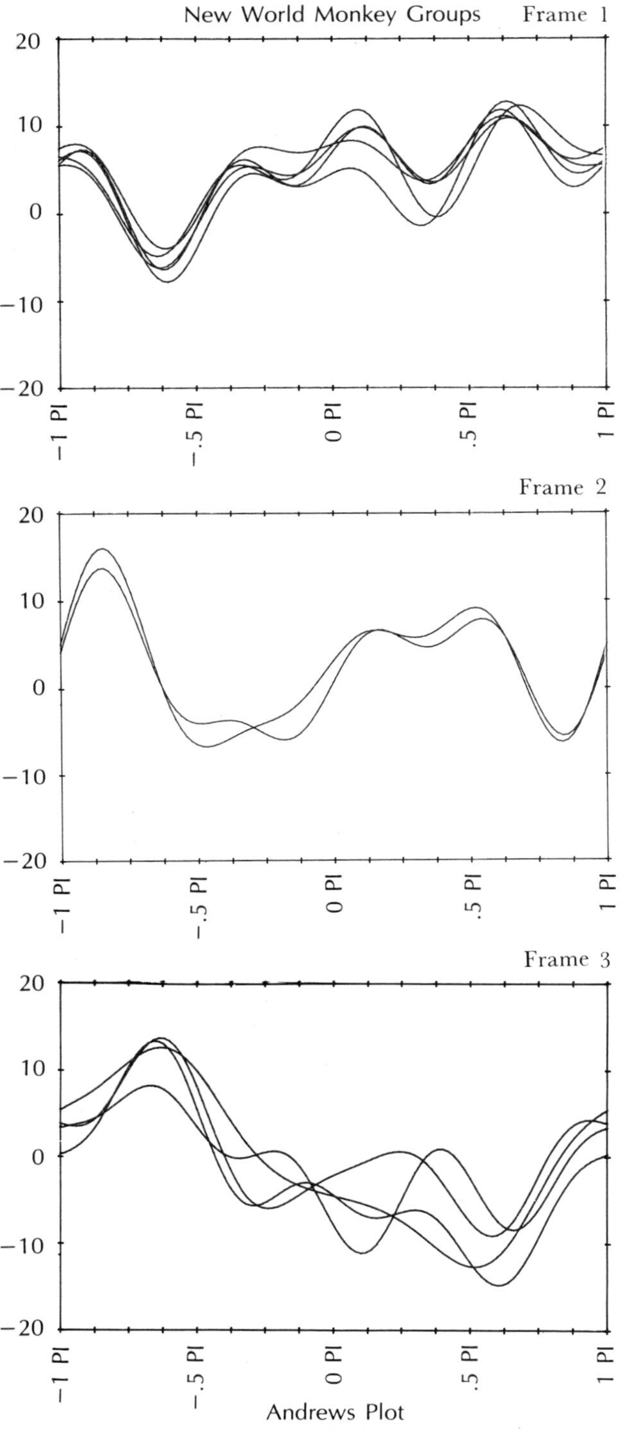

Fig. 8.27. The morphometric separations of the New World monkeys as displayed by the high-dimensional method applied to the study of detailed anatomical regions – the same three groups as in Fig. 8.26. exist. Upper group: Aotinae, Callithricidae and *Saimiri*, middle group – Pitheciinae, lower group – Atelinae, Alouattinae and *Cebus*.

following results for the Prosimii; in the case of the Prosimii problems exist for all modes of study – classical morphological, biomolecular or morphometric – because of inevitable limitations in materials available for study.

Prosimians

When, finally, we come to view arrangements within the Prosimii, much less clear alignments can be discerned. For although Fig. 8.28 posits an arrangement for prosimians resulting from classical morphological studies, it also notes that a great deal of argument and some widely differing opinions exist. There is argument as to whether or not tarsiers are prosimians — and whether or not tree-shrews are even primates.

And although that same Fig. 8.28 also posits an arrangement for the molecular and biochemical data, again there are problems. In this case the problems stem from the absolute paucity of data that are so far available for this group. For even the studies of Sarich and Wilson (1967), Dene, Goodman, Prychodko and Moore (1976) and Beard and Goodman (1976) are markedly deficient in coverage of genera.

And in exactly the same way, our own multivariate morphometric studies contain large gaps in species representation and small samples for at least some of the species that are included. It is true that in our results lemurs lie with lemurs (e.g. the conjunction of *Lemur* and *Varecia*, for instance), indriids lie with indriids (nearest neighbour relationships between *Propithecus*, *Indri* and *Avahi*, for example) and lorisines lie with lorisines (clusterings of *Perodicticus*, *Nycticebus*, *Arctocebus* and *Loris*.) But the overall sizes of the samples and the numbers of missing genera are such that we have to be very careful about reading too much into the analyses.

Yet, taking all this caution into account, Fig. 8.28 (results of morphometric study of overall proportions, Oxnard, 1981b) and Fig. 8.29 (results of summation of detailed anatomical regions, Oxnard, 1983c) suggest that our general thesis, that the multivariate statistical view aligns itself rather more with the biomolecular opinion than with the classical one, has a tentative reality even among prosimians.

Among the Prosimii, however, there are two major problems of a different type. Surely the validity of these groupings as representing anything to do with overall relationships falls down on

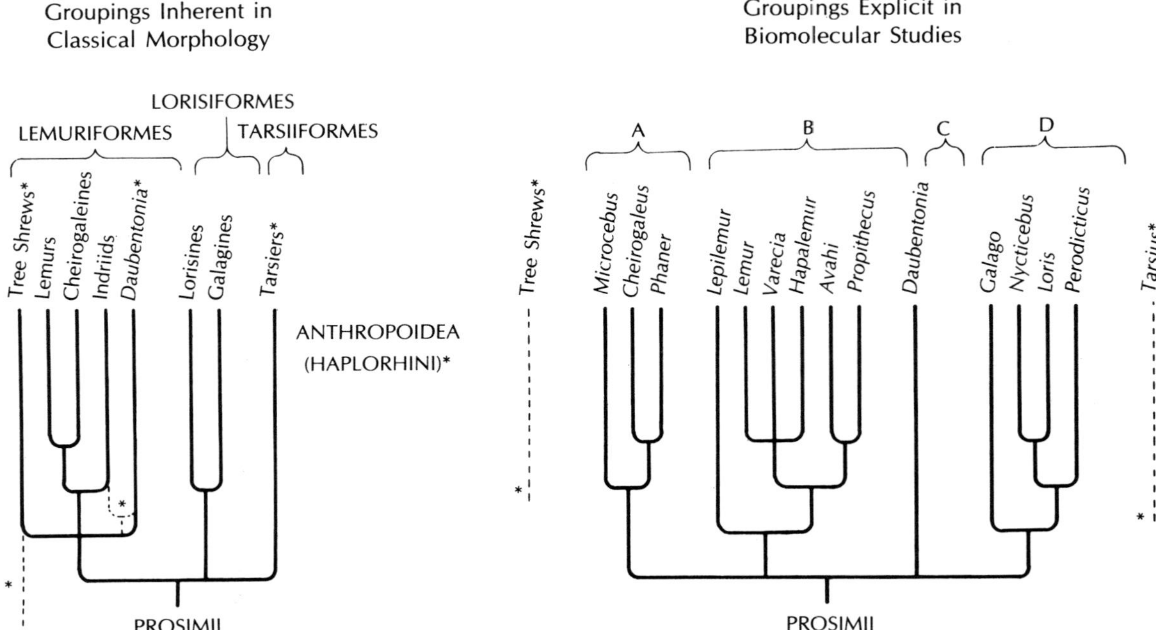

DISCORDANCES:
DETAILED DIFFERENCES
IN PROSIMIAN
GROUPING PATTERNS

Groupings Explicit in
Morphometric Studies

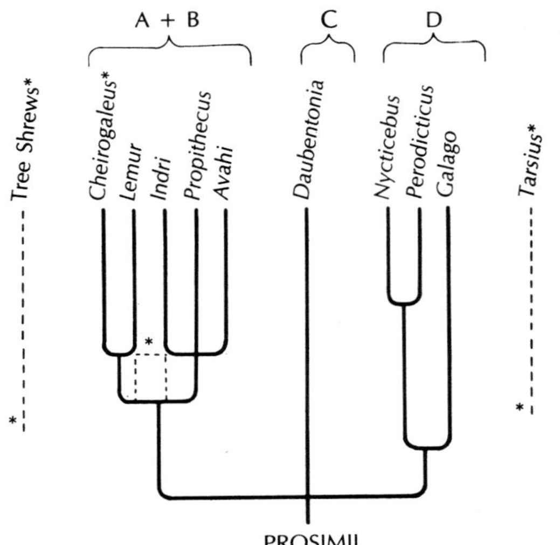

Fig. 8.28. The clusterings of prosimians by various methods. Here the studies of classical morphologists depend upon a very large base of data. The studies of molecular evolutionists and of multivariate morphometrics are much less strong. In these latter two cases the numbers of specimens examined in each group is much smaller, and many groups are totally unexamined to data. Yet even with these faults, there are certain agreements between the molecular and morphometric views that dissent from those of classical morphology. As the text indicates, the paucity of the new information means that we must be most cautious. Again, asterisks note alternative interpretations which are more numerous than among the higher primates.

two counts. Tarsiers are close to bush-babies, and aye-ayes are not close to indriids. In these two cases, the sizes of our samples are quite large so that this result is robust. Surely the results for these two particular groups are the evidence that denies our new hypothesis. It is, therefore, essential to look at these two cases in more detail.

Convergence and the spectral tarsier

Because of the lack of good descriptions, and because of a number of geographical and other errors, it is not entirely clear exactly when tarsiers were first discovered. Whether this is represented by Camel and Petiver's notes at the beginning of the

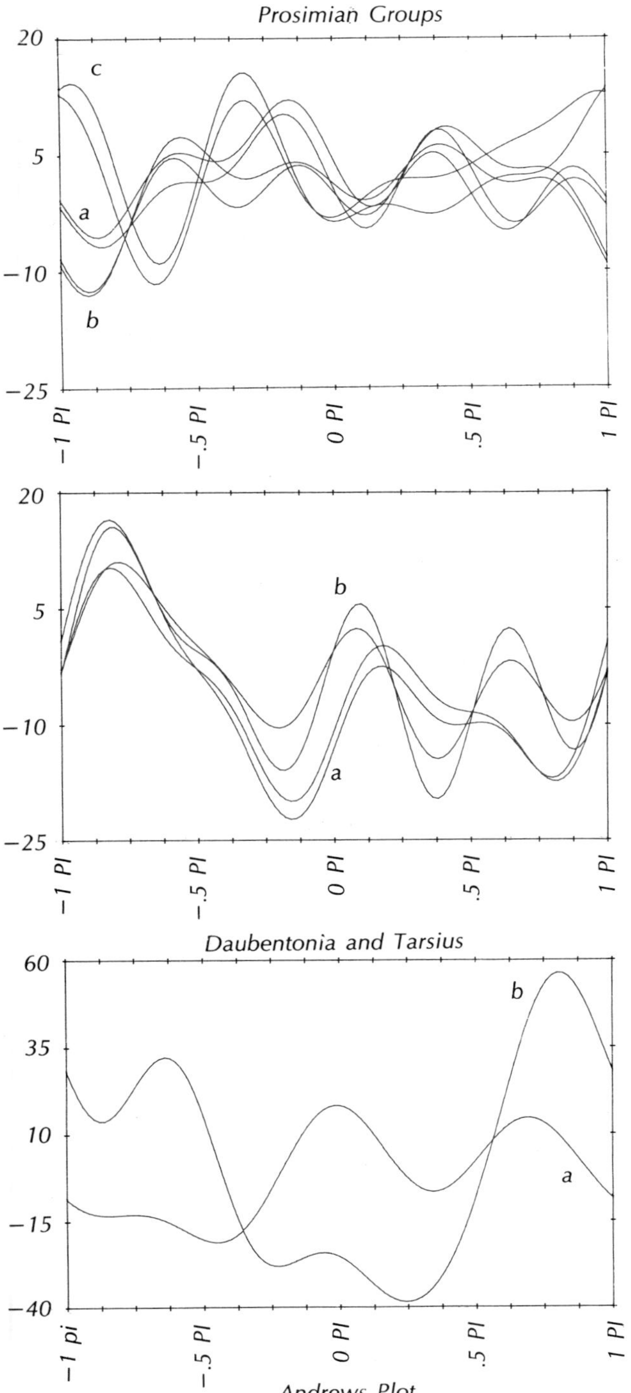

Fig. 8.29. The morphometric separations of prosimians as displayed by the high-dimensional method applied to the studies of detailed anatomical regions. Two groups can be readily identified: lemuriformes (and, within them, three subgroups: a) lemurs, b) cheirogaleines and c) indriids – upper frame), and lorisiformes (and, within them, two subgroups: a) lorises and b) galagos – middle frame). These groups exclude, of course, a) *Tarsius* and b) *Daubentonia* (lower frame – for more details see later).

eighteenth century, or whether the more certain description of Buffon later in that century (1765) is the more acceptable must remain open to some doubt. It is not in doubt, however, that this genus, found in Southeast Asia, is the sole living representative of a large group of primates that once flourished in dazzling variety and abundance.

***Tarsius:* the earlier views.** In comparison, for instance, with the doubts that surround *Daubentonia*, there seems no problem about the lineage of *Tarsius*: a clear fossil record goes back through the Miocene to the Palaeocene, with the main belly of the radiation being in the Eocene. It is, of course, the case that almost all of this exotic fossil history is based upon dental characteristics; a good deal less depends upon other cranial features; the post-cranial skeleton is scarcely known at all.

Yet this little animal, easily held in the palm of the hand, has for many years attracted the attention of comparative anatomists because it combines, in its anatomical organization, a number of remarkably basic characters together with an equally remarkable number of curious and highly specialized features.

The early history of the structure of *Tarsius* and speculations about its possible evolution show a range of opinions almost as wide as those exhibited about *Daubentonia*. As the aye-aye, so the tarsier was early allied with what are now believed to be non-primate mammals – to wit, the jerboas. And once recognized as a primate, *Tarsius* was variously allocated to the prosimian primates broadly defined, to lemur-like primates more specifically, to the anthropoid primates or to some primate group intermediate among them.

The consensus of opinion today recognizes *Tarsius* as a prosimian.

The modern tarsier is a crepuscular and almost entirely arboreal creature. It shows marked specializations in the relatively enormous size of the eyes and in the peculiar modifications of the limbs for leaping among the branches. The ears are huge; the structure of the nose and upper lip resembles that of New World monkeys; the tail is large and long with epidermal ridges, but is not prehensile in the usual sense of the term.

A more detailed survey of the various anatomical systems of *Tarsius* confirms that it possesses a remarkable combination of characters. Characteristics thought to be basic to early primates include molars with three simple cusps, the pattern of the digits of the hand, the unelaborated gut and the relatively smooth cerebral cortex. Characteristics thought to be highly specialized include the enormous eyes and orbits, and the modification of the hind limbs for leaping. Characters thought to be lemuroid include the well-defined fold under the tongue, the pattern of toes in the foot, the presence of toilet nails on some digits (though there are two of these and not one as in the lemurs), the large mobile ears and, inside the brain, a lateral geniculate nucleus of the inverted type. Characters thought to be anthropoid include the details of the upper lip, the restriction of the size of the nose, the partial separation of the orbits from the temples by an expanded bony shelf, the basal position of the foramen magnum, the differentiation of a macula in the retina and the enlarged occipital lobe in the brain with a highly complex visual cortex.

It is not surprising, therefore, that conflicting views have been advanced regarding the position of *Tarsius*.

Some authorities have been swayed by the lemuroid characteristics (particularly the resemblances with some of the lorisiformes, bush-babies, both in aspects of skull structure and in the modifications of the hind limbs for leaping) and have included them as prosimians. Others have been so impressed with the simian characteristics outlined above that the tarsier has been separated widely from the Prosimii and placed in a common taxonomic group with the monkeys, apes and man (Haplorhini). The proposition has been advanced that the Anthropoidea are derived from a tarsioid ancestry. Yet other investigators have placed the tarsier in a separate group occupying a position intermediate between the lower and higher primates.

The modern consensus. But it is really the accumulation of fossil evidence that make Simpson (1945) and Le Gros Clark (1959) willing to go along with the notion that the animal is infraordinally related to one (the Prosimii) of only two suborders of primates, thus emphasizing the basic characteristics of the fossils (Fig. 8.30). Their arguments are that the enormous radiation of the fossils and the great difficulties in differentiating between early lemuroids and early tarsioids are what determine the prosimian relationships of the entire group. This is notwithstanding the lack of modern lemur-like specializations in modern tarsiers.

This view suggests that the simian characteristics are illusory, the result of fortuitous resemblance, e.g. the reduction of the snout and the restriction of the nasal cavity may be only apparent, being concealed to a degree by expanded orbits; the reduction of the olfactory parts of the brain may be a secondary consequence of these peculiar alterations of the skull, the broad and rounded shape of the brain as a whole being perhaps attributable to the distortion and anteroposterior compression produced by the large eyes; the basal displacement of the foramen magnum and the existence of bone in the lateral wall of the orbit are possibly also secondary to the enlargement of the eyes. Thus, many of these features may not be morphologically equivalent to those apparently similar features in anthropoids.

But apart from these examples of possible convergence there remain a number of anatomical relationships which, taken together, really *do* seem to betoken an affinity between *Tarsius* and the higher primates: the differentiation in the eye of a retinal macula, the intrinsic structure of the cerebral visual cortex, the construction of the external parts of the upper lip and nose, the particular form (haemochorial) of the placenta and so on. It is, indeed, possible that these items may have been developed independently in *Tarsius*, but they may also betoken a degree of common ancestry or separate evolution from a prior stock with potentialities for development in similar directions.

ORDER **PRIMATES** linnaeus, 1758

Suborder **PROSIMII** illiger, 1811

Infraorder LEMURIFORMES Gregory, 1915
LEMURS, TREE-SHREWS

Infraorder LORISIFORMES Gregory, 1915
LORISES, GALAGOS

Infraorder TARSIIFORMES Gregory, 1915
TARSIUS

Suborder **ANTHROPOIDEA** Mivart, 1864

Superfam. Ceboidea Simpson, 1931
NEW WORLD MONKEYS

Superfam. Cercopithecoidea Simpson, 1931
OLD WORLD MONKEYS

Superfam. Hominoidea Simpson, 1931
APES and MAN

After Simpson, 1945

Fig. 8.30. The consensus of opinion of the relationships of the prosimians (after Simpson).

Recent challenges to the conventional view. In very recent years, a number of investigators have returned to the notion that the conventional position of *Tarsius* is to be challenged.

Professor A.J.E. Cave provides one of the pieces of this new evidence. At the same time as documenting many of the earlier suggestions that the basic separation among the primates is really such as to place the tarsier alongside the monkeys, apes and man in the Anthropoidea (Cave, 1973), he confirms that today most generally accepted classifications (e.g. Simpson, 1945) place *Tarsius* squarely among the Prosimii. But as Professor Cave goes on to investigate the architecture of the nasal fossa in a wide range of mammals, he arrives at a different conclusion. He points out that the nasal fossa is an anatomical region subserving a similar respiratory function in most mammals and he suggests that its own internal architectural make-up, apparently varying among terrestrial mammals without obvious equivalent internal functional differences, may thus be especially useful for distinguishing related creatures from those that are less closely linked. Professor Cave demonstrates that the architecture of the nasal fossa is uniform within the tarsiers, monkeys, apes and man, and that this group as a whole differs radically not only from all other mammalian orders but also from the remaining primates. The nature of the architectural uniformity that he observes (Fig. 8.31) is unlikely to have been determined by other specialized features such as enlarged orbits, the prior argument for rejecting information about the nose. The inference is drawn that perhaps the tarsiers, monkeys, apes and man indeed form a natural unit and that tree-shrews, lemuriformes and lorisiformes do not belong to that grouping.

In recent years, other studies have also suggested that clustering *Tarsius* with the monkeys, apes and man is a more likely reflection of phylogeny. Thus, Szalay (1975a, b), using data and arguments of a classical type, implies that the tarsiers, New World monkeys and Old World primates contrast with regular prosimians in sharing a series of characteristics of the ear region of the skull. He points to such features as an increase in relative size of the promontory artery as compared with the stapedial artery (two blood vessels within

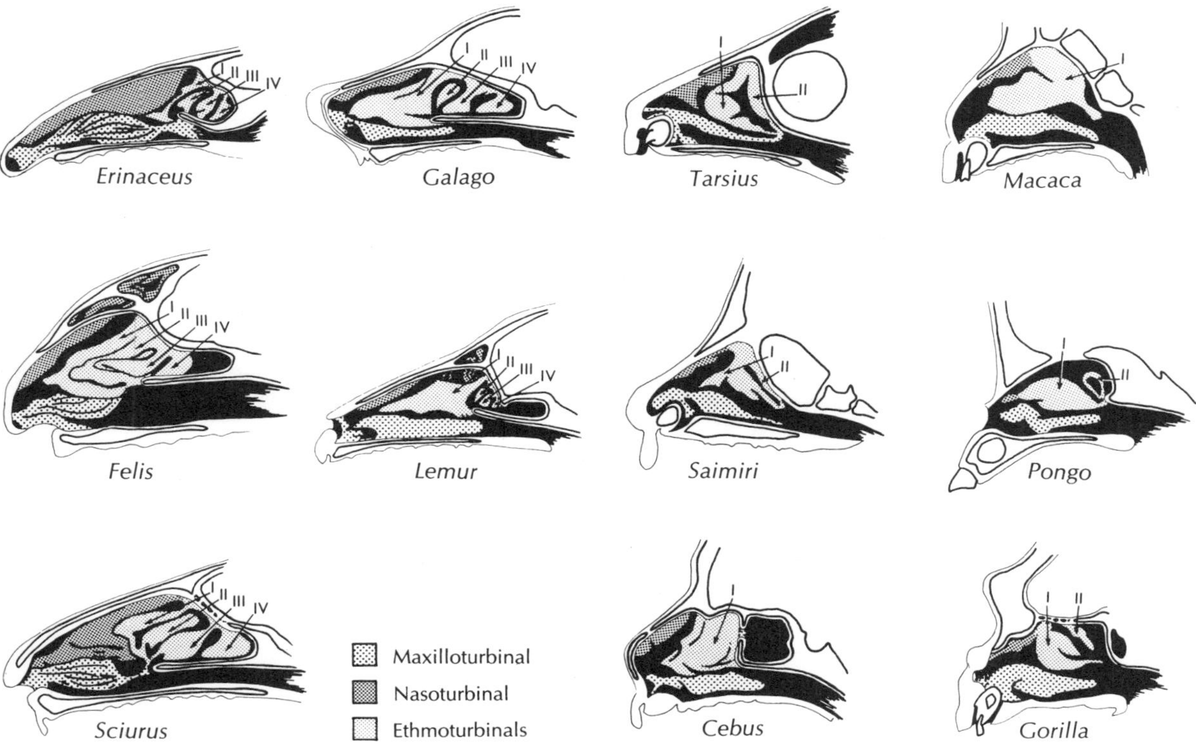

Fig. 8.31. An alternative view of the relationships of prosimians: the structure of the nose. Roman numerals indicate homologous anatomical parts. Tarsiers share the anthropoid pattern (redrawn after Cave).

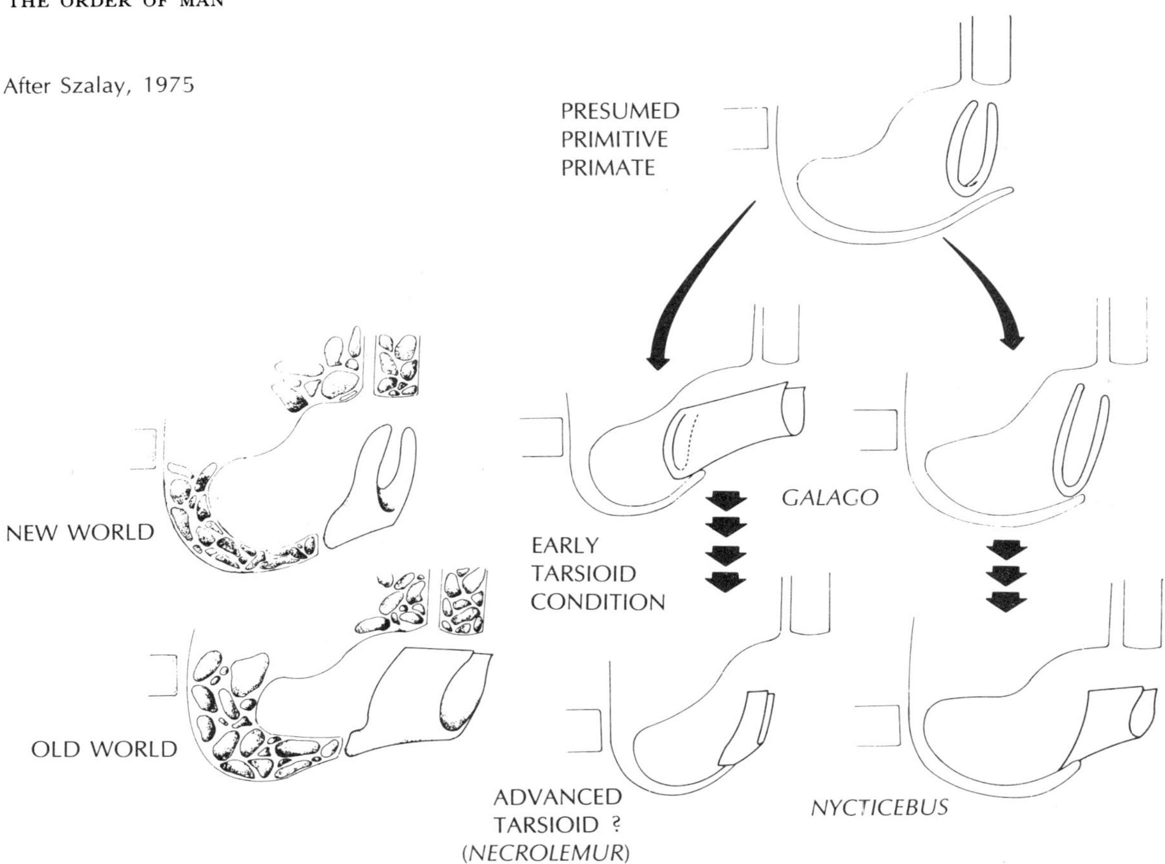

Fig. 8.32. An alternative view of the relationships of prosimians: the structure of the ear. Szalay believes that tarsiers approximate more to anthropoids (redrawn after Szalay).

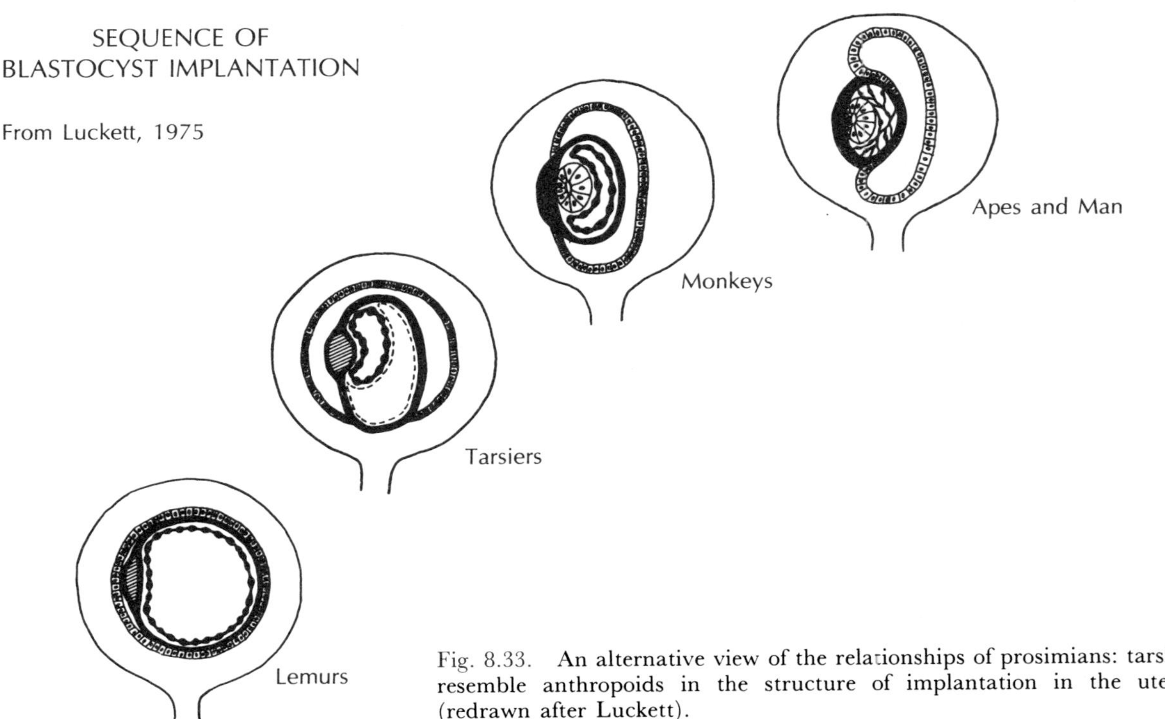

Fig. 8.33. An alternative view of the relationships of prosimians: tarsiers resemble anthropoids in the structure of implantation in the uterus (redrawn after Luckett).

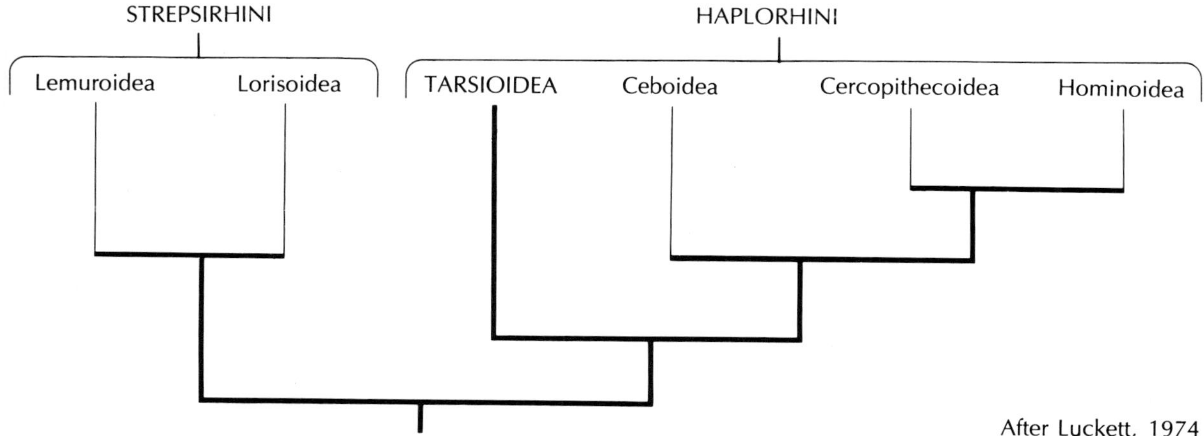

Fig. 8.34. An alternative view of the relationships of prosimians: 'Hennigian' clustering, using data from various reproductive structures, places tarsiers nearer anthropoids (redrawn from Luckett).

the ear); a medial shift of the entry of the carotid artery into the ear; and a change in the external configuration of the ear region particularly as epitomized in the living tarsier. Some of these features are summarized in Fig. 8.32.

Investigations of primate foetal membranes and placentae have led Luckett (1975) to postulate that the tarsiers and the anthropoidea are 'sister groups within the Haplorhini', the tarsiers being separate, therefore, from the remaining prosimians. His data are summarized in Figs. 8.33 and 8.34 and depend not only upon the use of physiological features relating to the reproductive system but also upon a line of argumentation derived from cladistic methods (Hennig, 1966).

And even data from molecular investigations, such as those provided recently by Dene, Goodman, Prychodko and Moore (1976) using immunodiffusion techniques, by Beard, Barnicott and Hewett-Emmet (1976) studying alpha and beta chains of haemoglobins, and by Boer and Boer-Van der Vlist (1973) investigating somatic chromosomes, or those given more than half a century ago on blood groups (Le Gros Clark, 1924) suggest that the closest affinities of *Tarsius* are with members of the Anthropoidea rather than with other primates (Fig. 8.35).

These various studies (and others: see Minkoff, 1974, and Groves, 1974) re-emphasize the question of the morphological assessment of the genus *Tarsius*. Is there other major evidence of a close relationship between tarsiers on the one hand and the monkeys, apes and man on the other, in the structural and functional affinities of these forms? Is evidence of a relationship of *Tarsius* to the other prosimians (*sensu* Simpson, 1945) absent? Certainly McKenna (1975) sees another picture (Fig. 8.36). It is possible that yet another approach, that taken in this book, may be applied to such questions.

Locomotor convergences in the limbs. A major problem in understanding the meaning of the anatomy of *Tarsius* is that it is so closely associated with the remarkable form of locomotion that the animal exhibits. So divergent is this locomotor pattern from that of any of the Anthropoidea that structures which may speak to an association of tarsiers with anthropoids may be overshadowed by functional adaptations to this form of leaping. Further, the extreme locomotor pattern of *Tarsius* seems very closely analagous to that of the bushbaby, undeniably a prosimian (Figs. 8.37 and 8.38). The degree of morphological convergence of *Tarsius* with such prosimians may hide morphological relationships speaking to its phylogenetic position.

An attempt is therefore made here to study the relationships of these creatures using the very large battery of data about the overall form and proportions of the body. The extent to which it is possible to 'dissect away' (to use an anatomical allusion) or to 'partition out' (to use a statistical metaphor) morphological convergence associated with an extreme locomotor pattern may provide another view of the affinities of the genus.

Again, the data used in these studies are (1) those of Professor A.H. Schultz derived from measurements of the overall proportions of 472

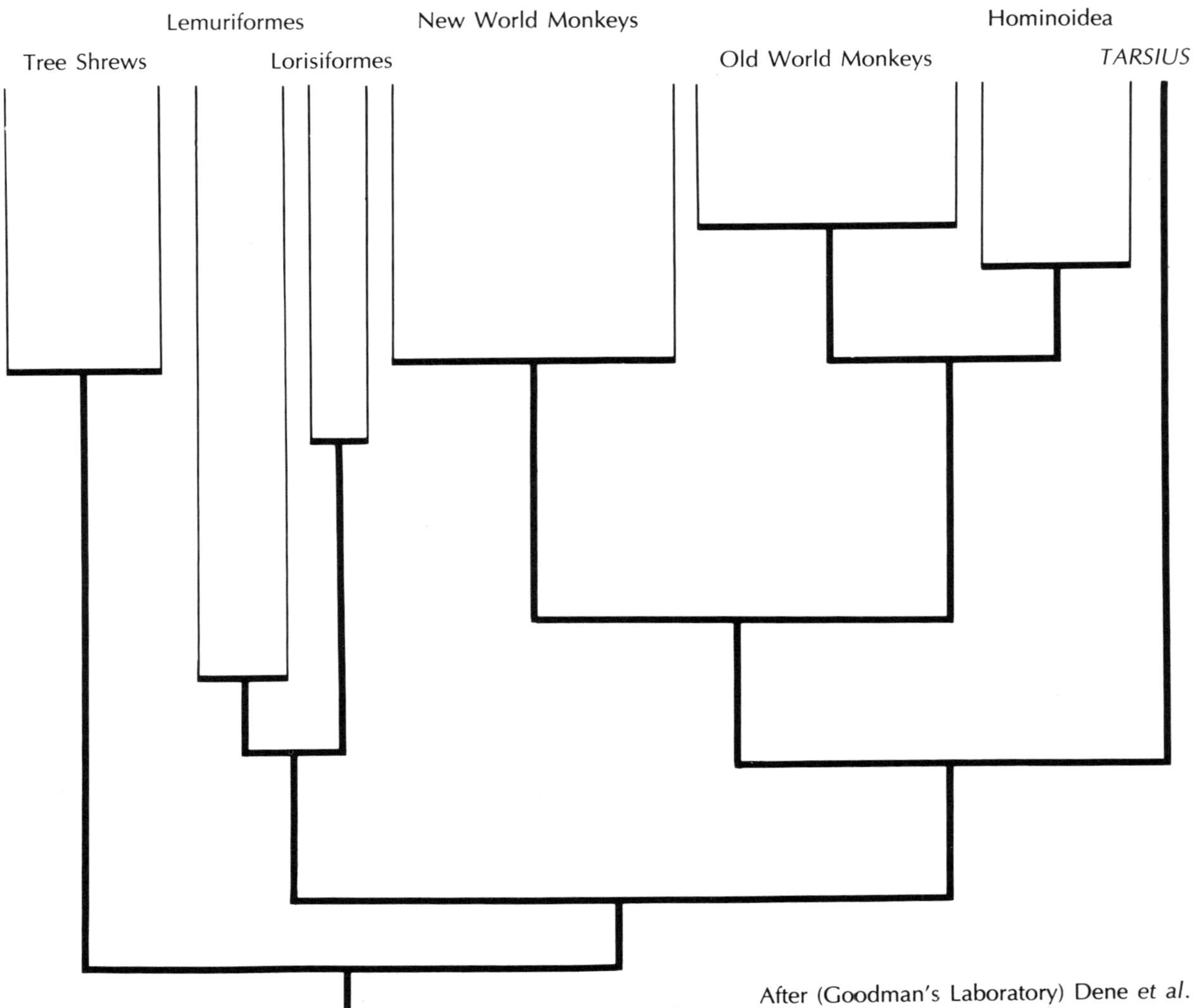

Fig. 8.35. An alternative view of the relationships of prosimians: immunodiffusion methods place tarsiers with anthropoids (redrawn after Dene, Goodman, Prychodko and Moore).

ORDER **PRIMATES** Linnaeus, 1758
Suborder **STREPSIRHINI** E. Geoffroy, 1812
LEMURS,
LORISES, GALAGOS
Suborder **HAPLORHINI** Pocock, 1918
Infraorder TARSIIFORMES Gregory, 1915
TARSIUS
Infraorder PLATYRRHINI E. Geoffroy, 1812
NEW WORLD MONKEYS
Infraorder CATARRHINI E. Geoffroy, 1812
OLD WORLD MONKEYS
APES and MAN

Based on M.C. McKenna, 1975

Fig. 8.36. Adding together alternative views: the relationships of primates, as described here, mirrors those suggested by McKenna, who also places tarsiers with the monkeys, apes and humans.

non-human primates representing 34 different primate genera. In addition, however, the data (2) of Professor J.E. McArdle derived from measurements of the hip and thigh from 289 specimens in 24 different prosimian species, and (3) of Dr. Francoise K. Jouffroy and the late Dr. J. Lessertisseur derived from measurements of the overall proportions of the limbs of 214 prosimians representing 17 different prosimian species are all available. Considerable numbers of specimens exist for most genera in these studies, but in 10 out of the 34 genera of Professor Schultz's data, less than four specimens are represented.

All measurements were made uniformly by the investigators in each study separately. The overall lists of dimensions studied are found earlier in this book. As explained in the original publications, the

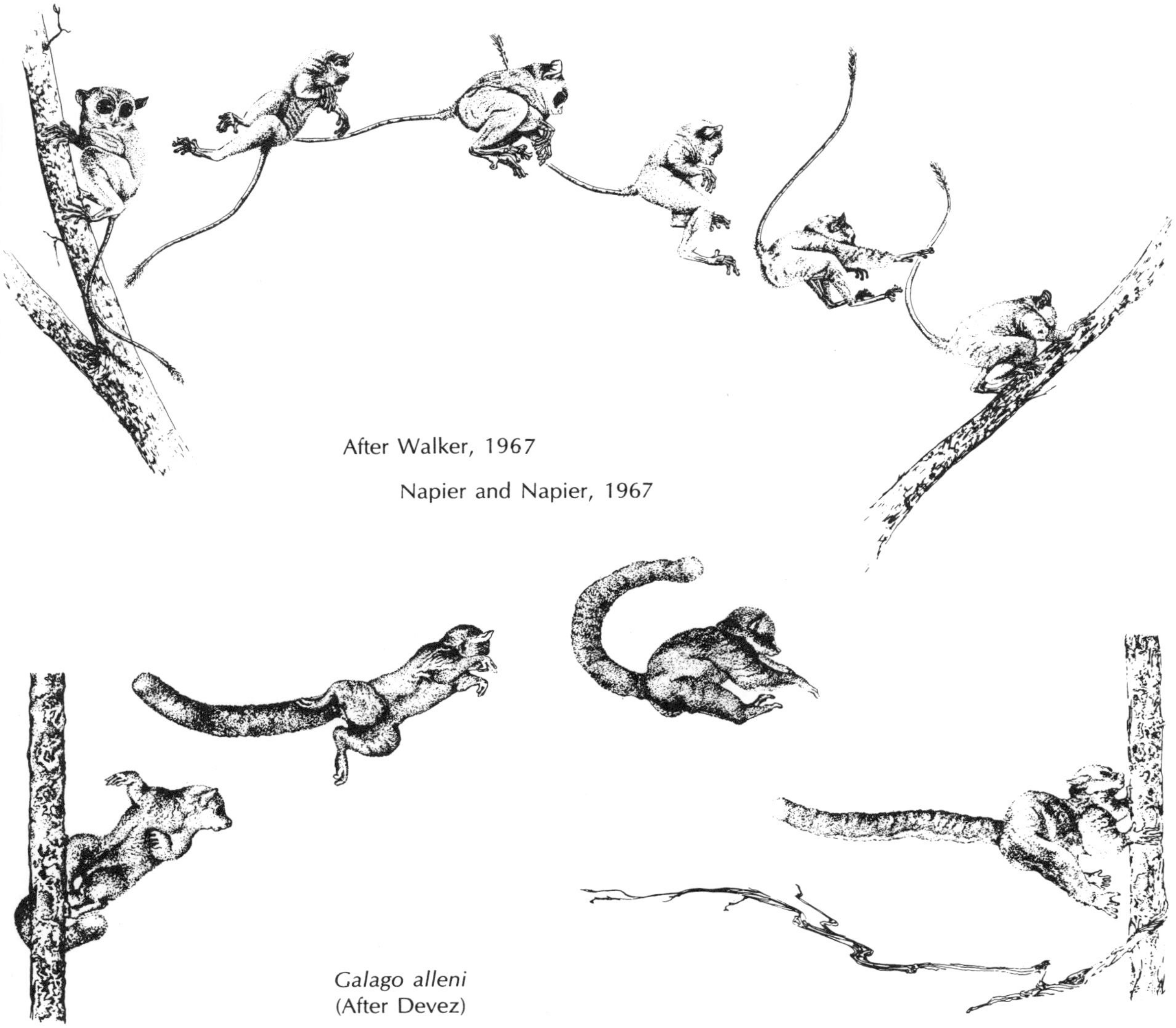
After Walker, 1967
Napier and Napier, 1967
Galago alleni
(After Devez)

Fig. 8.37. The extreme locomotor pattern of *Tarsius* and *Galago*: long leaps compared.

Tarsius spectrum
Galago alleni

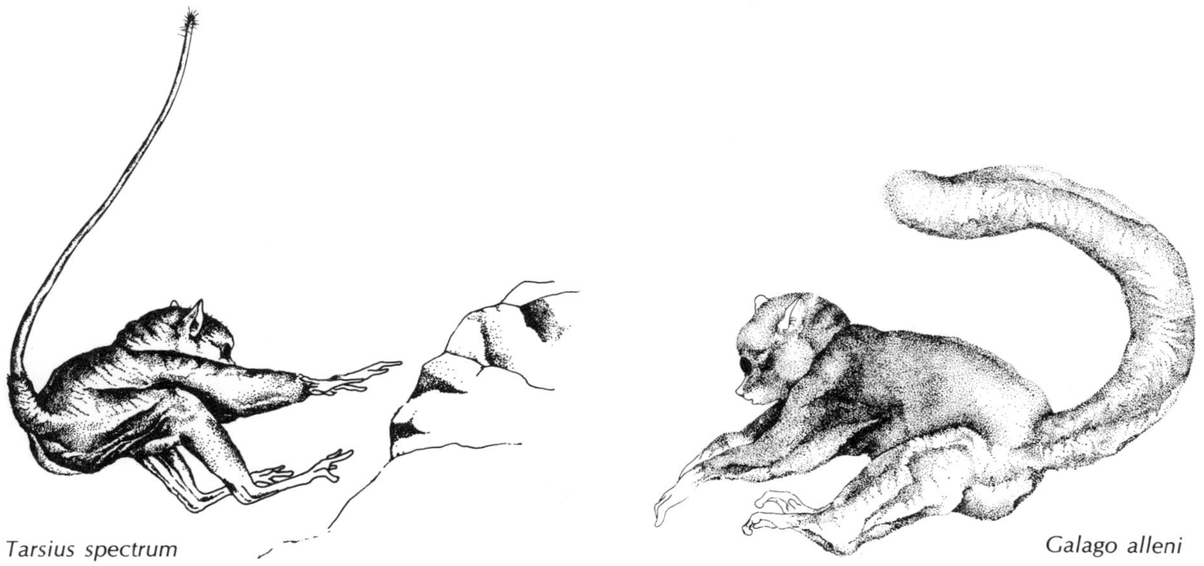

Fig. 8.38. The extreme locomotor pattern of *Tarsius* and *Galago*: mid-air positions compared.

scale of inaccuracy due to unavoidable mensurational inconsistencies is sufficiently small to render it unlikely that the results of the comparisons made here are affected by them.

The untransformed mean values of individual indices are given in the various original studies. Although no individual index makes effective separations among all primate groups, visual assessment of the entire range of indices suggests that the Hominoidea are distinguishable from the other Anthropoidea. The Prosimii (*sensu* Simpson, 1945) seem to be discrete from most Anthropoidea, but merge with some of them; within the Prosimii, there are obvious major similarities between tarsiers and bush-babies. But in most dimensions the general scale of these differentiations is quite poor, and univariate analyses of this type do not readily provide good quantitative differentiation.

The multivariate analyses of the various anatomical regions separately also show that tarsiers and bush-babies tend to fall near one another as outlined in Chapter 6. However, in each case, the distance between them is not overly close, being 11.5 generalized distance units in the upper limb study, 7.5 units in the lower limb study, 7.2 units for the trunk study and 4.5 units for the head study. It certainly cannot be said that the two species are morphologically identical or even nearly so in the individual anatomical regions; but they do appear to be closer to one another in these areas than to any other primates.

These arrangements in the studies of upper and lower limb dimensions seem to have biological relevance to the functions of the upper and lower limbs respectively as suggested in prior chapters.

Thus, the upper limb analysis grades the primates in a broad band related to degree of tension-bearing and upper limb raising. Bush-babies and tarsiers fit this interpretation well. Their limbs are not overly mobile and presumably bear heavy compressive stresses (relative to their size) in landing after long leaps; they lie appropriately near the end of the spectrum which they share, distantly, with other genera (e.g. baboons and patas monkeys) whose forelimbs bear weight by compression (although not because of great leaping, but because they are highly terrestrial).

In the case of the lower limb analyses, the model of the primates is a star with generalized quadrupedal animals in the centre and extreme species in the rays. Tarsiers and bush-babies, both extreme leaping forms, also fit this interpretation well. Both share a ray linked to the main body of quadrupedal species through the gentle and mouse lemurs which lie intermediately.

And similar findings are apparent in the studies of the overall proportions of the limbs of prosimians (Oxnard, German, Jouffroy and Lessertisseur, 1981; and Chapters 5 and 6) and of the dimensions of the prosimian hip and thigh (Oxnard, German and McArdle, 1981; and Chapters 5 and 6). Tarsiers in each study are outlying to bush-babies, which are themselves outlying among prosimian forms.

Whatever other information they contain, these arrangements seem to be evidence of functional convergence for leaping in the structures of the upper and lower limbs of bush-babies and tarsiers.

Studies of the whole body. The nature and degree of separations that appear in the analyses of the trunk and head differ from those in the limb studies. As explained earlier in this chapter, in the trunk and head separations of animals tend to be eclectic, individual genera throughout the primates being differentiated from one another by one or more of a wide variety of discriminant axes. Bush-babies and tarsiers tend to fall together as a pair, but they are accompanied by a variety of other primate genera in each instance.

Thus, although for example the first discriminant axis of the study of the head dimensions places these two species near one another at approximately + 11.0 generalized distance units, so also several other primates (indris, aye-ayes and mouse lemurs among the Prosimii; baboons and orang-utans among the Anthropoidea) occupy this approximate locus. And the third discriminant axis of the analysis of trunk dimensions, for instance, places bush-babies and tarsiers at +3.0 units; again, however, this position is shared with avahis and sifakas (but not indris) among the Prosimii, and patas and proboscis monkeys among the Anthropoidea.

These relationships with some of the other primates are clear-cut but do not lend themselves readily to any biological explanations. Conjunctions of genera such as *Tarsius*, *Galago*, *Indri*, *Daubentonia*, *Microcebus*, *Papio* and *Pongo* (first axis of the study of head dimensions) or such as *Tarsius*, *Galago*, *Lichanotus*, *Propithecus*, *Papio* and *Nasalis* (second axis of the study of trunk dimensions) are not readily interpretable.

And in all these separate analyses, even the proximity of the bush-baby to the tarsier is not as total as may appear from scanning early discriminant axes alone. In each analysis, separately, sufficient discrimination is built up in statistically significant higher axes that there are actually major distinctions between these two genera.

How different are the separations obtained in the overall analysis of the total set of dimensions! Although most of the usual arrangements among the primates are adhered to, the one major exception is the conjunction of tarsiers and bush-babies. These two genera do not appear to be as well differentiated as befits their separate infraordinal status within the Prosimii. They lie together on a side chain emanating from part of the main skeleton occupied by mouse lemurs.

However, further examination of the nature of the network of generalized distance linkages surrounding these two genera provides unexpected information. If we look beyond the minimum link between tarsiers and bush-babies and examine other neighbouring links, a different picture is provided. The bush-baby 'belongs' morphometrically with mouse lemurs, ring-tailed lemurs and mongoose lemurs. But tarsiers do not possess near minimum links with any other prosimian genus; their next nearest neighbours are tamarins, douroucoulis and squirrel monkeys from the New World and guenons and mangabeys from the Old World (Figs. 8.39 and 8.40).

Fig. 8.39. The morphometric view of the primates displayed by the linear minimum spanning tree with the link between the bush-baby (G) and the tarsier (T) included (upper frame) or ignored (lower frame). Bush-babies stay with prosimians; tarsiers move over to link with a variety of monkeys. Compare with Fig. 8.13.

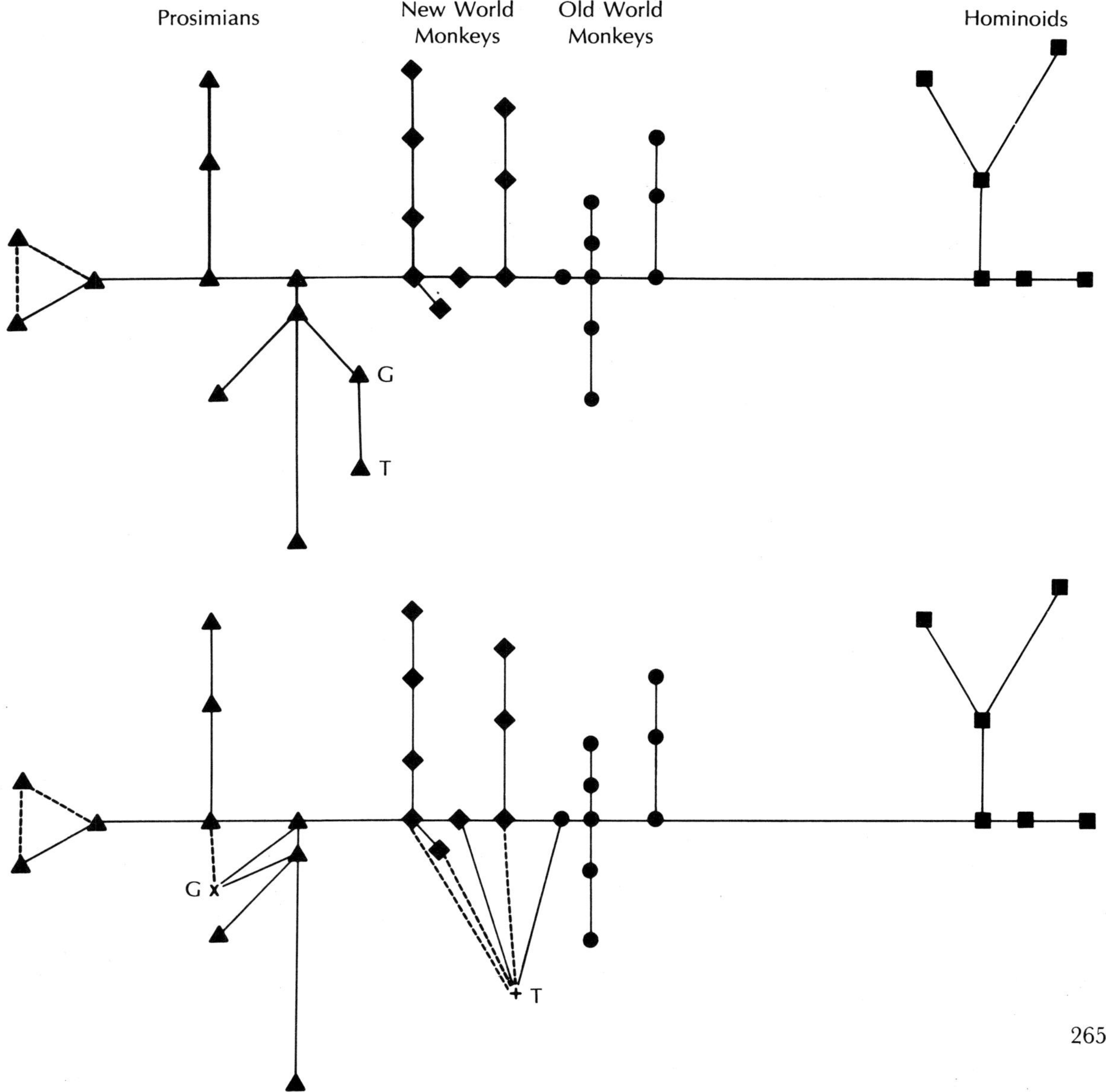

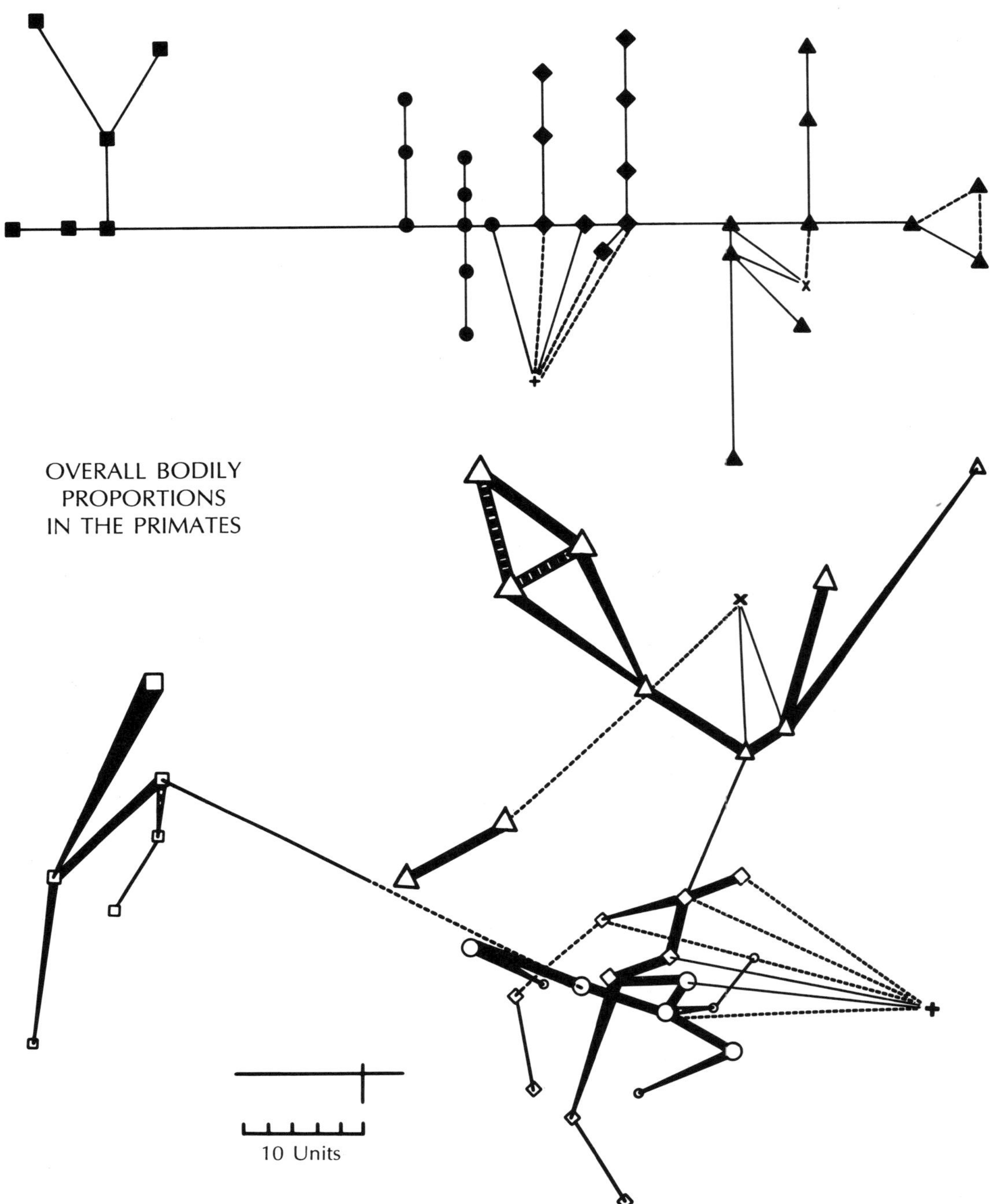

Fig. 8.40. This is seen even more clearly in the more complicated minimum spanning tree drawn in perspective. Although the general form of the tree is complex, the next nearest neighbours of tarsiers stamp them as even more closely related to anthropoids. Compare with Fig. 8.14.

Is it then possible that the minimum link between *Tarsius* and *Galago* results from convergent features of the data that are so heavy that they overshadow the taxonomic evidence in the data? Is it possible that removing this link between them dissects out convergence and allows another view of their morphological associations? The picture presented is not unconvincing. The morphological affinities of *Tarsius*, once we allow for convergence, may truly be with humans, apes and monkeys

(Anthropoidea); those of *Galago* are clearly with lemurs and lorises (Prosimii). This is further support for an anthropoid or intermediate location for *Tarsius* (Oxnard, 1978b; Jouffrey, Oxnard and German, 1982).

It might be thought that overlooking the minimum link between *Tarsius* and *Galago* in this manner is special pleading for a phylogenetic placement of *Tarsius* towards the Anthropoidea. It must, therefore, be pointed out that nowhere else in this analysis of the overall proportions of the primates does removal of a particular minimum link make any special difference to the morphological relationships of the genera. Removal, for instance, of the minimum link between orang-utans and siamangs still leaves a next close link between orang-utans and gibbons; removal, for example, of the minimum link between humans and gorillas still leaves as our next nearest neighbour chimpanzees. And it should also be noted that overlooking the minimum link between tarsiers and bush-babies does not provide this unique result when applied to the various sub-analyses relating to each individual anatomical region. Thus, for both the upper limb and lower limb analyses, the next nearest neighbour of *Tarsius* is another leaping prosimian, *Euoticus* (a bush-baby).

A second point that is of interest for this discussion relates to the more usual ways of making phylogenetic assessments for these species. The kinds of evidences upon which we usually depend for assessments are features such as patterns of tooth cusps, sutural arrangements in orbits, middle ear architectures and so on. The choice by Professor Cave, of characteristics of the nasal fossa, or by Professor Szalay of the architecture of the basicranium, are exactly such attempts to isolate anatomical arrangements that might relate less to 'habitus' and perhaps, therefore, more to 'heritage'. Such a weighting of biological characteristics of this type is a usual practice, even though the bases upon which it may be done may vary from one investigator to another. In the present study, ignoring the link between *Tarsius* and *Galago* and relying subsequently upon the picture provided by the next nearest minimum links with other species *may* be the equivalent of making such a weighting. However, in this case we are not weighting the individual characters themselves but rather certain features of their multivariate compound.

Given the hindsight provided by this study, we may ask: can we go back to the original dimensions and define among them those that appear to provide the information about the similarity of tarsiers with bush-babies (convergence) and those that seem to be speaking to the likeness of tarsiers to monkeys (phylogenetic proximity)? The answer appears to be that, in this case, we cannot. The contributions of the original dimensions to the multivariate separations are such that almost all dimensions contribute to almost all separations. Each original datum contains both convergent and phylogenetic information. Presumably, therefore, the way in which the original dimensions are associated with these two biological phenomena, convergence and phylogenetic proximity, is such as to lie not within the separations of species by the individual dimensions but within the complex pattern of covariation among the dimensions. Thus, the multivariate statistical approach is necessary to reveal the full picture.

This discussion leads in turn to the final point. For although a form of weighting has been carried out in this study in order to see more deeply into the nature of the relationships of *Tarsius* and *Galago*, no such weighting has been necessary for the great bulk of primate species. Presumably this is because, although there are many anatomically localized parallels and convergences among the other primates, summation of information from many anatomical regions in other primates overpowers the individual adaptive contents from the separate regions and provides an overall picture relating to phylogeny. It is only in the case of *Tarsius* and *Galago* that the adaptive information in the data is so powerful that, even when information pertaining to relationships is summated in the overall analysis, it is still hidden except by this special investigation.

The problem for our wider thesis about all the primates that is posed (and now answered) for *Tarsius* also applies to *Daubentonia*, another case we must therefore examine in more detail.

Divergence and *Daubentonia*?

For a long time after its first discovery on the island of Madagascar, a very considerable difference of opinion existed as to the zoological position of the aye-aye.

***Daubentonia:* the early confusion**. Sonnerat, the original discoverer, notes (1789) that 'although the Aye-aye resembles a squirrel, yet it differs therefrom by some essential characters, being also allied to the Lemur and Monkey' (quoted from Owen, 1866). Later workers emphasize the rodent-like aspects of its structure, although the problems of the classification of some other species at the same time emphasize that even these investigators are aware that such a placement is not necessarily all that it appears. Thus, though Buffon (1789) describes the aye-aye's resemblance to squirrels, he also finds similarities with tarsiers; but believing the tarsier to be a jerboa, he has no problem with the placement of the aye-aye in the Rodentia. In this Buffon is followed by Gmelin (1790) and Cuvier (1798) at the end of the century.

Just after the turn of that century, however, Shaw (1800) re-emphasizes the similarities of the new genus to the lemurs in an investigation that places greater weight than had those of previous investigators upon features of the limbs. And Owen's (1866) description of the further history of the placement of the genus shows that different workers continue to bounce it between the Rodentia and the then equivalent of the Prosimii. But Owen believes his own study demonstrates conclusively that it is a lemur.

The initial century of doubt depends upon the existence of those many curious features in which the aye-aye does truly differ from most other primates. These include its possession of (a) superficially rodent-like dentition with continually growing upper and lower incisors and large gaps behind the incisor teeth, (b) special architectural features of the skull, especially, of course, the rounded and foreshortened brain case and the marked downward bending of the facial skeleton upon the cranial base, and of course (c) the peculiar features of the extremities, especially their enormous length relative to the size of the animal, the filiform appearance of the uniquely mobile third finger of the hand and the existence of claws with double matrix on most digits of both hands and feet.

The modern position. Since Owen's extensive work, however, the prosimian characteristics of the aye-aye have scarcely been questioned so that today not only is it recognized clearly as a prosimian, but by most students as lemuriform, and by many as very closely related to the indriids. This view has been accepted notwithstanding arguments about the names and levels of higher categories of prosimians and especially about whether or not the aye-aye should be in its own separate genus, family, superfamily or infra-order (e.g. Elliot, 1913; Simpson, 1945; Osman Hill, 1953; and, most recently of all, Tattersall and Schwartz, 1975, and Hershkovitz, 1977).

The change in opinion depends upon adherence to the concept that the many morphological features listed above which mark the aye-aye as aberrant are very recent evolutionary developments closely related to specialized aspects of the animal's behaviour and replicated among fossil forms in only a single sub-fossil species very much like the living aye-aye, but larger. This opinion has been further adumbrated by most emphasis being placed upon the features of the aye-aye that are shared with other lemurs and especially the indriids: features such as the detailed form of the temporal arterial circulation, some specializations of the premolar and molar teeth and some of the developmental characteristics of the dentition. Support for this view is also said to stem from the complete lack of any fossil history for the genus prior to the Pleistocene fossil *D. robustus*.

A relatively few modern students (e.g. Abel, 1931) place the aye-aye and its supposed allies in a separate suborder. This is largely based upon the belief that the aye-aye has had a distinct lineage since the Paleocene. But Simpson (1945) thinks that some of this evidence is invalid and the rest equivocal, and he writes, 'It now seems more probable that the peculiarities of *Daubentonia* are rather superficial and that it is a true lemur, perhaps even of rather late origin from the typical lemurs. At any rate, its subordinal distinction is not warranted by present [i.e. 1945] data.'

New doubts, the shoulder and the pelvis. Recently, however, there has been renewed interest in the morphological features of the aye-aye. Thus, Groves (1974), in a somewhat neglected and little cited review of the taxonomy and phylogeny of prosimians, expresses his 'uneasiness' with the generally accepted fit of *Daubentonia* with other Malagasy lemurs and, indeed, with the lemurs as a whole. And Jouffroy (1975), through her own original work with specimens of *Daubentonia*, turns our attention once again to the remarkable features of the hands that so clearly differentiate it from other prosimians. And to this she allies a number of other features scattered across the anatomical spectrum (for example, she points to the occipital origin of the trapezius muscle, a feature common among many anthropoids but unknown to other prosimians). Without re-entering discussion of the taxonomic or evolutionary position of the species, she emphasizes its very strong dissimilarity with the, by comparison, very homogeneous and restricted group of lemurs.

This renewed interest is also evident in certain new kinds of structural information that have become available in recent years from our own laboratories. In most of these studies, although our usual aim is a survey of the Order as a whole, there is often reference to the aye-aye. In many cases the number of representatives of the genus is very small: seven in some studies, two or three in others; in one investigation only a single specimen is available. But the results of the various studies are so clear that even with such small samples it is possible to be rather confident that they stand up.

The position of the aye-aye in the study of the quantitative data on the shoulder bones using multivariate statistics seems quite unequivocal. It leads Ashton, Healy, Oxnard and Spence (1966) to speculate that 'the quadrupedal nature of the genus is corroborated, although peculiarities of its locomotor pattern are not reflected in any unusual placement'. In that study, the aye-aye appears to fall fairly and squarely within a mixed bag of genera of both the Anthropoidea and Prosimii that have in common only that they are believed to be generally quadrupedal in habit.

But in this first multivariate study, because there is only a single specimen of *Daubentonia*, the data are entered after the analysis thus using, for the determination of the position of the aye-aye, the mathematical coefficients derived from an analysis to which the aye-aye itself had not been allowed to contribute. We cannot, therefore, be certain that its relationships are defined correctly.

A subsequent study (Ashton, Flinn, Oxnard and Spence, 1971), in which two specimens are available, allows a different and better approach. In this new study, even although two specimens is still a very small number, the genus is entered as a group into the initial analysis and thus contributes, together with all other genera, to its final placement. This procedure seems to confirm superficially the placement of the shoulder of the aye-aye as squarely among those of other quadrupedal genera, irrespective of taxonomic group.

But a sub-study is available from this second investigation in which an attempt is made to define the positions of various superfamilies of primates. This reveals, quite unexpectedly, that the Daubentonioidea (as represented by the aye-aye) is morphologically distinct from all other primates.

Further investigation of this problem by Oxnard (1973a, 1975a) reveals the underlying difficulty. Whether examined through its association with the various higher taxonomic groups (superfamilies), or whether examined in its own right as a single genus among all the other primate genera, the aye-aye actually lies uniquely separate from any other primate genus. This had not previously been recognized because this separation is either contained within a combination of small, and therefore not obvious, separate parts of individual discriminant axes (which, in summation, provide a distinct and statistically significant placement for the genus from all other primates) or is included as a very marked and statistically significant outlier in axes much higher than those usually examined. These two results are equivalent depending upon the precise ways in which the analyses are carried out. They imply that the shoulder of the aye-aye is morphometrically unique from that of all other primates (Fig. 8.41).

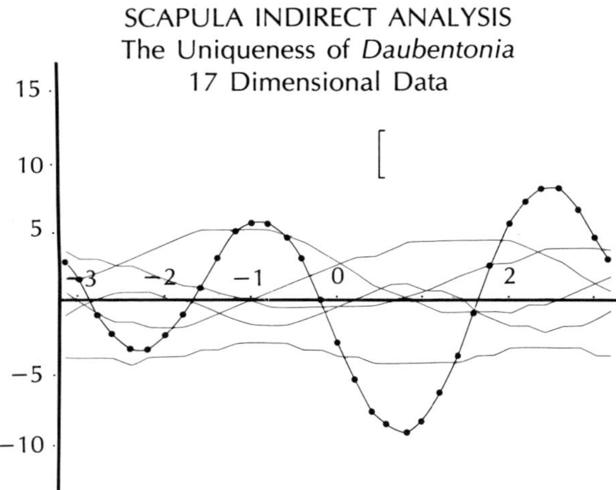

Fig. 8.41. High-dimensional display of the morphometric result for the shoulder stamps the aye-aye as unique. The flat continuous curves are the patterns for all other primates: the dotted curve for the aye-aye is very wavy and stands out from among all primates.

Oxnard (1973a, 1975a) therefore suggests that, although detailed information about peculiarities of the use of the shoulder in this animal are not available, the curious placement may be due to some functional factor not fully understood. When Oxnard (1973a) attempts to return from the multivariate statistical description of the shoulder to the observational description of the scapula of the aye-aye, he is able to see the confluence of the two results. The actual scapula of the aye-aye is as unique as its multivariate description. It is markedly different from that of any other primate due especially to its twistedness in a plane at right angles to that of the scapular fossae. This peculiarity of the scapula had, indeed, been noticed years before by Owen (1866) and is figured recently by Jouffroy (1975).

The shape of the mammalian scapula is now known to be tied so heavily to the action of the various muscles that attach to it, to the movements of which it is capable and to the stresses which it bears that it is almost inconceivable that this structural peculiarity of the aye-aye is not also of strong functional import (e.g. Oxnard, 1968a; Roberts, 1974; Bacon, Bacon and Griffiths, 1979). And in recent years, new studies of the aye-aye in the field and in captivity (e.g. Petter and Petter, 1967; Petter and Peyrieras, 1970) show that its habits are, indeed, curious compared with those of most other primates. For instance, the aye-aye's foraging propensities associated with the peculiar uses of the filiform third digit must have some impact upon shoulder function and, therefore, shape; the animal's ability, much better known now than before, for hanging by its lower limbs and foraging with its upper limbs in this reversed position may well affect the functional morphology of the shoulder; and its peculiarities in quadrupedal movement (Fig. 8.42), e.g. climbing on the under-surfaces of branches (Fig. 8.43), and especially the manner of its leaping (Fig. 8.44), may also have effects at the shoulder joint. Yet it must also be acknowledged that no study has been made which is detailed enough to demonstrate biomechanical associations between these peculiar

Fig. 8.43. Underbranch movement of the aye-aye.

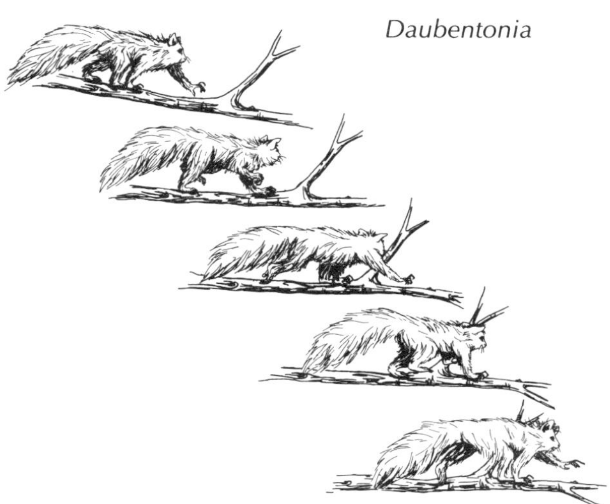

Fig. 8.42. Quadrupedal movement of the aye-aye.

Fig. 8.44. Leaping of the aye-aye.

behavioural features and the actual curious structures of the muscles, joints and bones of the animal.

When, therefore, it became possible to study another anatomical region, the hip, with a similar methodology, the relationships of *Daubentonia* were especially carefully assessed. In the case of the hip, however, the placement of the aye-aye is truly squarely among the various prosimian and monkey quadrupeds (Fig. 8.45). In this case also, measurements parallel morphology. A return to the original biological object in the form of the pelvis of the aye-aye reveals no very special osteological architecture for this bone.

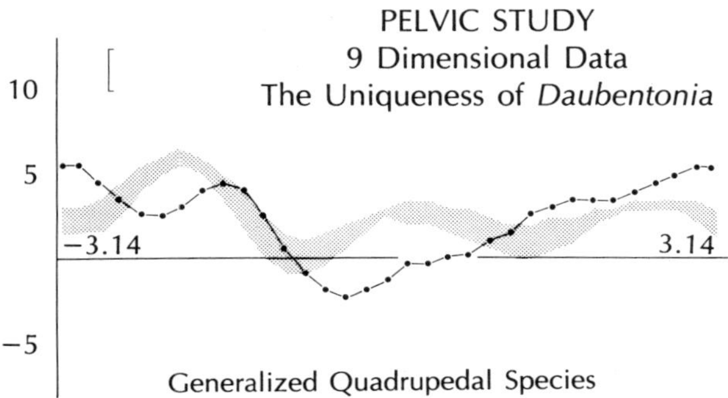

Fig. 8.45 High-dimensional display for the pelvis stamps the aye-aye (single line) as not far distant from regular quadrupeds (shaded envelope).

It is almost as though the form of the scapula mirrors the uniquely daubentonian features of the animal such as the teeth, skull and extremities, while the form of the pelvis mirrors the regular prosimian features of the animal such as the presence of nails, arterial arrangements in the skull and developmental patterns of the teeth.

New assessments of the structure of the aye-aye. In addition to these studies of the shoulder and hip, the whole series of investigations of the form of the body and its individual anatomical regions using the multivariate statistical approach speak to the relationships of the aye-aye. An attempt is made here to draw together the information about these features (some of which have appeared in the work of Oxnard and colleagues over the last two decades) and to assess them from the strictly parochial viewpoint of the placement, among the primates, of the aye-aye.

Some information stems from studies of the primates as a whole: the shoulder summarized in Oxnard, (1973a, 1975a), the arm and forearm (Ashton, Flinn, Oxnard and Spence, 1976), the pelvis (Zuckerman, Ashton, Flinn, Oxnard and Spence, 1973) and the overall proportions collected by Professor Schultz (Ashton, Flinn and Oxnard, 1975; Oxnard, 1977). In a more detailed manner, and aimed at prosimians alone with much bigger samples and at the species level, other information comes from investigations of the hip and thigh (Oxnard, German and McArdle, 1981) and from the overall proportions recorded by Lessertisseur and Jouffroy and analysed by Oxnard, German, Jouffroy and Lessertisseur (1981). The entire problem has been summarized in recent studies by Oxnard (1981a, 1982).

To what degree do these various researches add to our knowledge of how similar or dissimilar to other prosimians the aye-aye may be? To what extent can they pinpoint, in further detail, its underlying adaptations? How do they speak to the uniqueness of this peculiar genus?

The materials and methods. The details of the materials in each of those studies are given in the respective publications. However, in summary, the numbers of specimens of *Daubentonia* that have been examined include three each for the shoulder, the pelvis, the arm and forearm, one specimen in the studies of overall proportions and nine in each of the more detailed studies of the prosimian hip and thigh, and prosimian upper and lower limb proportions, respectively. Although these numbers are small and provide little information about variation within the genus, the scale of differences among the various primates are such that they do allow an assessment of the location of *Daubentonia* in each study. Given, then, that assessments are available from many anatomical regions, it is possible to provide an excellent overall comparative view of the morphological relationships of the genus.

In each case the comparisons are made with large numbers of specimens and genera of primates as outlined in earlier pages. And in each case the specimens were measured by individual sets of investigators only, again, as outlined earlier. The measurements for each investigation are thus derived by uniform techniques from single individuals or from two individuals working together (with appropriate inter-observer tests).

Appropriate tests of techniques have been carried out on all these data and they are outlined in the original publications and summarized in table 3.4 of Chapter 3. These tests make it highly unlikely that the precision of measurement has been affected by inaccuracies due to methods of measurement. In no case have measurements been pooled in blocks that contain data from different workers.

A summary of the dimensions used in the various investigations are presented in this and in prior chapters. Detailed descriptions of these are available in the original papers. In each case, however, sufficient measurements are taken so as to describe a major portion of the structure of each specimen at that particular anatomical level.

In the earlier studies only indices are available for analysis. But in the later studies from Chicago and the Chicago-Paris collaboration, measurements are also analysed. However, the results of those investigations show clearly that (a) analysis of measurements gives little information about most species and genera over and above that rather obviously related to overall sizes of specimens and that (b) analysis of indices provides far more sensitive indicators of differences among various species and genera with particular information that goes far beyond that of mere difference in overall size.

Study of the univariate differences between the aye-aye and the many other primate genera in each of the many different studies is indeed possible and is given in the preceding original studies. But the enormous bulk and complexity of such information, together with the fact that a large amount is redundant because of intercorrelations among the characters, renders univariate assessment virtually impossible.

Hence the data are analysed through the multivariate approach. And even that produces a large amount of information. In each study, the position of *Daubentonia* is determined relative to other anthropoid and prosimian genera through (a) bivariate plots of pairs of axes resulting from discriminant function analysis, (b) minimum spanning trees and three-dimensional models resulting from study of the matrix of generalized distance connections among the various groups of animals and (c) through the medium of Andrews' high-dimensional display in those cases where the number of discriminant axes required to position a group accurately is too big for a good picture to be obtained from the use of two-dimensional plots or three-dimensional models.

The multivariate statistical studies have been carried out with two notions in mind. *First*, the position of *Daubentonia* as a genus is examined in relation to the position of all other neighbouring genera and species. This allows some assessment of *Daubentonia* as a primate and as a prosimian. *Second*, however, recognizing that much of the data relate to the functions of the different anatomical parts, they are studied comparing *Daubentonia* (which is assumed to be quadrupedal) with other groups of quadrupedal forms.

The results: upper limb analyses. The results of the various analyses of different upper limb parts are shown in table 8.4 and Figs. 8.46 and 8.47.

Fig. 8.46. A three-dimensional model for the overall proportions of the primate upper limb demonstrates the separation of *Daubentonia* from all primates.

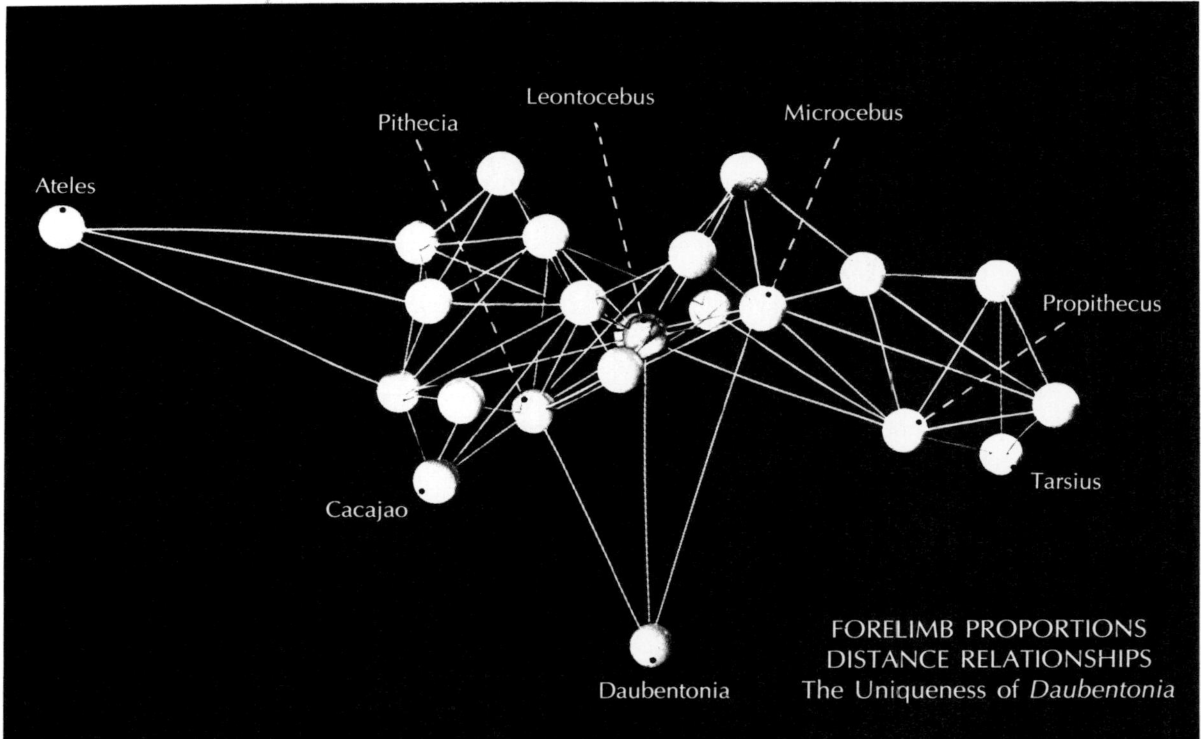

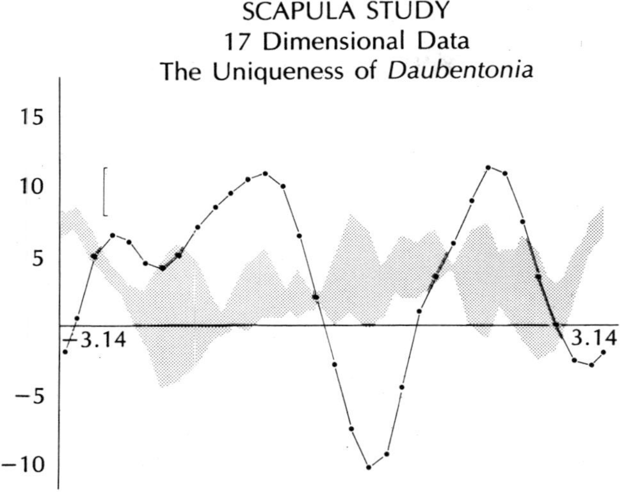

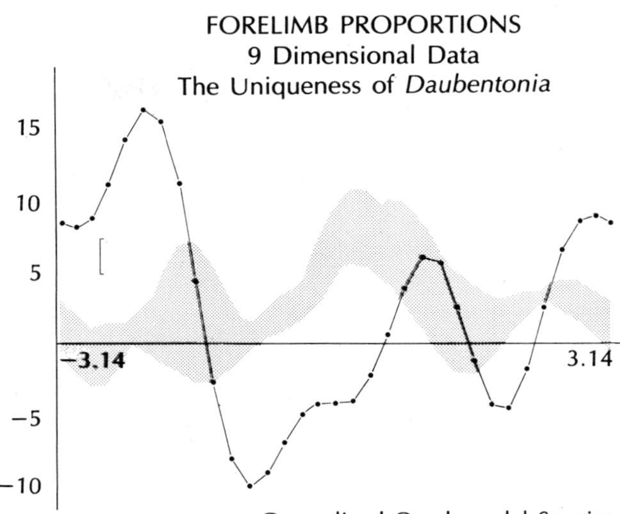

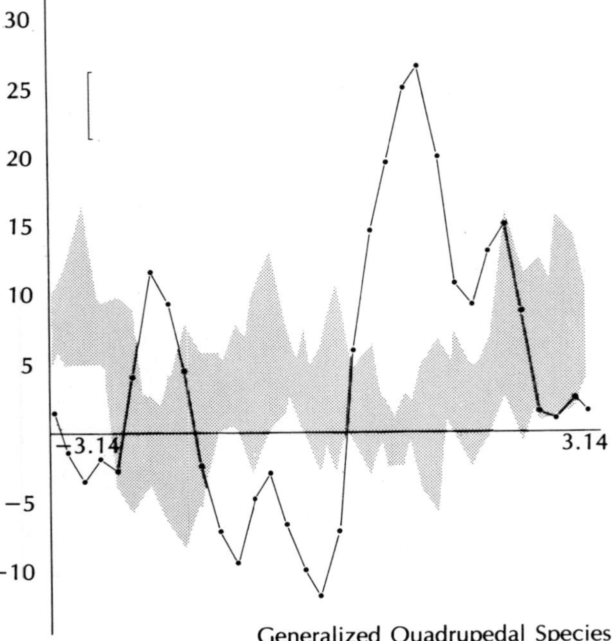

Fig. 8.47. High-dimensional displays for various studies of the upper limb demonstrate the unique separation of the aye-aye (single curve) from generalized quadrupedal species (shaded envelope).

The studies of the shoulder, arm, forearm, shoulder-arm-forearm combination and upper limb proportions all show that *Daubentonia* is further distant from its nearest primate neighbour than is any other primate. It lies at one edge of the universe of all primates. This is summarized in Fig. 8.46 which provides a three-dimensional model of generalized distance connections for prosimians and New World monkeys (apes and man lie yet further distant and are not figured). Table 8.4 provides the appropriate generalized distances for the investigations and it confirms this peripheral location.

Fig. 8.47 demonstrates that the aye-aye is especially different from those groups of prosimian and New World monkey genera that are generally thought to be quadrupedal (Pithecinae, Aotinae, Callithricidae, most Lemuridae, Tupaioidea) in the shoulder, the arm, the forearm, the shoulder-arm-forearm combination and in the overall proportions of the forelimb. Although far distant from these quadrupedal species, the aye-aye is even further distant from all more specialized species, such as the leaping indriids.

The results: lower limb studies. In the single less detailed study of the hip in the entire Order Primates (Zuckerman, Ashton, Flinn, Oxnard and Spence, 1973), *Daubentonia* does not differ from any other prosimian (table 8.4) nor from any regular quadrupedal primate.

But the more detailed studies of the hip and thigh of Prosimii and of the hindlimb proportions both of the entire Order and of the Prosimii alone demonstrate once again a picture largely similar to that stemming from the various forelimb studies (table 8.4, and Figs. 8.48 and 8.49).

Again, studies comparing *Daubentonia* with regular quadrupedal primates (Pithecinae, Aotinae,

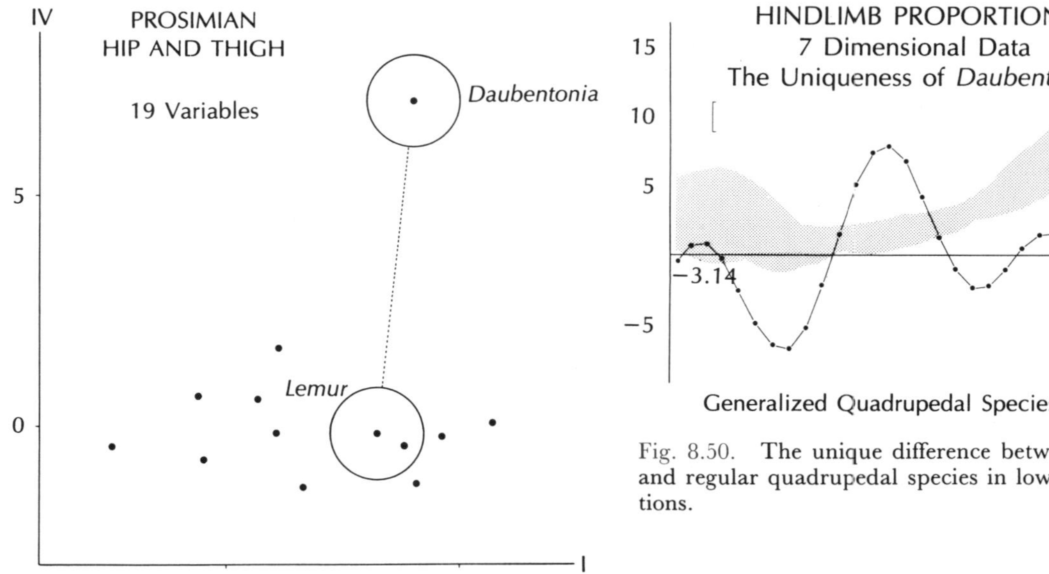

Fig. 8.48. The uniqueness of *Daubentonia* as seen in the study of the prosimian hip and thigh.

Fig. 8.50. The unique difference between the aye-aye and regular quadrupedal species in lower limb proportions.

Callithricidae, most Lemuridae, Tupaioidea) show *Daubentonia* to be markedly different (Fig. 8.50). *Daubentonia* is further still from the other more specialized forms, such as the leaping indriids.

Analyses of combinations of anatomical regions. Study of the combination of upper and lower limb dimensions of the entire primate Order shows (table 8.4) that *Daubentonia* is more distant from its next nearest neighbour than is any primate genus from any other. This is especially obvious in the study of combined upper and lower limb

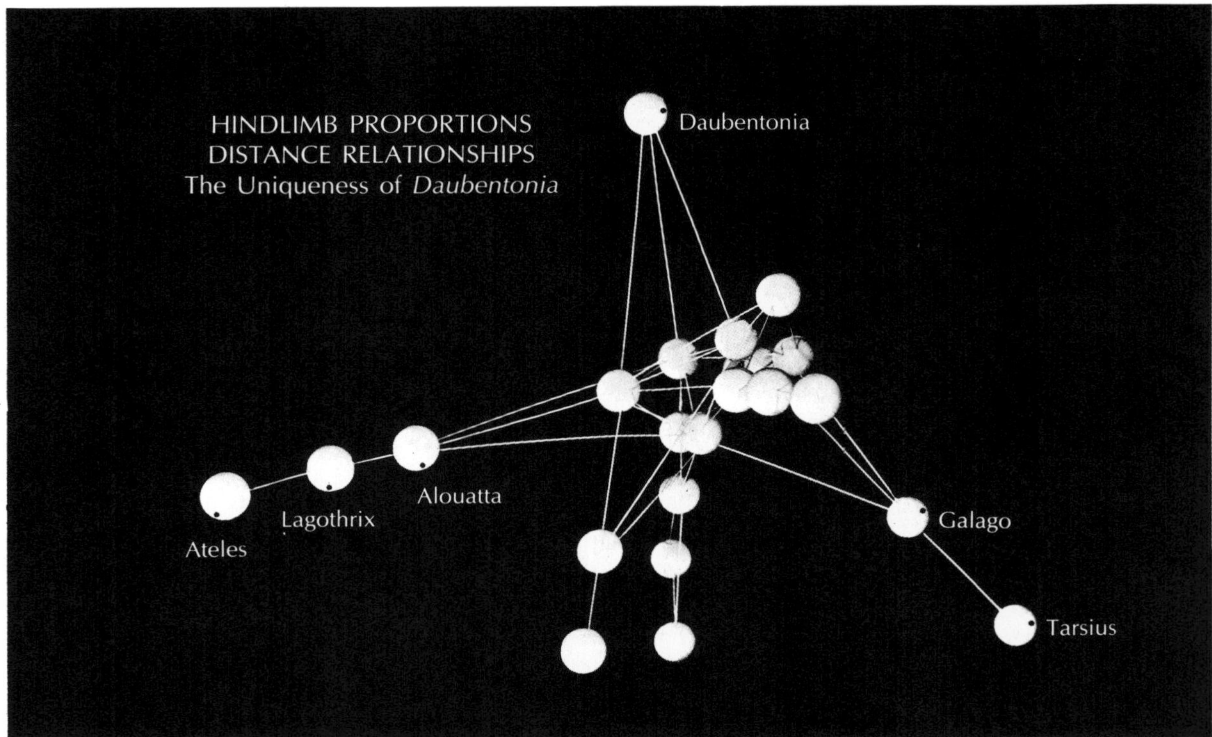

Fig. 8.49. The separation of *Daubentonia* as seen in the study of lower limb proportions.

Table 8.4

SQUARED GENERALIZED DISTANCE VALUES FOR *DAUBENTONIA* COMPARED WITH THE AVERAGE MINIMUM SQUARED DISTANCES FOR ALL OTHER SPECIES IN TWELVE DIFFERENT DISCRIMINANT FUNCTION STUDIES

Study	No. of variables	No. of specimens of *Daubentonia*	Minimum D-2 link for *Daubentonia*	Mean minimum link for all species
Shoulder	17	2	73.1	24.8
Arm and forearm	19	2	77.1	23.5#
Shoulder, arm and forearm	36	2	207.9	82.7
Primate upper limb	6	1	107.9	17.2
Prosimian upper limb	9	7	140.5	19.6
Primate pelvis	9	2	5.7*	6.9*
Prosimian hip and thigh	17	7	60.6	17.1
Primate lower limb	6	1	128.0	21.3
Prosimian lower limb	8	7	160.0	22.3
Combination of fore and hind limbs, primates	13	1	151.8	27.4
Combination of fore and hind limbs, prosmians	19	7	352.3	62.1
Overall bodily proportions in primates	23	1	772.9	67.9

* This study is the only one in which the minimum link for *Daubentonia* is less than the average minimum link throughout the primates.

\# In this study, though the minimum link for *Daubentonia* is indeed much larger than the average minimum link throughout the primates, it is not the absolutely greatest minimum link; it is exceeded by the length of the link between *Hylobates* and its nearest hominoid neighbour. In all other studies except the primate pelvis (footnote * above) the minimum link for *Daubentonia* exceeds that of all other genera and usually by a very large margin.

proportions for the Prosimii alone (Fig. 8.51). The very large separation that obtains for *Daubentonia* in that figure shows that the separations in each of the individual parts (the lower limb and upper limb) must be largely independent of one another and thus summate to a very large difference indeed.

The uniqueness of *Daubentonia* in comparison with regular quadrupedal species as outlined previously is also most marked in the combination of dimensions of the upper and lower limbs of the Prosimii (Fig. 8.52); and perhaps it is most marked of all in the study of the overall bodily proportions of the entire primate Order (Fig. 8.53).

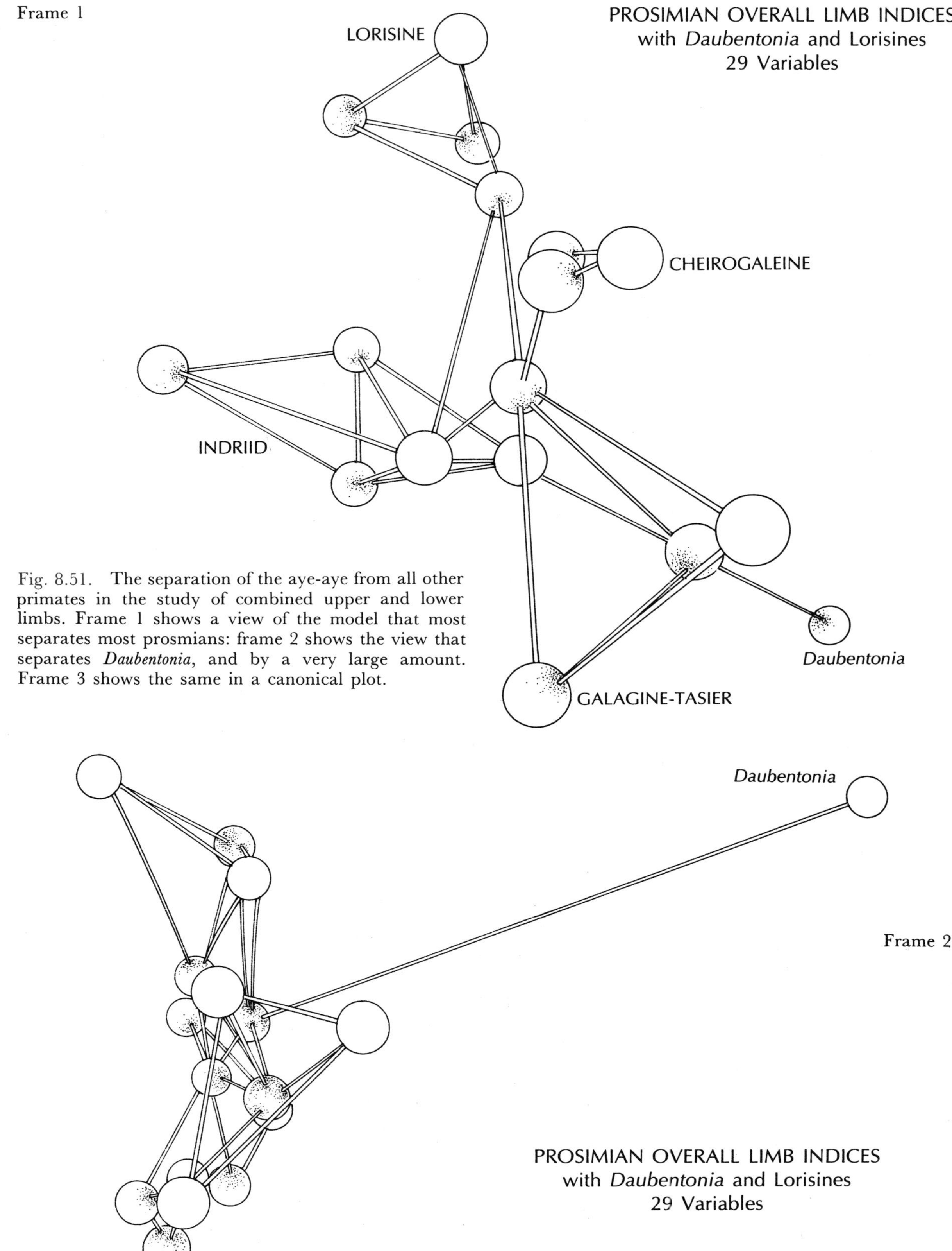

Fig. 8.51. The separation of the aye-aye from all other primates in the study of combined upper and lower limbs. Frame 1 shows a view of the model that most separates most prosmians: frame 2 shows the view that separates *Daubentonia*, and by a very large amount. Frame 3 shows the same in a canonical plot.

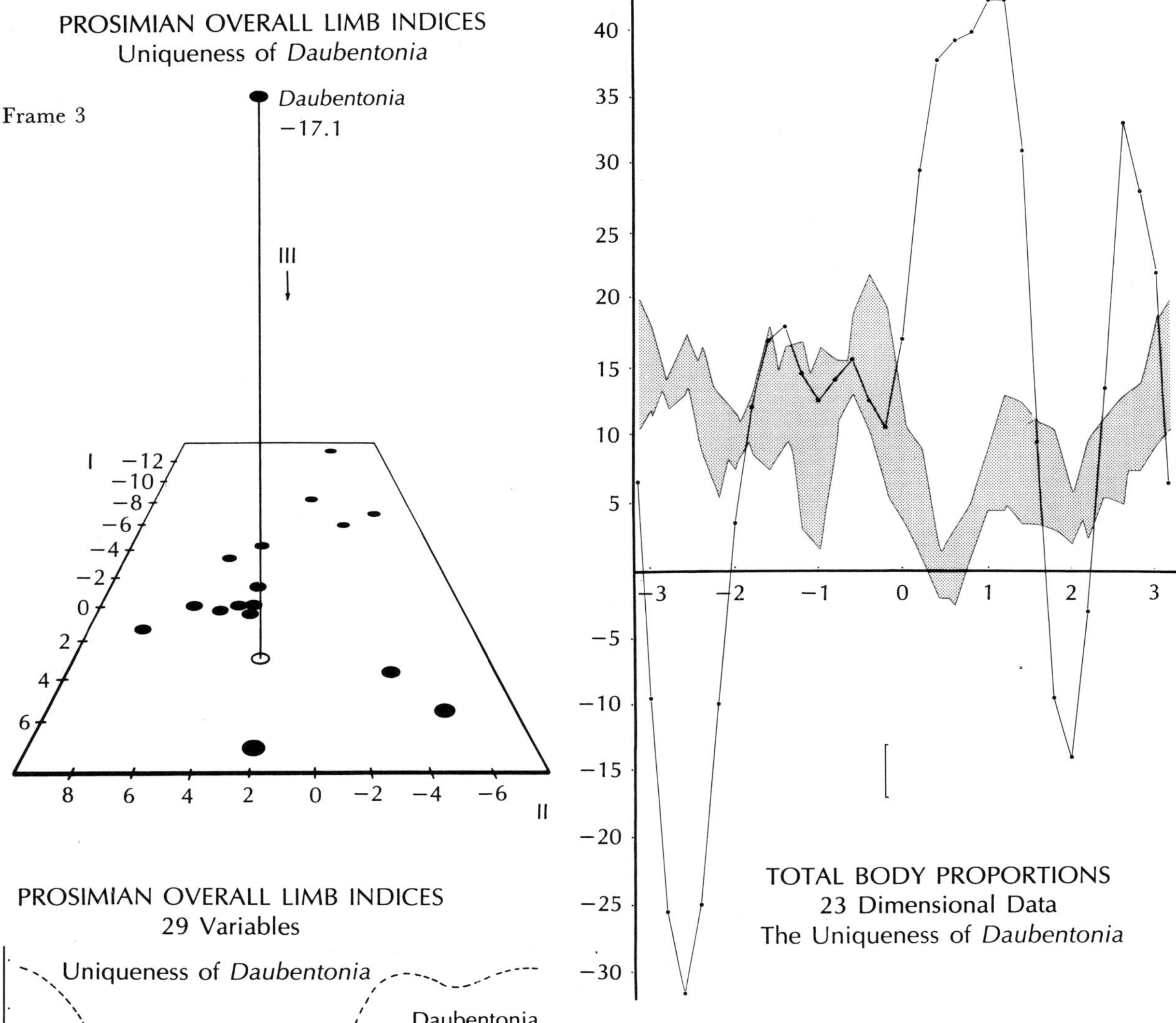

Fig. 8.53. The most marked separation of all is achieved in the morphometric study taking all dimensions of the primates together.

Fig. 8.52. The separation of the aye-aye from regular quadrupedal species in the study of upper and lower limbs.

The implications of these results. Although much early work emphasizes the differences of the aye-aye from other Prosimii, as indicated earlier in this chapter, a considerable body of more recent data emphasises its similarities with many prosimians and especially with the lemuriformes and, among them, indriids. These data, together with a lack of any long fossil history for *Daubentonia*, result in the lemuriform placement of the genus supported by most recent authors. The differences are said to be merely superficial.

This assessment may also have resulted, however, in some neglect of the more curious features of the genus. Our scan of the skeleton as evidenced through multivariate statistical study of dimensions of the shoulder, arm, forearm, upper limb as a whole, pelvis alone, pelvis and thigh, lower limb as a whole, upper and lower limbs together and overall bodily proportions, including not only the limbs but also the trunk and head, indicate unequivocally the total separation of this genus. In only a single study (that of nine dimensions of the pelvis of primates) does *Daubentonia* not appear completely separate; and even in that case, further study of the pelvis of prosimians utilizing a larger number of measurements provides a unique placement (table 8.4).

This wide separation is not only evident when the genus is compared with its various supposed close taxonomic relatives, but it is also true when the genus is compared with all those other prosimian and monkey genera that are believed to be quadrupedal. Although not illustrated here, it also follows that differences from other species with more extreme locomotor habits (for instance, extreme leapers of various kinds, such as indriids, galagines and tarsiers, or extreme arm-swinging types such as woolly monkeys, spider monkeys, gibbons and siamangs) are even greater. Thus, though the aye-aye is not at all similar to the regular quadrupedal species, it is these species from which it is least distant morphometrically.

So little is known about the habits of *Daubentonia* that it is difficult to interpret these results functionally. The animals are quadrupedal. But Petter and Petter (1967), Petter and Peyrieras (1970) and Petter, Albignac and Rumpler (1977) provide a number of descriptions and pictures that indicate that its quadrupedality may be very different from that of regular quadrupedal primates (however different these may be among themselves). Thus, although Fig. 8.42 demonstrates regular slow quadrupedal walking, Figs. 8.43 and 8.44 imply that climbing and leaping may possibly be quite curious, and the peculiar foraging habits of the aye-aye may well be involved in this curious morphology.

If these results are truly of functional significance, they may mean that the aye-aye is functionally uniquely different from other prosimians; and presumably some actual notion about its uniqueness might appear (a) if more extensive knowledge of the biomechanical import of its locomotion existed and (b) if other forms that happen to show parallel differences could be included in the analyses (perhaps these latter would include both nonprimate placental and marsupial species).

Conclusions for the uniqueness of Daubentonia. In the study of overall bodily proportions, the conjunction between tarsiers and bush-babies seems to result from a degree of functional *convergence* so heavy that it may be thought of as overshadowing the classificatory content of the data. In like manner, this result for the aye-aye may be speaking to a degree of functional *divergence* from all other primates that could also be thought of as overshadowing its true primate relationships.

But in this case the divergence is so fundamental, including nearly every part of the body and presumably therefore, thousands of characters, that we are bound to infer a much older and more divergent phylogenetic placement than generally believed in recent years.

We should, therefore, hold as most tentative any placement of the genus within current primate groups such as the Lemuroidea or Indriidae. Until more is known about how this particular type of data impacts upon classification, we should probably keep totally open minds about the genus. Searches for further resemblances and differences, searches for more convincing functional interpretations, searches especially for new fossil materials, all seem to be indicated. While for pragmatic reasons we may wish to leave classification where it is (with *Daubentonia* as an aberrant form separated from the indriids only a short time ago), those interested in the evolution and functional anatomy of prosimians may need, for the present, to operate as though the phylogenetic position of this genus within the Prosimii were unknown.

A return to the entire Order: functional data, taxonomic result

Let us, then, return to the study of the entire Order. Though the regional structural information (the material of Chapters 5, 6 and 7) speaks primarily to the functions of the individual anatomical regions, the overall information (the material of this chapter) speaks to the general relationships of the animals. This viewpoint is clear, as we have seen, when we consider the various groups of the Anthropoidea. It is somewhat less obvious for the Prosimii, in part because of smaller samples and fewer representative groups that it has been possible to examine. Another part of the difficulty with the study of the Prosimii relates to the superficially quite wrong assessments for two of the animals: tarsiers and aye-ayes.

However, as we have just seen, these assessments can be queried. Our new findings for *Tarsius* and *Daubentonia* actually strengthen the hypothesis, that there are links between multivariate morphometric structures of the primates and overall relationships of the primates.

The assessment of the aye-aye likewise, seems not to meld with conventional views about the genus. In this case, our detailed studies have shown that every individual anatomical region separately, and all of them in any and all combinations, are indeed enormously different for the aye-aye than for any other primate. Its uniqueness is so marked compared to any other primate, and so widespread throughout its own entire anatomy, that perhaps the conventional view about the aye-aye is wrong. Yet there are still problems.

One of these relates to the comparison of relationships as expressed through many years of studies of multitudinous data of a classical morphological type and those as expressed by relatively few years of investigation of data of a biomolecular kind. Although there are many disagreements among different morphologists (Fig. 8.17), and though the various biomolecular workers do not always agree entirely with one another (Fig. 8.18), there is, in general, considerable underlying agreement between these two groups about primate taxonomy. Thus, at the highest taxonomic levels (such as the primary division into prosimians and anthropoids as shown in Figs. 8.13, 8.14, 8.15 and 8.16) the same overall groupings are derived. And at the lowest taxonomic levels (such as in the study of geographic and other variants of individual species and species groups), again, the two methods are gradually coming together in focusing upon the same patterns (Albrecht, 1979). But the whole series of investigations (Figs. 8.19, 8.22, 8.25 and 8.28) reviewed here demonstrates unequivocally that, when we come to look at detailed patterns at intermediate levels, there are consistent and marked differences between the results of these two modes of study.

It may be thought by some that the classical structural approach provides the better picture simply because it depends upon such a huge volume of very different investigations carried out by many workers over the years, indeed over the centuries; on this count, the limited data stemming from the biomolecular studies can scarcely compare. However, this will not long remain the case: day by day, large laboratories and teams of investigators are adding to the biomolecular viewpoint. In comparison, work of a classical morphological type currently attracts few new investigators.

In contrast, others may think that the biomolecular approach is, in the end, the better because it deals more directly with the very stuff of evolution. It is far less diluted by non-genetic, developmental, biotic, environmental and other factors.

Function and relationship: three questions. Whatever may be the realities here, given that we have identified a difference between these two types of data, these two modes of study, we are forced to ask the question: *why should the new morphological data, when summarized using the multivariate statistical technique, provide a result similar to that obtained from the biomolecular mode and different from that given by classical morphology?*

The answer to this question may depend upon the answer to a prior question: *why should the mere addition to morphology of the multivariate statistical approach produce an answer different from that obtained from classical morphology alone?*

And this question, in turn, is based upon the answer to a third: *why should the morphological data studied by morphometric methods on individual anatomical regions, that contain clearly functional information, provide, when added together, a picture of overall primate relationships?*

Obviously we must take the questions in reverse order.

Discordance between morphometric parts and wholes. Why should several local functional results add up to an overall classificatory one?

One answer runs something like this. The information that stems from each anatomical region is truly functional, and the mere addition of many different functional regions helps to mimic the addition of many different characters such as is normally attempted in most studies of animal relationships. Though we may think that only a single biological system, the skeletal, is being examined, in fact the measurements of bones and limb segments that comprise the primary data must also be reflecting many other aspects of the function of the animals. These include especially, of course, functions of bones, muscles and joints; certainly, too, those of motor and sensory aspects of the nervous system; and even those relating to the environmental biology and ecology of the organisms. This widespread representation follows simply because locomotion is such an all-pervasive activity of most animals. We can thus readily understand why functional information apparently only from a few restricted skeletal regions may add up to be far more widely representative of the totality of the organism. In some ways, it is even possible that these morphometric data may better summarize the entire organism than the data of classical morphology, for these latter are very heavily biased towards skulls and teeth.

But there is also another possible answer. This resides in the fact that some small amount of taxonomic information is truly contained within each anatomical region. Though the structure of the shoulder as rendered morphometrically does seem to reflect mainly the function of the shoulder, some separations that coincide with taxonomy can be distinguished (Ashton, Healy, Oxnard and Spence, 1971). Though the architecture of the arm and forearm reflect the function of the upper limb, it is also possible, to a degree somewhat greater than in the shoulder, to distinguish groups that make taxonomic sense (Ashton, Flinn, Oxnard and Spence, 1976). In the same way, though the main features of pelvic form when rendered morphometrically are towards the function of the pelvis, some clusterings that resemble those of classification can be discerned (Zuckerman, Ashton, Flinn, Oxnard and Spence, 1973; Ashton, Flinn, Moore, Oxnard and Spence, 1981).

None of this is particularly surprising; for, though these post-cranial parts are extremely sensitive to function, there is no doubt at all that they must also reflect, to some degree or other, hereditary mechanisms, some of which will be separate from function. Thus, when we add many areas together, the following may occur. Functional components may be averaged in a way that tends to hide them because the functions of the various parts differ. In contrast, the various taxonomic components, not especially obvious in each individual anatomical region, may gradually summate to greater and greater degrees because each, though statistically independent points, biologically in the same direction as the others.

Probably there is some truth in both of these ideas.

Discordance between classical and biometric morphologies. We are thus compelled to ask the second question: why are the results of the multivariate morphometric and classical morphological approaches different?

When we look at the multivariate morphometric approach, we perceive one immediate way in which the result should differ from that of classical morphology. For, though the creative mind of the classical morphologist is able to take account of a great deal of interesting detail in the complex forms and patterns that are studied, that mind is not able to cope so easily with quantitative relationships and scarcely at all with associations that stem from statistical properties like variance and covariance, auto-correlation and cross-correlation. It is not that these statistical elements do not exist within classical morphological data — they surely do — it is merely that the methods applied in classical morphological studies have not, until recently, been capable of identifying and allowing for them.

The various methods used in multivariate and other analytical methods of analysing structures are exactly designed to take account of these phenomena. This is the simple major difference between the classical and multivariate morphometric approach; this, then, may be the major part of the answer to our second question.

Concordance between morphometrics and biomolecules. Let us move, then, to the primary question: why should the multivariate morphometric result resemble most closely that obtained from biomolecular structure? It cannot be because the initial data of the multivariate studies are as close to the stuff of evolution as the biomolecules; being morphological, whether quantitative or not, these data are presumably as far from the genetic materials as the classical morphological data. Is it possible that the reason for the similarity with the biomolecules is the same as the difference between the multivariate morphometric and classical approaches, i.e. that they depend upon there being no correlation?

In biomolecular studies many characters are examined and summated. But because of the way in which molecular phenomena have been discovered and examined, because of our still very restricted understanding of the biomolecular picture of whole organisms, because the coverage of biomolecular phenomena is extremely spotty, it is highly unlikely that there will be any very special intercorrelations among the various biomolecular datum items. There will, of course, be some correlations, but these are likely to be little more than those necessarily resulting from the almost random selection of these data.

The classical morphological data, in contrast,

display correlations (in addition to those that are accidental) for a multitude of causal reasons. This is because the coverage of the whole organism by morphological data is so complete. Especially obvious examples are the correlations that are due to underlying patterns of development and underlying mechanisms of inheritance.

Now, I do not wish to imply that such correlations do not exist in molecular data; clearly, they do. There must be DNA sequences that relate to specific RNA formations, and particular RNA formations must correlate closely with, say, particular serum proteins; these in turn must relate to specific physiological functions of, say, the blood. Such functional links will result in marked intercorrelations among molecular datum items.

But it is not, in fact, at all common, in choosing characters of the biomolecular type in evolutionary studies, for investigators to deliberately select such obviously linked systems. Indeed, such linked systems are so far scarcely known, save in a few special experimental species.

Almost the opposite is likely to be the case: those systems that are chosen for biomolecular evolutionary studies are much *less* likely to be correlated (for reasons other than accidental) than are any of the data used in morphological studies. The molecular data are rather likely to be almost totally independent one from another, although this will become less and less the case as we learn more about the molecular biology of whole organisms.

Such an independence is almost never the case for morphological features. For though we might think, superficially, that there will be almost no correlation (except for that accidentally present) between, for instance, the form of the head and the shape of the little finger or between, for example, the structure of the eye and the lengths of the digits, phenomena such as Down's syndrome and Marfan's disease remind us immediately that even such apparently remote anatomical areas can be most closely linked. And, though these particular examples do not operate, presumably, in normal individuals, many other equivalent and perhaps more subtle examples undoubtedly do.

Of course, once we know a great deal more about molecular biology, we will undoubtedly find many similar sets of linkages there; but at the moment we have little more than a small number of biomolecular examples, and it is rather likely that these may be quite isolated from one another.

We can now see that morphological data, when treated by the multivariate statistical technique, may mimic biomolecular information. Biomolecular data are generally rather independent; morphometric data, though as dependent as other morphological information, contribute to the final result only after the degrees of dependence have been eliminated by the multivariate statistical technique. Presumably, if it were possible to weight classical morphological data to allow for the intercorrelations that must be present among them, we would also obtain the biomolecular result.

The implication for fossils. All this is a most exciting conclusion, for it allows us to obtain results about the relationships of animals that concur with the biomolecular approaches even when biomolecular studies themselves cannot be carried out. For many zoologists, this means the ability to assess rare specimens for which skeletal materials are available in museums but fresh biological materials do not exist or cannot be obtained. For paleontologists this especially means improving the evaluation of fossils.

It is true that some fossils are available that can be assessed by biomolecular methods. But at present these are limited to uncommon examples such as preservation of actual materials by freezing (e.g. mammoths), natural mummification through dessication and within a matrix (the tar of the La Brea species, or the amber of certain insect finds).

But, in the majority of cases, the information that is available for extinct creatures is purely that about form and pattern determined from fossils. These can only be studied by methods that assess morphology; and though in the past this has mainly meant classical morphology, in the future it should include multivariate morphometric evaluations that may, if the above hypothesis is not too far awry, better mirror the biomolecular assessment.

Of course, there are many cautions and difficulties. The present studies have particularly emphasized that multivariate statistical studies of limited anatomical regions do *not* provide much information useful in studying the relationships of whole animals. It is, therefore, necessary that the fossil remains be complete enough that relatively full suites of characters can be obtained. And it is surely absolutely necessary, before any such assessments be attempted, that we have a very full understanding of the multivariate morphometrics of the most closely related living species. Finally, the statistical problems that result from missing data and from small samples or even single specimens obtrude into such investigations. But then none of this is new; such cautions apply equally to other modes of study, especially the classical. It is only that these cautions become more obvious when we are able to be quantitative and analytical.

Summary. First, we can see that analysis of the combination of upper and lower limb dimensions results in a signet ring-shaped structural spectrum. Although no functional estimates had been made prior to obtaining this result, with hindsight it appears to make a great deal of functional sense. The vast majority of animals that have equal usages of the limbs in locomotion all occupy the seal of the ring. Those few species heavily dependent upon upper limb function within locomotion occupy one haft leading away from the seal (certain New World monkeys, lesser apes, orang-utans); those few species heavily dependent upon lower limb function within locomotion occupy the other (indriids, bush-babies, tarsiers).

African apes and humans are the species that complete the ring and they, too, lie in positions that make sense in terms of the relative contributions of their upper and lower limbs to locomotion, especially when viewed in the light of their presumed evolutionary histories. Thus, locomotion in the African apes is currently mainly knuckle-walking which involves all four limbs; but their positions within the signet ring suggest that they have evolved from creatures that were much more upper limb dominant (as they, themselves, are on those occasions when they do enter the trees, and as are all other extant apes and some New World monkeys). Locomotion in humans is, of course, entirely lower limb dominant and, accordingly, humans are closest (though still far distant from them) to those other primates that are also heavily lower limb dominant in their locomotion (e.g. indriids). But the curious position of humans also indicates that their prior history may have involved ancestors that were much more upper limb dominant, similar to some form of ape or other, although not necessarily to any particular ape extant today.

Secondly, we note the combination of all transverse dimensions. This produces no sensible arrangement of individual genera as far as can be judged; certainly nothing that makes functional sense is apparent. It does separate the sexes. The way in which this is achieved suggests that sexual dimorphism in the primates cannot be described by one picture (such as increased size and robusticity in males). It may not, therefore, be due to one single mechanism or process. Though a single pattern does seem to account for sex differences in many of the genera, this is by no means always the case. In particular, each hominoid displays a different pattern of sexual dimorphism (not merely similar patterns to different degrees). This finding may, therefore, if replicated by more extensive study, affect the overall study of the evolution of secondary sexual differences in primates generally, indeed possibly in other animals even more widely. In the more restricted case of hominoid fossils, in which sexual dimorphism would seem to be especially important, it may also have major implications. In particular, using just one extant hominoid as the model for studying sex differences in the fossils may be most dangerous.

However, the third and principal result of the work described in this chapter stems from the analyses of the totality of bodily proportions and the overall summation of detailed anatomical regions. In their generality, they mirror the current taxonomy of the primates. However, in their details, the morphometric results differ in a number of ways from classical taxonomy.

For instance, though these analyses define, generally, the well-recognized group of the hominoids, they do not recognize pongids and hominids. The results obtained here place humans as closely similar to the African apes, and markedly different from the Asian apes. Though both New and Old World monkeys are generally detected as their individual super-family clusters, the next level of clustering in these results is not into the family and sub-family groupings of classical taxonomy. Though prosimians are generally recognized as a major cluster, some of the sub-clustering is not into the traditional groups. In particular, the associations of *Tarsius* and *Daubentonia* do not concur with majority taxonomic views. The result obtained here suggests that tarsiers are more closely related to the Anthropoidea not the Prosimii, and that aye-ayes are very far distant from any living primate.

A review of biochemical and biomolecular studies of primates demonstrates that, in exactly each case in which the morphometric result differs from classical taxonomy, so the biomolecular data differ from classical taxonomy. In contrast, the degree of concordance between molecules and morphometrics is remarkable.

CHAPTER 9

Whole Anatomies – Their 'Dissection' by Primates

Abstract. In previous chapters we have seen that whenever we use the multivariate statistical method to look at the biometric form of anatomical regions we obtain arrangements of the animals that are most easily interpreted in terms of the function of the anatomical region. And in previous chapters we have also seen that when we look at combinations of all anatomical regions we obtain arrangements of the animals that follow, broadly, those of the current classification of the animals. In each case the question that we have tried to answer has been: how do these data arrange the various animals?

A final test of the validity of results like these comes from asking the reverse question: how do these data arrange the various anatomical dimensions?

An answer is attempted in this chapter. First, the clusterings of the dimensions are obtained for each of the analyses involving anatomical regions, both smaller regions such as the shoulder and the foot, and somewhat larger regions such as the entire upper limb and the entire lower limb. Second, the clusterings of the dimensions are obtained for larger combinations: upper and lower limbs combined for both the primates as a whole and for the Prosimii alone, and of course, finally, for all bodily proportions of all the available primates.

The answers to such questions about the arrangements of anatomical dimensions are seen to bear strongly upon the functional and systematic answers that come from study of the arrangements of the animals.

How anatomies 'arrange' primates

The main bulk of the multivariate morphometric studies (described in this book in Chapters 5, 6 and 7) have been aimed at analysing a variety of individual parts of primates. The question that is being asked in each case is: *how do these anatomical regions arrange the different primate genera?* In each case the answer seems to relate to function.

Anatomical regions and animal functions. In summary, previous chapters have shown that upper limbs (a) arrange the various primates in a band-like spectrum: those animals that use the upper limbs most in tension lie at one end of the spectrum and those that use the upper limbs most in compression at the other. Lower limbs (b) arrange the various species in a star-shaped spectrum. Species using the lower limbs in some extreme way (leaping, bipedalism, lower limb hanging are some examples) lie in the points or rays of the star; those animals that have no especial locomotor characteristics of the lower limb, save that they are regular quadrupeds, lie in the centre or nucleus of the star. Upper and lower limbs taken together (c) locate the primates in a signet ring-shaped spectrum. Those primates that use the limbs approximately equally (whatever their actual mode of locomotion) lie in the seal of the signet ring; those species that are lower limb dominant all lie within one haft of the signet ring; those that are upper limb dominant lie within the other haft of the ring. Finally, the results in every anatomical region (d), taken separately, show that quadrupedal species not displaying any of the more extreme locomotor activities are also arranged in ways that fit with what is known about the lesser differences in locomotion among them.

It can thus be said with considerable certainty that the structures of the individual anatomical regions correlate rather well with what is known about the main functions of those regions in the particular animals studied in each investigation. This is the case for every anatomical region that has so far been examined, even though undoubtedly some require further investigation with better biometric characterizations of the anatomy, better representation of the many different species of primates and, of course, better understanding of the behaviours and hence the functions in each individual species.

Anatomical wholes and animal relationships. A second large bulk of multivariate studies (described in Chapter 8) has been aimed at attempting to add together the various individual anatomical regions in order to see how whole animals are arranged by this approach.

In summary, these results parallel most closely the overall relationships of the primates. In those situations where classical morphological relationships differ from biomolecular ones, it is with the biomolecular ones that the multivariate results are most similar. This has led us to speculate that it may be the special characteristics of the multivariate approach to morphometric data that produces the similarity to the biomolecular view rather than the classical morphological picture.

How primates 'arrange' anatomies

It is, of course, also possible to analyse these various data sets in order to ask the reverse question: *how do the various genera of the primates arrange the anatomical parts?*

For, though we may be interested in 'dissecting' the order of the primates into their component animal sub-groups, we are also interested in dissecting their bodies into component anatomical sub-units (if they exist). Such dissection may allow us new insights into the structures of the animals.

Previous essays into anatomical 'dissection'

Before proceeding with this matter, it is worth reviewing other attempts that have been made to discover if 'fundamental units' form a basis for the structures of organisms. As usual, Aristotle was early in the day with his search for an 'essence' defining natural groups of objects. Thus, he posited the existence of a 'natural system' in which items were placed in appropriate groups by reference to a comparatively small number of characters (the 'essence') based upon some physical classification of form. Such an essence would, in modern parlance, be a set of characters strongly correlated with one another within the objects of interest. We understand, of course, that such a definition is merely a way of defining objects geometrically and does not necessarily speak to the biological meaning, if any, of the 'essence'.

Another ancient attempt to define structural units of animal anatomies depends simply upon defining what seem to be obvious anatomical parts of organisms. For instance, a limb forms an anatomical unit within the structure of most land animals, the head another, the trunk and perhaps its sub-units, the thorax and abdomen, yet others. But such modes of anatomical dissection are useful for little more than the convenience of descriptive anatomy; indeed, they form the basis of the organization of the teaching of much medical anatomy to this day.

The chief attempts to organize the sub-units of structure revolve rather more around underlying scientific ideas. The most powerful in this regard is the recognition of structural units through knowledge of developmental processes, patterns and fields, and multiple effects of single genes, gene complexes and gene modifiers. Such studies have been especially fruitful and sometimes bring together elements of structure whose essential unity might never otherwise have been recognized or suspected.

Many examples exist. The most obvious, perhaps, are those where there is an anatomical contiguity, for instance the existence of increasingly complex patterns of molarization of teeth along the tooth row in many mammals. In other cases, however, the contiguity may be biochemical and genetic rather than anatomical; such examples are much more difficult to assess until we have knowledge of the underlying biochemistry and genetics. Who would have thought, for instance, that a combination of long thin extremities, pigeon chest, high-arched palate, subluxation of the lens of the eye and deformation of the heart would have a single underlying association? This was not understood until the discovery of the generalized biochemical defect of connective tissue that stems from the inheritance of the single dominant trait with variable expression of Marfan's syndrome.

Another most powerful idea underlying the unity of some structures is function, and there have been many attempts to define anatomical units on such a basis. It has been shown that basic units can be discerned through experimental methods that help elucidate the functions of different structural elements, both individually and when taken in combination. A recent example here is the work of Karel Liem (1970) on the functions of the jaws in leaf fishes. He has shown, using a combination of electromyographic, radiological and surgical methods, that apparently separate characters within the jaws of these animals are all part of a single mechanical unit. But it is not easy to carry out such studies within an evolutionary context in many different species and in taxonomic groupings as large as classes or orders; consequently, such investigations usually provide examplars that suggest to us overall possibilities within a group but not the differences among all the members of such groups.

Thus, the method that may be of most use in situations where we would understand relationships among many different animals, where organisms are so rare that it is not justifiable to perform experimental studies upon them, and where the necessary comparisons are so numerous that logistical constraints prevent the necessary numbers of experiments from being carried out, is the 'morphological fishing expedition'. If we study many different structural elements in order to understand particular anatomies in many animals, it is possible that underlying sub-units among those anatomies will stand revealed. Certainly, we may ask this question of the voluminous quantitative data defining the various anatomical regions of the many representatives of the primate order included in our studies.

Morphometric 'dissection' of anatomies

Answering questions about the sub-units of anatomical structures when the method of study is measurement and multivariate statistical analysis has not often been carried out. However, one of our earliest studies, the analysis of a series of 17 dimensions of the primate shoulder, presents a very clear example (summarized in Oxnard, 1973a). In that case three discriminant axes contain the great bulk of the information about the separations of the various primates. The method itself determines that the information contained within each axis is statistically independent from that contained within each of the others. In addition, however, the particular separations of the animals achieved in that study seem also to suggest that there is independent biological information reflected in each discriminant axis.

A first try: interpretation of discriminant axes. An attempt was therefore made to try to discern the particular combination of anatomical variables that contributed most to each discriminant axis. This was done by studying the combination (the product) of (a) the loading factor for each variable in a given axis and (b) the amount contributed by each variable to the separations of the animals in that axis. It is interesting to note that a large loading factor alone in a given axis does not necessarily mean that a given variable makes a big contribution to separations in that axis (for the variable may provide almost no separation of the genera at all). And a large separation of the genera produced by a particular variable likewise does not necessarily mean that the variable contributes largely to the given axis (for the loading factor for the variable for that axis may be very small). It is their combination in a single axis that is important.

The result of carrying out this procedure on the

shoulder was clear-cut. The combinations of variables contributing to the different discriminant axes were not random or meaningless. Quite the contrary, the combinations of variables suggested that, in each axis, *first*, the variables were actually measuring morphologically identifiable sub-units of the entire structure, *second* the particular sub-units made sense when viewed in the light of the functions of the shoulder girdle, and *third*, the functional deductions were supportable by independent biomechanical experiments (Oxnard, 1968a, 1969, 1973a).

Thus, the first axis of that study contains mainly contributions from a group of variables that are measures of the cranio-lateral twistedness of the shoulder complex: engineering estimates of the mechanical efficiency of the shoulder show that this twistedness is related to the degree to which the upper limb is capable of being used in a variety of tension-bearing activities consequent upon arboreal acrobatic movements; it is thus with pleasure that we find that this first axis indeed separates primates and mammals on that functional basis. Those species (sloths, orang-utans for instance), that readily perform tension-bearing activities of the upper limbs are well separated by these variables from others (patas monkeys, ground squirrels, for example) whose limbs are mainly compression-bearing (Oxnard, 1973a).

In a similar way, study of the second discriminant axis provides another suite of dimensions as major contributing factors. In this case, again, their combination seems to be morphologically meaningful: they are all measures of the amount of medio-lateral shortening or lengthening of the shoulder girdle; from the biomechanical point of view this fits in well with the notion of differential degrees of three-dimensional mobility at the shoulder; and again, therefore, it is with extreme interest that we realize that this particular discriminant axis indeed separates arboreal mammals (e.g. giant tree squirrels, uakari monkeys) with more mobile shoulders from more terrestrial species (e.g. ground squirrels, vervet monkeys) with less mobility at that joint (Oxnard, 1973a).

The third discriminant axis speaks, among the primates, solely to the unique separation of humans; but among the mammals as a whole, it refers in a reverse manner to all the gliding and flying mammals, whether placental or marsupial (Oxnard, 1968a, 1973a).

These various combinations of variables depend upon the statistical computations. But the basic morphology that they reveal is rather clearly displayed by using Thompsonian Cartesian coordinate deformations to provide some estimate of the differences in shoulder shape between pairs of animals that are separated by a single discriminant axis alone (Figs. 9.1, 9.2 and 9.3).

Fig. 9.1. The morphology of the shoulder represented by the information contained within the first discriminant axis displayed through the deformation of Cartesian coordinates in two species, gorillas and baboons, that differ mainly in that axis. The overall difference is a cranio-lateral twist in gorillas as compared with baboons.

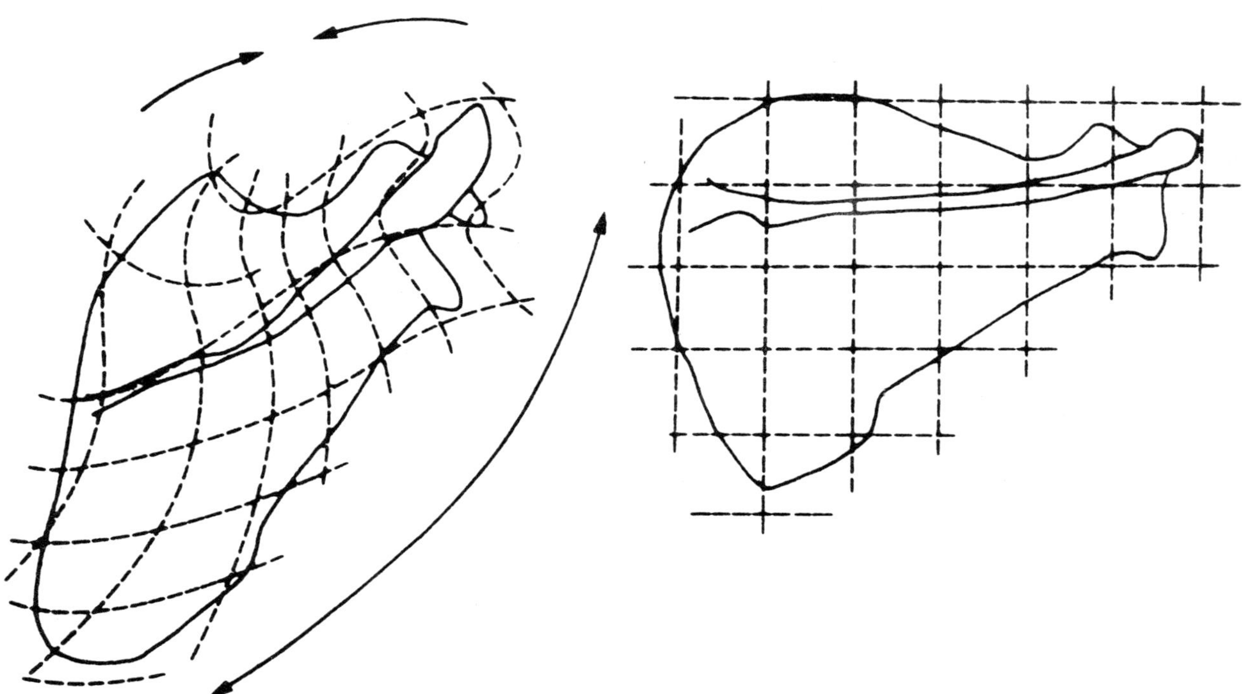

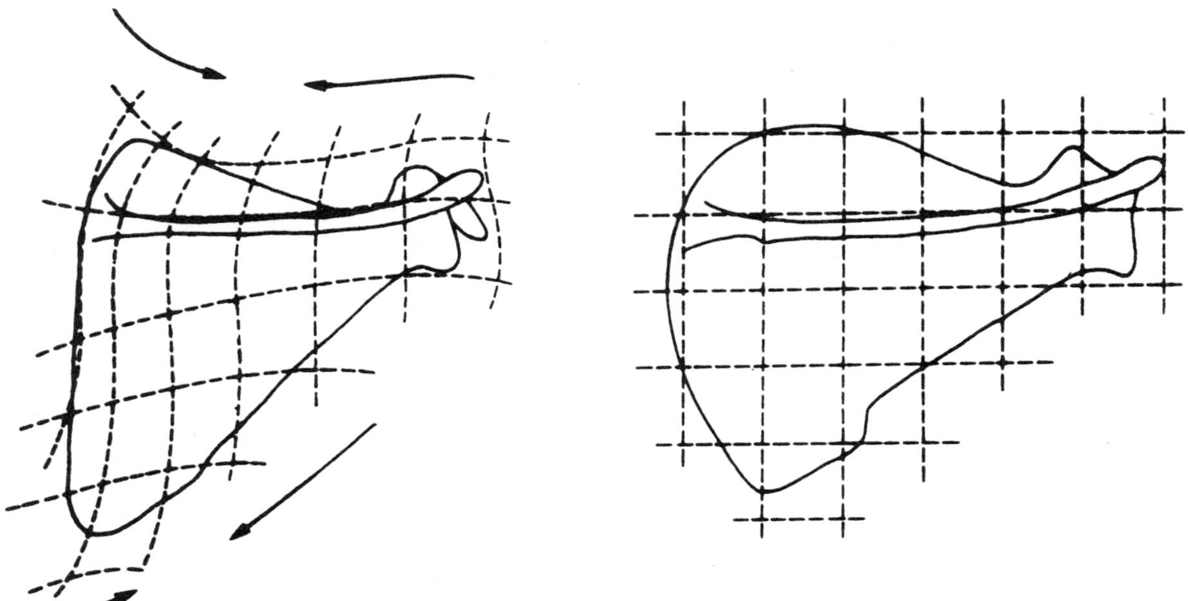

Fig. 9.2. The morphology of the shoulder represented by the information contained within the second discriminant axis displayed through the deformation of Cartesian coordinates in two species, uakari monkeys and macaques, that differ only in that axis. The overall difference is a medio-lateral narrowing in uakari monkeys as compared with macaques.

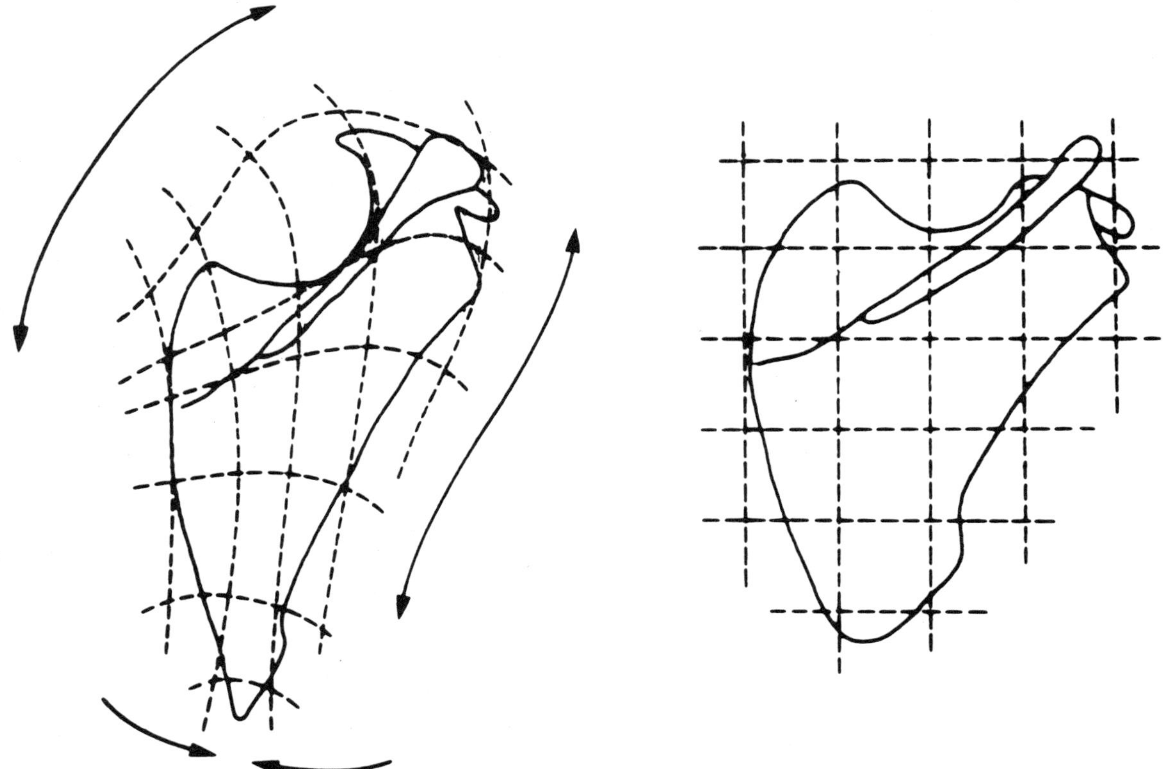

Fig. 9.3. The morphology of the shoulder represented by the information contained within the third discriminant axis displayed through the deformation of Cartesian coordinates in two species, 'flying lemurs' and *retroussé*-nosed langurs that differ only in the third axis. The overall difference is a cranio-lateral lengthening in 'flying lemurs' as compared with *retroussé*-nosed langurs.

The marked success of this particular study in revealing sub-units of shoulder structure has never been fully replicated in any other investigations. At first I was inclined to suppose *either* that we had been extraordinarily lucky in devising original variables that were particularly important functionally *or*, because the shoulder girdle is so completely suspended by muscles and its shape therefore influenced much more by mechanics, that any suite of measurements of the shoulder would reflect function more than measurements of most other anatomical regions. Probably there is some truth in both possibilities.

However, study of many other anatomical regions has now made it clear that the special contributions of the variables to the statistically determined discriminant axes are not necessarily the contributions that are biologically important. The original finding for the shoulder is not necessarily true for other anatomical regions.

That is, though by definition individual discriminant axes must contain independent statistical information, there is no *a priori* reason why the axes should be reflecting underlying biological information that is independent. Indeed, the opposite is more likely to be the case. The biological information is unlikely to be in independent blocks. Thus, rotated or oblique axes may better reflect any underlying 'meaning' that the data may hold.

A good example of this has already been encountered (Chapter 5) in the studies on prosimians where, though the discriminant separations are the ones provided by the statistics, the chief separations of the groups of animals are in directions at other angles (Fig. 9.4). Another example can be seen in studies of the talus in higher primates, where an axis best representing the overall sizes of the specimens can be rotated around by a series of quite minor manipulations of the original data (Fig. 9.5; and Lisowski, Albrecht and Oxnard, 1974).

A better attempt: factor analysis clusters. Thus, in studies that are more complex than was the case in the shoulder, and in studies that include, at one and the same time, discriminations among many different groups of animals (see the complexities of some of the three-dimensional models based upon generalized distances), then it is all the more likely that biologically important combinations of variables will *not* be reflected by single discriminant axes, but rather by combinations of them. Accordingly, then, it has now seemed to make considerable sense to search actively for combinations of variables using factor (principal components) analysis.

This is a mode of multivariate statistical study that asks a different question of the data: *what combination or combinations of variables are responsible for producing the separations of the groups of animals that we have observed?*

Even here, factor (principal components) analysis does not provide, alone, the readily recognizable picture. For, of course, it is likely that groups

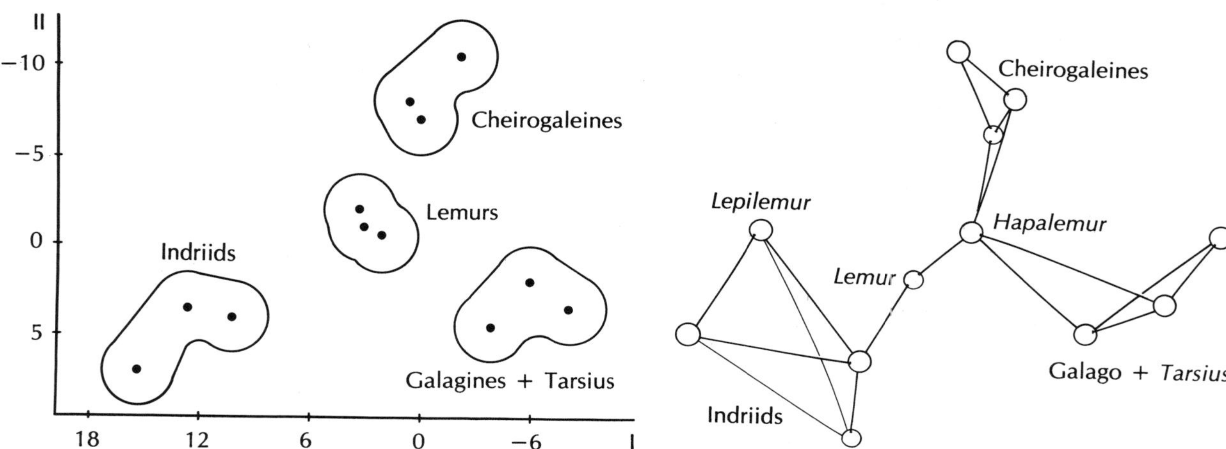

Fig. 9.4. The morphometric result of analysing dimensions of the limbs of prosimians. In the plot of the first two discriminant axes the separations of the various groups of animals are in directions at angles to each of the first two axes. This is seen even more clearly in a three-dimensional model of the same result. Conventions as in similar prior figures.

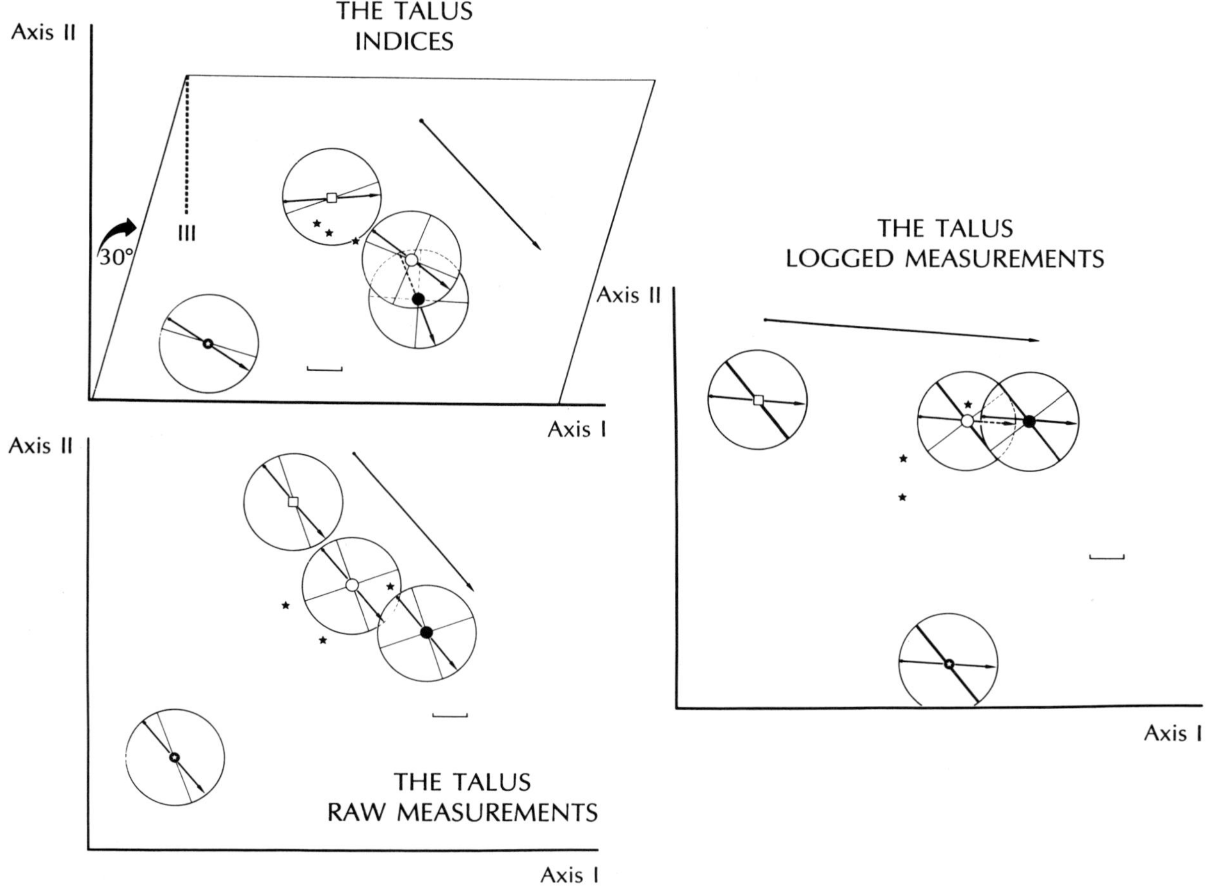

Fig. 9.5. The morphometric result of analysing dimensions of the talus in higher primates. Various manipulations of the data are made (analysis of measures, analysis of logarithmically transformed measures, analysis of indices). In each case similar separations are obtained. For the logarithmically transformed measures, the major differences among groups are directly aligned with the first axis; for the measures, the separations among the groups are rotated to lie at angles to both the first and second axes; for the indices, the relationships among the groups are further rotated to lie at angles to the first three axes.

of variables will be reflected in several different factors. What we need to know is: which groups of variables are providing similar contributions within the entire matrix of factor scores? This is exactly similar to our prior problem of trying to decide which groups of animals are separated in similar ways within the entire matrix of discriminant scores. Accordingly, then, the problem can be answered by using the same display: Andrews' high-dimensional technique.

In the previous cases in which we were interested in discovering groups of organisms that were similarly placed among a large number of discriminant axes, the discriminant coefficients for each organism were embedded in a sine-cosine curve as explained in Chapter 3. We were able, then, to recognize similarities among the animals by demonstrating similarities in the curves that represent them. In exactly the same way in the present cases we can embed the factor scores for each individual variable in a sine-cosine function and thus obtain a curve that represents the variable. Groups of variables that have a similar factor structure will then be displayed as curves that are all rather similar to one another. This allows us to identify sub-groupings of variables that underly the whole anatomy.

Our test example: biometrical 'dissection' of the shoulder

In order to see how this mode of analysis works, it is perhaps best to test it out on the scapular data that we already know so well. Accordingly then, factor (principal components) analysis was performed on the 17 dimensions of the primate shoulder. The various tests outlined in Chapter 3 indicate that the method is viable. Several factors (or components) are identifiable as being statistically significant and factor scores are obtained for each of the variables. Using these factor scores, Andrews' high-dimensional curves are plotted for each variable. It proves relatively easy to sort the variables on the basis of these curves.

The first frame of Fig. 9.6 shows curves for the group of variables that contribute most to the first factor; it demonstrates just how closely similar they are to one another. The second frame shows another set of closely similar curves for the second factor. The third frame displays the similar curves for the variables contributing most to the third factor. Comparisons among these frames shows just how distinct are these special groupings of variables.

The key question, then, is: *what are these groups of variables?* Are they at all similar to the groups that were realized in the prior discriminant function studies? The answer is that the two sets of groupings are almost identical. Table 9.1 provides lists of the main contributing variables to each discriminant axis in the earlier discriminant function analysis and compares them to the groups of variables with similar internal factor structure in the new factor analysis. The concordance between the two is almost perfect. This at least confirms that the new factor method provides groupings of shoulder variables similar to those given by the earlier discriminant study that we understand so well.

Thus, again, the overall features that these groups of variables seem to be measuring are the cranio-lateral twistedness of the shoulder girdle (first axis), its medio-lateral lengthening and shortening (second axis) and an oblique lengthening and shortening in the shoulder (third axis).

The precise variables differ slightly between the two studies. And there is a reversal of axis number between the two investigations (for instance, the second axis of the first study is comparable with the third axis of the second and vice versa). Notwithstanding these slight differences, the factor and discriminant investigations are providing the same information about clusters of variables. In the discriminant study this is merely a fortuitous similarity because of the special concordance between variable clusters and animal clusters. But in the factor study this would be the case whatever were the relationships between the structures of the variables and the structures of the animals. As a test of this last statement let us proceed to another simple example.

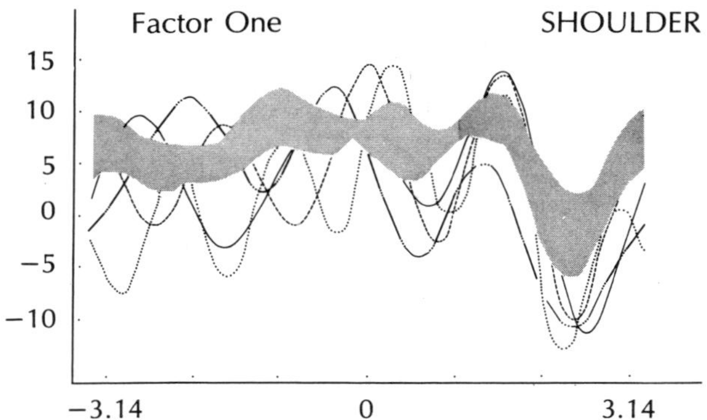

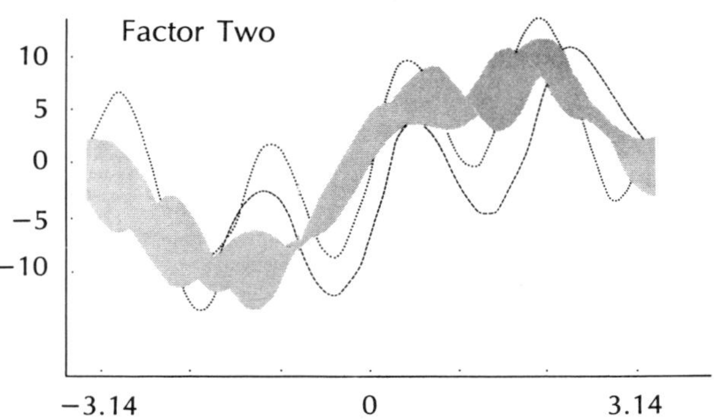

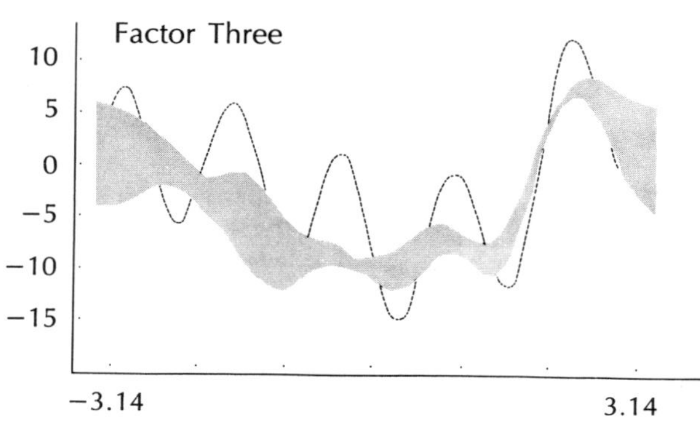

Fig. 9.6. Clustering of variables by factor analysis as they make contributions to the separations of the groups of animals in the study of the primate shoulder. The groups are displayed using the high-dimensional technique. In each case the groups are very clearly demarcated as shown by the shaded envelopes. Individual dotted curves are of variables somewhat peripheral to the group. This convention is followed in the remaining diagrams.

Table 9.1

COMPARISON OF KEY VARIABLES IN DISCRIMINANT AND FACTOR ANALYSES OF THE PRIMATE SHOULDER

Variable	Canonical axis 1	Factor axis 1	Canonical axis 2	Factor axis 2	Canonical axis 3	Factor axis 3
1	*	*				
2	*	*				
3			*	*		
4			*	*		
5	*	*	*	*		
6	*	*				
7						
8	*	*				
9	*	*	*	*	*	
10		*				*
11					*	*
12					*	*
13						
14						
15	*	*				
16			*		*	*
17					*	*

These three pairs of canonical and factor axes have virtually identical patterns of contributions of variables. The first pair (as judged by the variables) seem to measure the cranio-lateral twistedness of the shoulder girlde and separate the animals according to the degree of compression or tension in the shoulder during locomotion. The second pair (as judged by the variables) seem to measure the medio-lateral shortening of the shoulder and separate the animals according to the degree of arboreal activity in the shoulder. The third pair (as judged by the variables) seem to measure mostly minor details of shape and separate mainly prosimians from anthropoids.

Another simple case: the talus in higher primates

The talus in higher primates is another relatively simple investigation in which both the number of variables and the number of species represented are quite small. These data, outlined in more detail in Chapter 6, were studied by Lisowski, Albrecht and Oxnard (1974, 1976). They include eight measurements taken upon eight genera of non-human primates and four groups of humans. In addition, data are available from 21 individual fossil specimens.

That data set, though relatively simple because of the small numbers involved, has proven controversial because it demonstrates that those fossil tali from Olduvai and Southern Africa known as australopithecines (including both robust and gracile species and the individual specimen known as *Homo habilis*) are not at all like humans. The data suggest therefore that one bone of the foot, the talus, from the fossil known as Olduvai Hominid 10, is unlikely to have been from a foot adapted for bipedality in the manner of man. And, though it is likely indeed that the foot of that creature was capable of participating in bipedal locomotion (even the feet of the present day apes are *capable* of such an activity, as are the feet of many other primates) it seems as though an ability for climbing in some manner or other was also among the locomotor repertoires for which it was well adapted.

Controversial as those findings were when they were first made, they are now confirmed both by further studies of our own upon the arches of the entire foot (Oxnard and Lisowski, 1980; Oxnard,

1980b) and by new studies of the joint complexes of the feet of many living mammals as well as fossil primates by Professor O.J. Lewis (1980a, b, c, 1981). The talar assessments, therefore, however surprising they may have appeared to many workers, are thus confirmed; and it is of especial interest to restudy those data using the new 'biometrical dissectional method' in order to see what aspects of the shape of the talus may have been most responsible for the result.

Accordingly, then, that data set is here analysed using factor analysis in which a vector of factor scores for each variable is obtained. The patterns of clustering of the variables are displayed using Andrews' high-dimensional sine-cosine curves. We would not normally expect any very startling result because of the small size of the study; there are only eight variables, nine species and a total of 343 specimens. It is thus highly likely that every individual measure contributes quite largely to the entire result; certainly, in the corresponding prior discriminant analysis, almost every variable does contribute, in differing combinations, to each of the first three axes. However, Fig. 9.7 and table 9.2 demonstrate that, indeed, there are three rather clear groups of variables.

One suite includes variables whose similar factor scores produce very flat curves, mainly on the positive side of zero (first frame, Fig. 9.7). These variables are all measures of the length, breadth and height of the talus, and seem, therefore, to reflect the overall proportions of the bone. This may be a particular reason why human tali are so markedly separate from those of all the non-human primates; for big as they are, even the tali of gorillas are nowhere near the robusticity of those of humans, the only members of the extant series of animals in whom the entire body weight is borne by the talus and is not shared at all by components of the upper limb.

A second group of variables displays a pattern of high dimensional curves that is partly negative and that has a marked minimum and maximum, quite different, therefore, from the prior set (second frame, Fig. 9.7). These variables are the ones measuring the different angulations and twists in the talus. This presumably is related to the functions of the talus in those foot complexes that involve the flexibility of the foot. For, though most human feet can indeed be inverted and everted, the feet of most primates, especially those that are more highly arboreal, are far more mobile and flexible. They are not only capable of more marked inversion and eversion than the foot of man, but also of much greater change in proximo-distal curvature when the foot is grasping a tree limb; and, of course, such feet display increased twisting during the action of the grasping big toe that is not possessed by man. It is this suite of variables that is associated with differentiation of the orang-utan and most other arboreal primates from the African apes and man (though the African apes share these features to some smaller degree). And it is thus of especial interest to us to note that it is this same suite of characters that especially distinguishes the foot of various fossils, most notably that of Olduvai Hominid 10, a supposed biped, from both man and the terrestrial African apes.

Finally, there is yet a third grouping of variables that forms a cluster, but in this case one less clearly defined (third frame, Fig. 9.7). These high-

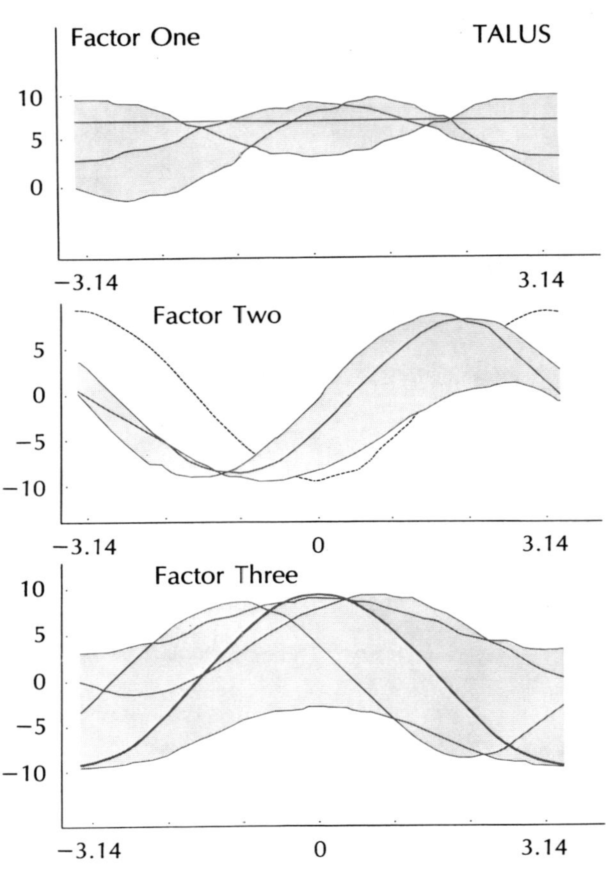

Fig. 9.7. The variables clustered as they make contributions to the separations of animals in the study of the primate talus. Again, the groups are displayed through the envelopes resulting from the high-dimensional technique. The groups are less clearly demarcated than is the case for the shoulder; they are nevertheless still quite clear. Conventions as in Fig. 9.6.

Table 9.2

FACTOR ANALYSIS OF DATA ON THE TALUS: CLUSTERS OF VARIABLES

Clusters	Anatomical features	Overall description
Factor axis 1	Maximum medial height Maximum lateral height Maximum length	General talus proportions
Factor axis 2	Angle between neck and body Torsion between head and body Curvature of trochlear facet	General twists of talus
Factor axis 3	Maximum medial height Maximum lateral height Posterior trochlear breadth – Angle between neck and body – Torsion between head and body	Overall proportions positive, twists negative

dimensional curves share a form in which there is a single central maximum. The group includes variables from each of the two prior groups but combined in such a way that variables from the first associate with variables from the second in a reverse fashion. That is variables that contribute negatively to one axis (say variable 7 to axis 2, or variable 3 to axis 3) contribute positively to other axes (both variables 7 and 3 contribute positively to axis 1). This combination of variables suggests that the interaction of overall proportions with twisting is also important to the separation of the species.

This example particularly demonstrates the folly of assuming that simple combinations of variables will always be defined by statistical axes.

Primarily, however, the study shows that the talus can not only be biometrically defined, it can also be biometrically dissected.

A complex example: biometrical 'dissection' of arm and forearm

We may now turn to the arm and forearm for a definitive study because the most valuable usage of this technique is where the number of defining variables is rather large. In this study there are a total of 19 variables taken on 525 specimens representing as many as 39 different primate forms (as outlined in Chapter 5). Though discriminant function analysis provides information about the separations of the various primates from one another in the first axis, the remaining axes, when taken individually, do not seem to be especially useful.

Accordingly, then, factor (principal components) analysis was performed on this data set and Andrews' high-dimensional curves generated from the factor scores for each variable. Several different groupings of variables are obvious. Some include variables that are linked as belonging to specific factor axes, but others include variables that are linked because they display similar patterns of contribution to each of several factor axes. It is information of this type that is difficult to obtain without the high-dimensional display.

These various groupings are shown in Figs. 9.8 to 9.10. Their usefulness depends upon whether they make some functional-anatomical sense or whether they are merely nonsense combinations.

The first group of variables, forearm flexion. The first group of variables defined by this technique is shown in Fig. 9.8 (the entire arrangement of variables is outlined in table 9.3). We may ask the question: does this group of variables, all similar enough as to fall within the shaded envelope of Fig. 9.8 (except for the single curve noted) make any specific functional or anatomical sense? The answer is that it does: every one of these variables measures some anatomical feature localized around the elbow (for instance, the relative projec-

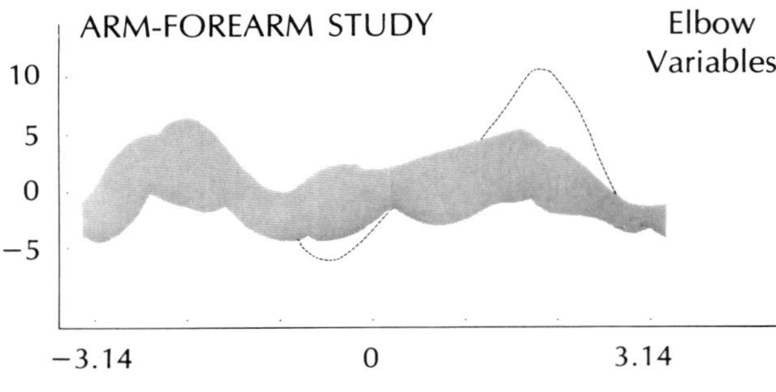

Fig. 9.8. A first group of variables in the more complex study of the primate arm and forearm. These variables are very similar to one another in their contributions to the analysis. They all relate to the elbow joint.

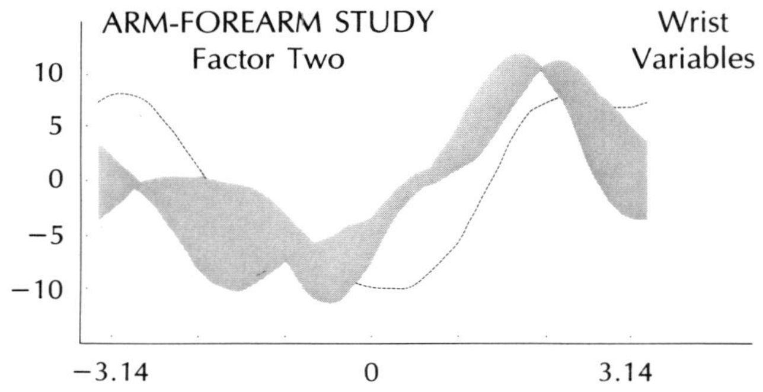

ARM-FOREARM STUDY
Factor Two
Wrist Variables
10
5
0
−5
−10
−3.14
0
3.14

Fig. 9.9. A second group of variables in the more complex study of the primate arm and forearm. These variables are also very similar to one another in their contributions to the analysis. They are all located around the wrist.

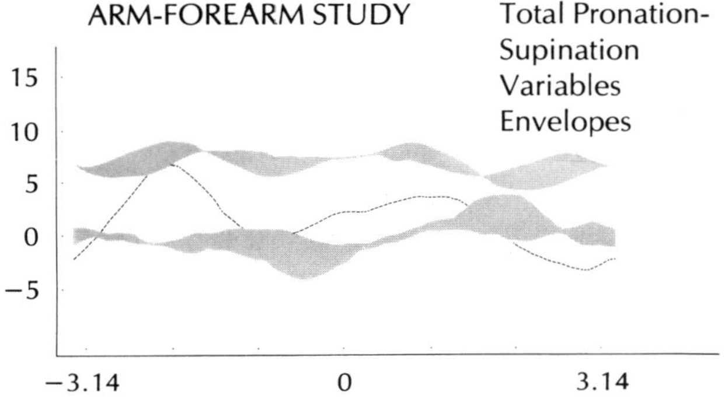

Fig. 9.10. The third group of variables is similar in displaying very little curvature in the high-dimensional plots. They are themselves divided into two blocks. The upper block relates to proximo-distal extents of insertion of pronating-supinating muscle attachments; the lower block relates to the central-peripheral extents of attachments of pronating-supinating muscles. The entire group relates, of course, to the movements of the 'radio-ulnar joint complex'.

Table 9.3

FACTOR ANALYSIS OF DATA ON ARM AND FOREARM:
CLUSTERS OF VARIABLES

Clusters	Anatomical features	Overall description
Shared factor pattern	Facet on humerus for ulna Facet on ulna for humerus Projection epicondyles Position radial tuberosity Insertion of triceps	Measures at elbow
Factor axis 2	Projection of ulnar styloid Relative sizes, both styloids Projection of radial styloid Relative widths, radius and ulna	Measures at wrist
Shared factor pattern	Distal insertion of pronator Distal insertion of biceps Lateral bowing of radius Angle of interosseous ridge Maximum bowing of radius	Measures relating to pronation and supination

tion laterally on the humerus of the bony attachments of many forearm muscles, the angle on the ulna of the articular surface for the humerus and the position on the radius of the attachment of the biceps muscle). It is particularly fascinating that the variables in this group include measures taken upon each of the three bones implicated at the elbow joint and, furthermore, measures confined to the elbow ends of these bones. This truly makes functional and anatomical sense.

A second group of variables, hand movements. The second group of variables is demonstrated by this technique in Fig. 9.9 Does this grouping make any functional sense? Indeed, it does. Each variable here (table 9.3) is a measure of some localized anatomical structure located at the wrist (for example, the relative widths of the distal ends of the radius and the ulna, and the relative lengths of the distal bony [styloid] processes of the radius and ulna measured separately).

An excellent test of this result would have existed if we had taken measurements on the proximal carpal bones. If these, too, had fallen with this set of variables, then we could have been absolutely certain that the factor analysis had identified an anatomical complex associated with the wrist joint and presumably, therefore, with the structures permitting the movements of wrist flexion and extension, and radial and ulnar deviation.

The third cluster of variables, forearm rotation. The third cluster of variables shows very little curvature in the high-dimensional plots (Fig. 9.10 and table 9.3). The curves fall into two clearly defined sub-groups, with a single variable that does not seem to be able to make up its mind, as it were, to which group it would like to belong.

It turns out that the variables in the upper envelope of Fig. 9.10 relate to measures of the proximo-distal extent of insertion in the forearm of muscles (such as biceps and pronator teres) that are responsible for producing pronation and supination of the radius and ulna. They thus relate to the longitudinally directed component of the lever arm of these pronating and supinating muscles.

The variables in the lower envelope of Fig. 9.10 (representing, therefore, a similar pattern of factor scores except for a difference in the first factor) are also variables that seem to do with pronation and supination (as defined in the original publication of Ashton, Flinn, Oxnard and Spence, 1976). They differ from the first set in being measures of the amount of bowing between the two forearm bones; this, among other things, affects the circularly directed components of the lever arms for the insertions of the pronator and supinator musculature.

The apparently intermediate variable shown in Fig. 9.10 does not belong, either anatomically or functionally, with these features; but then it is rather likely that it is truly displaying a quite different pattern of factor scores; it has merely been included in the discussion here because, of all the other variables, it is the one closest of this particular group.

It is especially interesting (a) that the technique has isolated for us a single joint system, lying between the radius and the ulna and extending throughout the entire length of the forearm and (b) that it has separated the longitudinal and circular components of this joint system.

The overall meaning of the arm and forearm clusters. Now in some ways we may think that this study has not achieved very much for us. Any competent anatomist would have said that, of course, the elbow, wrist and radio-ulnar complex are the three important functional units in the arm and forearm. The interest for us, however, is that in this case we have not assumed that knowledge; we have gone on an anatomical 'fishing expedition'; and we are most heartened to see that the biometrical definition of the structure of the arm and forearm is indeed easily dissected into anatomically recognizable and biomechanically acceptable subunits (table 9.3).

It is especially important that mere anatomical contiguity is not the result. For instance, some of the measures associated with pronation and supination are anatomically much closer to the wrist, others to the elbow; yet the analysis has placed them in a single, radio-ulnar group.

Another complex case: the hip and thigh in prosimians

The various studies of the hip and thigh in prosimians are especially useful in providing functional insights into differences among prosimians. This is particularly so in suggesting that there exist different forms of leaping with different implications for hip and thigh structure (Chapter 5). Again, however, in that study the individual discriminant axes do not themselves appear to possess any special anatomical relevance, though by definition they are especially important statistically. Most of the variables seem to contribute to all of the axes that perform the largest separations. Indeed, the relative biological *un*importance of the discriminant axes is apparent from the fact that it is the combination of axes, as in the high-dimensional and generalized distance methods of display, that seems to be of most value in separating the different groups of animals.

The same procedure (together with the same precautions, Chapter 3) is therefore carried out on this data set in order to obtain factor scores and high-dimensional displays for each variable. There are 23 variables taken on 289 specimens representing 20 individually recognizable species and genera of prosimians outlined in more detail in Chapter 6.

The most clear-cut groupings, muscle attachments. Some of the of variables identified in this study are presented overall in table 9.4. But their reality is clearest in the various separate figures. Thus, one most obvious grouping is represented by factor 5 which comprises a very narrow envelope (Fig. 9.11). These variables must therefore be making very similar contributions to the analysis. They are measures at the upper end of the femur; the relative positions of the lesser and third trochanters, bony processes to which are attached the large and functionally important iliopsoas and gluteus superficialis muscles respectively.

Another grouping of variables forms factor 4, again as a very narrow envelope that implies that the contributions of these variables must be remarkably similar (Fig. 9.12). They are measures of central points on the pelvis: the pubis to which is attached the adductor musculature, and the anterior inferior iliac spine which gives origin to the rectus femoris muscle. Osteometrically these points are rather close together; functionally they are associated with muscles capable of both flexion and adduction of the thigh.

I have described these two higher factors first because they are tightly defined and clear. Of much greater interest are the combinations of variables included in the earlier factors.

Other clusters of variables: pelvic form. Thus, Fig. 9.13 shows the envelope of variables that forms the first factor. This comprises the combination of positive contributions from cranial pelvic parts (the relative length of the ilium and the relative position of the superior iliac spine, also on the ilium) together with negative contributions from caudal pelvic parts (the relative lengths of the ischium and pubis). All are, of course, pelvic measures. They are further united in contrasting the cranial muscle mass attachments (the iliopsoas and gluteal muscles) with the more caudal muscle attachments (the hamstrings and adductor musculature). They are related to the three developmental parts of the pelvis which are fused at the centrally located acetabulum.

Table 9.4

FACTOR ANALYSIS OF DATA ON PROSIMIAN HIP AND THIGH:
CLUSTERS OF VARIABLES

Clusters	Anatomical features	Overall description
Factor 1	Iliac length Position iliac spine Ischial length negative Pubic length negative	Cranial muscle versus caudal muscle attachments
Factor 2	Pubic length Ischial length Iliac length	Cranial plus caudal measures
Factor 3	Presacral iliac length Pubic length Ischial length Ratio pubic and ischial lengths	Measures relating to all three regions of pelvis

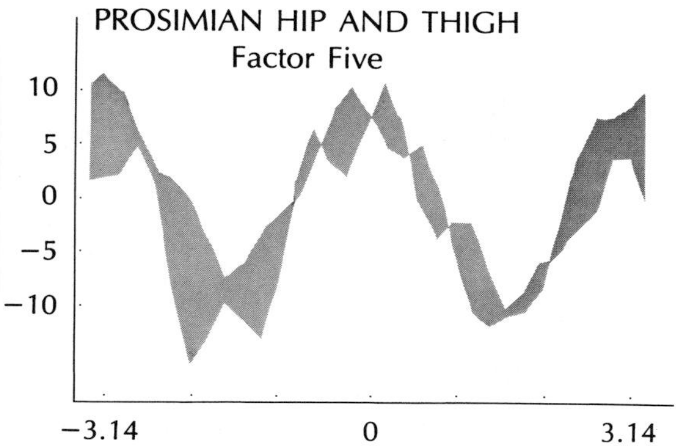

Fig. 9.11. The very narrow envelope of variables in the fifth factor of the study of the prosimian hip and thigh. These are variables measuring muscle attachments at the upper end of the femur.

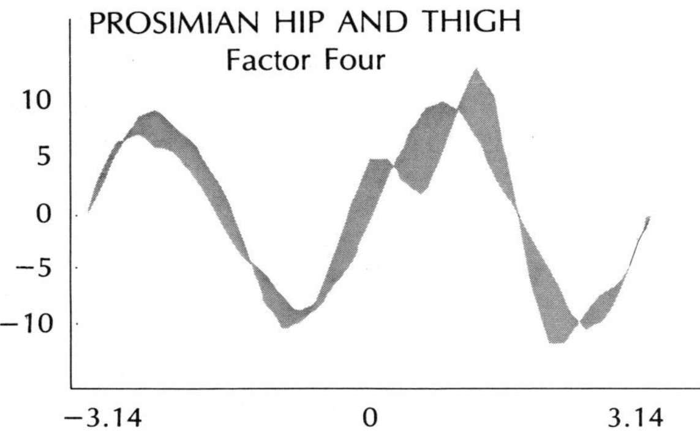

Fig. 9.12. The very narrow envelope of variables in the fourth factor of the study of the prosimian hip and thigh. These variables measure muscle attachments ventrally on the pelvis.

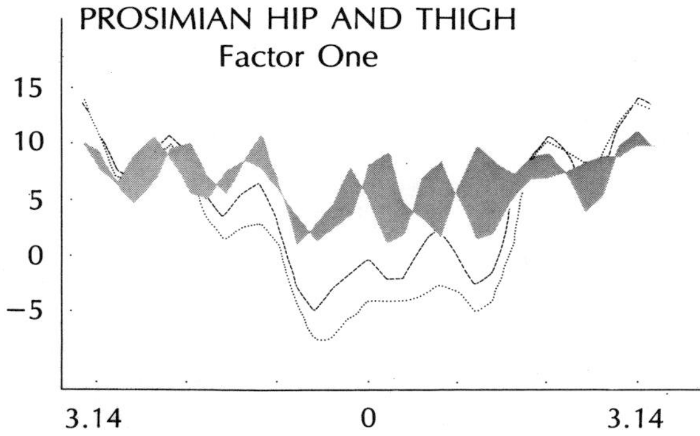

Fig. 9.13. The envelope of variables in the first factor of the study of the prosimian hip and thigh. They comprise the relative proportions of the three main parts of the pelvis.

The second factor has, as its principal contributors, three measures that are also related to all three regions of the pelvis (pubis, ischium and ilium, Fig. 9.14). These are relative pubic length, relative ischial length and one of several measures of relative iliac length.

The third factor also consists primarily of features that relate to all three regions of the pelvis (Fig. 9.15). In this case the particular dimensions are presacral length of ilium, relative pubic length, relative ischial length and the ratio between pubic and ischial lengths.

Apart from the fact that the measures in each of these three factor axes are all pelvic, and especially that they all seem to relate to the three main developmental and functional arms of the pelvis, there is little that it seems we can say. The same variables contribute from one factor to the next, and the difference is principally that the signs of the contribution (whether positive or negative) change

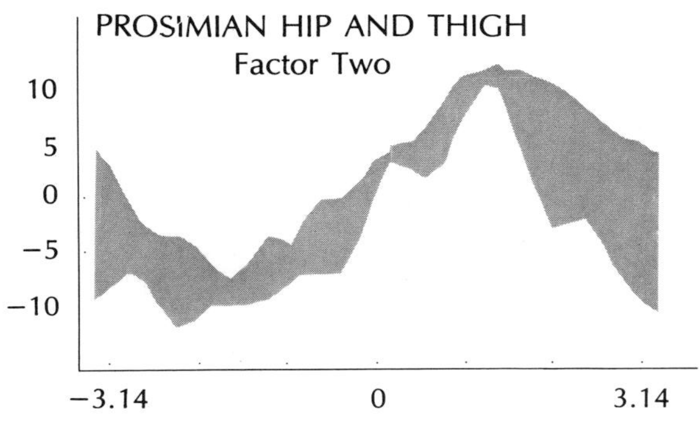

Fig. 9.14. The envelope of variables in the second factor of the study of the prosimian hip and thigh. Again, they comprise the relative proportions of the three main parts of the pelvis, but in a different combination.

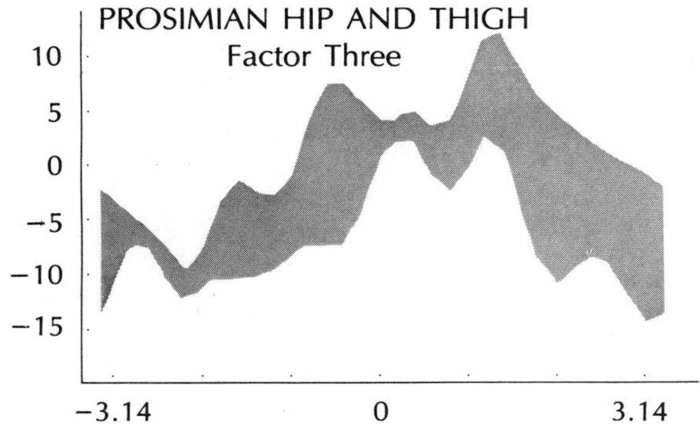

Fig. 9.15. The envelope of variables in the third factor of the prosimian hip and thigh. Once more, they comprise the relative proportions of the three main parts of the pelvis in a third combination.

from one axis to another. For example, in the first factor the relationship between the ischium and the pubis contributes negatively, whereas in the third factor it contributes positively. It is thus especially clear that it is not just groups of individual variables that are important but rather the different types of interactions among them.

We have therefore attempted to view the variables contributing to all three axes in a single picture. This is shown in Fig. 9.16 and it demonstrates that there is indeed an overall relationship between the variables; for though the left-hand part of the plot places them all as lying within a rather narrow envelope (narrow, that is, considering just how many variables are represented here) the right-hand side of the plot splays them out rather like a hand of playing cards.

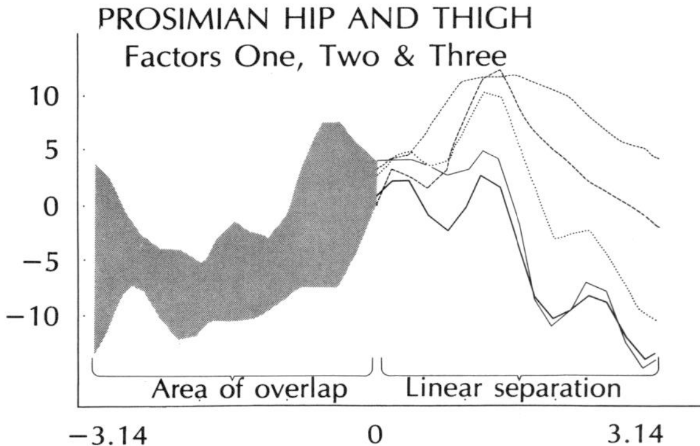

Fig. 9.16. The envelope of variables of the first three factors of the prosimian hip and thigh combined. The high-dimensional display demonstrates that for half of the plot these variables are all similar. For the other half of the plot they are separated vertically in a linear fashion. This means that this separation is a function of a first axis difference combined with that of other axes. This suggests that, in some complex way, size may be implicated in the separation.

The variables proceed, from above downwards, from an ilio-pubic comparison, through a pubo-pelvic ratio, an ischio-pelvic index, an ischio-pubic ratio, to a second ischio-pubic relationship. In other words, although these variables are aligned into three separate factors, there *is* an overall relationship among them that places the entire suite together as measures of each of the three arms of this tri-radiate bone. Not a single one of the femoral measurements is included here, nor any of the measurements of the smaller pelvic details such as the individual variables on the sacral articulation or the acetabulum.

Is any explanation possible for this arrangement? It is of interest that the linear arrangement that encloses all three of these groups of variables seems to relate to the overall proportions of the pelvis. Pelvic dimensions that are not included here measure only restricted parts, for instance, the relative size of the sacral articular area or the relative size of the greater trochanter. The linear arrangement of these major variables is a vertical displacement of the curves (Fig. 9.16); this relates only to differences in the first factor though these differences do interact with those in other factors; and this, in turn, suggests that the arrangement may relate to overall size.

Certainly, as we view the results of the study on the hip and thigh (Chapter 5), it is clear that size plays some part in the separations of the animals even though it is also clear that size is not implicated in the same way for each group of animals. The above result may be displaying for us some of that complexity: that three different groups of variables show different clusterings in their arrangements while yet sharing amongst themselves an arrangement that may be in part due to size.

Further groupings of variables. Of course, in this extensive study there are far more than the five factors just described. A sixth factor represents mainly the shape of the femoral head, how spherical it is, how much it departs from a sphere. A seventh factor relates principally to the features of the sacral articulation, its relative length and breadth. An eighth factor relates primarily to the form and length of the contact between the two pubic bones. A ninth is associated very largely with the relative height of the greater trochanter of the femur. All of these are statistically significant factors, but in each case the individual features that are isolated are much smaller details of the hip and thigh complex than are the major features in the earlier factors.

Presumably it would be useful to make an even bigger study of the hip and thigh including an even wider range of variables, especially more measures of these smaller anatomical regions. It would then be easier to see if the features agglomerated into the blocks suggested by these anatomical details or if they were somehow related with other more anatomically distant variables. Pending such further studies, however, we must remain most pleased at the anatomical dissection that it has been possible to achieve with this 'fishing expedition'. In every case the results provide futher insight into structures; in no case are the anatomical assemblages nonsense.

Biometric 'dissection' of overall proportions of primates

All of the above investigations are of localized anatomical regions in which there is an obvious relationship between form and function, in which the separations of the animals are most related to the functions of the specific anatomical parts in contention (Chapters 5, 6 and 7). It is not overly surprising, therefore, that sub-units of the different anatomical regions are defined that also seem to be allied to function. It is true that this is most strong for the shoulder complex, but it does indeed exist in every one of the regions studied here.

But, in the material of Chapter 8 on studies of how overall bodily proportions separate the animals, our findings relate not to functional differences between animals but to overall relationships between animals. It is therefore with especial interest that we await the application of this type of analysis to those data. Can we expect, even if the overall study provides groupings of animals that seem to be taxonomic, that the clustering of the variables will still be functional? Or will we find some different clustering of variables with other biological implications? Or none at all? Let us proceed.

Factor (principal components) analysis and Andrews' high-dimensional display have therefore been carried out on Professor Schultz's data (outlined briefly in Chapter 8) on the overall proportions of the primates. It is a fairly complex situation with 23 variables taken on each of 472 specimens of primates representing 34 primate genera. Again, we remind ourselves that, of all the data sets, this is the one most deficient through containing a rather larger number of smaller samples than we would like. However, the various prior studies outlined in Chapter 3 have indicated that, at the scale of separation existing among genera, these small samples do not seem to be a major problem.

The clusters of variables. The various clusters of variables are shown in table 9.5.

The first cluster (factor 1) comprises all the transverse dimensions of the trunk. These are the same ones responsible for the major part of the separations between the sexes of the primates (Chapter 8). The remarkable similarity in these variables is shown by the very narrow envelope in Fig. 9.17.

The second cluster (factor 2) involves variables that define the form and proportions of the upper limb (Fig. 9.18). In this case some of the variables are contained within a fairly narrow envelope; others are somewhat outlying as indicated by dotted lines in the figure. Nevertheless, it is apparent that there is a common factor structure among them.

The third cluster (factor 3 – Fig. 9.19) comprises a very narrow envelope that contains all of the variables resulting from the form and proportions of the face alone.

This anatomical relationship is to be contrasted with a fourth cluster of variables (but, again, not a factor) defining the rest of the head (Fig. 9.20). Face and cranium are evidently clearly demarcated clusters of variables.

The fifth cluster (factor 4) is a very narrow envelope containing the variables that define the hand and foot together (Fig. 9.21).

This grouping of hands and feet is contrasted with a sixth cluster of variables (but, again, not a single factor) that comprises both upper and lower limbs taken together (Fig. 9.22).

Interpretations. Once again, the groupings of the variables make considerable anatomical sense. Faces separately from crania, upper limbs separately from lower limbs, transverse measures separately from longitudinal ones are all reasonable anatomical units. And combinations of hands with feet and upper limbs with lower limbs are reasonable anatomical interactions.

Nowhere among the clusters of variables do we find groupings or interactions that put a measure of, say, the ear together with measures of the thigh, forearm and big toe; nowhere do we find a combination as peculiar as, say, a transverse measure of the chest, a longitudinal measure of the thumb and a ratio involving both upper and lower limbs; nowhere do we find nonsense groupings such as these. Our competent anatomist would continue to derive satisaction from saying, 'I told you so.' And we, too, continue to derive satisfaction from finding that the results of our 'fishing expedition' make considerable anatomical sense even though they are somewhat constrained because Schultz's set of variables provides only a few rather coarse measures for any given region.

A test of the study. Another way of testing these particular combinations of dimensions is to compare them with the equivalent subset of the data of Jouffroy and Lessertisseur, chosen so that they mimic, as far as possible, the variables taken by Professor Schultz.

These more restricted data, when investigated using factor analysis and Andrews' high-dimensional display, also reveal several clusterings of variables. A first involves upper limb variables alone, much as the second factor for Schultz's data in Fig. 9.18. A second comprises only hand and foot variables, taken together, similar to the fourth factor in Schultz's data as in Fig. 9.21. A third contains combinations of variables of upper and lower limbs, and this replicates the group of variables in Fig. 9.22 of Schultz's data. Table 9.6 defines the groupings of variables in each study indicating especially the congruences between the two. Such a concordance between these two different data sets, taken by different investigators on different specimens, indicates most strongly that there is something very real in the particular combinations of variables obtained.

Table 9.5

FACTOR ANALYSIS OF SCHULTZ'S DATA ON OVERALL PROPORTIONS: CLUSTERS OF VARIABLES

Clusters	Anatomical features	Overall description
Factor 1	Relative chest circumference Relative shoulder breadth Relative hip breadth Chest index	Transverse dimensions of trunk
Factor 2	Relative upper limb length Intermembral index Brachial index Relative hand length Relative thumb length Relative hand breadth	Upper limb variables
Factor 3	Relative face height Relative upper face height Interocular index	Face variables
Common factor pattern	Relative head size Cephalic index Relative head diameter	Head variables
Factor 4	Relative hand length Relative foot length Relative thumb length	Hand and foot variables
Common factor pattern	Intermembral index Brachial index Crural index Relative lower limb length	Upper and lower limb variables

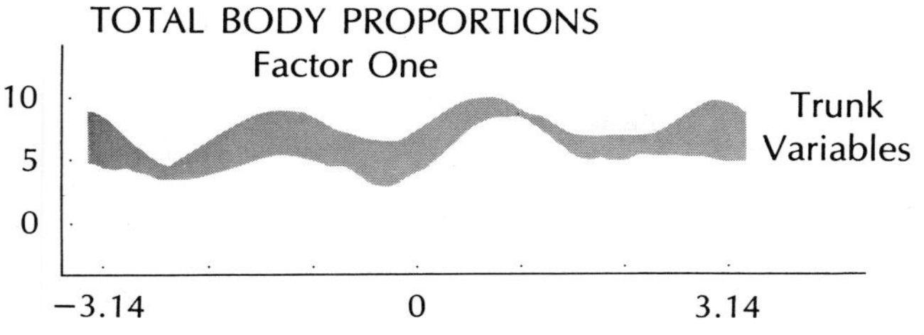

Fig. 9.17. Study of the overall proportions of primates. High-dimensional display shows that trunk dimensions taken together make a marked contribution in factor 1.

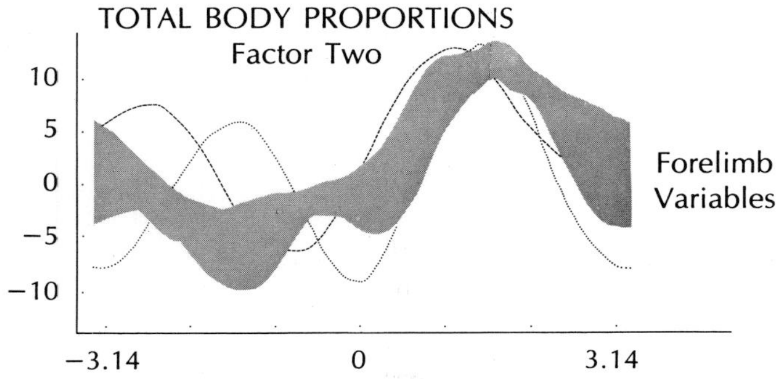

Fig. 9.18. Study of the overall proportions of primates. Upper limb variables all fall very close together in the high-dimensional display through similarities in factor 2.

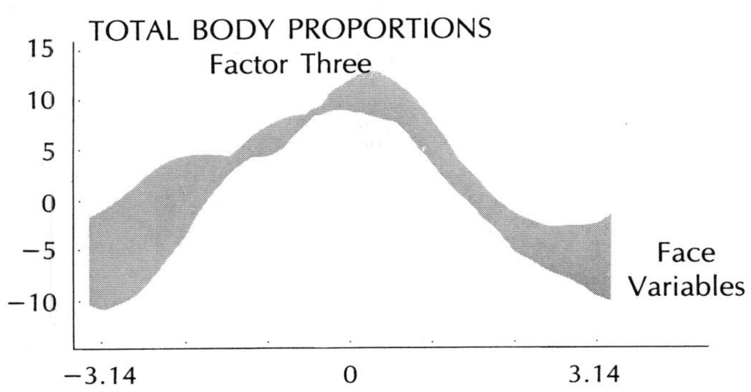

Fig. 9.19. Study of the overall proportions of the primates. Variables defining the form and proportions of the face are clustered by high-dimensional display through their combination in factor 3.

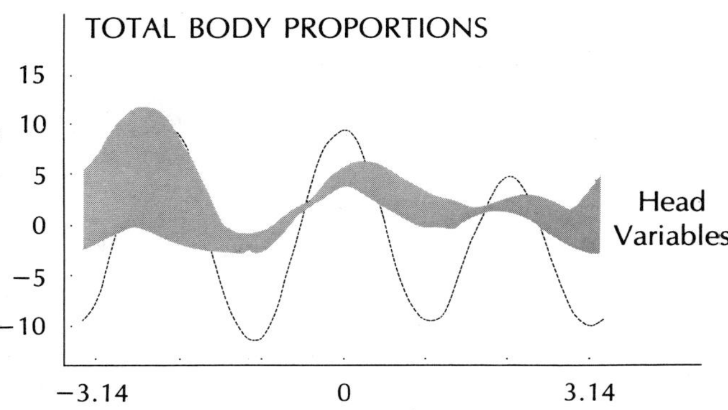

Fig. 9.20. Study of the overall proportions of the primates. Variables defining the form and proportions of the skull are revealed by high-dimensional display but not because of their combination in a single factor axis; rather is this through each possessing a similar pattern in each of several factor axes. Contrast with Fig. 9.19.

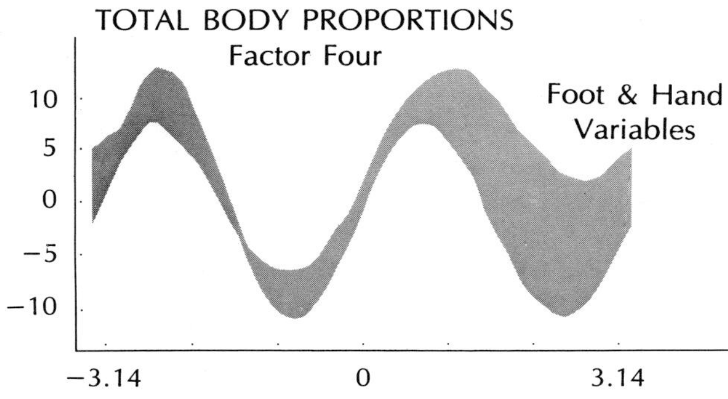

Fig. 9.21. Study of the overall proportions of the primates. Variables defining the form and proportions of the hand and foot taken together appear in the high-dimensional display; these are grouped in a single factor axis – four.

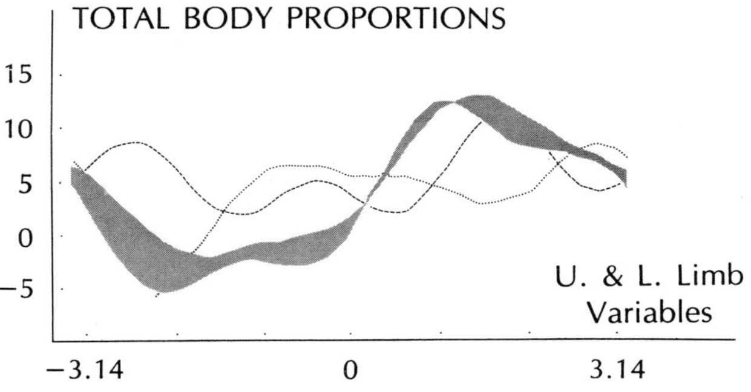

Fig. 9.22. Study of the overall proportions of the primates. Variables defining the form and proportions of upper and lower limbs taken together are demonstrated by the high-dimensional display. As with Fig. 9.20, the clustering of variables results not from a single factor axis but from a common pattern of contributions in several factor axes.

Table 9.6

COMPARISONS BETWEEN EQUIVALENT FACTOR ANALYSES OF SCHULTZ'S AND JOUFFROY AND LESSERTISSEUR'S DATA

Schultz's data: groups of variables	Jouffroy and Lessertisseur's restricted data: groups of variables
Trunk measures: figure 9.17	*
Upper limb measures: figure 9.18	Upper limb measures
Face: figure 9.19	*
Foot and hand together: figure 9.21	Foot and hand together: factor 4
Head measures: figure 9.20	*
Upper and lower limbs together: figure 9.22	Upper and lower limbs together: factor 1

* variables not available for study of Jouffroy and Lessertisseur

Biometric 'dissection' of prosimian proportions

However, the complete data of Jouffroy and Lessertisseur turn out to be far more interesting; they contain many more variables in each limb and, of course, have bigger samples of a wider representation of prosimian species and genera. In fact, the suite of dimensions is large enough that it is possible to examine each limb separately, as well as the entire assemblage taken together.

The main clusters of variables. Analysis of the upper limb variables produces three main factors (Fig. 9.23). The first consists of variables that represent every longitudinal segment of the upper limb. The second comprises variables that refer only to the fourth digit, measures of both metacarpals and phalanges for that digit. The third contains data from the hand alone but, in this case, for metacarpals and phalanges from the three middle digits. These are not unreasonable combinations of variables and we will discuss them further later.

Analysis of the lower limb reveals, in the same way, two main factors (Fig. 9.24). The first factor is very similar to factor 1 in the upper limb, comprising, as it does, variables from every longitudinal segment of the lower limb. The second factor is very similar to factor 3 in the upper limb because it includes measures on the phalanges and metatarsals of the three middle digits. There is no axis comprising measures on the fourth pedal digit alone and in this regard the lower limb differs from the upper limb.

The addition of these two anatomical regions is the most interesting. There are five major factors in addition to a number of highly statistically significant factors with single variable contributions.

The first factor includes many variables all lying in a very narrow envelope and all, therefore, very similar except as noted by the outlying dotted curves in Fig. 9.25. These variables include four measures of the lengths of the longitudinal segments of the upper limb, four measures similarly of the lower limb lengths, five measures of various rays in the hand and five measures of similar rays in the foot. This combination is very similar to the first factors of each of the upper and lower limb studies.

The second factor (Fig. 9.25) contains a group of variables that pertain only to digit four in the hand (like the second factor of the upper limb study).

The fourth factor (Fig. 9.26) contains a group of variables that make it very similar to both factor

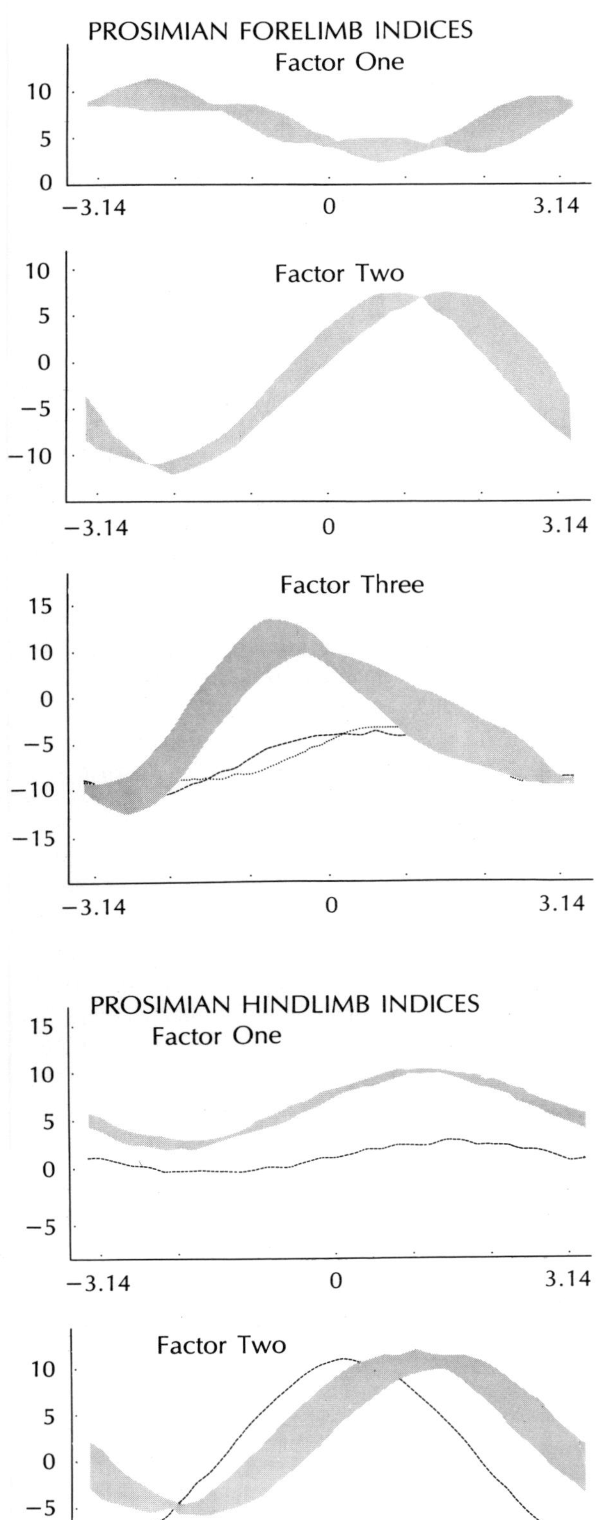

Fig 9.23. Study of the upper limb in the prosimians using the high-dimensional display. There are three main factors: the upper limb as a whole, each of the elements of the fourth digit and all of the elements of each of the middle three fingers (including palm bones).

◀

Fig. 9.25. Study of the upper and lower limbs of prosimians combined using the high-dimensional display. A first group of variables contains four measures of upper limb length, four of lower limb length, five measures of hand length and five of foot length. A second group of variables pertains only to the fourth finger. A third group of variables represents forearm and lower leg (shin), together with hand and foot.

▼

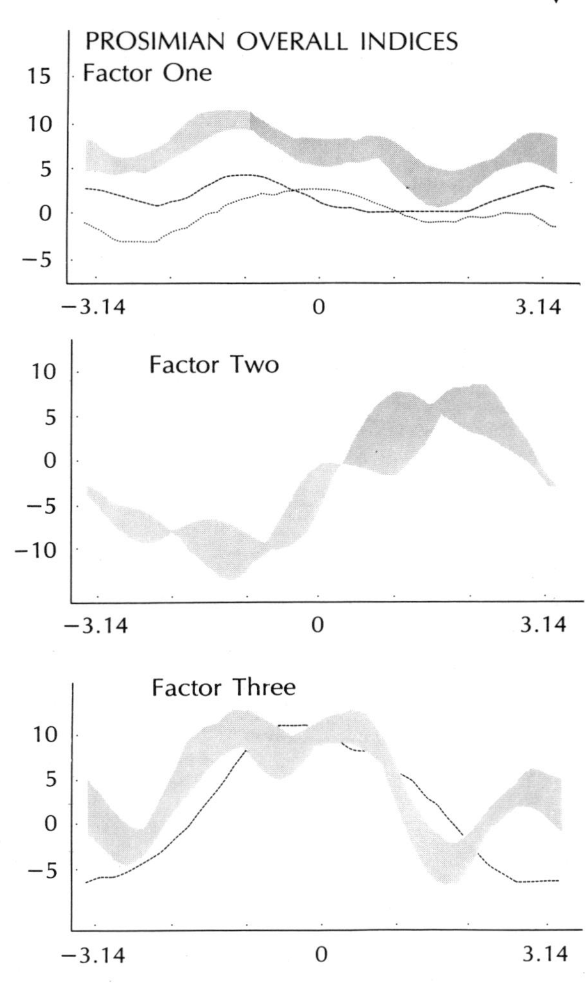

◀

Fig. 9.24. Study of the lower limb in the prosimians using the high-dimensional display. There are two main groups of variables: the lower limb as a whole and the middle three toes (including instep bones).

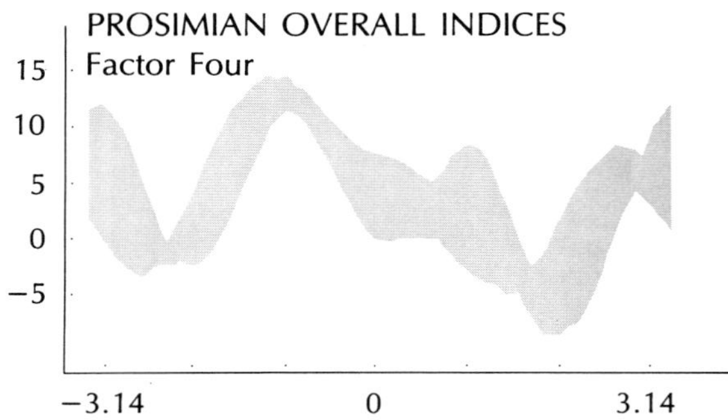

Fig. 9.26. Study of the upper and lower limbs of prosimians combined by high-dimensional display, continued. A fourth group of variables pertains to the three middle digits of both the hand and foot.

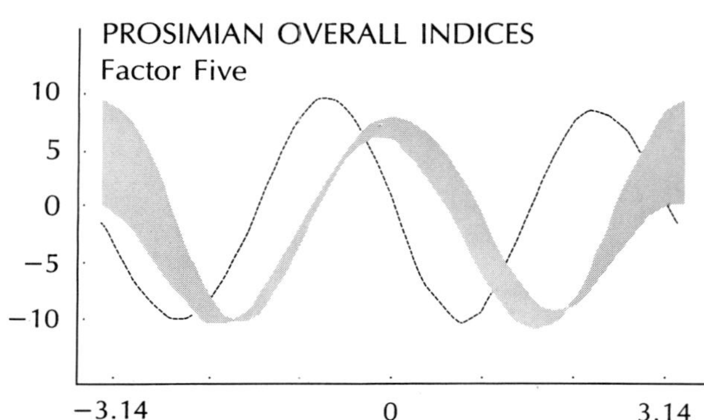

Fig. 9.27. Study of the upper and lower limbs of prosimians combined by high-dimensional display, continued. A fifth group of variables pertains to the upper arm and thigh, together with the hand and foot. Note the mirror relationship here with the variables in the third group of Fig. 9.25 (the forearm and lower leg together with the hand and foot).

3 of the upper limb and factor 2 of the lower limb: i.e. it contains variables defining the three middle digits of both hands and feet and including both fixed and free elements of the digits.

Factors 3 (Fig. 9.25) and 5 (Fig. 9.27) comprise variables from both upper and lower limbs. Factor 3 includes forearm and leg together with hand and foot; factor 5 contains arm and thigh together with hand and foot (a most interesting set of mirror relationships).

It is thus possible to see, in the entire study, (a) combinations of variables that entirely reflect those found in each anatomical region alone, (b) combinations of variables that add together equivalent variables from both upper and lower limbs, (c) combinations of variables that include interactions between alternate variables of the upper and lower limbs, together with (d) the special clustering based upon the variables of the fourth manual digit alone (table 9.7).

Possible interpretations. Our competent primate anatomist will have expected at least some of the groupings that have emerged. Perhaps, however, he will be somewhat excited about the second factor of both the upper limb and combined limb studies, because he will know that the fourth manual digit holds a special place in prosimian anatomy. It is the contribution of the fourth manual digit in the digital formula (which expresses the relative lengths of the digits) that is responsible for displaying a major difference between prosimians and anthropoids. In anthropoids, the third digit is the longest; in prosimians, the fourth. And this difference is not merely one of relative length: the long fourth digit of most prosimians is also a central functional element around which are arranged all the other structures of the hand such as the muscles and joints, and, functionally, the various postures and movements. In anthropoids it is the third manual digit that is so distinguished.

There is yet further information, however. Were we to be consulting competent embryologists and evolutionists, they too would recognize some of the groupings of variables that have emerged. For the first factor of the study combining both limbs includes variables proceeding proximo-distally down each limb; this may be reflecting something that our experts know exists during development: a proximo-distal developmental gradient down the limbs. And the fourth factor contains a cluster of variables that combines measures of the limbs taken cranio-caudally, from the pre-axial to the post-axial border of each limb, and evident, therefore, most clearly, in the hands and feet. This, our embryological experts would recognize, reflects the embryological segments associated with the development and evolution of mammalian limbs.

Finally, the third and fifth factors, reflecting specifically parallel and alternate combinations of arms with thighs, forearms with legs and hands with feet, would possibly be recognized by our embryologist-cum-evolutionist expert as mirroring something of that old concept, serial homology, the repetition of 'parts' in different 'segments' of the body.

Conclusions for statistical 'dissection'

The studies of Chapters 5, 6 and 7 provide fascin-

Table 9.7

FACTOR ANALYSIS OF DATA ON LIMB PROPORTIONS OF PROSIMIANS: CLUSTERS OF VARIABLES

Clusters	Anatomical features	Overall description
Factor 1	Eighteen variables: Four upper limb lengths Four lower limb lengths Five hand lengths Five foot lengths	Proximo-distal dimensions of the limb
Factor 2	Length of fourth metacarpal Lengths of fourth phalanges Length of fourth digit	Variables pertaining to digit four
Factor 3	Length of forearm Length of leg (shin) Length of hand Length of foot	Mirror image serial elements (see factor 5)
Factor 4	Lengths of middle metacarpals Lengths of middle metatarsals Middle phalanges of hand Middle phalanges of foot	Pre-axial to post-axial in hand and foot
Factor 5	Length of upper arm Length of thigh Length of hand Length of foot	Mirror image serial elements (see factor 3)

ating information as to how localized anatomical regions 'arrange' the primates. In every case our studies have shown that the basis of the arrangements seems to be the localized functions of those anatomical regions in the particular animals under study. In total contrast, the investigations of Chapter 8 have analysed these data in order to understand how entire anatomies arrange the primates. In this case the basis of the arrangements seems to be the evolutionary relationships of the entire order.

The studies of this chapter have investigated a series of reverse matters: how the different primate groups arrange the variables characterizing the anatomies.

We have found that localized anatomical regions consist of groups of variables, anatomical sub-units if you like, that make good sense when viewed in the light of the functions of the particular anatomical regions under study.

And we have also found that whole anatomies are equally easily broken down into sub-units; in this case the sub-units make good sense when viewed in the light of developmental and evolutionary phenomena such as developmental gradients, segmental origins and serial homologies.

Certainly, none of these groupings of variables makes arrant nonsense. Whatever may be the processes underlying their existence, they are highly unlikely to have been the result of chance alone. Our 'morphological fishing expeditions' have yielded information not at all surprising to anyone well-versed in anatomical, functional, developmental and evolutionary concepts of the vertebrates.

These findings have an important implication that relates to animal conservation. The principal way of discovering anatomical sub-units, with respect to function, development and so on, is through experimental investigations. Such studies are vitally important and must be done if we are to truly understand modern approaches to animal morphology. But they cannot be performed on a profusion of rare animals; they can only be easily undertaken on common laboratory animals. The methods described above can produce the equivalent information for endangered species by using materials already in museums without the need for further reductions of already precariously situated animals. Experimental work on common laboratory animals will always be required because that is the only way that the inferences of morphometrics can be corroborated.

Of course, the particular studies detailed above are not ideal. In the main, that is because the data were not collected with all these ideas in mind. Given, now, that we are sensitised to the additional possibilities, it seems that new studies, with variables specifically chosen to elucidate historical ideas as well as functional ones, may permit even clearer answers. This is a most exciting part of the challenge for a new generation of morphologists.

Summary. We have seen, from the analyses of this chapter, that sensible anatomical sub-units are clearly definable in every study so far undertaken, as judged by the clustering of the anatomical dimensions. We first look at anatomical regions: those parts of the analyses that seem to arrange the animals on functional bases.

The case that has been most completely studied and is clearly understood is the primate shoulder. Here, the clusters of variables fit very well with the functional interpretations that stem from the clusterings of animals observed in prior chapters. The clusters of variables make marked functional sense. The examination of the other anatomical regions: in order of anatomical complexity, the talus, the arm and forearm and the hip and thigh are next examined. In each case, clusters of variables are defined, and in each case it is not difficult to see the functional underpinnings of the particular clusters. Certainly, not a single cluster of variables is found that is a nonsense combination of variables.

We next look at those studies that combine large anatomical regions: upper and lower limbs, trunk, and head and neck. These are the data that, when we use them to arrange the animals do not provide functional information but rather information about primate relationships.

Again, sensible clusterings of anatomical dimensions are obtained. But these clusters seem to relate less to direct functional concepts (function within locomotion, that is) and more to phenomena like proximo-distal gradients, medio-lateral arrays and serially homologous elements. These are exactly the types of anatomical groupings that we might expect to see in data pointing to the overall relationships of animals.

Such anatomical dissection provides a powerful corroboration of the animal dissection that we have already carried out. It has major implications for studies of fossils and provides impetus for a wholly new set of investigations of the Order.

CHAPTER 10

Human Fossils – The New Revolution – Revisited

Conventional studies of australopithecines – New studies of australopithecines
First question – corroboration, or otherwise, by old methods
Second question – meld, or otherwise, of old data
Third question – new fossils, conventional methods
Fourth question – acceptance or otherwise of new views
Australopithecines (humans and locomotion – hominoids and systematics
investigators and convention) – Broader implications for human evolution.

The preceding studies have yielded a great deal of information about the living primates in their own right. Indeed, that is the main reason for their being undertaken. In addition, however, such a body of knowledge is an indispensable background to anyone wishing to study fossil primates by these same methods. That background can especially be used to study the history of the primates whenever reasonably large collections of fossils and large portions of the entire skeleton are available.

Happily, this is indeed the case for some parts of primate evolution. Some excellent materials exist: virtually whole skeletons of many specimens for *Notharctus*, for example, have been known for decades. Even for higher primates, relatively complete skeletons and several subjects are available, for instance, newly found fossils from the Afar Valley in Ethiopia.

In most cases, however, these methods have to deal with data with many missing parts. Though many specimens are known, say, for the ramapithecines as a group, the particular remnants are primarily confined to bits of jaws and teeth. And, in the case of, for example, the lower end of the humerus from Kanapoi, a single specimen of a tiny anatomical part, the elbow, is all that we have. Yet even with these limited pieces, the new methods for investigating structure, together with the regular observational techniques, can produce useful insights.

It is, therefore, not unreasonable to return to the material of the first chapter and see how these new tools, chiefly of course, multivariate morphometrics, are making impacts on our ideas about human evolution.

Conventional studies of australopithecines

For some time after their first discovery, there was a great deal of controversy about those fossils from Southern Africa and Olduvai that were ultimately designated australopithecines. Some workers assessed them as more related to the living apes, others as more closely linked with man. For many years now, however, the general consensus has been that these fossils are very close to the human lineage and that particular sub-groups, such as the gracile species or individual specimens like so-called *Homo habilis*, are direct human ancestors. This is the conventional idea of the single lineage indicated in the first chapter of this book.

Of course, the matter has never been as simple as that. Many different lineages have indeed been proposed over many years. Many different patterns of human evolution thus exist in the literature. But, in the main, they all conform to the notion of a single principal lineage. It is just that, at particular times, different specimens have been made to occupy that allegedly key position, the root and direct lineal ancestor of man.

Piltdown man, for example, when first discovered and for many years afterwards, was the source of the most twisted evolutionary contortions by investigators attempting to find this special place for that fossil. The discovery that it was a fraud buried those efforts.

The Taung child from Southern Africa played a similar role in many lineages developed after its discovery filtered into the consciousness of anthropologists. In time, however, this fossil too ceased being placed in the key position.

The even later discoveries of australopithecines at Olduvai, first *A. africanus* and later those particular specimens labelled by some *Homo habilis*, became, in their turn, the main focus of the human lineage, only to be displaced by yet newer finds.

And this process of discovering 'the missing link' is not finished; today the discoverers of the Afar specimens have purported evolutionary lineages that place those specimens as the ones critical to the human line. (Of course, the very concept of the missing link is non-biological; there is no such thing as the missing link; or there are thousands of missing links).

The major part of the evidence for the above views has come from several centuries of study of skulls, jaws and teeth. It depends upon the position that these particular anatomical parts provide special information relating fairly directly to the phylogenetic relationships of the primates.

It is only in more recent decades that information has started to flow from studies of post-cranial fragments. And it is becoming more generally recognized that such studies speak much less directly to phylogenetic position, though they have been interpreted in this way in the past. Most workers now believe that study of post-cranial fragments tells much more about the behaviours that shape, in both evolutionary and developmental senses, these particular anatomical parts. Study of post-cranial regions has therefore allowed assessments to be made about the possible behaviours of the australopithecines, although it is true that some of the post-cranial investigations have been read as confirming phylogenetic speculations advanced on the basis of the cranial and dental evidence.

Thus, earlier studies of the australopithecine pelvis found at Sterkfontein have been used to suggest that this is a pelvis involved in human bipedal locomotion. And, although there is some slight disagreement as to precisely how similar to human locomotion that may have been, there is general agreement that what we are assessing here is indeed a pelvis leading to the evolution of human bipedality (e.g. Clark, 1959). Many subsequent studies of the pelvis have been read as confirming that view (e.g. Lovejoy, Heiple and Burstein, 1973).

Similarly, initial investigations of a series of hand bones from Olduvai gorge seemed to suggest that the hand of this australopithecine is essentially similar to the human hand (e.g. Napier, 1962). Even now, some investigators believe that this is so (e.g. Day, 1977).

Similarly, again, a major part of the foot of *Homo habilis* discovered at Olduvai has been examined, rearticulated, reconstructed and pronounced the foot of a creature that is bipedal in the manner of man (Day and Wood, 1968). It is supposed even today that this is the foot of a human ancestor (Day, 1976, 1977).

Thus, we can see that, although the method of study of post-cranial fragments has been sometimes somewhat different from that of skulls and jaws, the overall opinion remains the same: that these creatures are closely involved in human ancestry and that some of them, the more gracile forms and especially the australopithecine that has been called by some *Homo habilis*, are directly ancestral to humans.

Many of these studies have been carried out using humans as models for comparison; the idea that the relationship is to say human bipedality is an already assumed underlying notion that the studies purport to support. But the fossil fragments may also be viewed against the background of the diversity of locomotor patterns that are known to exist among non-human primates. For it is likely that at many times during the evolution of the primates there have been species variously adapted for quadrupedal activities: running, leaping and climbing in the trees. And it is equally likely that at many times additional specializations with emphasis on the lower limb (e.g. such as is involved in extensive leaping: bush-babies, colobus monkeys) and on the upper limb (e.g. in acrobatic climbing: woolly monkeys, orang-utans) have broadened the basically quadrupedal primates into the diversity of complex arrangements described in this book. It must be from one part or another of this locomotor diversity that human bipedalism orginally evolved.

The question, then, that we are now interested in asking about any post-cranial fragment of an australopithecine, stems from the widely comparative discussion of post-cranial parts of living primates on prior pages. It is the following: to what extent does the morphology of a particular fragment indicate (a) close functional similarity with the corresponding part of extant man (i.e. similarly uniquely bipedal), or (b) some functional similarity with the equivalent part of any of a wide range of essentially quadrupedal forms as represented by the living non-human primates, or yet (c) a pattern of functional associations that point to intermediacy between man and one or another part of the non-human spectrum, or even, a question much less frequently posed, (d) a pattern of functional associations unique unto itself? Answers to any part of this complex question provide useful and interesting data for primate and human evolution, although for the australopithecines we must bear in mind that the first possibility is generally believed to be the case.

New studies of australopithecines

The results of applying some of the new methods to this particular group of fossils is now suggesting a picture that is rather different from the conventional. The new studies do not totally deny the conventional story. The australopithecines have many human resemblances that not only imply *some* genetic relationship but also betoken functional characteristics, such as *some* form of bipedality for instance, and *some* manipulative abilities for example.

But the new studies also show many features of australopithecine fragments reminiscent of those found in some of the arboreal apes and monkeys. They reflect quadrupedal and arboreal activities such as are not possessed by humans today. Most importantly of all, many of these studies are providing information about individual features or combinations of features that actually render the australopithecines markedly different from all living hominoids, humans and apes alike. These differences are often every bit as great, indeed often much greater, than those already existing between humans and African apes (e.g. as summarized in Oxnard, 1975a and b, and table 10.1). It is in this sense that the australopithecines may be

Table 10.1

DISCORDANCE BETWEEN CONVENTIONAL AND MULTIVARIATE MORPHOMETRIC VIEWS OF VARIOUS ANATOMICAL PARTS OF AUSTRALOPITHECINES

GRACILE AUSTRALOPITHECINES (including *Homo habilis*)

	Conventional view	Morphometric view	
	Position on human-African ape axis		
Anatomical part	Human ... African ape	Arboreal ape	Unique
Shoulder blade	.......*............................	*	
Clavicle	.......*............................	*	
Arm bone	..........*..........................		*
Elbow	..............*......................		*
Finger bones	..........*..........................	*	
Pelvis	....*................................		*
Ankle bone	.....*...............................	*	
Foot arches	.*..................................	*	
Toe phalanx	.*..................................		*

ROBUST AUSTRALOPITHECINES

	Conventional view	Morphometric view	
	Position on human-African ape axis		
Anatomical part	Human ... African ape	Arboreal ape	Unique
Arm bone	..........*..........................		*
Elbow	.............*.......................		*
Palm bone	..........*..........................		*
Ankle bone	.....*...............................	*	

characterized as unique among hominoids.

It is possible, of course, that such findings may be referring to some mode of behaviour of australopithecines that is totally different from anything that we see today or, indeed, have so far guessed. It is more likely, however, that the morphological distinctness of the australopithecines rests not in a totally new behaviour but rather in a combination of activities *no one of which is special by itself but of which the combination in a single animal is truly unique*. Thus, these creatures may have been capable of a form of bipedality far more efficient and habitual than that of any ape or monkey at the present day, combined with a form of quadrupedalism (perhaps including abilities in a climbing medium) far more sophisticated than that of which any human is capable. The various studies that have been performed seem to allow us to go into these matters in some detail. We can look at the same series of anatomical regions as before.

Though some new investigations of the Sterkfontein pelvis show features consonant with the conventional view, apparently representing adaptation to upright posture and movement, other features from that same investigation are clearly reminiscent of those of different ape pelves that may betoken various quadrupedal and climbing abilities (Zuckerman, Ashton, Flinn, Oxnard and Spence, 1973; Ashton, Flinn, Moore, Oxnard and Spence, 1981). In its totality, the picture from the new studies of the pelvis indicates a structure unique among hominoids (Fig. 10.1, and Oxnard, 1975a).

New investigations of some of the Olduvai and

Frame 1

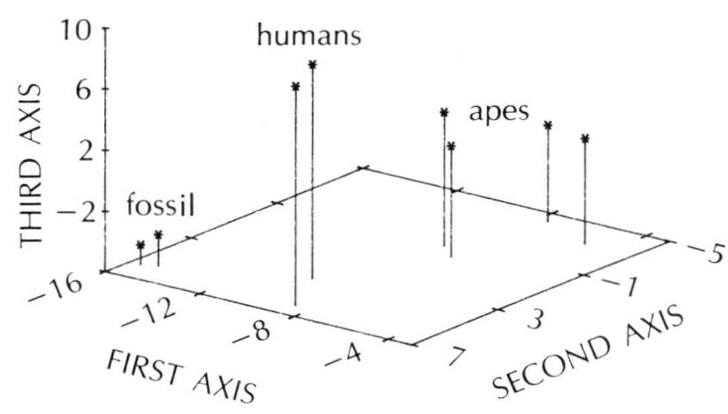

Fig. 10.1. Morphometric studies of the pelvis displayed using one three-dimensional and two high-dimensional plots (first three frames). The model shows that the fossils are not intermediate between humans and apes in the first three axes. The Andrews' high-dimensional curves confirm this for all canonical axes. The lower two curves in each high-dimensional plot represents two reconstructions of the fossil. The upper two curves in the second frame are the two African apes; the upper two in the third frame are modern and neanderthal man. This uniqueness of the australopithecine pelvis from Sterkfontein compared with humans and great apes concurs with the visual impression (fourth frame).

Frame 2

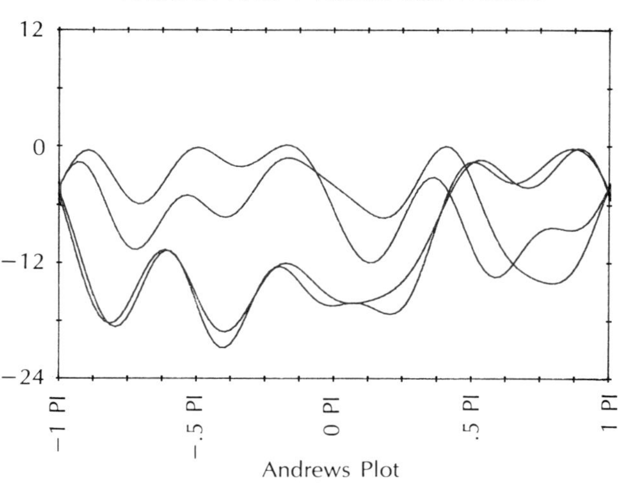

Frame 3

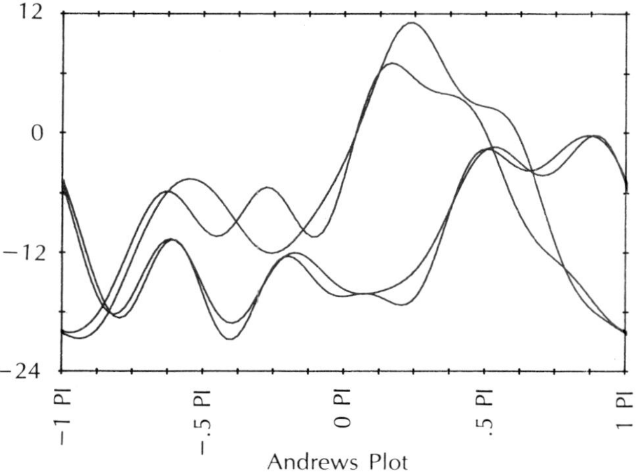

Frame 4

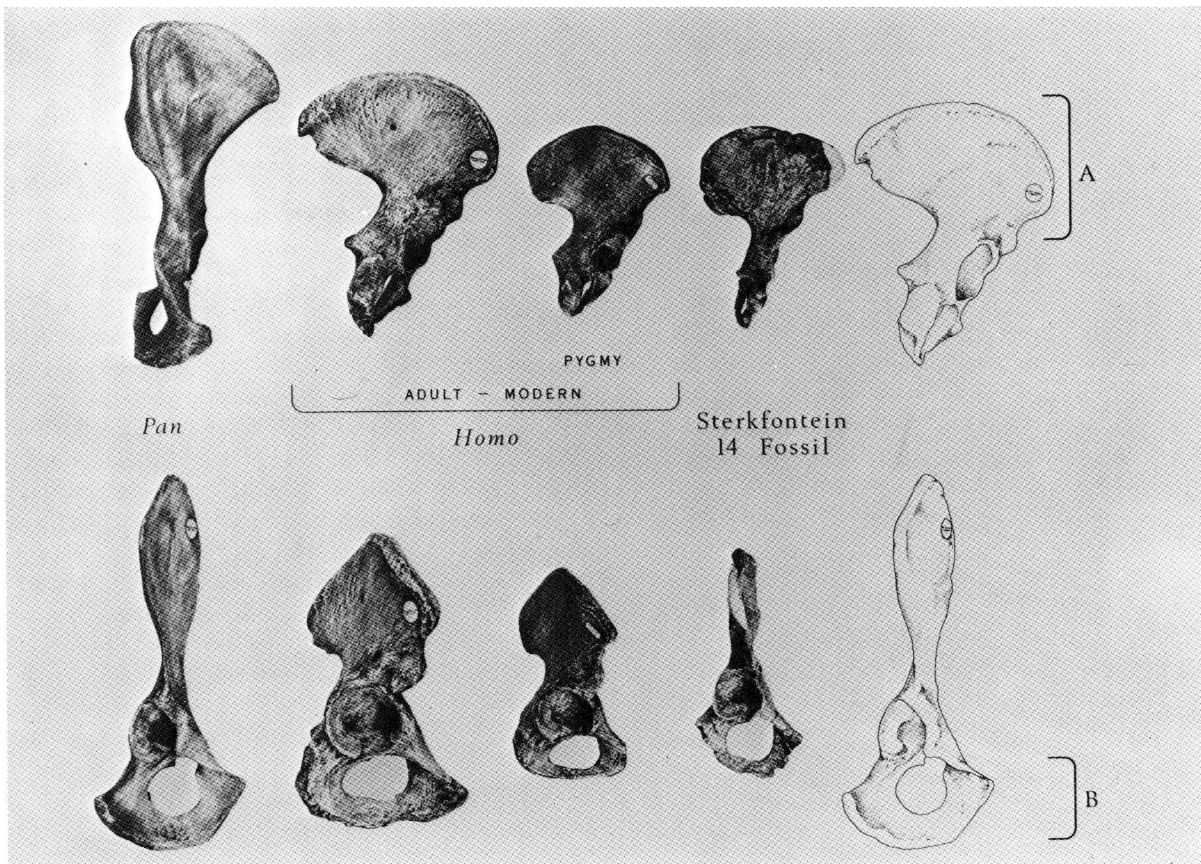

Southern African hand bones show a few features compatible with the conventional view that they are close to those of humans. But the major part of new morphometric and biomechanical investigations demonstrate that not only the robust but also the gracile specimens are actually quite different from those of humans (Figs. 10.2, 10.3, 10.4 and 10.5). Some of the evidence merely suggests that they are totally different from any living hominoids (Rightmire, 1972; Oxnard, 1975a, Fig. 10.2). Other evidence seems to relate to abilities for grasping with power reminiscent of what we find in the orang-utan (Figs. 10.3, 10.4 and Oxnard, 1973a), though it is necessary to exercise care here because these hand bones do not form a single assemblage. Indeed, some of the bones have already been removed from consideration as human ancestors in part because of these findings. Of those bones that now remain (Day, 1978) some are curved enough that they must have operated in this arboreal-grasping mode (Oxnard, 1973a, 1975a; Susman, 1979; Susman and Creel, 1979; Susman and Stern, 1979).

As a final example, new morphometric studies of

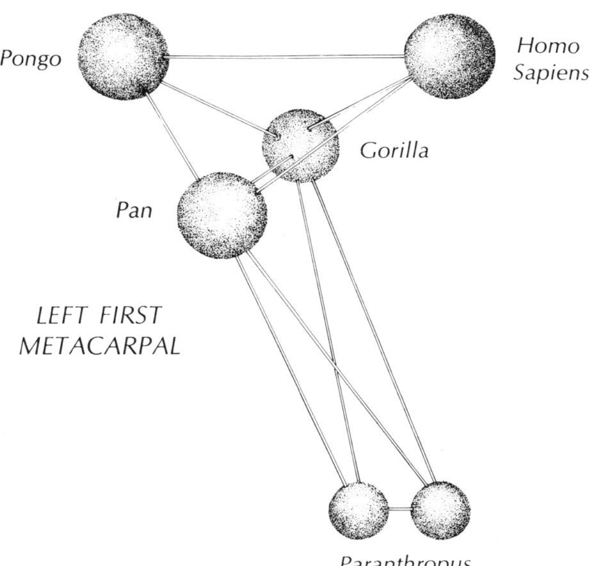

Fig. 10.2. A reinterpretation of the morphometric relationships of metacarpals of a robust australopithecine; it is clearly different from modern humans and great apes.

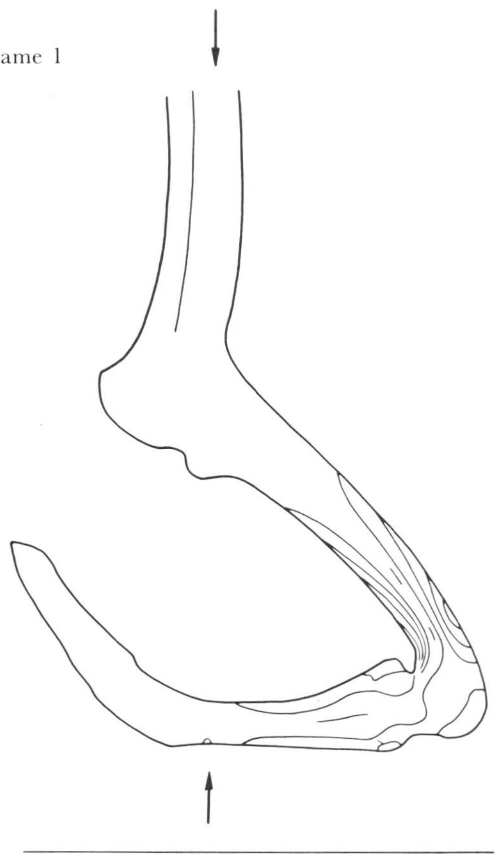

one of the foot bones, the talus or ankle bone, from both robust and gracile australopithecines, demonstrate how different they are from those of the present-day humans and African apes. These australopithecine ankle bones find their closest morphological analogues with other fossil ankle bones from Africa. Some of these are from those same eras, others from very much earlier times. The ankle bones from Olduvai and Kromdraai have also some similarities with the ankle bone of a living species that is principally arboreal — the orang-utan (Fig. 10.6 and Lisowski, Albrecht and Oxnard, 1974, 1976). Even the remaining bones of the Olduvai foot resemble those of man very much less than was once thought. For the foot bones of the fossil have now been articulated with one another in the light of information from dissections of modern apes and humans. When this is done, they display arches like those in modern day apes (Figs. 10.7 and 10.8, and Oxnard and Lisowski, 1979). Yet further evidence on this score comes from very detailed studies of the joints of the feet of a wide range of mammals (Lewis 1980a, b, c). Professor Lewis also shows some similarity be-

Fig. 10.3. An engineering analysis of a phalanx of an gracile australopithecine (bottom right). It is *inefficient* in a knuckle-walking simulation (containing very many isochromatic lines) as is the orang-utan (bottom left) but not, of course, the African great ape (top, containing many fewer isochromatic lines).

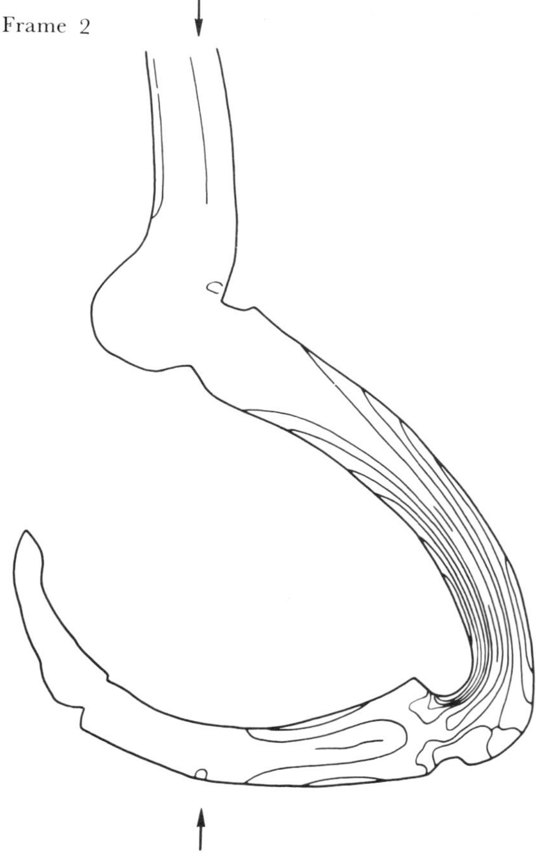

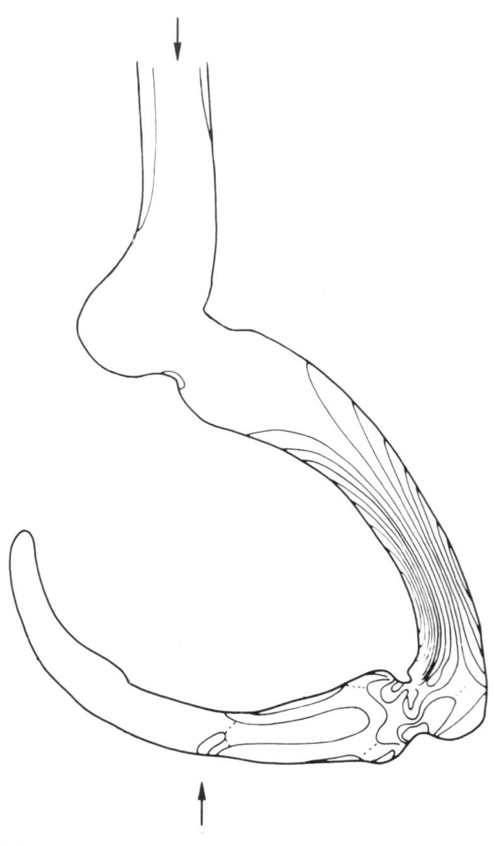

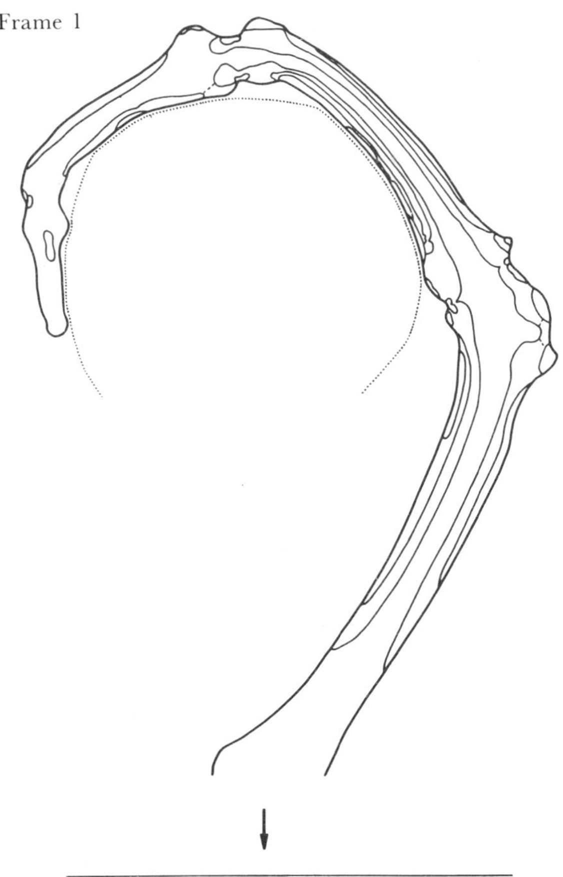

tween the Olduvai foot bones and those of living apes and monkeys. The fossil does not resemble humans at all (for overall review see Oxnard, 1980b).

Altogether now, we have carried out a number of new studies of post-cranial parts of these fossils. Some of them imply functional resemblances with primates that climb well (such as orang-utans and howler monkeys). Others indicate that australopithecine anatomies are totally novel. Both deny the conventional assessment as near human. An overall summary of these investigations was provided earlier in table 10.1. In addition, a number of other investigators using multivariate statistical methods are starting to recognize unique differences between australopithecine fossils and both humans and apes (e.g. the pelvis, McHenry and Corruccini, 1975; the femur, McHenry and Corruccini, 1978; the ulna, Feldesman, 1979).

One way to look at this problem is through multivariate statistical studies of all the data for the extant species combined as in the previous chapters. Thus, sets of 10 measurements could be taken upon each of possibly 20 anatomical regions. These could be measured upon each of 30 specimens that are available for each of, perhaps, 40

Fig. 10.4. Another part of the analysis of the phalanx of the fossil (bottom right). It is *efficient* (few isochromatics) in the hanging-climbing mode as also is the orang-utan (bottom left) but inefficient (many isochromatics) in the African ape (top).

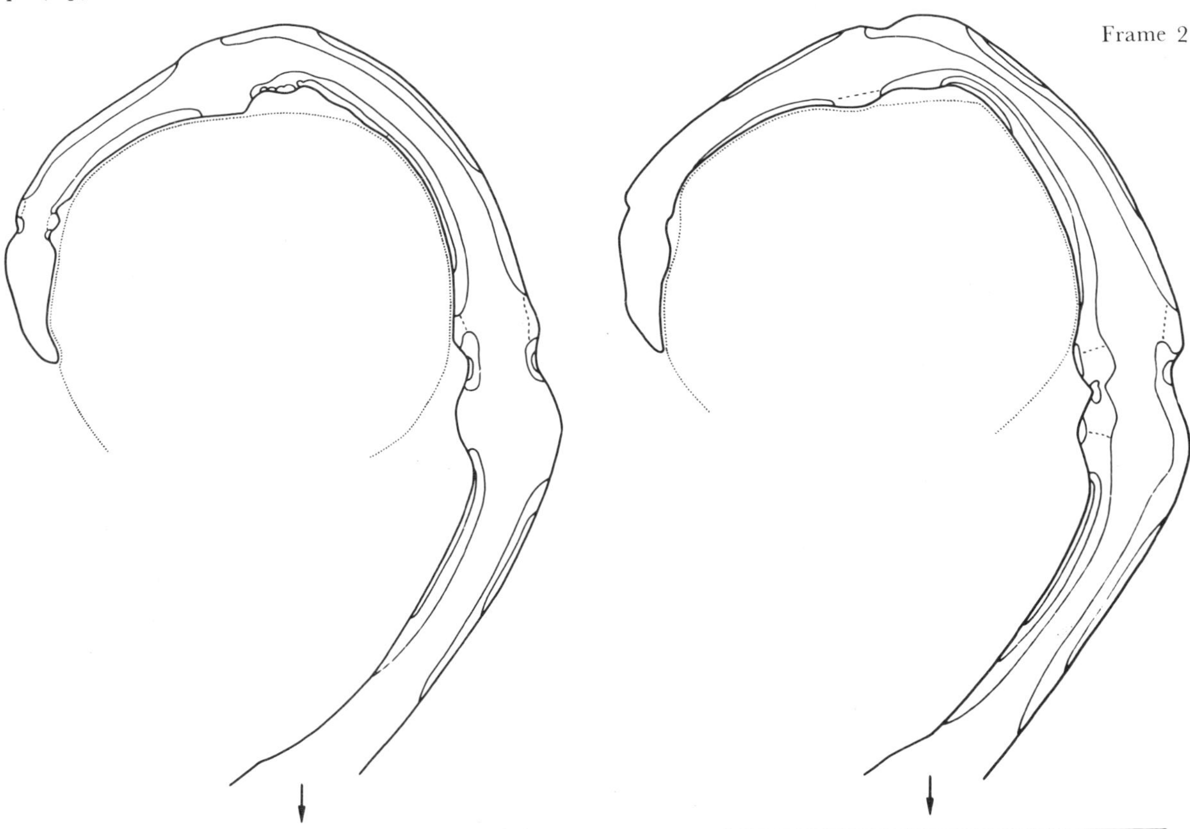

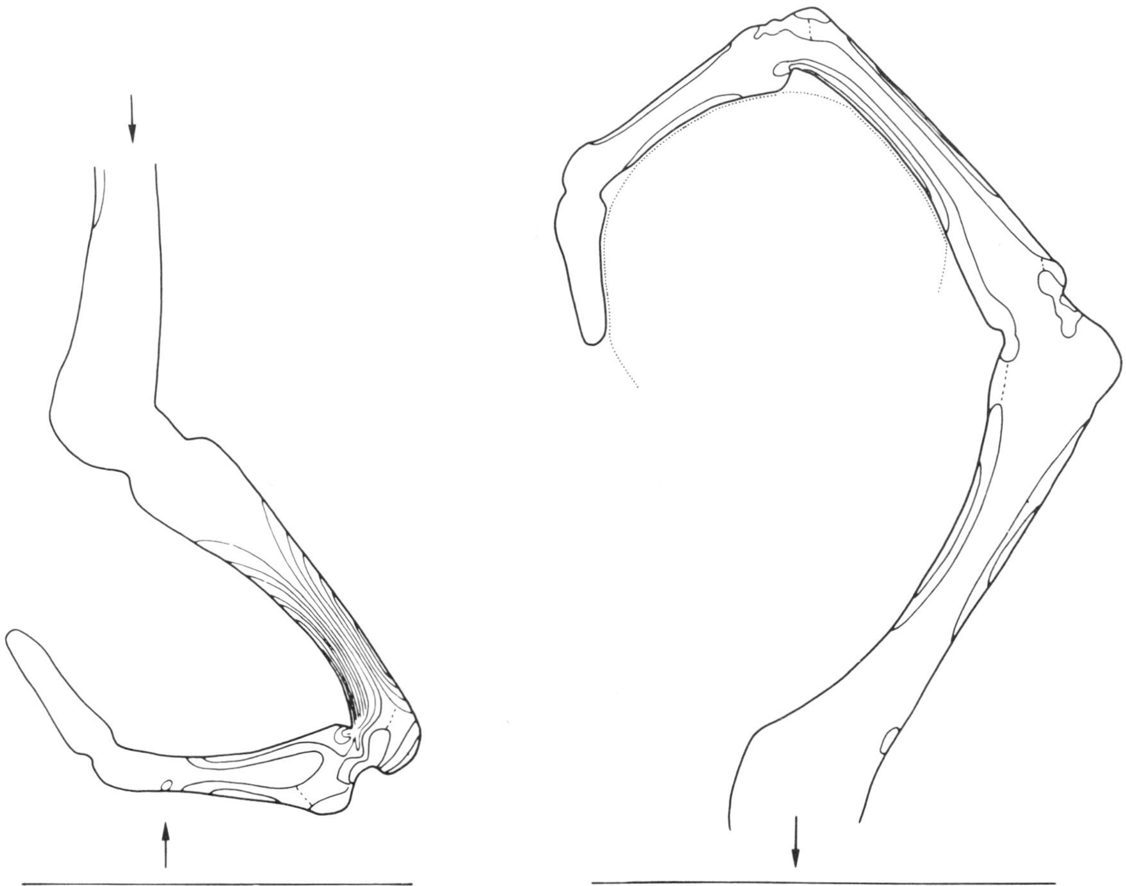

Fig. 10.5. The human phalanx is inefficient in both modes (although on a comparative basis, less so in hanging) and therefore also differs from the phalanx of the fossil.

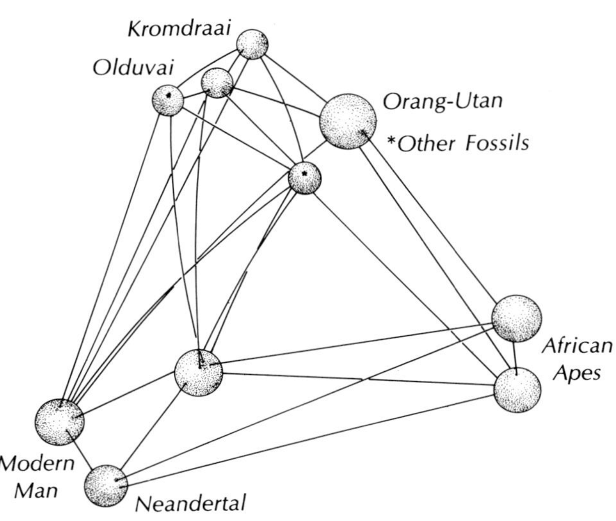

Fig. 10.6. A morphometric study of ankle bones of both gracile and robust species. They are similar to each other; they differ more from humans and African apes than these latter do from each other; they are closest, but not too much so, to the orang-utan.

genera of extant primates. A research plan such as this would certainly be a major study, yet in the near future it could be achieved.

Of course, when it comes to the matter of trying to interpolate the various fossil fragments, they would have to be combined as though they represented information about the approximate mean values for fossil groups, species or even genera. Inevitably it would be necessary to pool information from fossil remnants that might well, in truth, belong to separate forms. However, this limitation is not new and is with us all the time. It is a limitation that applies to any study of primate fossils, whether involving mathematical analyses or simple observation. The finds that have been made in the Afar Valley, where more than a dozen individuals seem to be represented, are a major hope for getting around this particular problem.

For the moment, therefore, we must look for other ways to test the new findings. Testing is especially important, for many of the new ideas

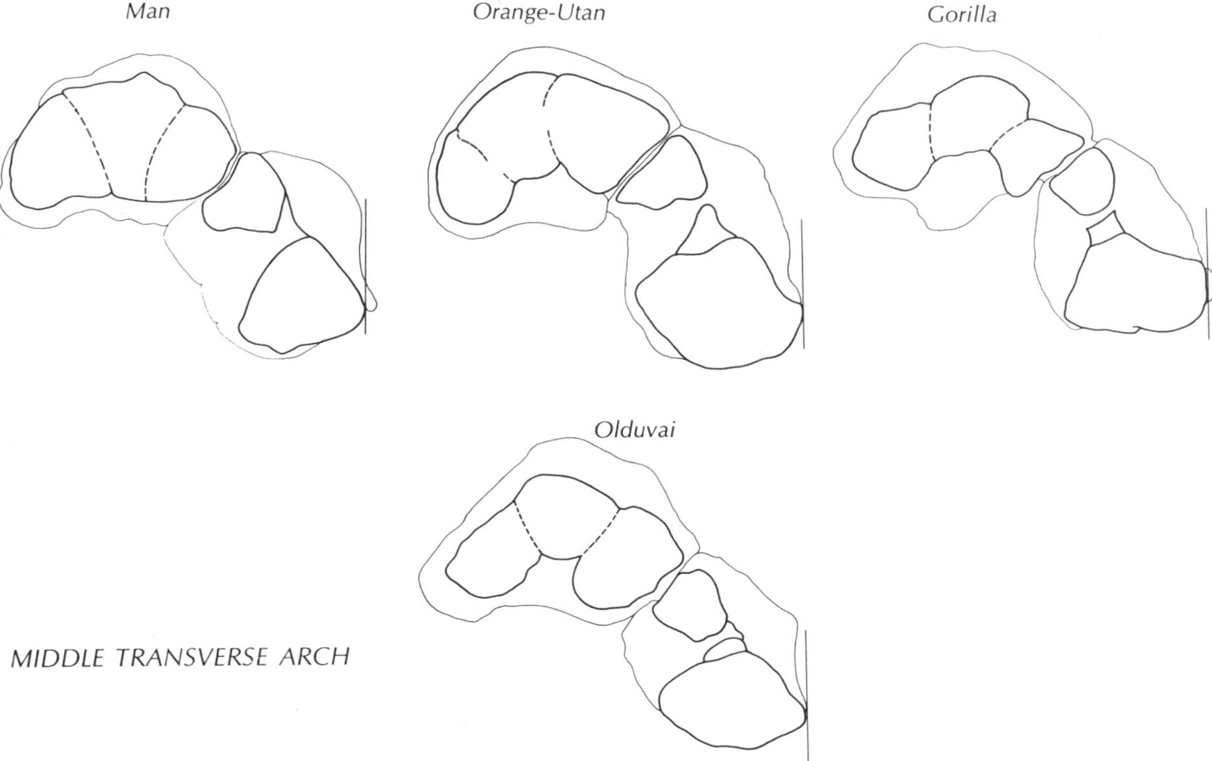

Fig. 10.7. Re-articulations of bones in primate feet. A transverse view across the arch of the foot near to the ankle. When articulated correctly, the fossil resembles various apes.

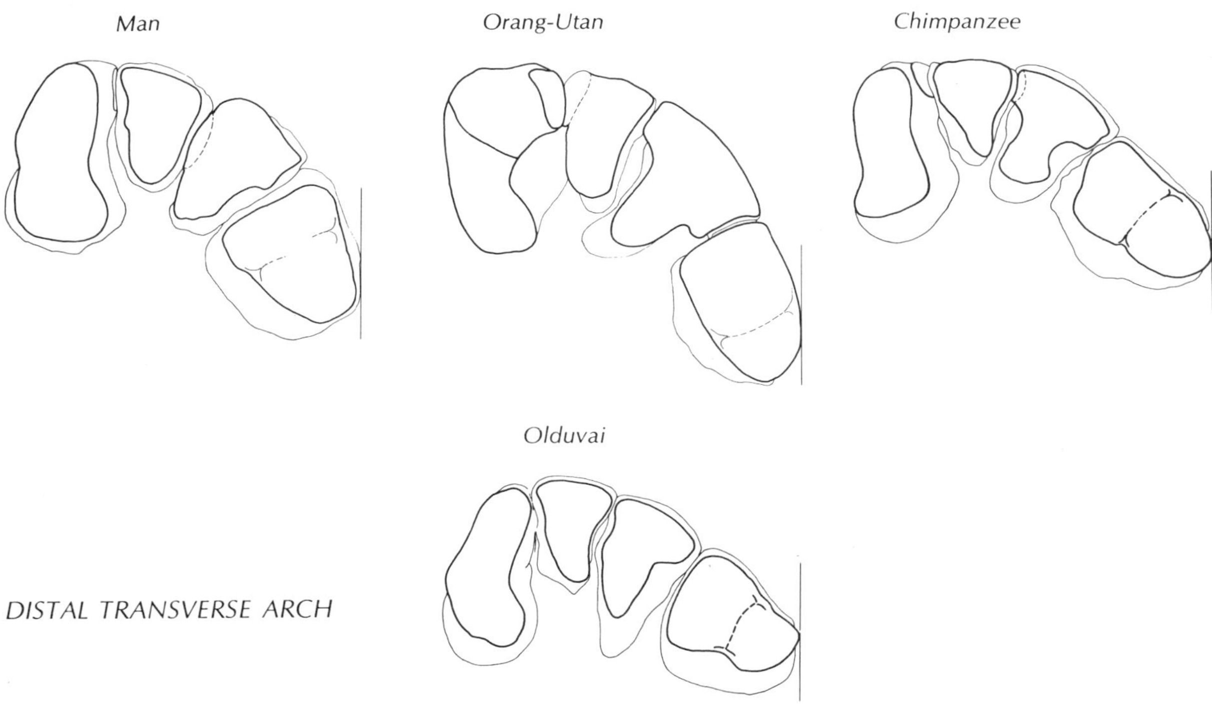

Fig. 10.8. Re-articulation of bones in primate feet. A transverse view across the arch of the foot nearer to the toes. When articulated correctly, the fossil again resembles some of the apes.

carry little weight with conventional anthropologists (a) because they are expressed through a relatively new technology, (b) because they are controversial to those who do not accept the findings and (c) because they are not readily accessible to those who do not themselves utilize these techniques.

Rapid advances sometimes occur in areas of science when totally new data suddenly open up new realms of insight and knowledge. During such phases, the new data may be so voluminous that, rather than the new being fitted into the picture provided by the old, we may require that the old be interpolated into the framework suggested by the new. Likewise, at such times, the use of new techniques may provide a view, even of the old information, that results in a major reassessment of our entire understanding. When either of these situations occurs, vigorous controversy may be engendered and considerable time may pass before the subject settles down into a new mode. The study of the place of the Olduvai and Sterkfontein hominids in human evolution is undergoing just such a period of saltatory change. In this case both the new methods and the new data have arrived almost together.

As a result we must look to four matters. *First*: do the older methods, in this case visual and mental assessment, provide any evaluations that, with hindsight, support the new possibility? *Second*: do the older data, in this case studies of fossil parts over the last half century or more, provide any information that, with hindsight, supports the new picture? *Third*: do new data studied by old methods give any credence to the new view? And *fourth* and finally, and perhaps most importantly: how are the new views being assimilated by the profession in general?

First question: corroboration, or otherwise, by old methods

We have therefore made attempts to provide appropriate visual comparisons of some individual bones (Oxnard, 1975a). These seem to confirm that there are visual aspects of bony structure that have not generally been taken into account in the conventional assessments.

Thus, classical study of, for instance, the scapula, the upper and lower ends of the humerus and the upper ends of the radius and ulna suggests to some workers that these fragments in australopithecines are more like those of humans than of apes; in the same way, classical studies of the pelvis, upper and lower ends of the femur and tibia, the entire talus and a number of other foot bones have been similarly assessed (e.g. Day, 1977).

Yet simple visual comparisons (Fig. 10.9) of differences in proportion of articular surfaces between upper and lower limbs in apes, humans and the fossils show most convincingly that human lower limb articular surfaces are large in comparison to human upper limb articular surfaces. This befits their bipedal status in which the lower limb takes all the body weight.

This same visual comparison (Fig. 10.9) reveals, in contrast, that ape upper and lower limb articular surfaces have more of an equal relationship. This presumably relates to a habitus for these creatures in which both limbs participate in bearing the body weight (and the upper limbs somewhat more than the lower, however that may be, whether through quadrupedal knuckle-walking on the ground or through quadrumanual climbing in the trees).

It is of extreme interest that, in such a set of comparisons, the little information that is available for the fossils show that the fragments resemble most, among living primates, the equivalent parts from apes (and among the apes, the orang-utan) more closely than they do humans (Fig. 10.9). These facts should be set alongside the comment of Richard Leakey (1973a), who reports that preliminary indications point to a relatively short lower limb and a longer upper limb for australopithecines.

Presumably none of this implies any close evolutionary relationship between these fossils and apes. Everyone believes that, of all extant forms, humans are closest to these fossils in an evolutionary sense. But, given that anatomical parts speak most to the functions of those parts, these findings raise doubt about functional adaptation in the fossils. Such findings must make us wonder whether the australopithecine pattern of bipedal adaptation really reflects a transitional phase to man. We can only come to the conclusion that, however able these creatures were at walking on two legs, they were also convincing quadrupeds and perhaps excellent climbers, feats denied to man today. It is therefore likely that, irrespective of how close these fossils are to human evolution, they must have been upon some side-path that did not lead to human-like functions.

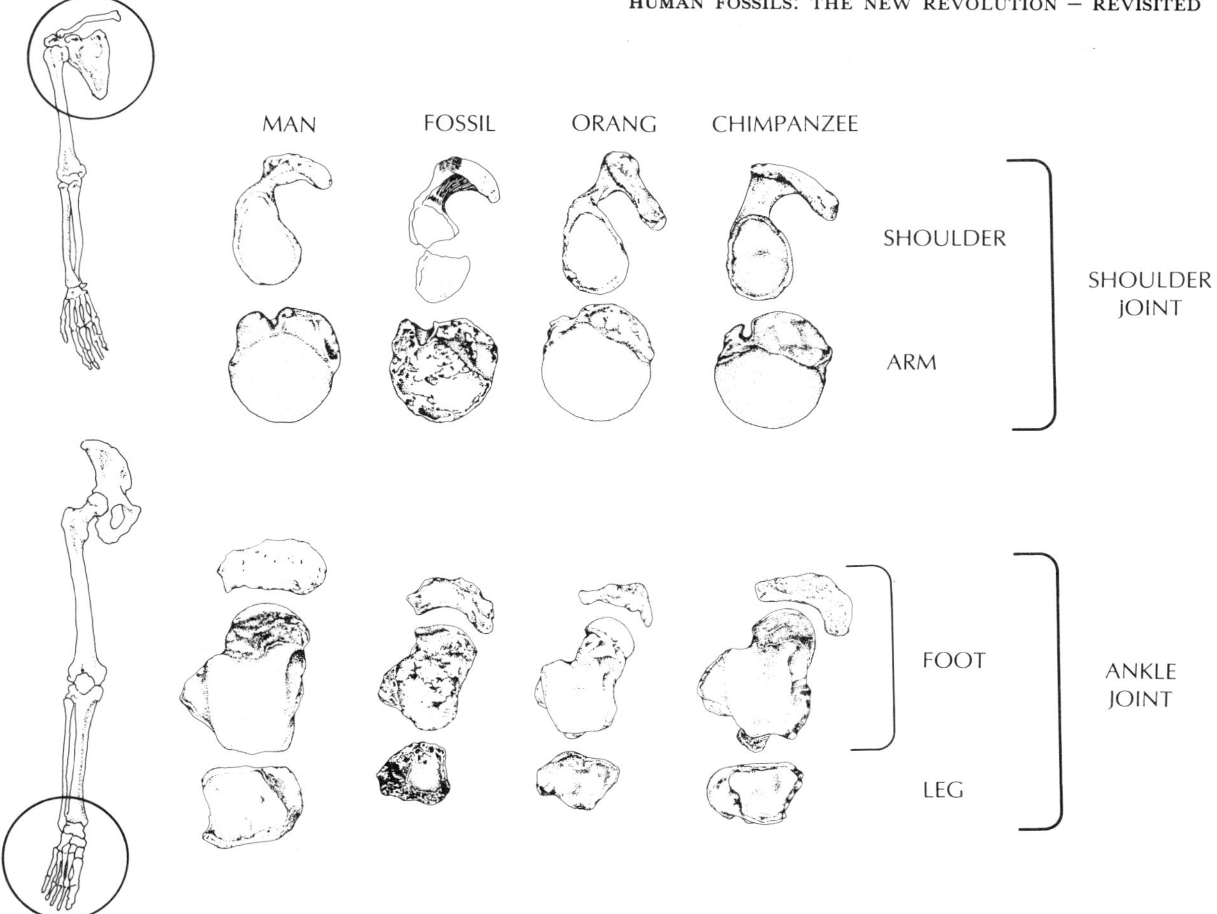

Fig. 10.9. Differences in the proportions of lower limb joint facets with upper limb ones. In each case the upper limb scapular facet is used as the standard. The fossils most resemble the orang-utan in these proportions. This is confirmed in studying other joints of the upper and lower limbs in the same way.

A final series of studies made by other workers using the classical methods is also available. One of these is the australopithecine ear ossicle studied by Rak and Clark (1979), which demonstrates a morphology indicating that they were unlikely to have been human ancestors. Another is the form of the enamel prism patterns on the teeth (Vrba and Grine, 1978) which do not implicate them as human ancestors any more than apes. A third includes simple odontometric studies (Hansinger, 1976), especially showing differences between these australopithecines and humans that suggest they are not human ancestors.

The old methods, then, including simple inspection, do seem to support the new position.

Second question: meld, or otherwise, of old data

Although, prior to the present spurt of fossil finding, the actual number of fossils known was rather small, the total number of investigations performed upon these fossils and reported in the literature is indeed voluminous. They include a morass of older studies of jaws, teeth and crania, and, in the main, it is upon these that the conventional picture of the australopithecines is based. It is necessary that we understand the degree to which the results of these older studies actually meld with, or are antagonistic to, the new views. Do they, in truth, negate the new ideas?

There are good reasons for examining these results once again. For the earlier controversies may have driven the earlier participants into extreme, perhaps untenable, positions. In the uproar, at the time, as to whether or not these creatures were near ape or human, the *opinion* that they were human won the day. This may well have resulted not only in the defeat of the contrary *opinion* but also in the burying of *that part of the evidence* upon which the contrary opinion was based. If this is so, it should be possible to unearth this *other part of the evidence*. This evidence may actually be more compatible with the new view; it may help open the possibility that these particular australopithecines are neither like African apes nor humans, and certainly not intermediate, but something markedly different from either. Let me present a few examples.

The various controversies over the shapes and sizes of the different australopithecine teeth are a case in point. Although many studies have been carried out, the general consensus is that the evidence clearly supports their near human status. What, however, are the facts? Some studies show that in some features (incisors, canines and lower first premolars) the australopithecines are indeed man-like, and this has been emphasized in the literature. But in other features (remaining premolars and molars) other studies show that they are ape-like (Ashton, Healy and Lipton, 1957). Assessment, with hindsight, of some of the data (e.g. for the milk canine) shows that some features are neither man-like nor ape-like but absolutely different from each. In the light of the foregoing discussion of the post-cranial fragments, the totality of these older results for the teeth is not incompatible with the idea that these creatures possess a combination of dental features rendering them distinct among the hominoids (Fig. 10.10).

Another example of such controversy results from the various studies of the basi-cranial axis, an anatomical feature of the base of the skull. The fact that the spheno-ethmoidal angle of this axis (an angle related to the anterior end of the base of the skull and therefore to the placement of the face) is markedly similar in both australopithecines and humans has been seized upon in many prior arguments and is much published. The fact that the foramino-basal angle (a posterior angle on the cranial base related to the form of the back of the cranium) is markedly similar in both fossils and apes has been neglected (table 10.2). Recent restudy of this area shows that, in combination, these characters render the fossils unique (Ashton, Flinn and Moore, 1975).

Yet another example includes the facial area and its contained infra-orbital foramina, the openings of canals whence issue some of the nerves to the upper lip. The fact that there is only a single infra-orbital foramen in both humans and fossils is well publicized. The other part of the evidence is that, in the position of the foramina upon the face, the fossils are much more like apes (table 10.3; Oxnard, 1955; Ashton and Oxnard, 1958; Ashton and Zuckerman, 1958). Taking these two facts together, and others which they mirror, the fossils exhibit anatomical arrangements that render them different from both human and great ape patterns.

One final instance that may be cited includes a study (Adams and Moore, 1975) of the condylar-nuchal region of the skull, that part of the skull giving attachment to the extensor muscles of the

Table 10.2

HOMINOID BASICRANIAL AXIS:
REGRESSION ADJUSTED

Genus	Anterior limb	Posterior limb	Spheno-Ethmoidal angle	Foramino-Basal angle
Homo	57	33	**133**	131
Gorilla	60	27	172	**121**
Pan	57	29	156	**121**
Pongo	53	31	165	**116**
A. africanus (sts 5)	51	29	**129**	**109**
A. robustus (Zinj)	64	25	121	128

(S.e. always < 2% of mean)

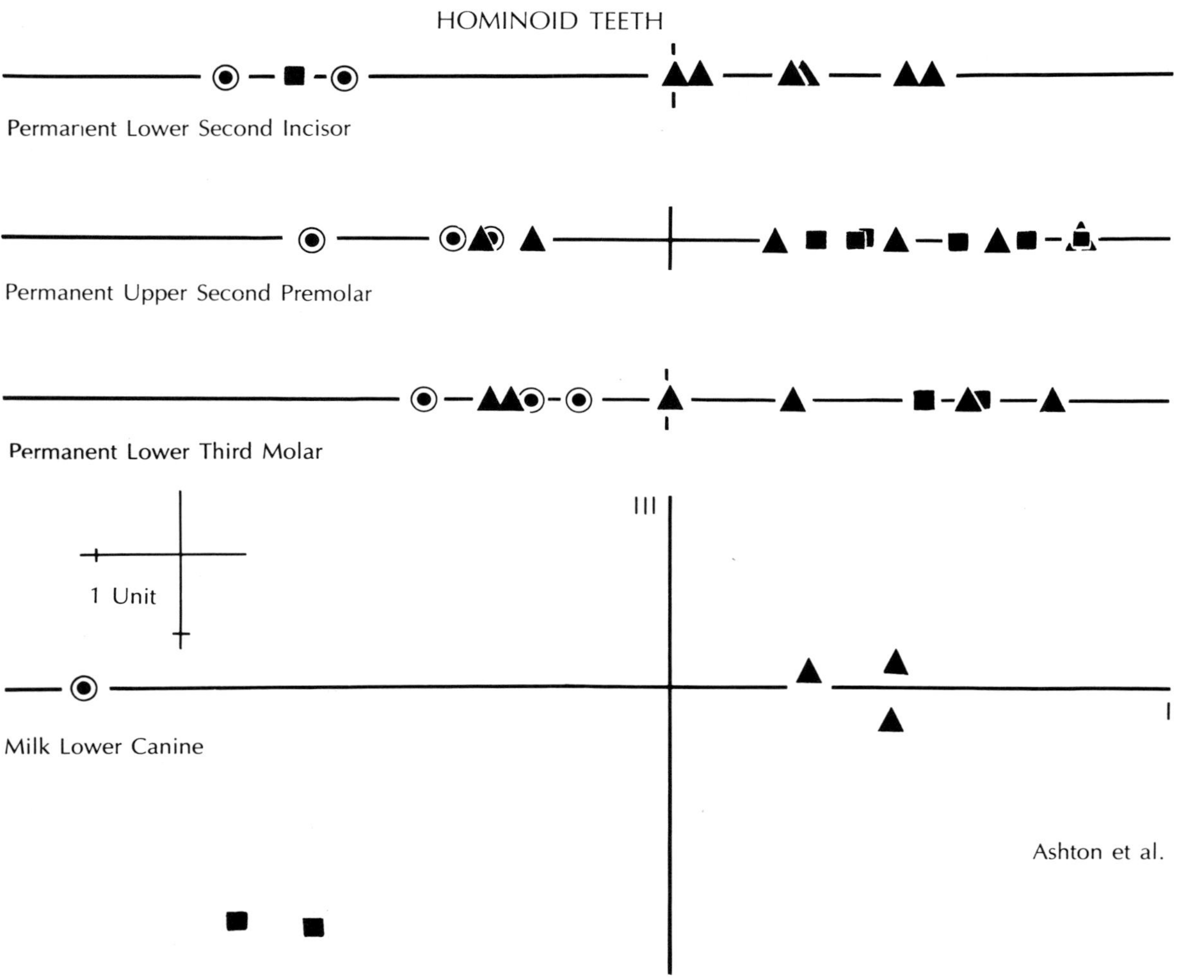

Fig. 10.10. An earlier morphometric study of teeth proportions often incorrectly cited. Note the different positions of the australopithecines (squares) as compared with various apes (triangles) and humans (circles). Redrawn after Ashton, Healy and Lipton.

Table 10.3

INFRA-ORBITAL FORAMINA

Genus	Number of Foramina	Position of Foramina*
Homo	**1.05**	21
Gorilla	1.60	**45**
Pan	2.40	**40**
Pongo	3.30	**35**
A. africanus	**1.00** (in 8)	**36** (Av of 2)
A. robustus	1.00 (in 4)	59 (Av of 3)

* %Distance from orbital to alveolar border

neck and providing for some of the joints that allow the head to move upon the spine. This study shows that the position of the occipital condyles in the fossils is convincingly ape-like; but it also shows that this feature is combined with an angulation of the condyles upon the skull that is markedly human (table 10.4). These facts may possibly relate to upright posture in a creature with an ape-like skull, certainly an unusual formation. Once again, the new position, uniqueness, is supported.

humans, but there will be an element of intermediacy in its morphological position. It will not be uniquely different from men and African apes.

The information presented above for the different fragments of australopithecines from Olduvai and Sterkfontein that are here in contention shows that, *compared with humans and African apes*, each fragment is far more different from the equivalent part of humans and apes than are these latter from each other; the fossils are indeed uniquely different from these extant hominoids. It is in this sense that I use the term 'unique' (Oxnard, 1979c). This uniqueness leads inexorably to the possibility that these fossils are not ancestral to either humans or apes.

The old data, then, do seem to support the new position.

Table 10.4

CONDYLAR–NUCHAL REGION

Genus	Condylar position index	Condylar angulation	Nuchal insertion index
Homo	74	**86**	106
Gorilla	**27**	62	39
Pan	**26**	65	44
A. africanus (sts 5)	**37**	**85**	67
A. robustus (Zinj)	50	92	62

(S.e. always < 2% of mean)

We need to be clear about the meaning we are applying to the term 'unique'. For today we are all heavily sensitised to the idea that each individual human being is a unique person. In a similar vein, many primatologists would think of each species of living primate as unique. Thus, some investigators interested in fossil primates (e.g. Howell, Washburn and Ciochon, 1978) have reacted to the notion of the uniqueness of the australopithecines with the idea that, of course, the australopithecines are unique. Anything, so goes the argument, that existed a million years earlier than something else, must be unique from it. The vital element, however, in our usage of the term 'unique' is the word 'compare'.

A prior human ancestor will be different from modern humans. But, *compared with the differences between African apes and humans* (and this is the key phrase), that human ancestor will be morphologically closer to both humans and African apes than humans are to African apes. It may not, of course, be absolutely intermediate between these apes and

Third question: new fossils, conventional methods

A criticism that can be made about these ideas relates to the fact that the number of anatomical regions and fossils actually examined with these techniques is too small to convert most paleoanthropologists. Most workers may say that the fossil record of human evolution is now too complete and too well studied for multivariate analyses of less than a dozen isolated fossil post-cranial bones to change many minds. Yet, although there have indeed been a great many papers published about the australopithecines in the years since their first discovery, the output of most of the earlier twentieth-century fossil hunters has not provided an overly large number of useful fragments for study.

It has only been within the last few years that really large and numerous finds have been made (e.g. by Richard Leakey at East Turkana; Mary Leakey at Laetoli; Johanson and Taieb in Ethiopia; Clark Howell in the Omo). Many of these new discoveries do not support the conventional views about the australopithecines from Olduvai and Southern Africa. In fact, many of these new discoveries, in that they suggest that man must be far older than previously realized, actually support the new views that have stemmed from the small number of multivariate studies.

Thus, just as the different behaviours postulated for australopithecines from Olduvai and Southern Africa by the new multivariate studies deny them a close link with the human lineage, so too do standard descriptions of some new fossil finds.

Thus, a skull (ER 1470 from East Turkana) has been found that is at least as old if not older than the Olduvai and Southern African fossils; this skull has a much bigger cranial capacity (about 800 ccs) than is usual for the geologically slightly younger australopithecines from Olduvai and Southern Africa at 400 to 550 ccs (Leakey, 1973b).

A talus, also from East Turkana, has been discovered that is at least as old if not older than the talus from Olduvai; yet this, too, is far more human than the Olduvai specimen (Fig. 10.11; Wood, 1973; Henderson and Wood, 1977). And the footprints recently discovered at Laetoli, very much older than the Olduvai foot, have been shown to be human in their contours (Day and Wickens, 1980) in a way denied to the reconstructed Olduvai remains (Oxnard and Lisowski, 1980; Lewis, 1981).

A humeral fragment has been found at Kanapoi that is almost five million years old yet also almost indistinguishable in shape from many modern humeri (Patterson and Howells, 1967; Oxnard, 1975a). Geologically much younger australopithecine humeri at one or two million years are vastly different from those of modern man (Fig. 10.12; McHenry, 1973; Oxnard, 1975a).

I equivocate about the actual dates of specimens

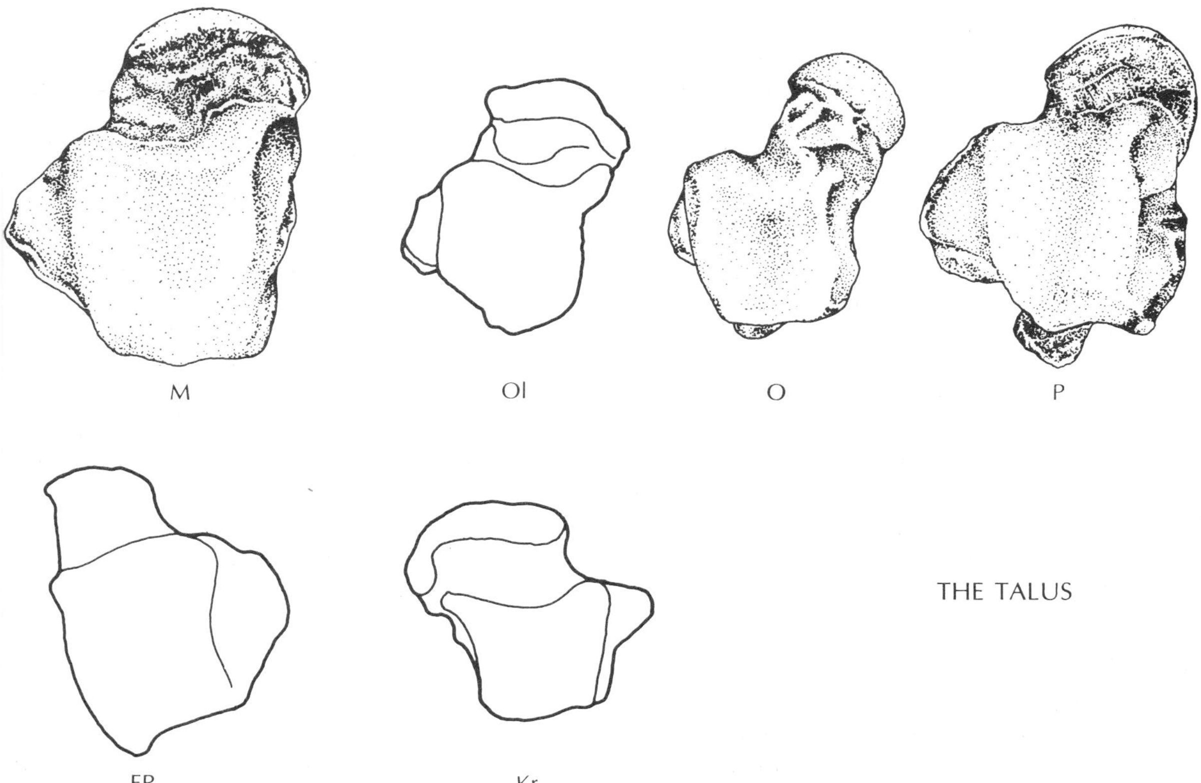

Fig. 10.11. The older, damaged, talus from East Turkana (ER) is much more like those of humans (M) than the more recent talus of so-called '*Homo habilis*' from Olduvai (Ol) or *Australopithecus robustus* from Kromdraai (Kr).

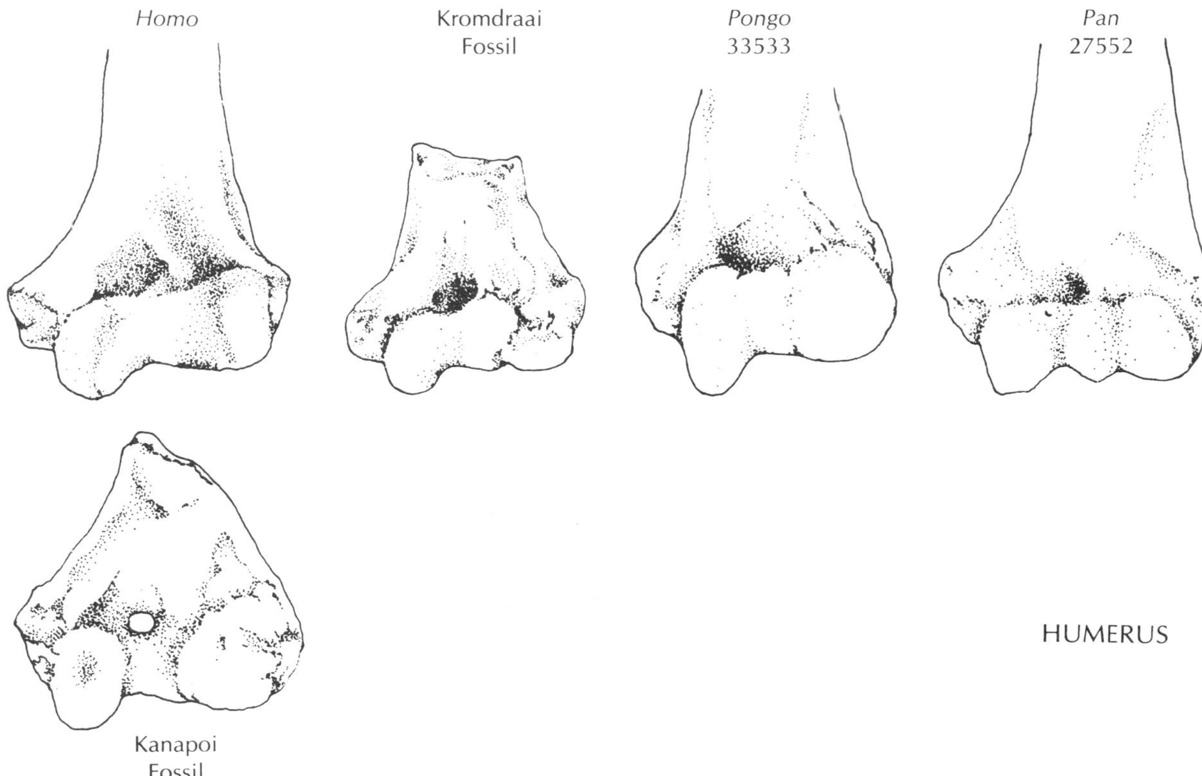

Fig. 10.12. The older humeral fragment from Kanapoi is much more like that of humans than are more recent humeri (e.g. that from Kromdraai) representing the conventional australopithecines.

such as ER 1470, because these change as the dating methods improve; but such changes in dates (e.g. from, say, 2.5 million years to 1.8 million years, as has occurred for the specimen ER 1470) while vitally important within the actual precise sequence at East Turkana, has no effect upon comparisons with such different australopithecines from Olduvai or Southern Africa. For whether these last are 0.75 million years or even 1.5 million years, with a cranial capacity of only a little more than half of that of ER 1470, it is unlikely that they have much to do with ER 1470 biologically. Even if they were marginally older than ER 1470, they could scarcely have given rise to it in such a short period of time.

Almost every year, it seems, are revealed new fossil specimens which are much more like man than the original australopithecines from Olduvai and Southern Africa, and which, at the same time, are at least as old geologically or even older than them. The Olduvai and Southern African australopithecines must have been, on the basis of this information alone, a long way from the lineage leading to man. The conventional wisdom would recognize the genus *Homo* at less than a million years and a common ancestor with chimpanzees and gorillas at one million, three million or five million years (all guesses put forward in the last two decades). The new finds confirm the morphometric suggestion that the human lineage itself must be much older than generally recognized. Although the new morphometric studies cannot say how old the human lineage must be, the new fossil finds can say that it must be at least as old as five million years, and they can suggest that a common ancestor with the African apes must be even before that.

The new finds, studied by old methods, thus support the new position.

Fourth question: acceptance or otherwise of new views

As might well have been expected, the new ideas have not arrived without fierce challenges. *First,* some investigators have directed attention to the enormous bulk of older data that apparently points the other way (Howell, Washburn and Ciochon, 1978). As we have just seen, however, much of the older data can indeed be interpreted alternatively, and this lessens the strength of arguments of that type. In addition, data exist that have been ignored in the prior interpretations. *Second*, such arguments depend quite overtly upon the strength of numbers. But though numbers may prevail in democratic institutions, in academia and in science they do not. One fact that controverts an hypothesis is more powerful than a thousand that support it.

It is, however, possible to try to deny the new views by *ex cathedra* statements saying that the new views are wrong. There have been some attempts in the recent literature to do exactly this. It is an important antagonistic method that we must examine.

What has been done is the following: when studies using complicated analytical methods, such as those of multivariate statistical analysis, provide assessments that are at variance with the conventional view, then the result of the new method is overridden by some simple *ex cathedra* statement that casts, apparently, utter doubt on the validity of the result. Let us look at some examples.

Many studies of the form of the shoulder have now been completed in which morphometric and biomechanical methods have been used. One final conclusion of all of these studies (of the shoulder of primates: Oxnard, 1967; of the shoulder of mammals: Oxnard, 1968a; of shoulder fragments in various australopithecine fossils: Oxnard, 1968b, c) is that the form of the human shoulder is morphologically closer to that of orang-utans, arboreal acrobats, than to terrestrial knuckle-walking chimpanzees and gorillas. This resemblance is not, of course, because of any genetic link between humans and orang-utans. It is rather because the form of the human shoulder is more likely to have stemmed from animals that climbed somewhat like arboreal apes, such as orang-utans, than from animals that knuckle-walked on all fours like chimpanzees and gorillas.

The fact upon which this idea rests, the similarity between the scapula of humans and orang-utans, depends upon a large body of information that has been swept away, apparently, by the simple statement by Tuttle (1974) 'that visual inspection clearly demonstrates that human shoulders are far more like those of African apes than like those of orang-utans'. This statement of opinion has been sezied upon by Corruccini (1975) and elevated to the status of fact.

Exactly how powerful is this *ex cathedra* statement? It is necessary to examine it because it is antagonistic to the morphometric view of the scapula. Fig. 10.13 shows the scapulae of the three great apes and man. All four are basically triangular. The three apes are similar in having the shoulder joint point upwards; in man it points laterally.

But Fig. 10.14 demonstrates unequivocally that there are a whole series of similarities between the scapulae of orang-utans and humans. In marked contrast to both humans and orang-utans are the African great apes; they differ markedly from both man and the orang-utan in almost every remaining aspect of the bone, especially in the strongly sloping spine with a deficient medial end.

In other words, even visual examination supports the statistical view; the cursory description is just not true.

The clavicle is another example. Various morphometric studies have demonstrated how important is the twist of the clavicle along its longitudinal axis. In particular this feature is important because, whereas in humans the clavicle is scarcely twisted at all, in the various apes, as in the Olduvai clavicle, it is heavily twisted. This particular feature does not fit with the idea that the fossils are functionally close to man.

But the entire finding has been dismissed by Day (1977) with the remark that, because the ends of the fossil clavicle are missing, the longitudinal torsion of the bone cannot be assessed. And this opinion, too, has been seized upon by other writers, (e.g. Vrba, 1978) as a fact.

Again, however, what are the facts of the matter? It is indeed true that absence of the bone ends prevents any measure of torsion in bones like, say, the femur and the humerus. In such bones, longitudinal torsion is merely the geometrical angle that can be measured between the placement of the two ends of the bone upon an essentially cylindrical shaft. But any anatomist knows that the torsion of the clavicle is quite different: the clavicle is a genuinely twisted bone with the twist extending throughout most of its length. In this respect it is rather like a propeller; even when propeller tips are missing, we can all still see that a propeller is twisted. Loss of the tips does not prevent a measure of the degree of propeller twist. This can again be demonstrated very clearly by means of a diagram (Fig. 10.15) which shows the lesser twisted state of the human clavicle, the more twisted form of ape clavicles and the large twist of the damaged fossil.

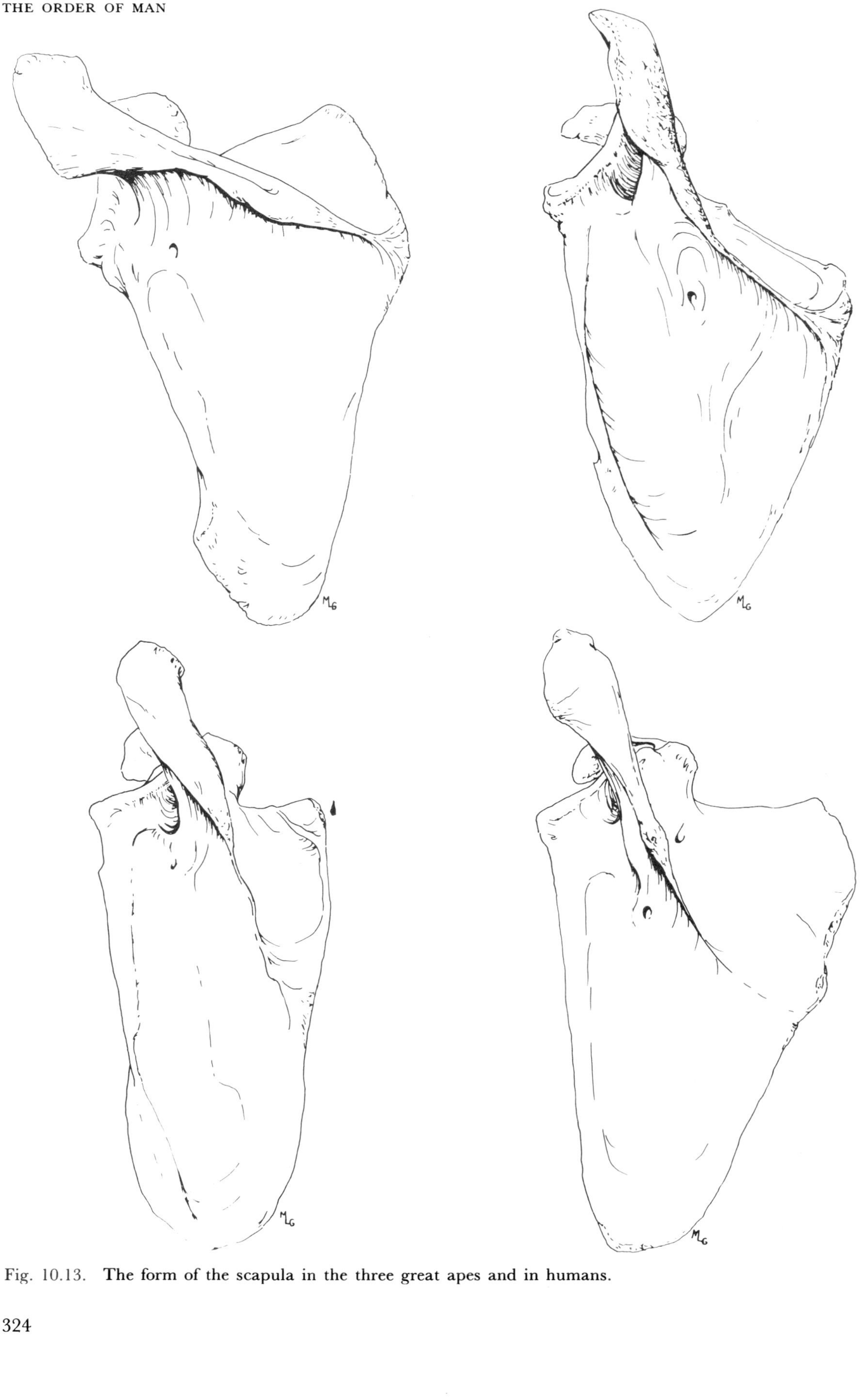

Fig. 10.13. The form of the scapula in the three great apes and in humans.

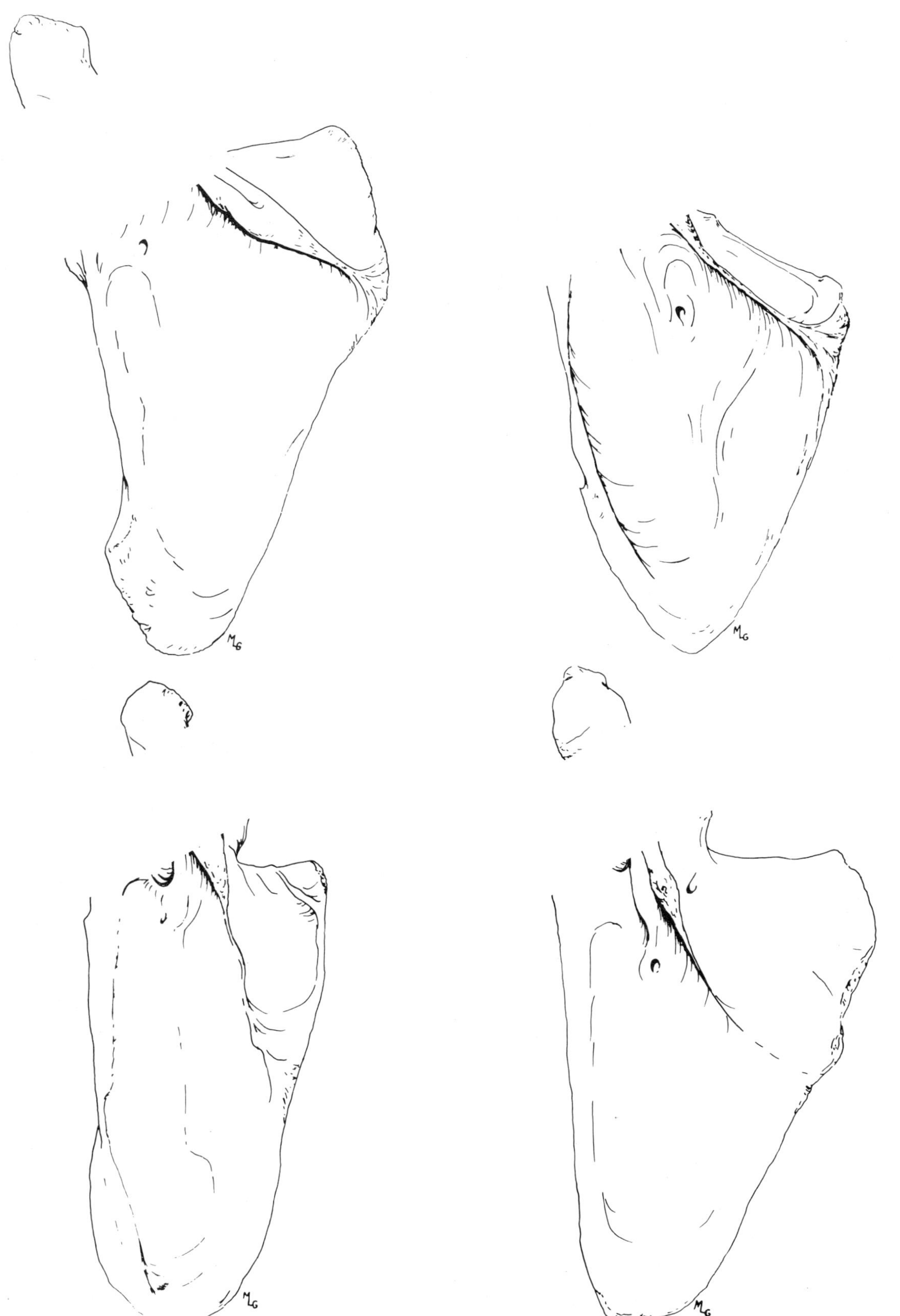

Fig. 10.14. The form of the scapula with the glenoid cavity and acromion removed. This emphasizes the general similarity between the gorilla and chimpanzee on the one hand, and the orang-utan and human on the other.

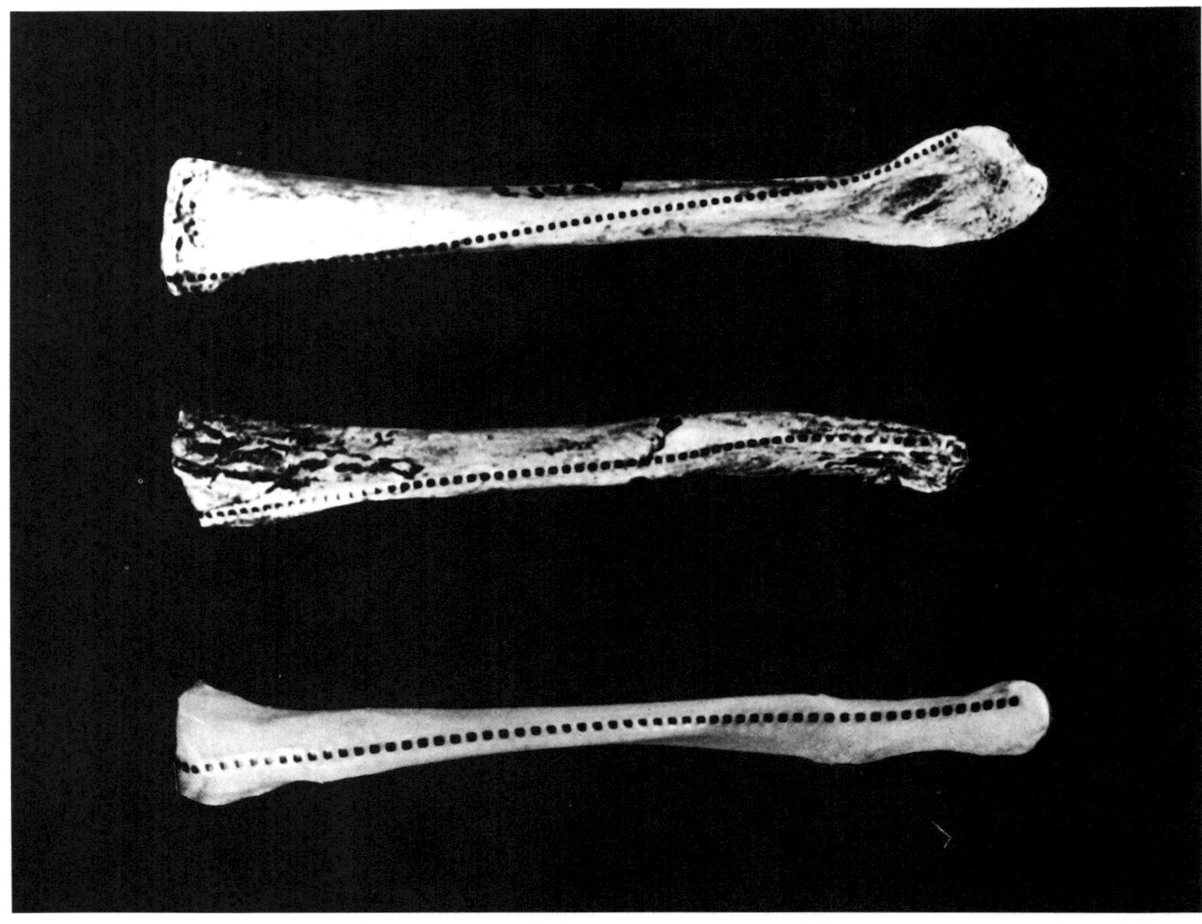

Fig. 10.15. The twist of the clavicle in apes (above), the Olduvai clavicle (centre) and humans (below). The dotted line on each photograph outlines the anteriorborder on each bone. It shows virtually no twist in humans, but a rather large twist in gorillas. These twists are quite evident from even a rather small portion of each clavicle. Enough of the Olduvai clavicle is present to indicate (a) that it has a rather larger angle of twist and (b) that in this feature it is more like a climbing or acrobatic ape or monkey than any human. This twist can be easily discerned without the whole bone being present.

This relationship is precisely what the morphometric approach tells us; the cursory comment is not correct.

Let me give a third example: the pelvis. Extensive studies of the pelvic musculature, osteology and osteometry (Zuckerman, Ashton, Flinn, Oxnard and Spence, 1973; since confirmed by Ashton, Flinn, Moore, Oxnard and Spence, 1981) have shown that the fossil (Sterkfontein) pelvis is quite different from both human and extant ape (see especially Oxnard, 1975a).

Yet it is still commonplace for this finding to be ignored as a minor point in the literature, such studies as those of Lovejoy (1979) pointing out that the human and Sterkfontein pelves are essentially similar. And the opinion is again taken up in reviews (e.g. Zihlman and Brunker, 1979) as a fact.

Again, what, pictorially, is the fact of the matter? We have investigated this by the very simple method of cutting special sections through both the relevant bones and casts of the fossil. Fig. 10.16 shows that the pelves of apes and monkeys (which have certain similarities because they are all quadrupedal animals) are such that single flat sections can be made that include both the hip joint and the sacral articulation. These are sections that pass, therefore, longitudinally through the bar of bone joining these two joints and carry much of the body weight. Similar sections can be made in human pelves, although in humans the shapes of the sections are quite different.

When, however, we attempt to make such sections for the Sterkfontein fossil we cannot do it unless we deliberately make a marked bend in the section (Fig. 10.16). The fossil pelvis is just different, completely different, from that of monkeys, apes or humans.

It is entirely possible that the human pelvis has evolved from one similar to that of an ape or monkey through the evolutionary development of two twists or bends: a backward bend of the ischium and a forward twist of the ilium. The first angulation may approximately 'correct' the second, so that the section in humans comes out flat and thus spuriously similar to the living non-human primates. But in the fossil, though the posterior bend of the ischium exists, the anterior twist of the ilium does not; hence in the fossil the bend introduced by a first development is not 'corrected' by that from a second. The net result is the marked distortion shown in the figure.

The fossil, despite all protestations to the contrary, is not like humans. And this is precisely what the multivariate approach tells us. Whether what is demonstrated by the multivariate statistic will be more easily accepted through this simple method of sectioning (though it is true that it is the hindsight of the multivariate study that helps us to make the

Fig. 10.16. Sections of the innominate bone of ape, human and Sterkfontein fossil. The section passes through the main load-bearing bar of bone from the ischial tuberosity giving attachment to the powerful hamstring muscles, through the main joint of load-bearing with the femur, the hip joint to the sacroiliac joint, the main weight-bearing connection with the vertebral column.

The frame below shows this section *en face* in human, (left) australopithecine, (centre) and orange-utan, (right). In this view the fossil shows both human-like and ape-like features and seems generally intermediate in form.

The frame (next page) shows a view of these bones at right angles to the plane of the section (at right angles, therefore, to the plane of the upper frame). In both orang-utan (upper bone) and human (lower bone) the section is more or less planar. In the fossil, however, the section had to be cut with a marked bend in order to include each element of the weight-bearing bar of bone (central bone). The fossil is clearly unique.

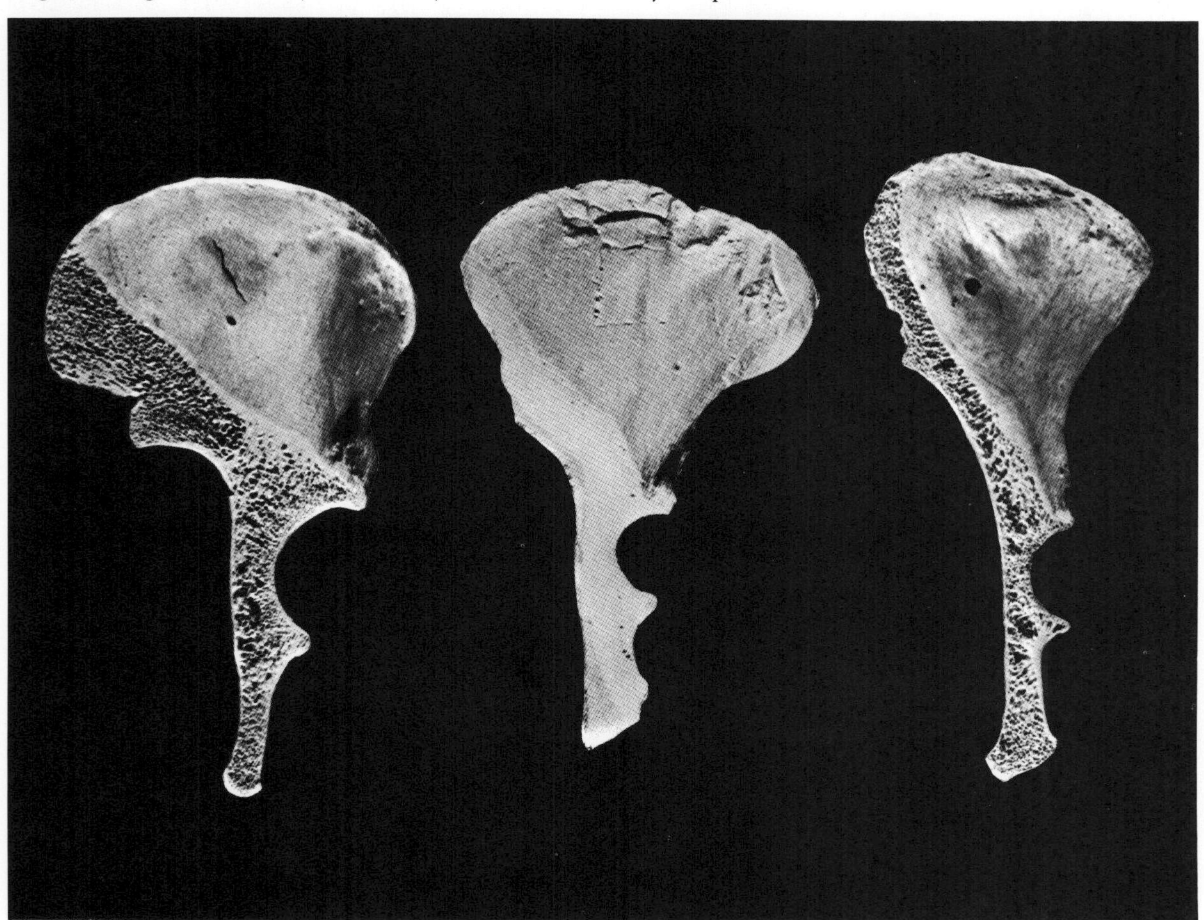

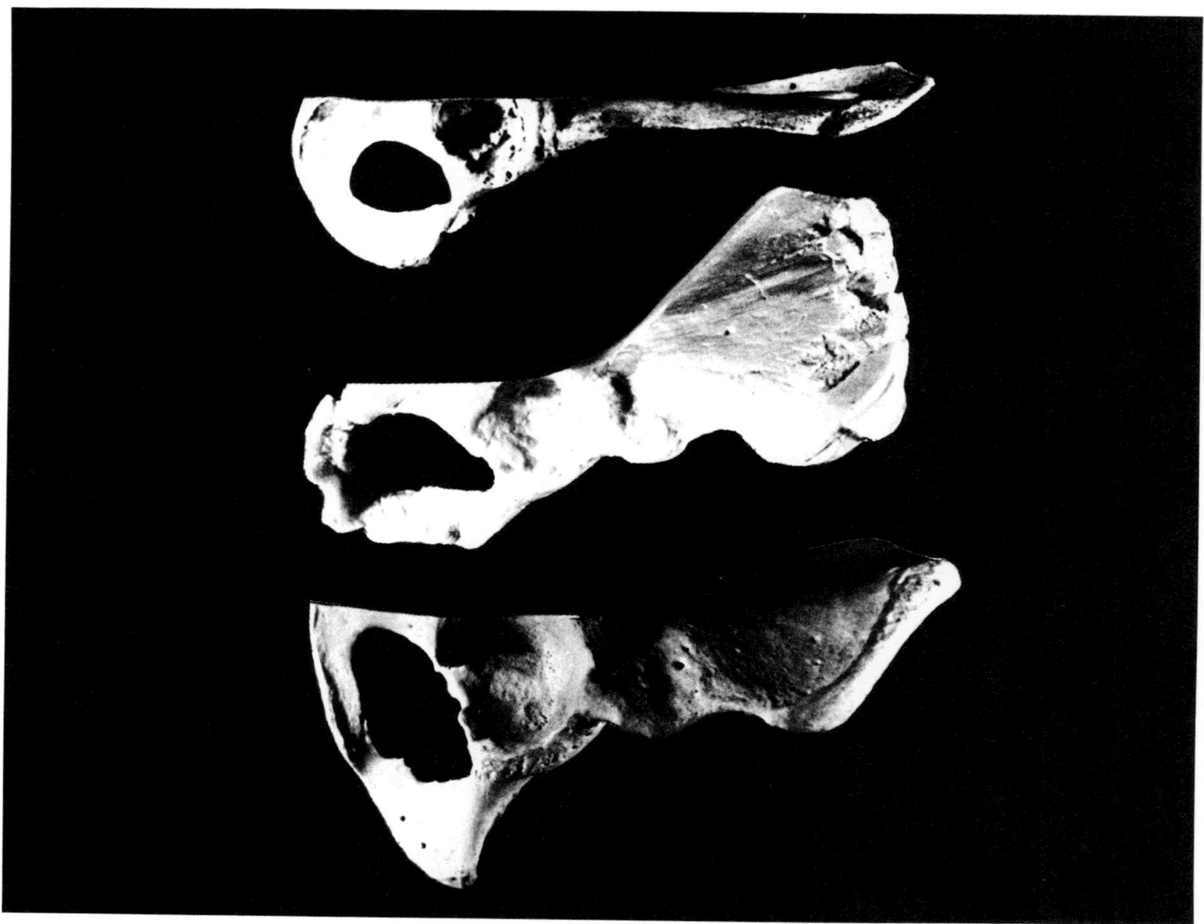

appropriate section), we will have to wait and see.

A final example stems from the studies on the Olduvai foot. Although multivariate studies of the talus deny close resemblances to bipedal man, these results have been cast aside with the opinion (Day, 1977) that the human arches possessed by the rest of the Olduvai foot are incontrovertible evidence of human bipedality. And this readily convinces those who write overviews of human evolution (e.g. Kennedy, 1980) that it is a fact.

Once again, careful visual examination readily demonstrates that this is just not so. Figs. 10.17 and 10.18 show that the highly arched foot of the Olduvai fossil is obtained through a series of osteological misalignments. More correct rearticulations of the foot bones do not produce high human-like arches (compare Figs. 10.7 and 10.8).

What I have attempted to do with these visual examples is not to hide the ways in which these fossils resemble humans, but to pinpoint quite clearly the ways in which they do not. And though it takes the multivariate morphometric method to be certain about this, the visual examples of Figs. 10.13 through 10.18 make excellent displays that deny the *ex cathedra* responses in the literature.

Australopithecines, humans and locomotion

The foregoing results of multivariate morphometric studies of australopithecines are the following: in terms of morphology, the various australopithecines are generally more similar to one another than any individual specimen is to any living primate. They are different from any living form to a degree greater than the differences between bipedal humans and terrestrial apes. Some of their similarities to living forms are especially reminiscent of the arboreal habitat.

The meaning of this is far more difficult to disentangle. It undoubtedly does not rest upon genetic propinquity as it is surely clear that, genetically, humans, the African apes and presumably

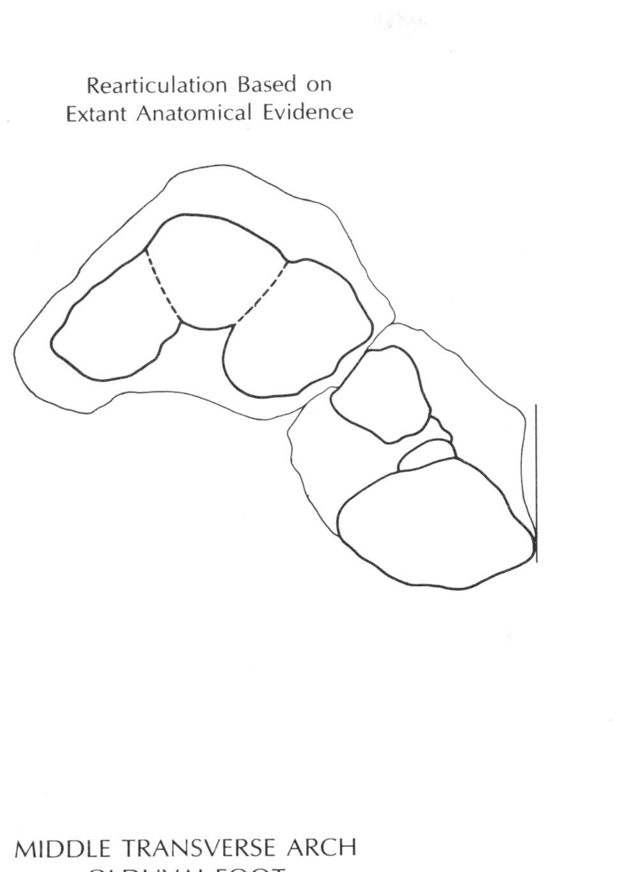

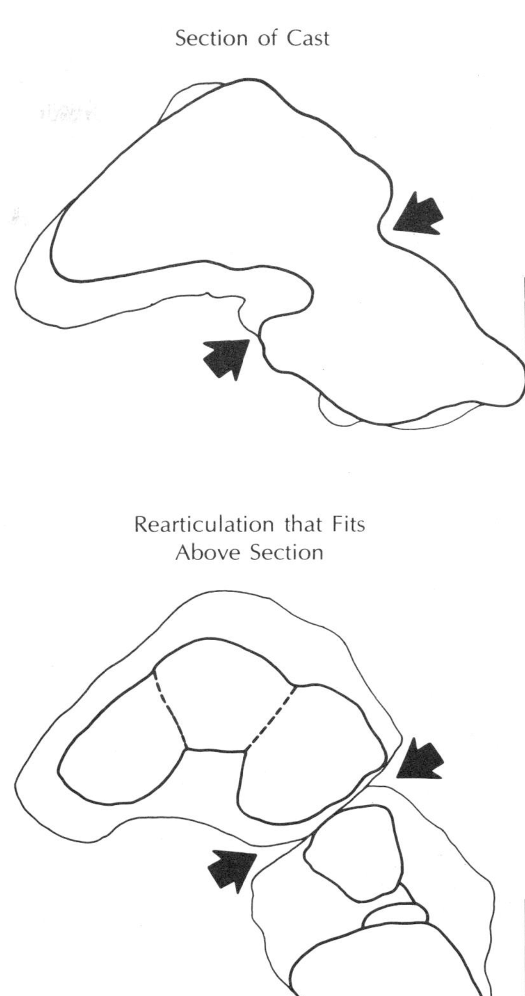

Fig. 10.17. Section (top right) near the ankle through a cast of the entire *Homo habilis* foot compared with equivalent transverse re-articulation (left) of the individual bones. Bottom right is the misaligned re-articulation that must be made in the original bones in order to reproduce the position in the cast. The arrow indicates the misalignment which makes the arch seem much deeper than it really is, and hence more human than it really should be.

the australopithecines are all closer to one another than are any of them to arboreal forms such as the orang-utan. It is presumably far more likely that the patterns of uniqueness and diversity that have been uncovered relate to biomechanical aspects of behaviour.

The results speak, therefore, to the following deductions. These australopithecines, in displaying uniqueness in morphology, may likewise have been functionally unique from all living hominoids. This may mean that they were not arboreal in the different manners of the Asiatic apes; it surely means they were not terrestrial in the special knuckle-walking mode of the African apes; they cannot have been solely bipedal like man. They therefore displayed *either* a totally new and unknown manner of locomotion which would be unique in its own right (and which we will judge unlikely), *or* they possessed such a mixture of locomotor abilities, therefore anatomical adaptations, and therefore, in turn, bony morphologies, as to be rendered unique through being curious functional and morphological mosaics.

The modes in which they are similar to man indicate propensities for a type of bipedality. But the differences indicate that this bipedal activity

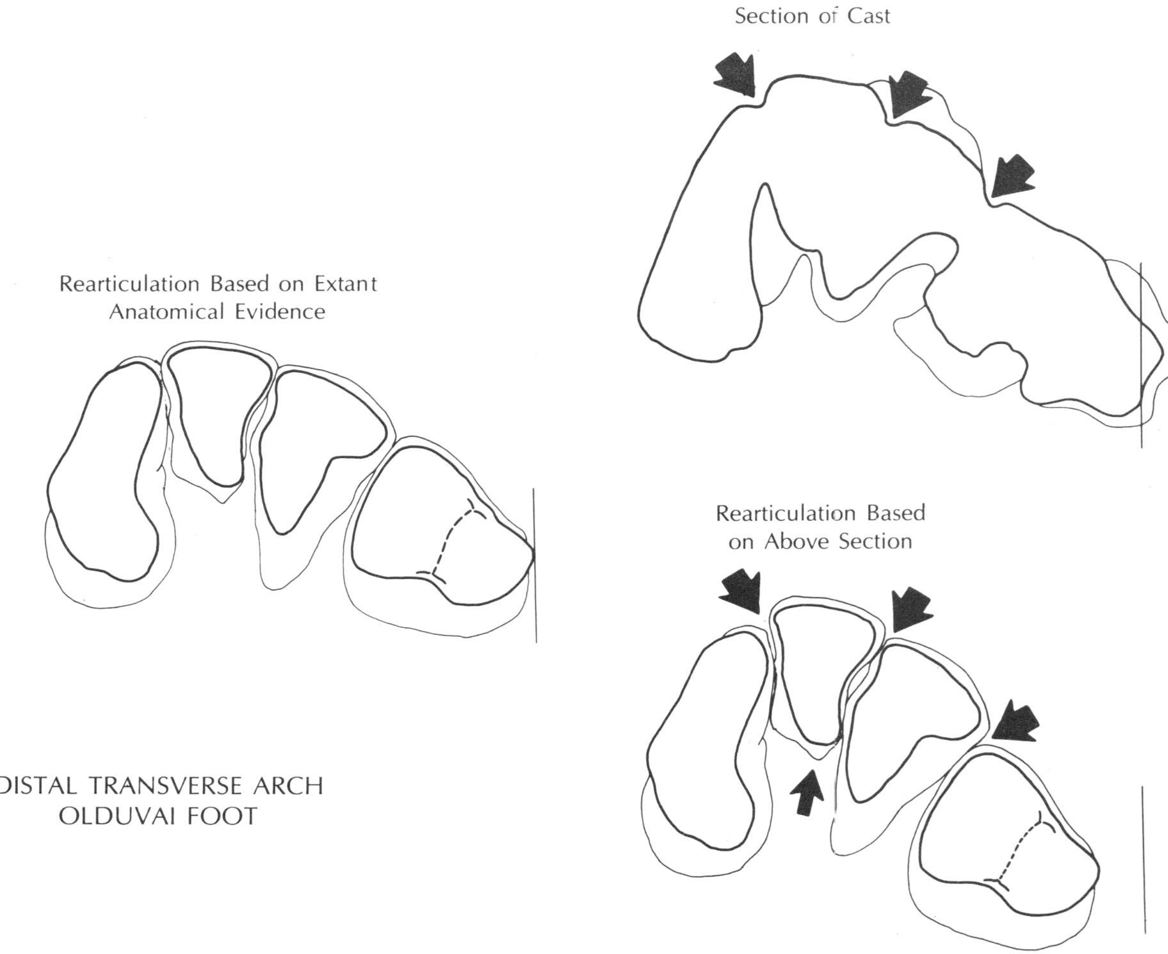

Fig. 10.18. Section (top right) near the toes through a cast of the entire *Homo habilis* foot compared with equivalent transverse re-articulation (left) of the individual bones. Bottom right is the misaligned re-articulation that must be made in the original bones in order to reproduce the position in the cast. The arrows indicate the misalignment which makes the arch seem much deeper than it really is, and hence more human than it really should be.

was not similar to that of man. The ways in which they are similar to the arboreal apes indicate abilities for quadrupedal movement, presumably in an arboreal environment and with a degree of acrobatic climbing. But, again, the differences indicate that this was not in the same manner as any present-day arboreal ape.

Let us be clear that these suggestions do not include merely an extension of human capabilities. Humans can climb using methods like those of apes rather than most monkeys, but our abilities to live in a climbing setting are almost non-existent. Likewise we can, and do, when infants, move on all four limbs; but, in fact, our quadrupedal facility is virtually zero. Our true abilities in both these directions cannot be said by anyone to be other than a total liability in any conceivable life framework where they would be required. Such was presumably not the case with these australopithecines, although we are not able, from these data, to disentangle the time relationships of the developments. The Olduvai and Sterkfontein australopithecines may display these morphologies because they had both sets of abilities, or because, while performing the one as a new acquisition, they had not yet lost the hallmarks of the other older mode.

Is it yet possible that the results are consonant with the australopithecines from Olduvai and Southern Africa being close to the pathway of the evolution of bipedality as expressed in the evolution of man? This seems unlikely for it would require that bipedality arose by some markedly complex process involving genetic mosaics that were like man at an earlier period, but were very much unlike man at a later time; it would require that all this occurred in a very short period of time; it would require some curious reversals from earlier specimens such as those of Kanapoi, Laetoli and East Turkana that were more like us to later ones as at Olduvai and Sterkfontein that were less.

The results are actually far more compatible with the simple idea that human bipedality was not the only experiment in this direction. The australopithecines are displaying for us another experiment, and, given that they are now extinct, one that failed. The reasons why this may have happened are speculative in the extreme at this point. We may well have to accept that human bipedality is far older than previously guessed, and that australopithecine locomotion included one or more parallel experiments in this direction. If the new claims that ramapithecines found in Yunnan, China, are bipedal and, at the same time, eight million years or so old, then bipedality is indeed ancient, and a radiation of different forms of bipedalism is indeed highly likely.

Australopithecines, hominoids and systematics

These results have implications that are systematic as well as biomechanical. Thus, though above we have interpreted the uniqueness of these australopithecines from Olduvai and Southern Africa as biomechanical, the very breadth of their uniqueness (involving not only every area of the postcranial skeleton that has so far been examined, but also every specimen studied and therefore every species of australopithecine from Olduvai and Southern Africa so far named, not only gracile, robust and hyper-robust forms, but *Homo habilis* and *H. africanus*) indicates a need to reassess the systematic and phylogenetic position of the group.

It is entirely possible that a more cautious assessment of these fossils is that, among the hominoids, they form a rather coherent group of specimens and species. It is entirely possible that they form an independent unit within the hominoid radiation, one that presumably became extinct. They may thus form (a) a third cluster of species that is parallel to (b) the African apes of the genus *Pan* and their relatives and progenitors and (c) the groups of humans and their relatives and progenitors of the genus *Homo*.

These comments relate, of course, only to the australopithecines from Olduvai, Sterkfontein, Kromdraai and Makapansgat, the ones that we have been able to study. They do not apply to the new fossils from the Afar, from East Turkana, from Laetoli and other recently examined sites, because these have not yet been studied with such hypotheses in mind. Though many of these latter are also called 'australopithecines' we must be aware that, although it is possible that some may be similar to the Olduvai and Sterkfontein australopithecines, others may be quite different. The claim (e.g. Johanson and White, 1979) that these new australopithecines are ancestral to both humans and later australopithecines may turn out to be correct. But the Afar fossils are new and have not yet been studied by independent laboratories to allow it to be corroborated. It is just as possible that the claim will turn out to be wrong. In the meantime, these new relationships of the Olduvai and Sterkfontein australopithecines may easily be hidden by references using the same nomen to other more recently discovered but geologically older fossils from such places as the Afar, Laetoli and East Turkana.

Australopithecines, investigators and convention

It is not to be expected that the biometrical studies here referred to will convince most workers that the conventional wisdom to which they subscribe should be qualified. The evidence that will do this will be that provided by the new fossil finds as they are made. I have indicated that some of these new finds have perhaps already been made in the Afar Valley, at Laetoli, in East Turkana and in the Omo. And we can confidently expect that Leakey and others will find yet more evidence of geologically earlier forms that are considerably more human than the fossils from Olduvai and Sterkfontein. We can especially expect to find such fossils in new, non-African sites as investigations proceed apace in Asia.

Nonetheless, the biometrical studies to which I have referred have been powerful enough to foreshadow what appears to be the impact of some of the emerging new finds. There will be confusion, of course, because these new fossils, themselves different from the australopithecines of Olduvai and Southern Africa, have also been tagged with similar names: the new find from the Afar Valley has been given the generic nomen, *Australopithecus* (Johanson and White, 1979).

Even though the existence of this form so much earlier in time supports the ideas presented here, it is entirely possible that we may eventually decide that this species, *Australopithecus afarensis*, is as lief upon a lineage leading only to the later australopithecine as upon other alternatives. If the small cranial capacity attributed to *A. afarensis* is correct, then this possibility is rather likely.

But, whatever turns out to be the case, and however much the changes are disguised by using the term *Australopithecus* for the new earlier fossils as well, the australopithecines known over the last several decades from Olduvai and Sterkfontein, Kromdraai and Makapansgat, are now irrevocably removed from a place in the evolution of human bipedalism, possibly from a place in a group any closer to humans than to African apes and certainly from any place in the direct human lineage.

Broader implications for human evolution

All of this should make us wonder about the usual presentation of human evolution in introductory textbooks, in encyclopaedias and in popular publications. In such volumes not only are australopithecines described as being of known bodily size and shape, but as possessing such abilities as bipedality and tool-using and -making and such developments as the use of fire and specific social structures. Even facial features are happily (and non-scientifically) reconstructed.

This new evidence relegating the conventional australopithecines to a side role in human evolution raises a series of rather fascinating possibilities.

First, these ideas mean that the evolution of the australopithecines may be a most interesting separate set of problems in their own right. We may be able to see more clearly the flowering (and withering) of this entire group of creatures once we are able to remove from our minds the idea that they were direct human ancestors. What were their relationships with the genus *Homo*, and what with the genus *Pan*? What were the actual behaviours of the various creatures that made up *Australopithecus*? We have already looked briefly at these possibilities. Now read Nota Bena following page 334.

Second, the simple biomechanical abilities of man that relate to bipedality may also have existed for far longer than we had previously thought. That, in turn, almost certainly means that these simple biological features may have evolved not once, not even twice, but perhaps several times. Such a multiple development, or radiation, would indeed closely resemble what zoologists readily recognize as evolution into a new adaptive zone or a new niche. It would also inevitably mean, however, that, in this new adaptive zone, the genus *Homo* may have been a part of a far larger and more complex radiation of species. Again, we have already looked briefly at this idea.

Third, it is rather likely that *Homo* as a genus has existed an order of magnitude longer than previously guessed. The new finds and new studies of old finds from Africa imply that this length of time may be five million years and even more, rather than the conventionally believed half a million years or so. The new finds from China (e.g. Wu, 1981) may possibly mean that some features previously thought to be human, but not necessarily of the genus *Homo* (e.g. bipedality) may be even older still: eight million years. Obviously, we need to await further judgements on these new Chinese finds, but it is well worth recording at this point that Professor Wu Ru Kang, of the Chinese Academy of Sciences, believes that these new finds could even implicate a ramapithecine-human transition as long ago as fourteen million years. The genus *Homo* must certainly have been contemporaneous with the australopithecines as we now know them and may well have extended a great dealer earlier.

Such time spans are so much greater than currently believed to be the case that they allow a very much longer period for the psycho-social evolution of the genus. It is possible, therefore, that those features of humans in which they differ most markedly from animals — behavioural, cultural, intellectual and creative — may depend rather less upon prior genetically determined behaviours in animal forebears and much more upon this extra amount of psycho-social evolutionary time. Let us take this particular notion somewhat further.

Thus, the commonly held view about the biological elements of the evolution of *Homo* is that the one to two million year old australopithecines from Olduvai and Sterkfontein (most particularly, perhaps, *Homo habilis* from Olduvai) are on a direct, or almost direct, lineage leading to the genus *Homo* (and especially to *Homo erectus*, at a few hundred thousand years). This concept can be applied to help interpret the psycho-social phase of human evolution (to use the phrase coined by Juian Huxley). It necessarily implies that this phase was very short and was, therefore, strongly underpinned by the extensive biological, especially genetic, mechanisms responsible for biological evolution generally.

With such a view it is reasonable to believe that those features of humans that are especially associated with being human, for example our aggression, our social structure, our creativity, even our mysticism, may be closely related to prior animal characteristics such as animal aggression, animal social structures, animal creativity and animal precursors to mysticism (whatever they may be).

Such human features are thus usually thought to add up to *a very thin veneer, due to a short period of evolution of a psycho-social type superimposed upon a very thick bedrock of prior animal behaviour.* This is certainly the viewpoint of many who look closely to animals for the determinants of human behaviour (e.g. Wilson, 1978).

If, however, the new information from our biological studies of skeletal parts of australopithecines (including *Homo habilis*) from Olduvai and Sterkfontein, that australopithecines of one to two million years old are on sidelines parallel to those leading to *Homo* and that the genus *Homo* has been around for five million years and probably a great deal longer is, indeed, truly correct, then new possibilities open up. *If* a brain with a human-like volume of 800 ccs (on the order of that of *Homo erectus*) has existed for several million years rather than several hundred thousand; *if* human-like bipedality has existed for even longer – possibly three or four million years; *if* human-like upper limbs have been around for even longer still, possibly more than five million years; *if* bipedality of some form or other (not necessarily like modern humans) has figured on the evolutionary scene for as long as eight million years; *if* some kind of transition to a very early human-like form really occurred at fourteen million years; *if* at least some of these things, *then* there has been one or perhaps even two orders of magnitude more time for the period of the psycho-social evolution of the genus *Homo* than has been accepted by many.

Given that psycho-social evolution itself proceeds at far greater rates than mere biological evolution, then it may indeed be that enormous changes have taken place in the intervening millions upon millions of years. Rather than being a thin veneer, human psycho-social characteristics of this sort may in actuality be *an enormously thick, complex psycho-social sludge overlying what is, in comparison, only a very thin sliver of prior animal behaviour.*

As we think of what it is that makes us human, this idea does not seem so extreme. Although many investigators have tried to draw analogies between many human behaviours and many animal behaviours, we must admit that, in general, such efforts try our patience to the utmost.

Animal aggression may, indeed, be the *initial* determinant of human aggression. But let us look carefully at the enormous complexities of human aggression: nastiness of children to one another in the playground; robbery with violence of both children and adults; family violence: the beating, not only of children, but of spouses, of parents and of grandparents; assault and battery; rape, rape with violence, rape with torture, rape even with murder; murder as a crime of passion, murder during the act of committing some other crime, murder in cold blood; sadistically and sexually oriented torture and murder; torture and murder as a 'normal' way of carrying out some of the business of our 'civilization'; torture and murder to obtain information both in war and 'peace'; warfare involving hand-to-hand combat; warfare through long-range weapons, through pilotless machines and using push-button techniques; wholesale massacre, and attempted and successful genocide; even the vision of the death of a world. *There is almost no limit to the aggressive ends of human behaviour; it seems most unlikely that all this can be laid primarily at the door of animal aggression.*

We may look at the development of human laws, from the dooms of the kings and the customs of the tribes, through the complicated forms and fictions of the laws of ancient Rome, through the star chamber and the court of Torquemada, through trial by ordeal and *force peine et dur*, to the extreme complex legalities of modern business, criminal, civil, international, sea and space law, even our new 'laws of ethics'. *It seems highly unlikely that these human social mores have much more than very tentative beginnings in a prior animal social structure.*

When we view the enormous complexities of human creativity, from the making and using of simple tools and other implements, through the design of implements that are not only useful but pleasing to the eye and the touch, and delight in our own home-made household articles, the potter's art, architecture, sculpture, painting, through the creations of those artists and scientists who have been among the geniuses of man — the art of

Michelangelo, the science of Einstein, the performance, more ephemeral but none the less real, of a *Swan Lake* ballet — the many other artistic, scientific and humanistic creations of humans, when we look at all these things it seems a very far cry from a chimpanzee shaping a small twig to eat termites as described by Jane von Lawick Goodall (1971, but actually noted well before the Second World War as reported by Merfield, 1963). *On this score, human creativity seems so very, very far from such animal behaviours.*

Finally, we may look at the ethical, moral, religious and mystical creations of the human mind, even some of the perversions of these things as church dogma, religious torture and sectarian heresy, as astrology, satanism and black magic, as the modern cults and so on. *It seems that we cannot even begin to see what may be the prior animal precursors of such activities and thoughts and visions.*

It is, therefore, perhaps a truly strange quirk that the even newer work of Wilson (Lumsden and Wilson, 1981) should now be providing a possible key to these ideas. For though Wilson's earlier studies suggest close analogies between human and animal behaviours (and this is echoed by the investigations of many ethologists over the years), his new ideas imply very much the opposite, though he may not agree. Lumsden and Wilson (1981) confirm that animal behaviours are mostly, in the terms of the computer specialist, hard-wired that is, they are rather directly related to the hardware of evolution, the genetic mechanism. But their new views suggest that human behaviours seem to be mostly soft-wired, depending upon the software programs of epigenetic accidents and choices. These must be superimposed upon the genetic hardware. The length of time that the hardware has been around is short enough that little change can have occurred in it; the software changes almost from generation to generation so that even only short periods of geological time would suffice to produce new behaviours widely separating humans from any animal. This conforms to the notion suggested above.

Lumsden and Wilson, in departing from the idea that human behaviour mostly depends on the hardware, have started to provide us with ideas about the nature of a new software-hardware interaction (their co-evolution) that does not exist in animals but that could be at the root of the uniqueness of human behaviour. They thus think that the soft-wired epigenetic cultural choices might feedback to change the hard-wired genes of humans; if this can and has occurred, miniscule genetic changes could, in their turn, easily produce big changes in the epigenetic rules; and these, in their turn, could have a fantastically large effect on culture (which affects reproductive behaviour), and so the cycle continues. Though such ideas are a major challenge to our current views of human evolution, and though they are not more than tenuous at the present time, they fit well with the results stemming from the morphological studies in this book. They are almost certainly not the ultimate answer, but it seems to me that they may not be bad ones to build on at the present time. Certainly, these are among the more creative ideas that have come from the evolutionary thinking of the present day.

It is especially fascinating to me that the new investigative methods described in this book have proven capable of forecasting (correctly, it appears, from the new fossils) a much longer period for the psycho-social evolution of humans. The new methods are to be reckoned as a set of very powerful tools that have been added to the armamentarium of the evolutionary morphologist. Findings resulting from the application of such methods, once they have been thoroughly checked, should not lightly be discarded just because they are at variance with the conventional wisdom. Indeed, such controversial results may well be the source of more thoughtful assessments of human evolution, and they may help us to escape from the bonds that traditional thinking sometimes lays upon us.

NOTA BENE

Human Evolution. Grounds for Doubt? New Confirmations!

Two important developments have occurred since this book went to press; they greatly increase the information available to us about two of the fossil groups most critical for understanding early human evolution. The first of these stems from a large number of newly published studies of australopithecines, mostly about the recent finds from the Afar valley in Ethiopia. The second comes from a large amount of new data on ramapithecines resulting from the recent discoveries in China.

New investigations of australopithecines

It is true that there are still a number of new publications about australopithecines concluding that these forms show 'a fully developed adaptation to bipedality' (Latimer,1983) or 'a human-like characterization of the total morphological pattern' (Cook, Buikstra, de Rousseau and Johanson, 1983). In the main phrases, such as these, stem from the associates of those who made the initial reports on these fossils and in general these workers continue to support the older idea.

But many recent investigations of australopithecines are newly providing critically important support for the ideas and arguments proposed in this book. An increasing number of workers are at last recognizing that the australopithecines were not primarily human-like and that they were most likely capable of a wide range of non-human activities. It is, of course, still uncommon for these studies to acknowledge the 'grounds for doubt' (the very title of a paper in *Nature*, Oxnard, 1975b) that my colleagues and I have expressed over many years now. Nevertheless the idea that these doubts truly exist seems now not in doubt.

Thus, the studies of Oxnard (1968b and c) which suggest that the shoulder of australopithecines from Southern Africa and Olduvai are shaped and were therefore muscled in such a way as to imply powerful locomotor activities such as are involved in living in trees at least part of the time, have been complemented by a new study of the thoracic vertebral column (Cook, Buikstra, DeRousseau and Johanson, 1983). This investigation, though it concludes that various pathological features of the vertebral column point to a human-like total morphological pattern, is also forced by its own data, to note that these pathologies imply that 'climbing and acrobatic activities may have been proportionately more important than they are in modern humans'. As, of course, these activities are almost totally unimportant to modern humans, that is a most important caveat.

In addition, however, this study notes that 'Lucy', the Afar australopithecine, also possesses a number of non-pathological features, such as long cervical spinous processes and robust thoracic neural arches. These bony features are so long and so robust, respectively, that they are likely to have provided the stress bearing structures necessary to support the actions of very powerful shoulder muscles in climbing and arboreal acrobatics suggested by our prior studies of the scapula and clavicle of other australopithecines.

Again, the study of Oxnard (1968b), stating that the upward direction of the shoulder joint in the fragmentary Sterkfontein scapula suggests that this animal must have had behaviours where the arm was held above the head, has been supported by new work of Stern and Susman (1983a). They find a similar upwardly pointing shoulder joint facet on the scapula in 'Lucy'; they too, interpret this as a valuable feature if the arm were held overhead much of the time as it is in climbing and hanging in trees.

Recent studies of the elbow region in the Afar fossils by Senut (1981) and by Feldesman (1982a and b) corroborate the results of Oxnard (1975a) on the lower end of the humerus of other australopithecines. These new results also emphasize some features of the australopithecine elbow region that are similar to those in pongids and other characteristics different from both pongids and hominids. Both sets of features are most likely related to some form of arboreal locomotion.

New considerations of the skeleton of the hand likewise complement the work of Oxnard (1973a) suggesting that the curved finger bones in the Olduvai hand indicate activities of that hand in powerful grasping as in climbing. Thus, a whole series of new studies of both the Olduvai and Afar hand bones even more powerfully imply arboreal activities for at least a part of their life styles (Stern and Susman, 1983). Stern and Susman show that 'Lucy' possessed long, slender and curved finger bones with evidence of scars for the attachment of powerful grasping muscles, and many signs in the hand (e.g. in the various palm joints, various wrist bones, and so on) of a strong musculature for flexing the wrist. 'When all elements of the *afarensis* hand are taken into account (write Stern and Susman) one is struck by the morphological similarity to apes. All of these similarities, (say Stern and Susman) are hallmarks of a creature that used hook-like fingers and strong wrists for climbing in trees.'

It was, indeed, a similar set of facts that confronted Napier (1962) in his original examination of the hand of so-called *Homo habilis*, an australopithecine from Olduvai; but being convinced that he was looking at a pre-human hand that made tools, he interpreted three features in which that hand was similar to a human hand as more weighty than ten in which he found it was similar to those of apes.

The doubts raised by our investigations of the Sterkfontein pelvis (Oxnard, 1973a; Zuckerman, Ashton, Flinn, Oxnard and Spence, 1973; Ashton, Flinn Moore Oxnard and Spence, 1981) are supported by studies of the pelvis of the Afar creature. Our examination of the Sterkfontein pelvis clearly indicates that, whatever else that creature was doing (and it almost certainly did include a type of bipedal activity) it must also have been a capable climbing creature and probably quite good at quadrupedal activities. The new studies of Stern and Susman (1983a) indicate that the pelvis in 'Lucy' likewise supports such notions for its activities. They have noted that the socket of the hip joint is not as spherical as in humans, that the thigh is somewhat flexed in relation to the pelvis during bipedal walking, and that the iliac crest is orientated laterally as in the chimpanzee. These all suggest to Stern and Susman that 'Lucy' kept its balance more like a chimpanzee than a human female. Further aspects of the hip and thigh suggest to Stern and Susman that 'Lucy' had the ability to move the hips in a way that suits climbing in trees. And these new studies of 'Lucy' have been further followed by studies of pelvic bones in other australopithecine fossils. Thus Susman, Stern and Rose (1983) believe that features of the iliac blade in both Olduvai and East Turkana specimens are such that arboreality was not only possible but frequent in these forms; this is in marked contrast, of course, with *Homo erectus*, in whom arboreality was infrequent or even non-existent.

Though I have not yet studied the knee in detail, I have noted that the size of the knee joint in australopithecines relative to the sizes of various upper limb joints implies a sharing of body weight-bearing between them (Oxnard, 1975a and chapter 10). And I am, of course, aware that earlier studies by Johanson (1980) indicate that he believes 'Lucy's' knee to be essentially human in its form thus denoting human-like bipedality. The extent to which the leg is placed laterally at the knee (valgus form of the knee) is an important indication of bipedality because it enables the foot to be placed more directly under the center of gravity of the body giving better balance and more time for the free leg to swing forward during walking; 'Lucy' possesses such a valgus knee. Again, however, new argumentation (by Jack Prost, 1980) suggests an opposite view. Prost maintains that this angle reflects the ability to climb; circumstantial evidence in favour of this alternative view is that among monkeys and apes the greatest angle of valgus is found in the orang-utan and the spider monkey; these are extremely able arborealists that have similar valgus angles as humans.

Further studies on the knee in the smaller Afar hominids by Tardieu (1981) conclude that the form of the joint did not allow the leg to swing as far forwards as that of humans suggesting that the stride was short, that its locking mechanism was not developed implying that full extension of the leg in walking, a key point of human bipedality, was lacking, and that it allowed the foot and lower leg greater freedom to be rotated thus enabling 'Lucy' to place its feet in a wider variety of orientations than humans. All three of these features suggest to Tardieu that 'Lucy' spent a considerable period of time climbing in the trees.

Again, though I have not yet studied the lower leg in any detail, I did note (Oxnard 1975a and chapter 10) that the relative sizes of the lower leg joints in australopithecines were more compatible with sharing of weight-bearing between both upper and lower limbs. This is confirmed and much extended by new findings by Stern and Susman (1983a). Thus, though the shape of the facet on the tibia for articulating at the ankle faces forwards in bipedal humans, in 'Lucy' the ankle joint faces backwards as much as in many extant apes. Thus its foot would normally be bent backwards relative to its leg, which Stern and Susman (1983a) point out would have been useful in reaching for branches during climbing. Another feature in the lower leg, a groove on the fibula for the peroneal muscles that help evert the foot, is small in bipedal humans but large in 'Lucy', as it is also in other apes. This may have been used by 'Lucy' to help maintain an arched foot for bipedal walking but, again, actual studies of the way that humans, chimpanzees and gibbons use the peroneal muscles in walking have enabled Stern and Susman (1983b) to rule out the possibility that *Australopithecus afarensis* used its large peroneal muscles to support an arched foot for bipedal walking. The peroneal groove is more likely yet another indication of a grasping foot such as would be used by an animal that climbed trees.

Finally we come to the foot itself which certainly ought to tell us if the Afar fossils were exclusively bipedal.

Thus, the studies that we have carried out on the Olduvai talus and the foot (Lisowski, Albrecht and Oxnard, 1974, 1976; Oxnard and Lisowski, 1980; Oxnard, 1980b) that clearly implicate these anatomical regions in arboreal climbing activities, have now been complemented by new studies of the Afar foot, especially the curved phalanges, that equally imply arboreal activity of one kind or another.

It is true that the calcaneum or heel bone is larger and more human-like than are the heel bones of apes. There may thus well be evidence here of the beginnings of bony arches for supporting a bipedal foot. But, as Stern and Susman (1983) point out, the toes are long and slender, much longer than human toes and very similar to those of apes. They are curved, again like those of apes and unlike that of the human. There is a hint that the big toe could be opposed, like a human thumb, and, of course, like ape big toes. Latimer (1983), a colleague of Johanson, implies that these features in the foot of 'Lucy' mean it was a strong walker. The view of Johanson and Edey (1981) that 'elongated toes were of service in moving over rough stony ground or in mud, where gripping ability may have been important' is further explained by Latimer: the curvature of the phalanges may have been 'an adaptive response to large dorsoplantar bending moments'. But all of this suggests to Stern and Susman (1983a) that 'the foot and ankle remains reveal to us an animal that engaged in climbing as well as bipedality. There is no evidence whatsoever that any extant primate has long curved heavily muscled hands and feet for any purpose other than to meet the demands of full or part-time arboreal life'. Stern and Susman thus show that elements of 'Lucy's' foot resemble, to a degree, what Oxnard and colleagues have shown for the Olduvai foot.

Even the much vaunted footprints from Laetoli allegedly proving human-like bipedality show, when examined rather carefully and with comparable footprints of humans and apes available for comparison, features that imply otherwise. Thus Stern and Susman (1983a) concede that the big toe seems to be in line with the rest of the foot as in humans, but they point out that there is little evidence of a ball at the base of the big toe and that therefore this animal did not transfer its weight along the foot as we do. They also remark upon the variability of the foot impressions for the other toes: on some prints the big toe seems longer, on other prints the other toes seem longer. Chimpanzees often curl their long four lateral toes out of the way when walking bipedally thus shortening the impressions that they make, and it seems likely that so, too, on occasions, did these fossils.

There is even evidence from other regions of the body (some of it reviewed in Stern and Susman, 1983a). Thus a study of endocranial casts shows that though the large skull from East Turkana (KNM-ER 1470) has hominid patterns of external brain features in the speech area, the smaller skulls of some australopithecines (e.g. from East Turkana and from Southern Africa) do not. Though geologically younger than KNM-ER 1470, they demonstrate pongid-like features in that region (Falk, 1982).

The evidence that many of the recently studied australopithecines were proficient in the trees is now very good; that they were bipeds though in a manner different from humans is less so, but still persuasive. This is exactly what we have found in the australopithecines of relatively later date, those from Olduvai and Sterkfontein, Kromdraai and Makapansgat, at least some of which are contemporaneous with *Homo erectus*. Arboreality in these forms must have continued well beyond the period in which human-like bipedality had evolved in very early *Homo erectus*..

It is particularly significant that this rush of new studies is from several laboratories, some at least academically independent of those responsible for the earlier results that so strongly define australopithecines as fully bipedal, fully tool-using and -making, and essentially little different from humans. Though the standard idea is that some of the australopithecines are implicated in a lineage of human-like forms, the new possibility suggested in this book, a radiation separate from either humans or African apes, has received this powerful corroboration. It is now being recognized widely that the australopithecines are not structurally closely similar to humans, that they must have been living at least in part in arboreal environments, and that many of the later specimens were contemporaneous or almost so with the earliest members of the genus *Homo*. This means that the new possibility is now truly open for discussion.

New data for ramapithecines

The new findings on the ramapithecines are also critically important support for the arguments put forward in this book. In this case the new picture stems from study of the very large number of new fossils that have been found by Professor Wu Rukang and his colleagues in Lufeng county, Yunnan. The analysis of these data in a collaborative effort is now based upon fossil fragments that are almost two orders of magnitude more numerous than previously known (Wu and Oxnard, 1983a and b).

The current view of the ramapithecines depends upon little more than two dozen fragments, mainly of teeth and parts of jaws, that have been discovered since the first find reported on by G. Edward Lewis in 1934. The initial discovery prompted Lewis to recognize a new form that he called *Ramapithecus*. This was followed in later years by a handful of fossils that were each recognized as new forms and they were given a series of separate names (*Kenyapithecus, Graecopithecus, Rudapithecus, Sivapithecus*) based upon the geographical localities at which they were found. But in 1965 Simons and Pilbeam reviewed the entire series and held the view that all these forms really comprized two species groups. One of these, *Sivapithecus*, was basically ape-like and it was therefore put forward as an ape ancestor; the other, *Ramapithecus*, seemed to possess a number of hominid-like features and was therefore entered as an early hominid ancestor. This view was still extant in 1977 (Simons, 1977) but a series of more recent studies have cast doubt upon it.

Thus, Andrews and Cronin (1982) and Lipson and Pilbeam (1982) have all suggested that the non-Chinese ramapithecines are really only a single species or species group, that the two forms, (*Sivapithecus* and *Ramapithecus*), are really only the males and females of a strongly sexually dimorphic species group, that the features of the various parts that are now known (these now include a few somewhat more complete crania and jaws than were known previously) are basically more like orang-utans than anything else, and that consequently, the ramapithecines should be removed entirely from a place in human and African great ape ancestry. They suggest, indeed, that the entire group be recognized as only *Sivapithecus*, orang-utan ancestors.

Professor Wu and his colleagues (Wu, Xu and Lu, 1983) have examined the new finds from China (which include several almost complete crania and jaws, more than any non-Chinese site has yet yielded, see figure 1) and they have tentatively upheld this picture. They have reported that visiting scientists to China have, in general, agreed. In an even more recent publication, however, Professor Wu and his colleagues (Wu and Wu, 1983) retain the possibility that this view may not be correct.

One of the reasons for putting forward this new idea is an attempt to make these data conform to those suggested by the concept of the molecular clock. The molecular clock, assessing the time from common ancestry of two species using the notion that molecular evolution has taken place in a linear manner, suggests that humans and African apes had a common ancestor at five million years ago or even closer to the present time. If this were true it would be logically impossible for there to have existed prior ancestors of humans (ramapithecines date from 8 to 14 million years ago) that were more like humans than apes. The new views of the fossils have therefore concentrated on the ape-like features of ramapithecines. And of these, big sexual dimorphism is one of the most powerful, being found in every great ape known, but not markedly present in any species of the genus *Homo* so far identified.

Our analyses of the data on the lengths, breadths and heights of over 900 teeth from the Yunnan coalfields, have given us the opportunity to test the current views about these fossils (Wu and Oxnard, 1983a and b).

The various dimensions were first examined assuming that indeed only a single species was present, and one that was sexually dimorphic. A typical result is shown in figure 2: the breadth of the lower canine. There certainly are two peaks present in the frequency distribution histograms. But unlike those resulting from sexual dimorphism in any species so far known, the amount of separation between the peaks is such that, if it indeed represented sexual dimorphism, it would allow absolute certainty in sexing (for instance, on the basis of canine breadth alone).

Next, the assumption was made that each sex had indeed been misinterpreted, and the peaks truly represent two separate forms: a smaller animal and a larger one. The frequency distribution histograms were therefore derived separately for each and attempts made to fit these new distributions to normal curves. For a small number of tooth dimensions (20 out of 48 in the smaller of the two animals, 5 out of 48 in the larger) normal histograms exist (e.g. figure 3, although this does not necessarily imply that sexual differences are not present, only that they are not statistically separate – see Wu and Oxnard, 1983b. But for the great majority of tooth loci (28 in the smaller species, 43 in the larger) the distributions for the histograms are non-normal, usually bi-modal (e.g. figure 4) and this suggests that two sexes are present for each form.

That this bimodality is likely to be sexual dimorphism is further strengthened by the finding that it is greater for breadth measures of teeth rather than lengths (also figure 4), a finding similar to that reported in Chapter 8 for overall dimensions of the body. The data also suggest that the degree of sexual dimorphism is considerably greater in the larger animal than in the smaller (also figure 4).

And the further likelihood that these findings relate to sexual dimorphism is provided by particular findings for the canine teeth. These have a ratio of area (determined from the product of length and breadth) to the areas of various other tooth regions in the larger animal such as are found in more sexually dimorphic living species, and ratios in the smaller animal such as are generally found in species that demonstrate a small degree of sexual dimorphism (table 1). The relationship between canine height and the heights of other parts of the dentition in the larger animal show that the canines protrude well beyond the usual line of the tooth row, whereas in the smaller animal they do not (figure 5). Comparisons of canine heights in relation to incisor heights in extant forms provide yet further support to the idea that there are two groups here one of which is like those extant species that are dimorphic, the other like extant species that are less so (table 2). Histograms for this single dimension, canine tooth height, would be critical in studying canine sexual dimorphism; unfortunately, because of deficiencies in the materials this measure is only marginally useful statistically; yet even so it is suggestive of two different forms, a larger animal with a marked bimodal distribution, a smaller animal in which this can be less easily discerned (figure 6).

Yet other information is evident from study of these dimensions.

The ratios between the areas of the different functional regions of the dentition (table 3) suggest that the larger animal was possibly somewhat more herbivorous (proportionately smaller incisors relative to premolars and molars as in apes) in contrast to the smaller animal that may have been more omnivorous (smaller incisors in relation to premolars and molars as in humans).

The form of the bimodal distributions is such as to imply that the numbers of each sex present in the sample (ratio of females to males) was 2:1 to 4:1 in the larger animal, but 1:1 in the smaller. This appears to be the case from consideration of the number of specimens in each peak of the bimodal distributions. In the larger species it seems that the peak of larger specimens (males) is always between one half and one quarter of the size of the peak of smaller specimens; in contrast, in the smaller species, it seems that the peaks representing the two sexes are always approximately equal in size (figures 7 and 8). This in turn implies a socio-sexual arrangement in the larger animal that resembles a harem family structure, and in the smaller animal that is similar to a nuclear family organization.

All of this suggests very strongly (the size of the ramapithecine tooth sample is 955, almost two orders of magnitude greater than the number of prior known fragments, a very large sample upon which to be able to work) that there truly are two species (or species groups) present in Yunnan. One of these, the larger creature, (*Sivapithecus*), with larger dental sexual dimorphism, larger canine dimorphism, larger canine heights and areas, more herbivorous dentition and considerably smaller number of males than females has attributes that are matched by many of the apes. In contrast, the smaller creature (*Ramapithecus*) possess smaller dental sexual dimorphism, smaller canine dimorphism, smaller canine heights and areas, more omnivorous dentition and equal numbers of male and females and thus has attributes that would not deny it a place in a radiation of prehuman forms.

There are many implications that follow from these data, not the least being that much more study is now possible. But at least one implication is that we really are seeing here species that may have been involved, separately, in both human and ape like radiations very early in time. Unless several series of curious reversals have occurred, this may well mean that the molecular clock does not run in a linear manner and that dates cannot be derived from molecular data in the simple ways that have so far been suggested. At least another implication is confirmation of the idea that there may have been a series of radiations of species rather than a linear descent towards humans. And yet a third implication is an enormous strengthening of the idea that Africa was not necessarily the place where it all happened; Asia, though far less studied, may have been enormously important in higher primate evolution.

Conclusions

It is not common, in the sciences, that such massive confirmation as is evident from these new studies of australopithecines and ramapithecines, should follow so closely on the heels of controversial ideas; but this has occurred; it lends powerful confirmation of them; and it is powerful support, therefore, for the methods that have produced them.

References

Andrews, P. and Cronin, J.E. (1982). The relationships of *Sivapithecus* and *Ramapithecus* and the evolution of the orang-utan. *Nature,* **297**, 541–546.

Cook, D.C., Buikstra, J.E., DeRousseau, C.J. and Johnanson, D.C. (1983). Vertebral pathology in the Afar australopithecines. *Amer. J. Phys. Anthrop.* **60**, 83–102.

Falk, D. (1983). Oldest human-like sulcal pattern in the fossil record. *Amer. J. Phys. Anthrop.* **60**, 60–192.

Feldesman, M.R. (1982a). Morphometric analysis of the distal humerus of some cenozoic catarrhines: the late divergence hypothesis revisited. *Amer. J. Phys. Anthrop.* **59**, 73–76.

Feldesman, M.R. (1982b). Morphometrics of the ulna of some cenzoic 'hominoids'. *Amer. J. Phys. Anthrop.* **57**, 187.

Latimer, B. (1983). The anterior foot skeleton of *Australopithecus afarensis. Amer. J. Phys. Anthrop.* **60,** 217.

Lipson, S. and Pilbeam, D. (1982). *Ramapithecus* and hominoid evolution. *J. Hum. Evol.* **11,** 545–548.

Napier, J.R. (1962). Fossil hand bones from Olduvai Gorge. *Nature*, **196**, 400–411.

Prost, J. (1980). The origin of bipedalism. *Amer. J. Phys. Anthrop* **52,** 175–190

Senut, B. (1981). Humeral outlines in some hominoid primates and in plio-pleistocene hominids. *Amer. J. Phys. Anthrop.* **56**, 275-284.

Simons, E. and Pilbeam, D. (1965). Preliminary revision of the Dryopithecinae (Pongidae, Anthropoidea). *Folia Primatol.* **3**, 81–152.

Simons, E. (1977). *Ramapithecus. Sci. Amer.* **236**, 28–35.

Stern, J.T. Jr. and Susman, R.L. (1983a). The locomotor anatomy of *Australopithecus afarensis. Amer. J. Phys. Anthrop*, **60**, 279–318.

Stern, J.T. Jr. and Susman, R.L. (1983b). Functions of peroneus longus and brevis during locomotion in apes and humans. *Amer. J. Phys. Anthrop.* **60**. 256.

Susman, R.L., Stern, J.T. Jr. and Rose, M.D. (1983). Morphology of KNM-ER 3228 and OH 28 innominates from East Africa. *Amer. J. Phys. Anthrop.* **60**, 259.

Tardieu, C. (1981). Morpho-functional analysis of the articular surfaces of the knee-joint in primates. In A.B. Chiarelli and R.S. Corrucini (eds.) *Primate evolutionary biology*. (Berlin, Springer Verlag), pp. 68–80.

Wu, R. and Wu, X. (1983). Hominid fossils from China and their relation to neighbouring regions. In *The palaeoenvironment of East Asia from the mid tertiary*. Centre for Asian Studies, University of Hong Kong. In the press.

Wu, R. and Oxnard, C.E. (1983a). Ramapithecines from China: evidence from the dimensions of the teeth. *Nature*, Submitted.

Wu, R. and Oxnard, C.E. (1983b). *Ramapithecus* and *Sivapithecus* from China: some implications for higher primate evolution. *Amer. J. Primatol.* In the press.

Wu, R., Xu, Q, and Lu, Q. (1983) Morphological features of *Ramapithecus* and *Sivapithecus* and their phylogenetic relationships. *Acta Anthr. Sinica*, **2**, 1–14.

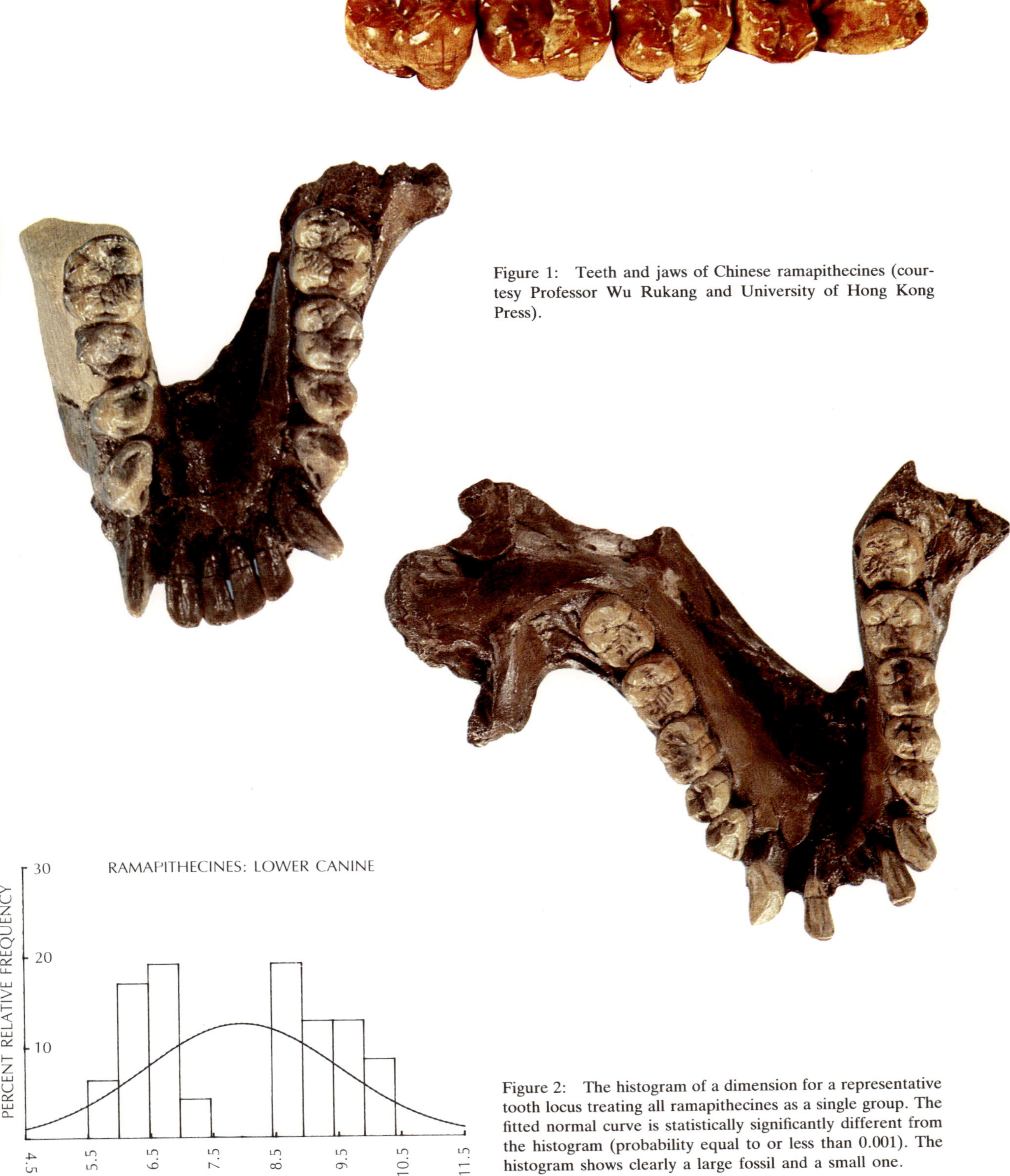

Figure 1: Teeth and jaws of Chinese ramapithecines (courtesy Professor Wu Rukang and University of Hong Kong Press).

Figure 2: The histogram of a dimension for a representative tooth locus treating all ramapithecines as a single group. The fitted normal curve is statistically significantly different from the histogram (probability equal to or less than 0.001). The histogram shows clearly a large fossil and a small one.

◀

Figure 3: An example in the two groups of fossils for a particular tooth locus of histograms that are not significantly different from their fitted normal curves (left-hand frame = larger fossil; right-hand frame = smaller fossil).

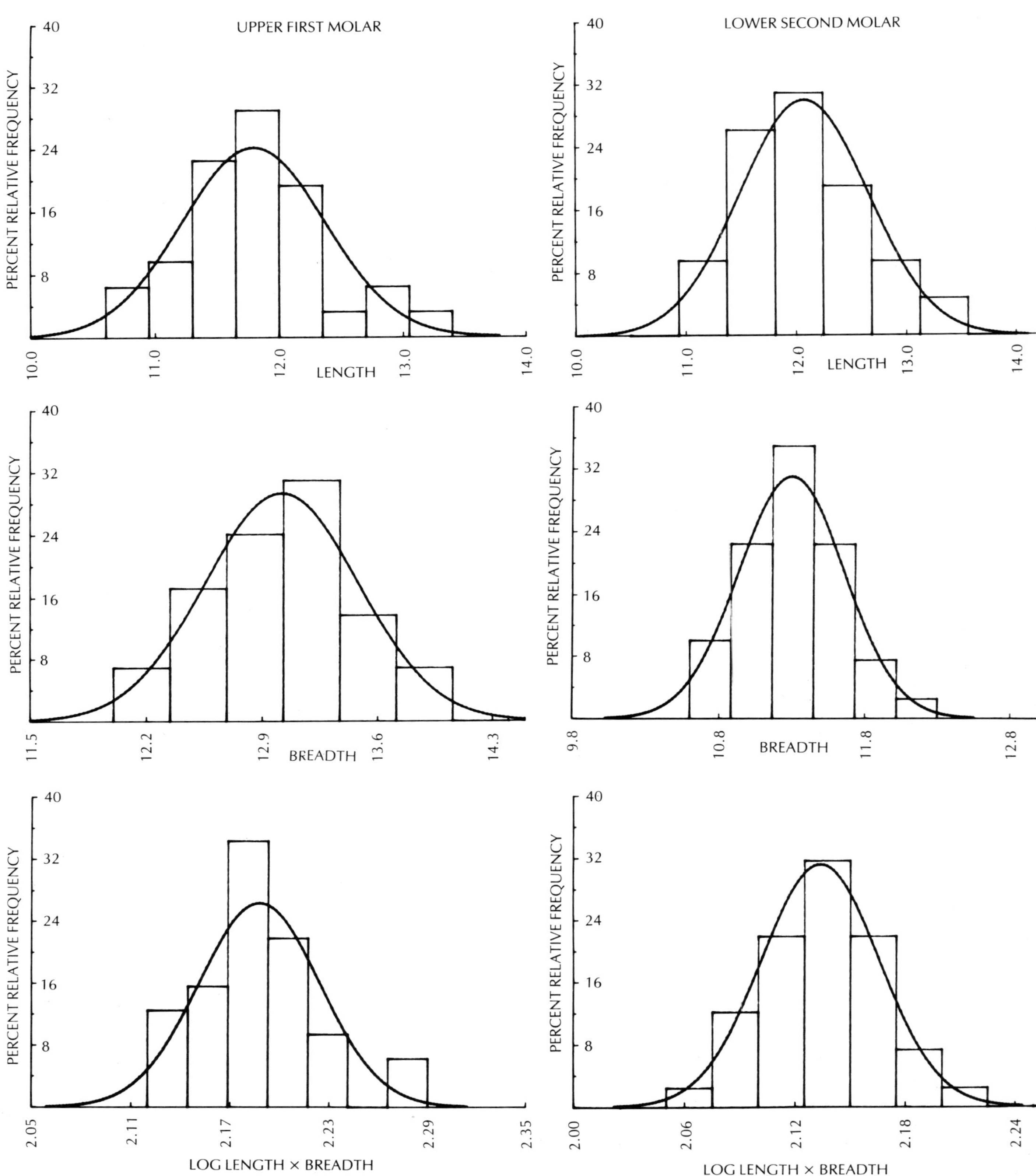

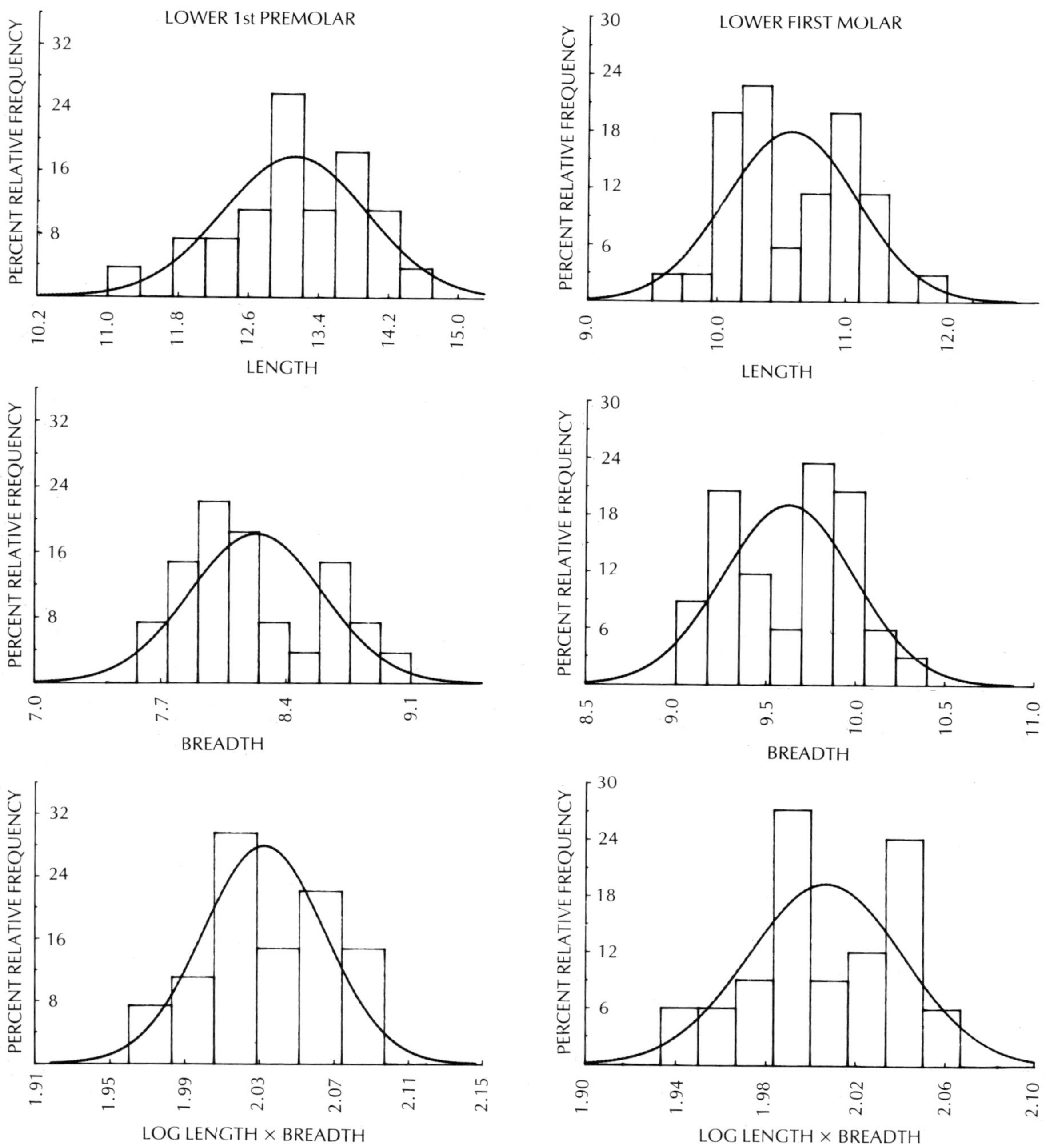

Figure 4: An example in the two groups of fossils for a particular tooth locus of histograms that are bimodal in form and statistically significantly different from their fitted normal curves (left-hand frame = larger fossil; right-hand frame = smaller fossil).

Table 1: RATIOS BETWEEN AREA OF CANINE AND AREA OF EACH OTHER DENTAL REGION (SEXES AVERAGED)

Ratio	Homo	Gorilla	Pan	Pongo	Small Fossil	Large Fossil
C'/I^{1+2}	0.72	1.29	0.84	0.79	0.74^{H}	1.35^{A}
C'/P^{3+4}	0.38	0.67	0.88	0.54	0.42^{H}	0.56^{A}
C'/M^{1+2+3}	0.13	0.31	0.40	0.30	0.15^{H}	$0.22^{A\text{-}H}$
$C,/I_{1+2}$	0.70	0.98	0.74	0.69	0.87^{ALL}	1.12^{Unique}
$C,/P_{3+4}$	0.40	0.74	0.90	0.71	$0.61^{H\text{-}G}$	0.73^{A}
$C,/M_{1+2+3+}$	0.14	0.35	0.42	0.36	0.25^{H}	0.31^{A}

H = Human-like, A = Ape-like.

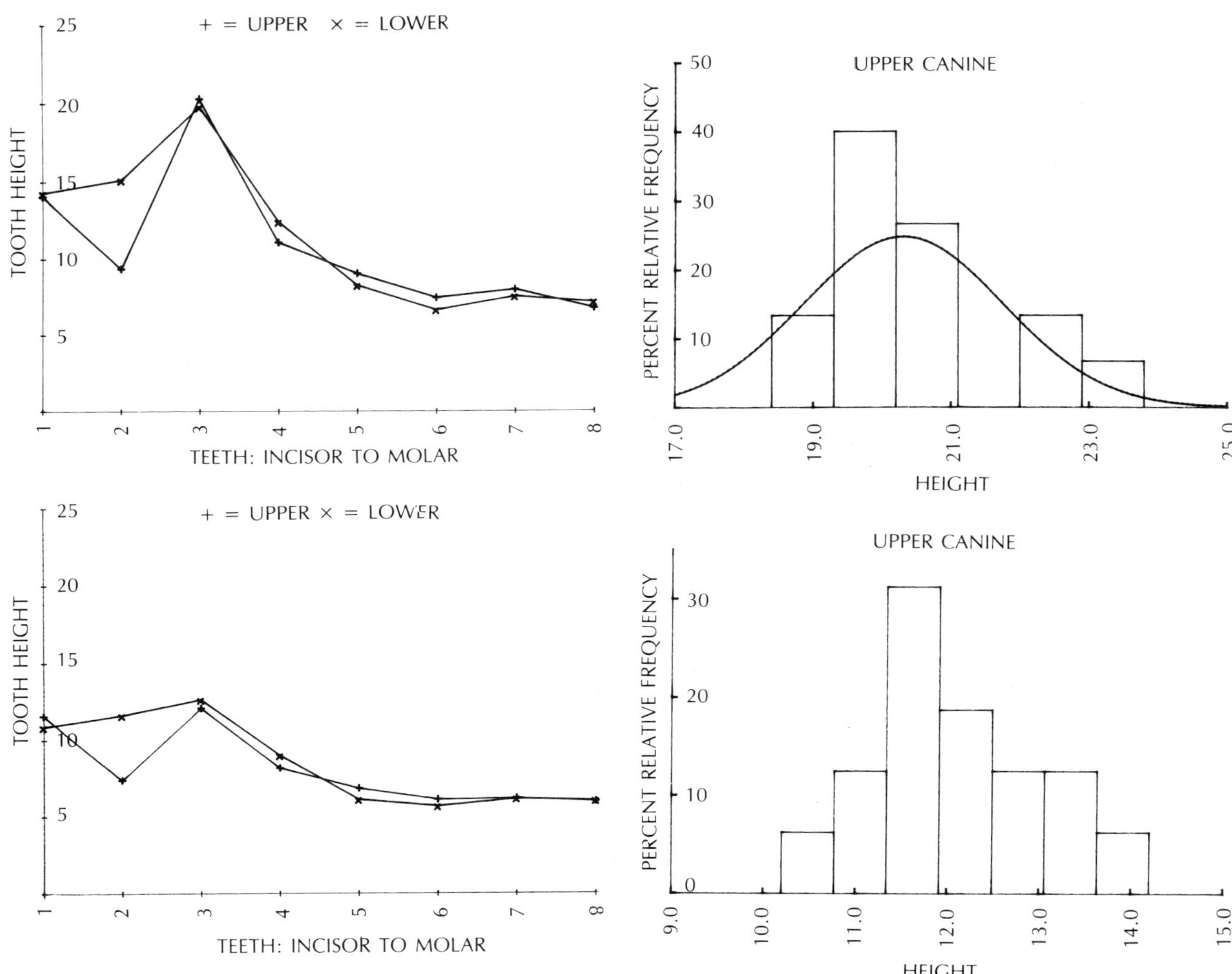

Figure 5: Mean heights for crowns of teeth along the tooth row in the larger fossil (above) and the smaller fossil (below).

Figure 6: Histograms and fitted normal curves (where computable) for canine heights in the larger fossil (above) and the smaller fossil (below).

Table 2: HEIGHTS OF CANINES COMPARED WITH HEIGHTS OF INCISORS IN VARIOUS HOMINOIDS (SEXES AVERAGED)

Ratio	Homo	Gorilla	Pan	Pongo	Small Fossil	Large Fossil
C'/I^1	0.96	1.92	1.47	1.30	1.13^H	1.42^A
$C_{,}/I_1$	1.05	1.67	1.45	1.71	1.10^H	1.39^A
C'/I^2	1.20	2.29	1.63	1.45	1.68^A	2.09^A
$C_{,}/I_2$	1.07	1.67	1.36	1.38	1.05^H	1.30^A

H = Human-like, A = Ape-like.

Table 3: RATIOS BETWEEN AREAS OF VARIOUS DENTAL REGIONS (SEXES AVERAGED)

Ratio	Home	Gorilla	Pan	Pongo	Small Fossil	Large Fossil
$I^{1+2} + C'/P^{1+2} + M^{1+2+3}$	0.65	1.00	1.35	1.22	0.73^H	1.17^A
$I_{1+2} + C_{,}/P_{1+2} + M_{1+2+3}$	0.69	1.01	1.48	1.22	$0.95^{H\text{-}G}$	1.02^A
P^{3+4}/M^{1+2+3}	0.34	0.45	0.46	0.55	0.37^H	$0.39^{\text{least distant G}}$
P_{3+3}/M_{1+2+3}	0.35	0.47	0.47	0.51	$0.41^{H\text{-}G}$	$0.43^{\text{least distant G}}$

H = Human-like, A = Ape-like, G = Gorilla-like,

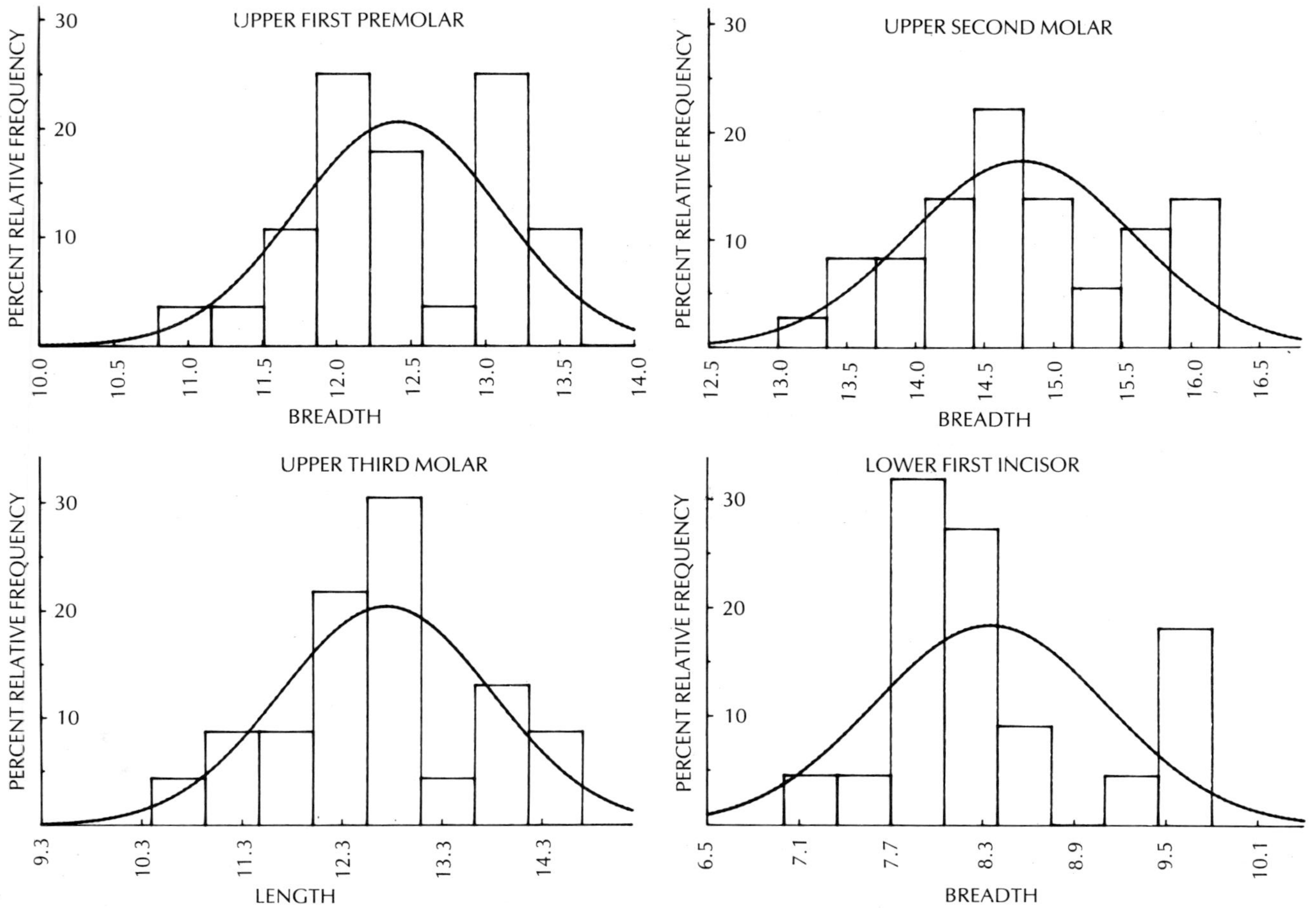

Figure 7: Several examples of bimodal histograms in the larger fossil. There are clearly two forms of the large fossil, a smaller form that is between twice and four times as frequent as the even larger form.

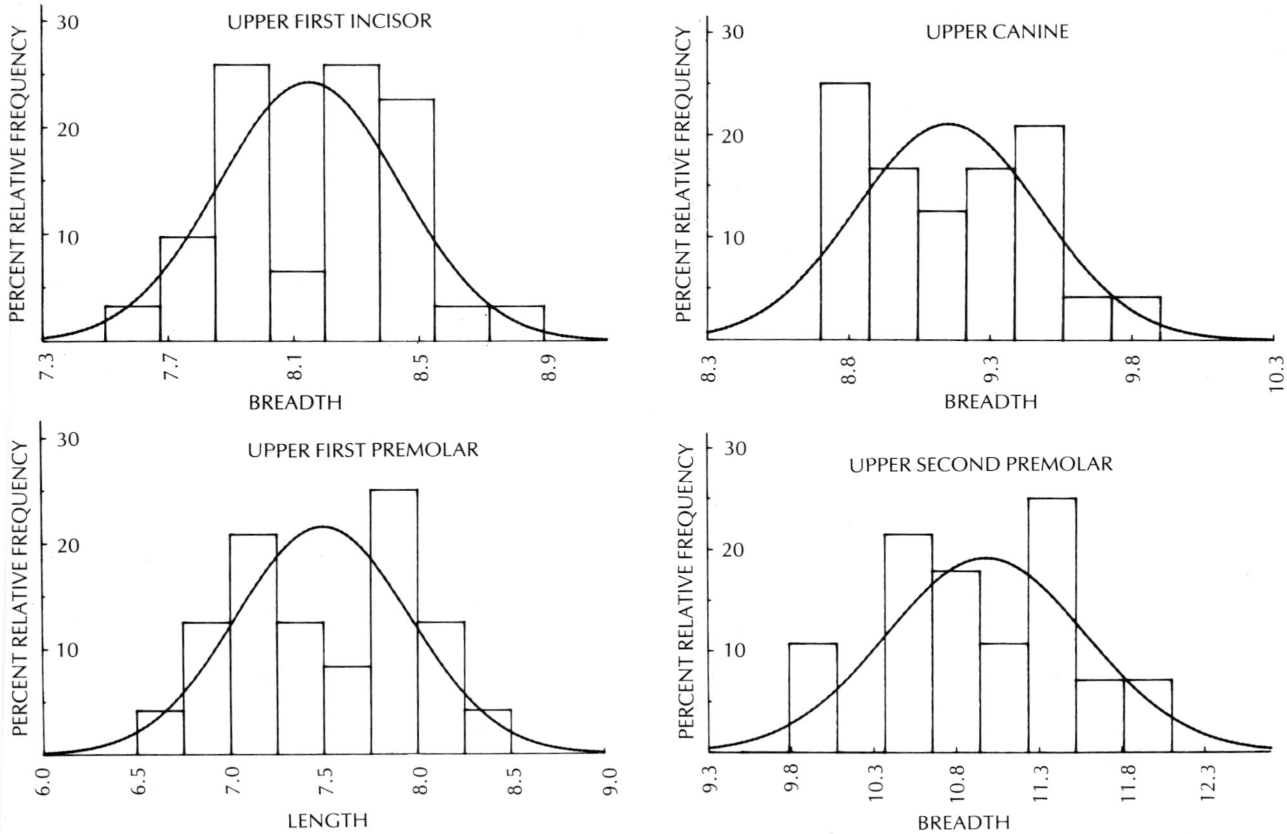

Figure 8: Several examples of bimodal histograms in the smaller fossil. There are clearly two forms of the small fossil, an even smaller form that is equally as frequent as the larger form.

References

Abel, O. (1931). *Die Stellung des Menschen im Rahmen der Wirbeltiere* (Jena, Fischer), pp. 1–398.

Adams, L.M. and Moore, W.J. (1975). Biomechanical appraisal of some skeletal features associated with head balance and posture in the Hominoidea. *Acta Anatomica* **92**, 580–94.

Agin, G.J. and Binford, T.O. (1976). Computer description of curved objects. *IEEE Trans. Comput.* **C–25**, 439–49.

Albrecht, G.H. (1978). The cranio-facial morphology of the Sulawesi macaques (multivariate approaches to biological problems). *Contr. Primatol.* **13**, 1–151.

Albrecht, G.H. (1979). The study of biological versus statistical variation in multivariate morphometrics: the descriptive use of multiple regression analysis. *Syst. Zool.* **28**, 338–44.

Albrecht, G.H. (1980). Multivariate analysis and the study of form, with special reference to canonical variate analysis. *Amer. Zool.* **20**, 679–93.

Alexander, R. McN. (1968). *Animal Mechanics.* (Seattle, University of Washington Press).

—— (1971). *Size and Shape.* (London, Clowes).

—— (1974). The mechanics of jumping by a dog (*Canis familiaris*). *J. Zool. Lond.* **173**, 549–73.

—— (1975). *The Chordates.* (London, Cambridge University Press)

—— (1977). Mechanics and scaling of terrestrial locomotion. In T.J. Pedley (ed.) *Scale Effects in Animal Locomotion.* (New York, Academic Press), pp. 93–110.

—— (1980). Forces in animal joints. *Eng. Med.* **9**, 93–97.

—— and Vernon, A. (1975). The mechanics of hopping by kangaroos (Macropodidae). *J. Zool. Lond.* **177**, 265–303.

Almir, L.U. and Shamir, J. (1976). Pattern recognition using one-dimensional Fourier transformations. *Optics Commun.* **18**, 304–306.

Andrews, D.F. (1972). Plots of high-dimensional data. *Biometrics* **28**, 125–136.

—— (1973). Graphical techniques for high-dimensional data. In T. Cacoullos (ed.) *Discriminant Analysis and Application.* (New York, Academic Press), pp. 37–39.

Andrews, H.C. (1970). *Computer Techniques in Image Processing.* (New York, Academic Press).

——(1972). *An Introduction to Mathematical Techniques in Pattern Recognition.* (New York, Wiley).

—— and Pratt, W.K. (1969). Transform image coding. In *Symp. Computer Processing in Communications* (Polytech. Instit. Brooklyn, April, 1969), pp. 63–84.

Anstey, R.L. and Delmet, D.A. (1972). Genetic meaning of zooecial chamber shapes in fossil Bryozoans: Fourier analysis. *Science* **177**, 1000–1002.

Appleby, R.M. and Jones, G.L. (1976). The analogue video reshaper – a new tool for paleontologists. *Paleontology* **19**, 565–86.

Arnold, J.S. 1966. External and trabecular morphologic changes in lumbar vertebrae in aging. In National Institute of Arthritis and Metabolic Disease *Progress in Methods of Bone Mineral Measurement.* (Bethesda, Md., U.S. Govt. Printing Office, pp. 352–411.

Ashton, E.H. (1949). A vice to facilitate the measurement of small skulls. *Amer. J. Phys. Anthrop.* **7**, 595–97.

——, Flinn R.M. and Moore, W.J. (1975). The basicranial axis in certain fossil hominoids. *J. Zool. Lond.* **176**, 577–91.

——, Flinn, R.M., Moore, W.J., Oxnard, C.E. and Spence, T.F. (1981). Further quantitative studies of form and function in the primate pelvis with special reference to *Australopithecus. Trans. Zool. Soc. Lond.* **360** 1–98.

——, Flinn, R.M. and Oxnard, C.E. (1975). The taxonomic and functional significance of overall body proportions in primates. *J. Zool. Lond.* **175**, 73–105.

——, Flinn, R.M., Oxnard, C.E., and Spence, T.F. (1971). The functional and classificatory significance of combined metrical features of the primate shoulder girdle. *J. Zool. Lond.* **,163**, 319–50.

——, Flinn, R.M., Oxnard, C.E., and Spence, T.F. (1976). The adaptive and classificatory significance of certain quantitative features of the forelimb in primates. *J. Zool. Lond.* **179**, 515–56.

——, Healy, M.J.R., Oxnard, C.E., and Spence, T.F. (1965). The combination of locomotor features of the primate shoulder girdle by canonical analysis. *J. Zool. Lond.* **147**, 406–29.

——, Healy, M.J.R. and Lipton, S. (1957). The descriptive use of discriminant functions in physical anthropology. *Proc. Roy. Soc. Lond. Ser. B* **146**, 555–72.

—— and Oxnard, C.E. (1958). Some variations of the maxillary nerve in primates. *Proc. Zool. Soc. Lond.* **131**, 457–70.

—— and Oxnard, C.E. (1963). The musculature of the primate shoulder. *Trans. Zool. Soc. Lond.* **29**, 553–650.

—— and Oxnard, C.E. (1964a). Locomotor patterns in primates *Proc. Zool. Soc. Lond.* **142**, 1–28.

—— and Oxnard, C.E. (1964b). Functional adaptations in the primate shoulder girdle. *Proc. Zool. Soc. Lond.* **142**, 49–66.

—— and Zuckerman, S. (1958). The infraorbital foramen in the Hominoide. *Proc. Zool. Soc. Lond.* **131**, 417–85.

Atchley, W.R., Gaskins, C.T. and Anderson D. (1976). Statistical properties of ratios. I. Empirical results. *Syst. Zool.* **25**, 137–48.

Attneave, F. and Arnoult, M.D. (1966). The quantitative study of shape and pattern perception. In L. Uhr. (ed) *Pattern Recognition* (New York, Wiley), pp. 123–41.

Baba, M.L., Darga, L.L. and Goodman, M. (1979). Immunodiffusion systematics of the primates. Part V. The platyrrhini. *Folia Primatol.* **32**, 207–238.

Bacon, G.E., Bacon, P.J., and Griffiths, R.K. (1979). Stress distribution in the scapula studied by neutron diffraction. *Proc. Roy. Soc. Lond* **204**, 355–62.

Baldwin, L.A., and Teleki, G. (1976). Patterns of gibbon behaviour on Hall's Island, Bermuda. A preliminary ethogram for *Hylobates lar.* In *Gibbon and Siamang* (Basel Karger) Vol. 4, pp. 21–105.

Barnicott, N.A. (1969). Some biochemical and serological aspects of primate evolution. *Sci. Progr. Oxf.* **57**, 459–93.

Barski, G., Butler, J.W., and Thomas, R.J. (1969). Computer analysis of animal cell movement *in vitro. Exp. Cell Res.* **56**, 363–68.

Basmajian, J.V. (1967) *Muscles Alive* (Baltimore Williams Wilkins).

Beard, J.M., Barnicott, N.A. and Hewett-Emmet, D. (1976). Alpha and beta chains of the major haemoglobin and a note on the minor component of *Tarsius. Nature* **259**, 338–41.

—— and Goodman, M (1976). The haemoglobins of *Tarsius bancanus* . In M. Goodman and R.E. Tashian (ed.) *Molecular Anthropology* (New York, Plenum), pp. 239–55.

Bearder, S.K. (1974). Aspects of the ecology and behaviour of the thick-tailed bushbaby *Galago crassicaudatus.* (Ph. D. thesis) University of the Witzwatersrand, Johannesburg).

—— and Doyle G.A. (1974). Ecology of bushbabies, *Galago senegalensis* and *Galago crassicaudatus*, with some notes on their behaviour in the field. In R.D. Martin, G.A. Doyle and A.C. Walker (eds.) *Prosimian Biology* (London Duckworth) pp. 109–30.

Becker, H.C., Meyers, P.H. and Nice, C.M. Jr. (1969). Laser light diffraction, spatial filtering, and reconstruction of medical radiographic images. *Ann. N. Y. Acad. Sci.* **157**, 465–86.

Benveniste, R.A. and Todaro, T.C. (1976). Evolution of type C viral genes: evidence for an Asian origin of man. *Nature* **26**, 101–108.

Blackith, R.E. (1963). A multivariate analysis of Latin elegaic verse. *Lang. Speech* **6**, 196–205.

—— and Reyment, R.A. (1971). *Multivariate Morphometrics* (London, Academic Press).

Blum, H. (1962). An associative machine for dealing with the visual field and some of its biological implications. In E.E. Bernard and M.R. Kare (eds.), *Biological prototypes and synthetic systems*, (New York, Plenum Press) pp. 224–60.

—— (1967). A transformation for extracting new

descriptors of shape. In W. Walthen-Dunn (ed.), Models for the perception of speech and visual form. *Proc. Symp. Data Sciences Lab. and Air Force Cambridge Research Lab.* (Boston, M.I.T. Press) pp. 362–80.

Bock, R.D. (1975). *Multivariate Statistical Methods in Behavioural Research* (New York, McGraw-Hill).

Boer L.E.M. de. and Boer-Van der Vlist, J. de. (1973). The somatic chromosomes and the idiogram of *Tarsius syrichta carbonarius* Heude, 1898 (Primates: Tarsioidea). *Genen. Phaenen* **16**, 65 –71.

Bonner, J.T. (1965). *Size and Cycle* (Princeton, Princeton University Press).

Bookstein, F.L. (1977a). The measurement of biological shape and shape change (Ph.D. thesis, Univ. Michigan, Ann Arbor).

—— (1977b). The study of shape transformation after D'Arcy Thompson. *Math. Biosci.* **34**, 177–219.

—— (1978). *Lecture Notes in Biomathematics* (Berlin, Springer-Verlag).

—— (1980). When one form is between two others: an application of biorthogonal analysis. *Amer. Zool.* **20**, 627–41.

Brehm, A.E. (1868). *La vie des animaux illustrèe, Les Mammifères* (Paris Baillière et Fils), vol. 1, pp. 130 and 140.

Brown, T. (1973). *Morphology of the Australian Skull: Studied by Multivariate Analysis* (Canberra, Australian Instit. Aboriginal Studies).

Bruce, E.J. and Ayala, F.J. (1978). Humans and apes are genetically very similar. *Nature* **276**, 264–65.

Buffon, G.L.L., Comte de. (1765). *Histoire naturelle, gènèrale et particulière* (Paris, L'imprimerie du Roi) **13**, 87–91.

—— (1789). *Histoire naturelle, gènèrale et particulière* (Paris, L'imprimerie du Roi) **7**, 268–71.

Butler, J.W., Butler, M.K. and Marczynska, B. (1969). Automatic analysis of 835 marmoset spreads. *Ann. N.Y. Acad. Sci.* **157**, 424–37.

Cable, L.E. and Van Haagen, R.H. (1967). Digital caliper. *Copeia* **3**, 683–85.

Calow, L.J. and Alexander, R. McN. (1973). A mechanical analysis of the hind leg of a frog. *J. Zool. Lond.* **171**, 293–321.

Campbell, B. (1937). The shoulder musculature of the platyrrhine monkeys. *J. Mammal.* **8**, 66–71.

Carpenter, C.R. (1934). A field study of the behaviour and social relations of the howling monkeys (*Alouatta palliata*). *Comp. Psychol. Monogr.* **10**, 1–168.

—— (1940). A field study in Siam of the behaviour and social relations of the gibbon (*Hylobates lar*). *Comp. Psychol. Monogr.* **16**, 1–212.

—— (1976). Suspensory behaviour of gibbons *Hylobates lar*: a photoessay. In *Gibbon and Siamang*. (Basel, Marger), Vol. 4. pp. 1–20.

Cartmill, M. (1970). The orbits of arboreal mammals: a reassessment of the arboreal theory of primate evolution (Ph.D. Thesis, Univ. Chicago, Chicago).

—— (1972). Arboreal adaptations and the origin of the order Primates. In R. Tuttle (ed.), *The Functional and Evolutionary Biology of Primates* (Chicago, Aldine-Atherton), pp. 97–122.

—— and Milton, K. (1977). The lorisiform wrist joint and the evolution of 'brachiating' adaptations in the Hominoidea. *Amer. J. Phys. Anthrop.* **47**, 249–72.

Cave, A.J.E. (1973). The primate nasal fossa. *J. Linn. Soc.* **5**, 377–87.

Chance, M.R.A., and Jolly, C.J. (1970). *Social Groups of Monkeys, Apes and Men* (New York, Dutton).

Charles-Dominique, P. (1971). Eco-ethologie des prosimiens du Gabon. *Biol. Gabon.* **2**, 121–228.

—— (1972). Ecologies et vie sociale de *Galago demidovii* (Fischer 1808, Prosimii). *Z. f. Tierpsychol. suppl.* **9**, 7–41.

—— (1977). *Ecology and Behaviour of Nocturnal Primates* (New York, Columbia University Press).

Chatterjee, S. (1974). A rhynchosaur from the upper triassic Maleri formation of India. *Phil. Trans. R. Soc. Lond. Ser. B* **267**, 209–61.

Cheng, G.C. (1969). Pictorial pattern recognition. *Pattern Recognition* **1**, 187–94.

Chernoff, H. (1973). The use of faces to represent points in *k*-dimensional space graphically. *J. Amer. Stat. Assoc.* **68**, 361–68.

Chiarelli, A.B. (1973). *Evolution of the Primates* (London, Academic Press).

Chmielewski, N. and Varner, J.R. (1969). An application of holographic contouring in dentistry. *Biomed. Sci. Instrum.* **6**, 72–79.

Chopra, S.R.K (1958). A 'pelvimeter' for orientation and measurement of the innominate bone. *Man* **58**, 126–128.

Chu, E.H.Y. and Bender M.A. (1961). Chromosome cytology and evolution in primates. *Science* **133**, 1399–1405.

Clark, W.E. Le Gros. (1924). Notes on the living tarsier (*Tarsius spectrum*) *Proc. Zool. Soc. Lond.* **1924**, 217–23.

—— (1959). *The Antecedents of Man.* (Edinburgh, University Press).

Claus, W.D. Jr., McManus, L.R., and Durand, P.E. (1974). Development of head forms for sizing infantry helmets. *CEMIE Tech. Rep.* **75:23** (Natick, Mass., Clothing Equip. Mat. Eng. Lab.), pp.1–26.

Clutton-Brock, T.H., Harvey, P.H. and Rudder, B. (1977) Sexual dimorphism, socionomic sex ratio, and body weight in primates. *Nature,* **269**. 797 –800.

Coimbra-Filho, A.F. and Mittermeier, R.A. (1973). New data on the taxonomy of the Brazilian marmosets of the genus *Callithrix* Erxleben, 1877. *Folia Primatol.* **20**, 241–64.

Cole, A.J. (1969). *Numerical Taxonomy* (New York, Academic Press).

Cooley, W.W. and Lohnes, P. R. (1971). *Multivariate Data Analysis* (New York, Wiley).

Coolidge, H.J. (1933). *Pan paniscus,* pygmy chimpanzee from south of the Congo river. *Amer. J Phys. Anthrop.* **18**, 1–59.

Coppens, Y., Howell, F.C., Isaac, G.L. and Leakey, R.E.F. (1976). *Earliest Man and Environments in the Lake Rudolf Basin.* (Chicago, University of Chicago).

Corruccini, R.S. (1975). Morphometric assessment of australopithecine postcranial affinities. *Syst. Zool.* **24**, 226–33.

Creel, N. (1976). A stereophotogrammetric system for biological applications. *Amer. J. Phys. Anthrop.* **44**, 173.

Creel, N. and Preuschoft, H. (1970). Hominoid taxonomy. A canonical analysis of cranial dimensions. *Proc. 3rd. Int. Congr. Primatol.* Zurich, 1970) **1**, 36–43.

Crompton, A.W. and Jenkins, F.A. Jr. (1973). Mammals from reptiles: a review of mammalian origins. In F.A. Donath, F.G. Stehli and W.G. Wetherill (eds.) *Annual Reviews of Earth and Planetary Sciences* (Palo Alto, Ann. Rev.) pp. 131–55.

Cutrona, L.J. (1965). Recent developments in optical technology. In J.T. Tippett, D.A. Berkowitz, L.D. Clapp, C.J. Koester and A. Vanderburgh, Jr. (eds.) *Optical and Electro-optical Information Processing* (Cambridge, MIT Press), pp. 83–123.

Cuvier (1798). *Tableau Elementaire de l'Histoire Naturelle.* Cited by Owen, 1866.

Davis, D.D. (1964). The giant panda: a morphological study of evolutionary mechanisms. *Field. Zool. Mem.* **3**, 1–339.

Davis, J.C. (1970). Optical processing of microporous fabrics. In J.L. Cutbill (ed.) *Data Processing in Biology and Geology* (London, Academic Press), pp. 69-87.

—— (1973). *Statistics and Data Analysis in Geology* (New York, Wiley).

Day, M.H. (1976). Hominid postcranial material from Bed I, Olduvai Gorge. In G.L. Isaacs and E. McCown (eds.), *Human origins* (Menlo Park, California, Staples Press), pp. 363–74.

—— (1977). *Guide to Fossil Man* (London, Cassel).

—— (1978). Functional interpretation of the morphology of postcranial remains of early African hominids. In C. Jolly (ed.) *Early Hominids of Africa* (London, Duckworth).

—— and Napier, J.R. (1964). Hominid fossils from bed I, Olduvai Gorge, Tanganyika: fossil foot bones. *Nature* **201**, 967–70.

—— and Pitcher-Wilmott, R.W. (1975). Sexual differentiation in the innominate bone studied by multivariate analysis. *Ann. Hum. Biol.* **2**, 143–51.

—— and Wickens, J. (1980). Hominid footprints, photogrammetry and bipedalism. *Proc. Anat. Soc. Grt. Brit. Ire.* **26**, 19–22.

—— and Wood, B.A. (1968). Functional affinities of the Olduvai hominid 8 talus. *Man* **3**, 440–55.

De Beer, Sir G. (1940). *Embryos and Ancestors* (Oxford, Clarendon Press).

Delson, E. (1977). Catarrhine phylogeny and classification: principles, methods and comments. *J. Hum. Evol.* **6**, 433–59.

—— and Andrews, P. (1975). Evolution and interrelationships of the catarrhine primates. In W.P. Luckett and F.S. Szalay (eds.) *Phylogeny of the Primates* (New York, Plenum), pp. 405–446.

——, Eldridge, N. and Tattersall, I. (1977). Reconstruction of hominid phylogeny: a testable framework based on cladistic analysis. *J. Hum. Evol.* **6**, 263–78.

Dene, H.M., Goodman, M., Prychkodko, W. and Moore, G.W. (1976). Immunodiffusion systematics of the Primates. *Folia Primatol.* **25**, 35–61.

Dobrin, M.B. (1968). Optical processing in the earth sciences. *IEEE Spectrum* **5**, 59–66.

——, Ingalls, A.L. and Long, J.A. (1965). Velocity and frequency filtering of seismic data using laser light. *Geophysics* **30**, 1144–78.

Duda, R.O. and Hart, P.E. (1973). *Pattern Classification and Scene Analysis* (New York, Wiley Interscience).

Dunn, J.C. (1975). Group averaged linear transforms that detect corners and edges. *IEEE Trans Comput.* **C-24**, 1191–1201.

Eccles, M.J., McQueen, M.P.C. and Rosen, D. (1977). Analysis of the digitized boundaries of planar objects. *Pattern Recognition* **9**, 31–41.

Ellefson, J. (1974). A natural history of the gibbons in the Malay peninsula. In D. Rumbaugh (ed.) *Gibbon and Siamang* (Basel, Karger) Vol. 3 pp.1–136.

Elliot, D.G. (1913). *A Review of the Primates* (New York, the American Museum of Natural History).

Elliot, S. B. and Morris, W.J. (1978). *Holographic Contouring of Fossil Teeth.* (Los Angeles, Occidental College), pp. 1–15 (mimeographed).

Erikson, G.E. (1963). Brachiation in New World monkeys and in anthropoid apes. *Symp. Zool. Soc. Lond.* **10**, 135–63.

Evans, F.G. (1957). *Stress and Strain in Bones* (Springfield, Thomas).

Feldesman, M.R. (1976). *The Primate Forelimb: Morphometric Study of Locomotor Diversity*, Univ. Oregon Anthrop. Papers No 10, pp. 1–154.

—— (1978). *Multivariate Morphometric Analysis of* Dryopithecus (Proconsul) africanus *Forelimb* (Oregon, Portland State Univ.) pp. 1–32 (mimeographed).

—— (1979). Further morphometric studies of the ulna from the Omo Basin, Ethiopia. *Amer. J. Phys. Anthrop.* **51**, 409–16.

Fienberg, S.E. (1979). Graphical methods in statistics. *The Amer. Stat.* **33**, 165–78.

Fisher, R.A. (1936). The use of multiple measurements in taxonomic problems. *Ann. Eugenics* **7**, 179–88.

Fitch, W.M. (1977). The phyletic interpretation of macromolecular sequence information: simple methods. In M. Goodman and R.E. Tashian (ed.) *Molecular Anthropology* (New York, Plenum), pp. 211–48.

—— and Langley C.H. (1976). Evolutionary rates in proteins: neutral mutations and the molecular clock. In M. Goodman and R.E. Tashian (ed.), *Molecular Anthropology* (New York, Plenum), pp. 197–219.

Fleagle, J.G. (1974). Dynamics of brachiating siamang [*Hylobates* (*Symphalangus*) *syndactylus*]. *Nature* **248**, 259–60.

—— (1976). Locomotor behaviour and skeletal anatomy of sympatric Malaysian leaf monkeys (*Presbytis obscura* and *Presbytis melalophos*). *Yrbk. Phys. Anthrop.* **20**, 440–53.

—— (1977). Brachiation and biomechanics: the siamang as an example. *Malay. Nat. J.* **30**, 45–51.

—— (1978). Locomotion, posture and habitat utilization in two sympatric Malaysian leaf monkeys (*Presbytis obscura* and *Presbytis melalophos*) In G. Montgomery (ed.) *The Ecology of Arboreal Folivores.* (Washington, Smithsonian Institution Press), pp. 243–251.

——, Kay, R.F. and Simons, E.L. (1980) Sexual dimorphism in early anthropoids. *Nature* **287**, 328–330.

Fooden, J. (1961). Urinary amino acids of non-human primates. *Zoologica* **46**, 167–80.

—— (1969). Taxonomy and evolution of the monkeys of the Celebes (Primates: Cercopithecidae). *Bibl. Primatol.* **10**, 1–148.

Frey, H. (1923). Untersuchungen uber die Scapula, speziell uber ihre aussere Form und deren Abhangigkeit von der Funktion. *Z. Anat. EntwGesch.* **68**, 277–324.

Friedenthal, H. (1900). Ueber einen experimentellen Nachweis von Blutsverwantschaft. *Arch. Anat. Physiol Lpz., Physiol. Abt.*, 494–508.

Gabrielson, V.K. (1977). Mesh generation for two-dimensional regions using a DVST (Direct view storage tube) graphic terminal. *Comput. Graphics* **2**, 59–66.

Gautier-Hion, A. (1975) Dimorphism sexuelle et organisation sociale chez les cercopithecines forestiers africains. *Mammalia* **39** 365–374.

Gingerich, P.D. (1981) Cranial morphology and adaptations in eocene Adapidae. 1. Sexual dimorphism in *Adapis magnus* and *Adapis parisiensis*. *Amer. J. Phys. Anthrop.* **56**. 217–234.

Gnanadesikan, R. (1977) *Methods for Statistical Data Analysis of Multivariate Observations* (New York, Wiley).

Gmelin (1790). *Systema Naturae.* Cited by Owen, 1866.

Goodall, J. (1965). Chimpanzees of the Gombe State Reserve. In I. DeVore (ed.), *Primate Behaviour* (New York, Holt, Rhinehart and Winston), pp. 425–77.

—— (1971). *In the Shadow of Man* (Boston, Houghton Miflin).

Goodman, J.W. (1968). *Introduction to Fourier Optics* (New York, McGraw-Hill).

Goodman, M. (1976). Towards a genealogical description of the Primates. In M. Goodman and R.E. Tashian (eds.), *Molecular Anthropology* (New York, Plenum), pp. 321–53.

—— and Tashian, R.E. (1976). *Molecular Anthropology* (New York, Plenum).

Gordon, W.J. and Hall, C.A. (1973). Construction of curvilinear co-ordinate systems and applications to mesh generation. *Int. J. Numer. Methods Eng.* **7**, 461–77.

Gould, S.J. (1977). *Ontogeny and Phylogeny* (Cambridge, Harvard University Press).

Gower, J.C. (1967). A comparison of some methods of cluster analysis. *Biometrics* **23**, 623–36.

—— and Ross G.J.S. (1969). Minimum spanning trees and single linkage cluster analysis. *Applied Stat. C.* **18**, 54–64.

Gray, Sir J. (1968). *Animal Locomotion* (New York, Norton).

Groves, C.P. (1974). Taxonomy and phylogeny of prosimians. In R.D. Martin, C.A. Doyle and A.C. Walker (eds.), *Prosimian Biology* (London, Duckworth), pp. 449–73.

Hakim, N.S. and King A.I. (1978). A three-dimensional finite element dynamic response analysis of a vertebra with experimental verification. *J. Biomech.* **12**, 277–92.

Hall, B.K. (1978). *Developmental and Cellular Skeletal Biology* (London, Acad, Press).

Hall-Craggs, E.C.B. (1965a). The analysis of the jump of the lesser galago *J. Zool. Lond.* **147**, 20–29.

—— (1965b). An osteometric study of the hindlimb of the Galagidae. *J. Anat. Lond.* **99**, 119–26.

Hansinger, M.J. (1976). Hominids in plio/pleistocene Africa: an odontometric approach to phylogeny (Ph.D. thesis, University of Florida).

Harvey, P.H., Kavanagh, M. and Clutton-Brock, T.H. (1978) Sexual dimorphism in primate teeth. *J. Zool. Lond.* **186**. 475–485.

Henderson, A. and Wood, B.A. (1977). The functional anatomy of the Olduvai (O.H.8) foot. *J. Anat. Lond.* **124**, 252.

Herrera, A.E., Lehman, H., Joysey, K.A. and Friday, A.E. (1976). Evolution of myoglobin amino acid sequences in primates and other vertebrates. In M. Goodman and R.E. Tashian (eds.) *Molecular Anthropology* (New York, Plenum), pp.289–300.

Hershkovitz, P. (1977). *Living New World Monkeys* (Platyrrhini) (Chicago, University of Chicago Press).

Hennig, W. (1966). *Phylogenetic Systematics* (Urbana, University of Illinois Press).

Hewett-Emmett, D., Cook, C.N. and Barnicot, R.A. (1976). Old World monkey haemoglobins: deciphering phylogeny from complex patterns of molecular evolution. In M. Goodman and R.E. Tashian (eds.) *Molecular Anthropology* (New York, Plenum), pp. 257–76.

Hildebrand, M. (1967). Symmetrical gaits of primates. *Amer. J. Phys. Anthrop.* **26**, 119–30.

—— (1974). *Analysis of Vertebrate Structure.* (New York, Wiley).

—— (1976). Analysis of tetrapod gaits: general considerations and symmetrical gaits. In R. M. Herman, S. Grillner, P.S.G. Stern and D. G. Stuart (eds.) *Neural Control of Locomotion* (New York, Plenum Press), pp. 203–36.

—— (1977). Analysis of asymmetrical gaits. *J. Mammal.* **58**, 131–55.

—— (1980). The adaptive significance of tetrapod gait selection. *Amer. Zool.* **20**, 225–67.

Hildich, C.J. (1969). A system of automatic chromosome analysis. In A. Graselli (ed.) *Automatic Interpretation and Classification of Images* (New York, Academic Press), pp. 1–434.

Hill, A.V. (1970). *First and Last Experiments in Muscle Mechanics* (Cambridge, Cambridge University Press).

Hill, W.C.O. (1953–1966) *Primates. Comparative Anatomy and Taxonomy* (Edinburgh, Edinburgh University Press).

Holeman, J.M. (1968). Holographic character reader. In L.N. Kanal (ed.) *Pattern Recognition* (Washington, Thompson Book Co.), pp. 63–78.

Howell, F.C., Washburn, S.L. and Ciochon, R.L. (1978). Relationships of *Australopithecus* and *Homo. J. Hum. Evol.* **7**, 127–31.

Howells, W.W. (1969). Criteria for selection of osteometric dimensions. *Amer. J. Phys. Anthrop.* **30**, 451–58.

—— (1973). *Cranial Variation in Man: A Study by Multivariate Analysis of Patterns of Difference among Recent Human Populations* (Harvard, Peabody Museum Of Archaeology and Ethnology).

Inman, V.T., Saunders, J.B. de C.M. and Abbott, R.C. (1944). Observations on the function of the shoulder joint. *J. Bone Jt. Surg.* **26**, 1–30.

Ishida, H., Kimura, T. and Okada, M. (1974). Patterns of bipedal walking in anthropoid primates. *Symp. 5th Congr. Int'l. Primat. Soc.* **5**, 287–301.

Jenkins, F.A. Jr. (1970). Limb movements in a monotreme (*Tachyglossus aculeatus*): a cineradiographic analysis. *Science* **168**, 1473–75.

—— (1971). Limb posture and locomotion in the Virginia oppossum (*Didelphis marsupialis*) and other non-cursorial mammals. *J. Zool. Lond.* **165**, 303–15.

—— (1972). Chimpanzee bipedalism: cineradiographic analysis and implications for the evolution of gait. *Science* **178**, 877–79.

—— (1974). The movement of the shoulder in claviculate and aclaviculate mammals. *J. Morphol.* **144**, 71–84.

—— (1975). *Primate Locomotion* (New York, Academic Press).

——, Dombrowski, P. J., and Gordon, E.P. (1978). Analysis of the shoulder in brachiating spider monkeys. *Amer. J. Phys. Anthrop.* **48**, 65–76.

Johanson, D. and Edey, M. (1980) *Lucy: the beginings of humankind.* (Simon and Schuster, New York.)

Johanson, D.C. and Taieb, M. (1979). Plio-pleistocene hominid discoveries in Hadar, Ethiopia. *Nature* **260**, 293–97.

—— and White, T.D. (1979). A systematic assessment of early African hominids. *Science* **203**, 321–30.

Jolly, C.J. (1978). *Early Hominids of Africa* (London, Duckworth). P. 431–434 b50.

Jouffroy, F.-K. (1975). Osteology and myology of the lemuriform post-cranial skeleton. In I. Tattersall and R.W. Sussman (eds.) *Lemur Biology* (New York, Plenum), pp. 149–92.

—— and Gasc J.P. (1974). A cineradiographical analysis of leaping in an African prosimian (*Galago alleni*). In F.A. Jenkins (ed.) *Primate Locomotion.* (New York Academic Press), pp. 117–42.

—— and Lessertisseur, J. (1978). Étude ecomorphologique des proportions des membres des primates et specialement des prosimiens. *Ann. Sci. Nat. Zool. Paris* **20**, 99–128.

—— and J. Lessertisseur (1979). Relationships between limb morphology and locomotor adaptations among prosimians: an osteometric study. In M.E. Morbeck, H. Preuschoft and N. Gomberg (eds.), *Environment, Behaviour and Morphology: Dynamic Interactions in Primates* (New York, Fisher), pp. 143–82.

——, Oxnard, C.E. and German, R.Z. (1982). Interpretation phylogenetique des proportions des members des tarsiers, par comparaison avec les autres prosimiens sauteurs. Analyse multivariée. *C.R. Acad. Sc. Paris.* **295**, 315–320.

Jungers, W.L. (1976). Hindlimb and pelvic adaptations to vertical climbing and clinging in *Megaladapis*, a giant sub-fossil prosimian from Madagascar. *Yrbk. Phy. Anthrop.* **20**, 508–24.

—— and Fleagle, J.G. (1980). Post-natal growth allometry of the extremities in *Cebus albifrons* and *Cebus apella*: a longitudinal and comparative study. *Amer. J. Phys. Anthrop.* **53**, 471–78.

——, Jouffroy, F.-K. and Stern, J.T. Jr. (1980). Gross structure and function of the quadriceps femoris in *Lemur fulvus*: an analysis based on telemetered electromyography. *J. Morphol.* **164**, 287–99.

—— and Stern, J.T. Jr. (1980). Telemetered electromyography of forelimb muscle chains in gibbons (*Hylobates lar*). *Science* **208**, 617–19.

Kaesler, R.L. and Waters, J.A. (1972). Fourier analysis of the ostracode margin. *Geol. Sci. Amer. Bull.* **83**, 1169–78.

Keith, Sir A. (1949). Foetalization as a factor in human evolution. *A New Theory of Human Evolution* (London, Watts), pp. 192–201.

Kempson, G.E., Spivey, C.J., Swanson, S.A.V. and Freeman, M.A.R. (1971). Pattern of cartilage stiffness in normal and degenerate human femoral heads. *J. Biomech.* **4**, 597–609.

Kennedy, G.E. (1980). *Paleoanthropology* (New York, McGraw-Hill).

King, M.C. and Wilson A.C. (1975). Evolution at two levels in humans and chimpanzees. *Science* **188**, 107–16.

Kingdon, J. (1971). *An Atlas of East African Mammals.* Volume 1 (New York, Academic Press).

Kinzey, W.G. (1976). Positional behavior and ecology in *Callicebus torquatus*. *Yrbk. Phys. Anthrop.* **20**, 468–80.

Kohne, D.E. (1975). DNA evolution data and its relevance to mammalian phylogeny. In W.P. Luckett and F.S. Szalay (eds.) *Phylogeny of the Primates* (New York, Plenum), pp. 249–61.

Kopp, R.E., Lisa, J., Mendelsohn, J., Pernick, B., Stone, H. and Wohlers, R. (1976). Coherent optical processing of cervical cytologic samples. *J. Histochem. Cytochem.* **24**, 122–37.

Kummer, B. (1966). Photoelastic studies on the functional structure of bone. *Folia Biotheoret.* **6**, 31–40.

Langston, W. Jr. (1953). Permian amphibians from New Mexico. *Univ. Calif. Publ. Geol. Sci.* **29**, 349–416.

Lanyon, L.E. (1971). Strain in sheep lumbar vertebrae recorded during life. *Acta. Ortho. Scand.* **42**, 102–12.

—— (1972). *In vivo* bone strain recorded from thoracic vertebrae in sheep. *J. Biomech.* **5**, 227–81.

—— (1973). Analysis of surface bone strain in the calaneus of sheep during normal locomotion. *J. Biomech.* **6**, 41–49.

—— (1974). Experimental support for the trajectorial theory of bone structure. *J. Bone Jt. Surg.* **56B**, 160–66.

—— and Smith, R.N. (1970). Bone strain in the tibia during normal quadrupedal locomotion. *Acta. Orthop. Scand.* **41**, 238–48.

——, Hampson, W.G.J., Goodship, A.G. and Shah, J.S. (1975). Bone deformation recorded *in vivo* from strain gauges attached to the human tibial shaft. *Acta. Orthop. Scand.* **46**, 256–68.

Leakey, R.E.F. (1973a). Further evidence of lower pleistocene hominids from East Rudolf, North Kenya, 1972. *Nature* **242**, 170–73.

—— (1973b). Evidence for an advanced plio-pleistocene hominid from East Rudolf, Kenya. *Nature* **242**, 477–80.

Lessertisseur, J. (1970). Les proportions du membre posterieur de l'homme comparées a celles des autres primates. *Bull. Mem. Soc. Anthrop. Pairs* **6** 227–41.

—— and Jouffroy, F-K. (1973). Tendance locomotrice des primates traduites par les proportions du pieds. *Folia Primatol.* **20**, 125–60.

Lestrel, P.E. (1976). Some problems in the assessment of morphological size and shape differences. *Yrbk. Phys. Anthrop.* **18**, 140–62.

—— and Brown, H.D. (1976). Fourier analysis of adolescent growth of the cranial vault: a longitudinal study. *Hum. Biol.* **48**, 517–28.

——, Kimbel, W.H., Prior W. and Fleischmann, M.L. (1977). Size and shape of the hominoid distal femur. *Amer. J. Phys. Anthrop.* **46**, 281–90.

Leutenneger, W. and Kelly, J.T. (1977). Relationship of sexual dimorphism in canine size and body size to social, behavioural and ecological correlates in anthropoid primates. *Primates.* **18**. 117–136.

Lewis, O.J. (1965). Evolutionary change in the primate wrist and inferior radio-ulnar joints. *Anat. Rec.* **151**, 275–86.

—— (1969). The hominoid wrist joint. *Amer. J. Phys. Anthrop.* **30**, 251–68.

—— (1971). Brachiation and the early evolution of the Hominoidea. *Nature* **230**, 577–78.

—— (1974). The wrist articulations of the hominoidea. In F.A. Jenkins Jr. (ed.) *Primate Locomotion.* (New York, Academic Press), pp. 143–69.

—— (1980a). The joints of the evolving foot. Part I. The ankle joint. *J. Anat. Lond.* **130**, 527–43.

—— (1980b). The joints of the evolving foot. Part II. The intrinsic joints. *J. Anat. Lond.* **130**, 883–57.

—— (1980c). The joints of the evolving foot. Part III. The fossil evidence. *J. Anat. Lond.* **131**, 275–98.

—— (1981). Functional morphology of the joints of the evolving foot. *Symp. Zool. Soc. Lond.* **46**, 169–88.

Liem, K.F. (1970). Comparative functional anatomy of the Nandidae (Pisces: Teleosteii). *Field. Zool. Mem.* **56**, 1–166.

Lipson, H. (1972). *Optical Transforms* (London, Academic Press).

Lisowski, F.P. (1967). Angular growth changes and comparisons in the primate talus. *Folia. Primatol.* **7**, 81–97.

——, Albrecht, G.H. and Oxnard, C.E. (1974). The form of the talus in some higher primates. *Amer. J. Phys. Anthrop.* **41**, 191–215.

——, Albrecht, G.H. and Oxnard, C.E. (1976). African fossil tali: further multivariate morphometric studies. *Amer. J. Phys Anthrop,* **45**, 5–18.

——, Van der Stelt, A., and Vis, J.H. (1961). Upright posture: an experimental investigation. *Acta F.R.N. Univ. Comen.* **5**, 127–36.

Lovejoy, C.O. (1979). Contemporary methodological approaches to individual primate fossil analysis. In M.E. Morbeck, H. Preuschoft and N. Gomberg (eds.), *Environment, Behaviour and Morphlogy: Dynamic Interactions in Primates* (New York, Fischer), pp. 229–44.

——, Heiple, K.G. and Burstein, A.H. (1973). The gait of *Australopithecus. Amer. J. Phys. Anthrop.* **44**, 489–506.

Lu, K.H. (1965). Harmonic analysis of the human face. *Biometrics* **21**, 491–505.

Luckett, W.P. (1974). Comparative development and evolution of the placenta in primates. *Contr. Primatol* **3**, 142–234.

—— (1975) Ontogeny of foetal membranes and placentae: their bearing on primate phylogeny. In W.P. Luckett and F. Szalay (eds.) *The Phylogeny of the Primates)* (New York, Plenum), pp. 157–82.

—— (1977). Ontogeny of amniote foetal membranes and their application to phylogeny. In M.K. Hecht, P.C. Goody and B. M. Hecht (eds.) *Major Patterns in Vertebrate Evolution* (New York, Plenum), pp. 439–516.

Lull, R.S. and Gray, S.W. (1949). Growth patterns in ceratopsia. *Amer. J. Sci.* **247**, 429–503.

Lumsden, C.J. and Wilson, E.O. (1981). *Genes, Mind and Culture: the Coevolutionary Process* (Cambridge, Harvard, Mass.).

Macgregor, A.R., Newton, I. and Gilder, R.S. (1971). A photogrammetric method of investigating facial changes following the loss of teach. *Med. Biol. Illustr.* **21**, 75–82.

MacKinnon, J. (1974). The behaviour and ecology of wild orang-utans (*Pongo pygmaeus*). *Anim. Behav.* **22**, 3–74.

Manaster, B.J.M. (1975). Locomotor adaptations within the *Cercopithecus, Cerocebus* and *Presbytis*, genera: a multivariate approach (Ph.D. thesis, The University of Chicago, Chicago).

—— (1979). Locomotor adaptations within the *Cercopithecus* genus: a multivariate approach. *Amer. J. Phys. Anthrop.* **50**, 169–82.

Martin, R. (1957–1966) *Lehrbuch der Anthropologie.* 3rd edn. (Stuttgart, Fischer).

McArdle, J.E. (1978). Functional anatomy of the hip and thigh of the Lorisiformes (Ph.D. thesis, University of Chicago, Chicago).

—— (1981). *Functional Morphology of the Hip and Thigh of the Lorisiformes* (Karger, Basel).

McHenry, H.M. (1973). Early hominid humerus from East Rudolf, Kenya. *Science* **180**, 739–41.

—— (1976). Early hominid body weight and encephalization. *Amer. J. Phys. Anthrop.* **45**, 77–83.

—— and Corruccini, R.S. (1975). Multivariate analysis of early hominoid pelvic bones. *Amer. J. Phys. Anthrop.* **46**, 263–70.

—— and Corruccini, R.S. (1978). The femur in human evolution. *Amer. J. Phys. Anthrop.* **49**, 473–88.

McKenna, M.C. (1975). Towards a phylogenetic classification of the mammalia. In W.P. Luckett and F.S. Szalay (eds.), *Phylogeny of the Primates* (New York, Plenum Press), pp. 21–41.

McMahon, T.A. (1975). Using body size to understand the structural design of organisms: quadrupedal locomotion. *J. App. Psych.* **39**, 619–27.

McQueen, C.M., Smith, D.A., Monk, I.B. and Horton, P.W. (1973). A television scanning system for the measurement of spatial variation of microdensity in bone sections. *Calcif. Tissue Res.* **11**, 124–32.

Medawar, P.B. (1944). The shape of the human being as a function of time. *Proc. Roy. Soc. Lond. Ser. B* **132**, 133–41.

—— (1945). Size, shape and age. In W.E. Le Gros Clark and P.B. Medawar (eds.) *Essays on Growth and Form presented to D'Arcy W. Thompson* (Oxford, Oxford University Press), pp. 157–87.

Meltzer, B., Searle, N.H. and Brown, R. (1967). Numerical specification of biological form. *Nature* **216**, 32–36.

Merfield, F.G. (1963). *Gorillas Were My Neighbours* (London, Longmans, Green and Co.).

Miller, R.A. (1932). Evolution of the pectoral girdle and forelimb in the Primates. *Amer. J. Phys. Anthrop.* **17**, 1–56.

—— (1943). Functional and morphological adaptations in the forelimbs of the slow lemurs. *Amer. J. Anat.* **73**, 153–83.

Minkoff, E.C. (1974). The direction of lower primate evolution: an old hypothesis revived. *Amer. Nat.* **108**, 519–32.

Mittermeier, R.A. (1978). Locomotion and posture in *Ateles geoffroyi* and *Ateles paniscus. Folia Primatol.* **30**, 161–93.

Mivart, G. (1867). Additional notes on the osteology of the *Lemuridae. Proc. Zool. Soc. Lond.* 960–75.

—— (1873). On *Lepilemur* and *Cheirogaleus*, and the zoological rank of the *Lemuroidea. Proc. Zool. Soc. Lond.* 484–510.

Mollison, T. (1910). Die Korperproportionen der Primaten. *Morph. Jhb.* 42–79–304.

Morbeck, M.E. (1976). Leaping, bounding and bipedalism in *Colobus guereza:* a spectrum of positional behaviour. *Yrbk. Phys. Anthrop.* **20**, 408–20.

—— (1979). Forelimb use and positional adaptation in *Colobus guereza:* integration of behavioural, ecological and anatomical data. In M.E. Morbeck, H. Preuschoft and N. Gomberg (eds.) *Environment, Behaviour and Morphology: Dynamic Interactions in Primates* (New York, Fischer), pp. 95–118.

Morton, D.J. (1924). Evolution of the human foot. Part II. *Amer. J. Phys. Anthrop.* **7**, 1–52.

Murray, P.D.F. (1936). *Bones. A Study of the Development and Structure of the Vertebrate Skeleton* (London, Cambridge University Press).

Muybridge, E. (1899). *Animals in Motion* (London, Chapman and Hall). Republished with minor changes, 1957, (ed.) L.S. Brown (New York, Dover).

Myers, R.H. and Shafer, D.A. (1980). Hybrid ape offspring of a mating of gibbon and siamang. *Science* **205**, 308–10.

Nagy, G. (1968). State of the art in pattern recognition. *Proc. IEEE.* **56**, 836–62.

Napier, J.R. (1962). Monkeys and their habitats. *New Scientist* **15**, 88–92.

—— (1964). The evolution of bipedal walking in the hominoids. *Extrait de Biol.* **75**, 673–708.

—— and Davis, P. (1959). *The Forelimb Skeleton and Associated Remains of Proconsul africanus.* Fossil mammals of Africa, No 16 (London, Brit. Mus. Nat. Hist), pp. 1–69.

—— and Napier, P.H. (1967). *A Handbook of the Living Primates* (London, Academic Press).

—— and Walker, A.C. (1967). Vertical clinging and leaping – a newly recognised category of locomotor behaviour of primates. *Folia Primatol* **6**, 204–19.

Niemitz, C. (1974). A contribution to the postnatal behavioural development of *Tarsius bancanus*, Horsfield, 1821, studied in two cases. *Folia Primatol.* **21**, 250–76.

Nissen, H.W. (1931). A field study of the chimpanzee. *Comp. Psychol. Monogr.* **8**, 1–122.

Owen, R. (1866). On the aye-aye (*Chiromys*, Cuvier). *Trans. Zool. Soc. London* **5**, 33–101.

Oxnard, C.E. (1955). The maxillary nerve in mammals (B.Sc. thesis, University of Birmingham, Birmingham).

—— (1963). Locomotor adaptations in the primate forelimb. *Symp. Zool. Soc. Lond.* **10**, 165–82.

—— (1967). The functional morphology of the primate shoulder as revealed by comparative anatomical, osteometric and discriminant function techniques. *Amer. J. Phys. Anthrop.* **26**, 219–40.

—— (1968a). The architecture of the shoulder in some mammals. *J. Morph.* **126**, 249–90.

—— (1968b). A note on the fragmentary Sterkfontein scapula. *Amer. J. Phys. Anthrop.* **28**, 213–17.

—— (1968c). A note on the Olduvai clavicular fragment. *Amer. J. Phys. Anthrop.* **29**, 429–31.

—— (1969). Mathematics, shape and function: a study in primate anatomy. *Amer. Sci.* **57**, 75–96.

—— (1970). Functional morphology of primates: some mathematical and physical methods. *Burg Wartenstein Symposium* **48**, 1–42.

—— (1971). Tensile forces in skeletal structures. *J.Morph.* **134**, 425–35.

—— (1972). The use of optical data analysis in functional morphology: investigation of vertebral trabecular patterns. In R.H. Tuttle (ed.) *The Functional and Evolutionary Biology of Primates* (Chicago, Aldine-Atherton), pp. 337–47.

—— (1973a). *Form and Pattern in Human Evolution: Some Mathematical, Physical and Engineering approaches* (Chicago, University of Chicago).

—— (1973b). Some problems in the assessment of skeletal forms. *Symp. Soc. Study. Hum. Biol.* **9**, 103–25.

—— (1973c). Some locomotor adaptations among lower primates. *Symp. Zool. Soc. Lond.* **33**, 255–99.

—— (1975a). *Uniqueness and Diversity in Human Evolution: Morphometric Studies of Australopithecines* (Chicago, University of Chicago).

—— (1975b). The place of the australopithecines in human evolution: grounds for doubt? *Nature* **258**, 389–95.

—— (1975c). Primate locomotor classifications for evaluating fossils: their inutility, and an alternative. In S. Kondo (ed.) *Proc. Symp. 5th. Congr. Internat. Primatol. Soc.* (Tokyo, Japan Science Press), pp. 269–84.

—— (1977). Human fossils: the new revolution. In R.M. Hutchins and M.J. Adler (ed.) *Great Ideas Today* (Chicago, Britannica Press), pp. 92–153.

—— (1978a). One biologist's view of morphometrics. *Ann. Rev. Ecol. Syst.* **9**, 219–41.

—— (1978b). The problem of convergence and the place of *Tarsius* in primate phylogeny. In D.J. Chiver and K.A. Joysey (eds.) *Recent Advances in Primatology, volume 3, Evolution* (London, Academic Press), pp. 239–47.

—— (1978c). Primate quadrupedalism: some subtle structural correlates. *Yrbk. Phys. Anthrop.* **20**, 538–54.

—— (1979a). Some methodological factors in-studying the morphological behavioural interface. In M.E. Morbeck, H. Preuschoft and N. Gomberg (ed.), *Environment, Behaviour and Mor-*

phology: Dynamic Interactions in Primates (New York, Fischer), pp. 183–207.

—— (1979b). The morphological behavioural interface in extant primates: some implications for systematics and evolution. In M.E. Morbeck, H. Preuschoft and N. Gomberg (eds.) *Environments Behaviour and Morphology: Dynamic Interactions in Primates* (New York, Fischer), pp. 209–27.

—— (1979c). The relationships of *Australopithecus* and *Homo*: another view. *J. Hum. Evol.* **8**, 427 –32.

—— (1980a). Contribution to the analysis of form. *Amer. Zool.* **20**, 619–26; 695–706; 721–22.

—— (1980b). Convention and controversy in human evolution. *Homo.* **30**, 225–46.

—— (1981a). The uniqueness of *Daubentonia Amer. J. Phys. Anthrop.* **54**, 1–22.

—— (1981b). The place of man among the primates: anatomical, molecular and morphometric evidence. *Homo* **3**, 23–53.

—— (1981c). *Beyond Biometry: Holistic Views of Biological Structure.* (Hong Kong, Hong Kong University Press).

—— (1982). Commentary on two misconceptions of phylogeny and classification. Amer. J. Phy. Anthrop. **59** 305–306.

—— (1983a). Multivariate statistics in physical anthropology: testing and interpretation. *Zeit. f. Morph. u. Anthrop.* **73**, 237–238.

—— (1983b). Sexual dimorphisms in the overall proportions of primates *Amer. J. Primatol.* **4**. 1–22.

—— (1983c). Anatomical, biomolecular and morphometric views of the living primates. In R.J. Harrison and V. Navaratnam, (eds) *Progress in Anatomy.* (Cambridge, Cambridge University Press).

——, German, R., Jouffroy F.-K. and Lessertisseur, J. (1981). The morphometrics of limb proportions in leaping .prosimians. *Amer. J. Phys. Anthrop.* **54**, 421–30.

——, German, R. and McArdle, J.E. (1981). The functional morphometrics of the hip and thigh in leaping prosimians. *Amer. J. Phys Anthrop.* **54**, 481–98.

—— and Lisowski, F.P. (1978). The Olduvai foot. *Anat. Rec.* **190**, 499–500.

—— and Lisowski, F.P. (1980). Functional articulation of some hominoid foot bones: implications for the Olduvai (Hominid 8) foot. *Amer. J. Phys. Anthrop.* **52**, 107–48.

—— and Neely, P. (1969). The descriptive use of neighbourhood limited classification in functional morphology: an analysis of the shoulder in primates. *J. Morph.* **129**, 117–48.

—— and Yang, H.C.L. (1981). Beyond biometrics: studies of complex biological patterns. *Symp. Zool. Soc. Lond.* **46**, 127–67.

Oyen, O.J. and Walker, A. (1975). Stereometric craniometry. *Amer. J. Phys. Anthrop.* **46**, 177–82.

Patterson, B. and Howells, W.W. (1967). Hominid humeral fragment from early pleistocene of northwestern Kenya. *Science* **156**, 64–66.

Patterson, T. (1979). The behaviour of a group of captive pygmy chimpanzees (*Pan paniscus*). Primates **20**, 341–54.

Pedley, T.J. (1977). *Scale Effects in Animal Locomotion* (New York, Academic Press).

Petter, J.J. (1962). Recherches sur l'ecologie et l'ethologie des lemuriens malgaches. *Mem. Mus. Hist. Nat.* **27**, 1–146.

——, Albinac, R. and Rumpler, Y. (1977). *Faune de Madagascar: Lemuriens* (Paris, Orstomcnrs).

—— and Petter A. (1967). The aye-aye of Madagascar. In S.A. Altmann (ed.) *Social Communication among Primates* (Chicago, Univ. of Chicago Press).

—— and Peyrieras, A. (1970). Nouvelle contribution a l'étude d'un lemurien malagache, le aye-aye (*Daubentonia madagascariensis* E. Geoffroy). *Mammalia* **34**, 167–93.

Pfeiler, M. (1969). Image transmission and image processing in radiology. In A. Graselli (ed.) *Automatic Interpretation and Classification of Images* (New York, Academic Press), pp. 399–415.

Philbrick, O. (1968). Shape description with the medial axis transformation. In G.C. Cheng, R.S. Ledley, D.K. Pollock. and A. Rosenfeld (eds.) *Pictorial Pattern Recognition* (Washington, D.C., Thompson).

Pincus, H.J. (1969). Sensitivity of optical data

processing to changes in rock fabric. Parts I, II and III. *Int. J. Rock Mech. Min. Sci.* **6**, 259–68, 269–72, 273–76.

——, Power, P.C. Jr. and Woodzick, T. (1973). Analysis of contour maps by optical diffraction. *Geoform.* **14**, 39–52.

Potts, A.M., Hodges, D., Shelman, C., Fritz, K.J., Levy, N.S. and Mangnall, Y. (1972). Morphology of the primate optic nerve. Parts I, II and III. *Invest. Ophthalmol.* **11**, 980–88, 989–1003, 1004–16.

Power, P.C. and Pincus, H.J. (1974). Optical diffraction analysis of petrographic thin sections. *Science* **186**, 234–39.

Prensky, W. (1971). Automated image analysis in autoradiography. *Exp. Cell. Res.* **63**, 388–94.

Preston, F.W. (1953). The shapes of birds' eggs. *Auk.* **70**, 160–82.

Preuschoft, H. (1973). Body posture and locomotion in some East African miocene dryopithecidae. *Symp. Soc. Study. Hum. Biol* **9**, 13–46.

Priemel, G. (1937). Die platyrrhinen Affen als Bewegungstype unter besondere Berucksichtigung der Extremformen, *Callicebus* und *Ateles*. Z. *Morph. Okol. Tiere.* **33**, 1–52.

Prost, J. (1965). A definitional system for the classification of primate locomotion. *Amer. J. Phys. Anthrop.* **67**, 1198–1214.

Prothero, J.W., Tamarin, A. and Pickering, R. (1974). Morphometrics of living specimens. A methodology for the quantitative three-dimensional study of growing microscopic embryos. *J. Microscop. Oxford.* **101**, 31–58.

Radin, E.L., Paul, I.L. and Rose, R.M. (1973). Mechanical factors in the aetiology of osteoarthrosis. *Ann. Rheum. Dis.* **34**, 1–12.

Rak, Y. and Clark, R.J. (1979). Ear ossicle of *Australopithecus robustus*. *Nature* **279**, 62–63.

Richards, O.W. (1955). D'Arcy W. Thompson's mathematical transformations and the analysis of growth. *Ann. N.Y. Acad. Sci.* **63**, 456–73.

—— and Kavenagh, A.J. (1945). The analysis of growing form. In W.E. Le Gros Clark and P.B. Medawar (eds.) *Essays on Growth and Form presented to D'Arcy Wentworth Thompson* (London, O.U.P.), pp. 108–230.

—— and Riley, G.A. (1937). The growth of amphibian larvae illustrated by transformed coordinates. *J. Exp. Zool.* **77**, 159–67.

Rightmire, P. (1972). Multivariate analysis of an early hominid metacarpal from Swartkrans. *Nature* **176**, 159–61.

Rink, M. (1976). A computerized quantitative image analysis procedure for investigating features and an adapted image processor. *J. Microscop. Oxford* **107**, 267–86.

Ripley, S. (1967). The leaping of langurs: a problem in the study of locomotor adaptation. *Amer. J. Phys. Anthrop.* **26**, 149–170.

—— (1976a). Gray zones and gray langurs: is the 'semi'-concept seminal? *Yrbk, Phys. Anthrop.* **20**, 376–94.

—— (1976b). Crossing the arbor-terrestrial rubicon: etic dimensions of a habitat boundary transition. *Yrbk. Phys. Anthrop.* **20**, 395–407.

—— (1979). Environmental grain, niche diversification and positional behaviour in neogene primates: an evolutionary hypothesis. In M.E. Morbeck, H. Preuschoft and N. Gomberg (eds.) *Environment, Behaviour and Morphology: Dynamic Interactions in Primates* (New York, Fischer), pp. 37–74.

Roberts, D. (1974). Structure and function of the primate scapula. In F.A. Jenkins, Jr. (ed.) *Primate Locomotion* (New York, Academic Press).

Robertson, A. (1972). Quantitative analysis of the development of cellular slime molds. *Lect. Math. Life Sci.* **4**, 43–47.

—— and Cohen, M.H. (1974). Quantitative analysis of the development of cellular slime molds. Part II. *Lect. Math. Life Sci.* **4**, 47–62.

Rodman, P.S. (1979). Skeletal differentiation of *Macaca fascicularis* and *Macaca nemestrina* in relation to arboreal and terrestrial quadrupedalism. *Amer. J. Phys. Anthrop.* **51**, 51–62.

Romero-Herrara, A.E., Lehman, H., Joysey, K.A. and Friday A.E. (1976). Evolution of myoglobin amino acid sequences in Primates and other vertebrates. In M. Goodman and R.E. Tashian (eds.) *Molecular Anthropology* (New York, Plenum), pp. 289–300.

Rose, M.D. (1974). Postural adaptations in New and Old World monkeys. In F.A. Jenkins, Jr.

(ed.) *Primate Locomotion* (New York, Academic Press), pp. 201–22.

Rosenfeld, A. (1969). *Picture Processing by Computer* (New York, Academic Press).

Rubin, J. and Friedman, H.P. (1967). *A Cluster Analysis and Taxonomy System for Grouping and Classifying Data* (New York, I.B.M. Corporation).

Rybicki, E.F., Simonen, F.A. and Weiss, E.B. (1972). On the mathematical analysis of stress in the human femur. *J. Biomech.* **5**, 203–15.

Sarich, J.M. and Cronin, J.E. (1976). Molecular systematics of the primates. In M. Goodman and R.E. Tashian (eds.) *Molecular Anthropology* (New York, Plenum), pp. 141–70.

—— and Wilson, A.C. (1976). Rates of albumin evolution in primates. *Proc. Nat. Acad. Sci. U.S.A.* **58**, 142–162.

Savara, B.S. (1965). Applications of photogrammetry for quantitative study of tooth and face morphology. *Amer. J. Phys. Anthrop.* **23**, 427–34.

Scheibengraber, K.J. (1974). A new holographic subtraction system. *Proc. Tech. Prog. Electro optical Systems Design Conf.* (San Francisco, Internat. Laser Exp.), pp. 58–62.

—— (1977). Coherent optical correlation: a new method of craniological comparison (Ph.D. thesis, Univ. Wisconsin, Milwaukee).

—— (1979). Coherent optical correlation: a new method of cranial comparison. *Amer. J. Phys. Anthrop.* **51**, 225–71.

Schon, M.A. (1968). The muscular system of the red howling monkey. *Bull. U. S. Nat. Mus.* **273**, 1–185.

—— and Ziemer, L. K. (1973). Wrist mechanisms and locomotor behaviour of *Dryopithecus* (*Proconsul*) *africanus*. *Folia Primatol.* **20**, 1–11.

Schultz, A.H. (1929). The technique of measuring the outer body of human foetuses and of primates in general. *Contr. Embryol.* 213–57.

—— (1969) *The Life of Primates* (New York, Universe Books).

—— (1970). The comparative uniformity of the Cercopithecidae. In J.R. Napier and P. Napier (eds.) *Old World Monkeys* (New York, Academic Press), pp. 39–52.

Seltzer, R.H. (1969). Computer processing of the Roentgen image. *Radiol. Clin. North Amer.* **7**, 461–471.

Shaw, D. (1800). Long-fingered Lemur. *General Zoology* **1**, 1–134.

Shelman, C.B. (1972). The application of list processing techniques to picture processing. *Pattern Recognition* **4**, 201–10.

Shulman, A.R. (1970). *Optical Data Processing* (New York, Wiley).

Simmons, D.J. and Kunin, A.S. (1979). *Skeletal Research* (New York, Academic Press).

Simons E. and Delson, E. (1978). Cercopithecidae and Parapithecidae. In V.J. Maglio and H.B.S. Cooke (eds.) *Evolution of African Mammals* (Cambridge, Mass., Harvard University Press), pp. 100–119.

Simpson, G.G. (1945). The principles of classification and a classification of mammals. *Bull. Amer. Mus. Nat. Hist.* **85**, 1–350.

—— (1953). *The Major Features of Evolution* (New York, Columbia University Press).

Singh, I. (1978). The architecture of cancellous bone. *J. Anat.* **127**, 305–10.

Smith, J.W. (1962). The structure and stress relations of fibrous epiphyseal plates. *J. Anat.* **96**, 209–25.

Smart, I.H.M. (1969). The method of transformed coordinates applied to the deformations produced by the walls of a tubular viscus on a contained body: the avian egg as a model system. *J. Anat.* **104**, 507–18.

Sneath, P.H.A. (1967). Trend surface analysis of transformation grids. *J. Zool. Lond.* **151**, 65–122.

—— and Sokal, R.R. (1973). *Numerical Taxonomy* (San Francisco, Freeman).

Sokal, R.R. (1966). Numerical Taxonomy. *Sci. Amer.* **215**, 106–15.

—— and Sneath, P.H.A. (1963). *Principles of Numerical Taxonomy* (San Francisco, Freeman).

Sonnerat M. (1789). *Voyage aux Indes orientales et à la Chine*, vol. 2, book 5 (Paris, Froule), pp. 137–39.

Srivastava, G.S. (1977). Optical processing of structural contour maps. *Math. Geol.* **9**, 3–38.

Steegman, A.T. Jr. (1970). Reliability testing of a facial contourometer. *Amer. J. Phys. Anthrop.* **33**, 241–47.

Stern, J.T. Jr. (1971). Functional myology of the hip and thigh of cebid monkeys and its implications for the evolution of erect posture. *Bibl. Primatol.* **14**, 1–319.

—— (1975). Before bipedality. *Yrbk. Phys. Anthrop.* **19**, 59–68.

—— and Oxnard, C.E. (1973). Primate locomotion: some links with evolution and morphology *Primatologia,* volume 4, lief 11 (Karger, Basel).

——, Wells, J.P., Jungers, W.L. and Vangor, A.K. (1980). An electromyographic study of serratus anterior in Atelines and *Alouatta*: implications for hominoid evolution. *Amer. J. Phys. Anthrop.* **52**, 323–34.

——, Wells, J.P., Jungers, W.L., Vangor, A.K. and Fleagle, J.G. (1980). An electromyographic study of the pectoralis major in Atelines and *Hylobates*, with special reference to the evolution of a pars clavicularis. *Amer. J. Phys. Anthrop.* **52**, 13–25.

Susman, R.L. (1979). Comparative and functional morphology of hominoid fingers. *Amer. J. Phys. Anthrop.* **50**, 215–36.

—— and Creel, N. (1979). Functional and morphological affinities of the sub-adult hand (O.H. 7) from Olduvai Gorge. *Amer. J. Phys. Anthrop.* **51**, 311–32.

——, Badrian, N.L. and Badrian, A.J. (1980). Locomotor behaviour of *Pan paniscus* in Zaire. *Amer. J. Phys. Anthrop.* **53**, 69–80.

—— and Stern, J.T. Jr. (1979). Telemetered electromyography of the flexor digitorum profundus and flexor digitorum superficialis in *Pan troglodytes* and implications for interpretation of the O.H. 7 hand. *Amer. J. Phys. Anthrop.* **50**, 565–74.

Szalay, F.S. (1975a). Phylogeny, adaptations and dispersal of the tarsiiforme primates. In W.P. Luckett and F.S. Szalay (eds.) *The Phylogeny of the Primates* (New York, Plenum).

—— (1975b). Phylogeny of primate higher taxa. The basicranial evidence. In W.P. Luckett and F.S. Szalay (eds.) *The Phylogeny of the Primates* (New York, Plenum), pp. 91–125.

Takasaki, H. (1970). Moiré topography. *Appl. Opt.* **9**, 1457–72.

Tappen, N. (1960). Problems of distribution and adaptation of the African monkeys. *Curr. Anthropol.* **1**, 91–120.

Tattersall, I. and Schwartz, J.H. (1975) Relationships among the Malagasy lemurs. The craniodental evidence. In W. P. Luckett and F.S. Szalay (eds.) *The Phylogeny of the Primates* (New York, Plenum Press).

Taylor, C.A. and Ranniko, J.K. (1974). Problems in the use of selective optical spatial filtering to obtain enhanced information from electron micrographs. *J. Microscop. Oxford* **100**, 307–14.

Tennent, J.E. (1861). *Sketches of the Natural History of Ceylon* (London, Longman Green).

Thompson, D'Arcy W. (1915). Morphology and Mathematics. *Trans. Roy. Soc. Edin.* **50**, 857–95.

—— (1917). *On Growth and Form* (Cambridge, Cambridge University Press).

Tobler, W.R. (1977). *Bidimensional Regression: a Computer Program* (Santa Barbara, Univ. Calif. Geol. Dept.), pp. 1–72 (mimeographed).

Tuttle, R.H. (1972). Relative mass of cheiridial muscles in catarrhine primates. In R.H. Tuttle (ed.), *The Functional and Evolutionary Biology of Primates* (Chicago, Aldine-Atherton), pp. 262–91.

—— (1974). Darwin's apes, dental apes and the descent of man: normal science in evolutionary anthropology. *Curr. Anthr.* **15**, 389–98.

—— (1975). Parallelism, brachiation and hominoid phylogeny. In W.P. Luckett and F.S. Szalay (eds.) *The Phylogeny of the Primates* (New York, Plenum), pp. 447–80.

—— (1976). Electromyography of pongid shoulder muscles and hominoid evolution. I. Retractors of the humerus and rotators of the scapula. *Yrbk. Phys. Anthrop.* **20**, 491–97.

—— (1977). Naturalistic positional behaviour of apes and models of hominid evolution, 1929–1976. In G.H. Bourne (ed.) *Progress in Ape Research* (New York, Academic Press), pp. 277–96.

—— and Basmajian, J.V. (1978a). Electromyography of pongid shoulder muscles. II. Deltoid, rhomboid and "rotator cuff". *Amer. J. Phys. Anthrop.* **49**, 47–56.

—— and Basmajian, J.V. (1978b). Electromyography of pongid shoulder muscles. III. Quadrupedal positional behaviour. *Amer. J. Phys. Anthrop.* **49**, 57–70.

Udupa, K.J. and Murphy, I.S.N. (1977). Machine visualization of three-dimensional objects via skeletal transformations. *IEEE Trans. Syst. Man. Cybern.* SMC-7, 424–34.

Underwood, E.E. (1970). *Quantitative Stereology* (Reading, Mass., Addison-Wesley).

Vangor. A.K. (1978). Electromyography of gait in non-human primates and its significance for the evolution of bipedality (Ph.D. thesis. State Univ. New York, Stony Brook).

Vienot, J.-C., Duvernoy, J., Tribillon, G. and Tribillon, J.-L. (1973). Three methods of information assessment for optical data processing. *Appl. Optics* **12**, 950–60.

Volkov, T. (1903-1904). Les variations squelettiques du pied chez les primates et dans les races humaines. *Bull. Mem. Soc. Anthropol. Paris,* series 5, **4**, 623–708, **5**, 1–50 and 201–331.

Vrba, E. (1978). Further contributions to the study of scapular form. *Amer. J. Phys. Anthrop.*

—— and Grine, F.E. (1978). Australopithecine enamel patterns. *Science* **202**, 890–92.

Waddington, C.H. (1977). *Tools for Thought* (New York, Basic Books).

Wainright, S.A., Biggs, W.D., Curry, J.D. and Gosline, J.M. (1976). *Mechanical Design in Organisms* (New York, Wiley).

Walker, A. (1967). Locomotor adaptations in recent and fossil Madagascan lemurs (Ph.D. thesis, University of London).

—— (1969). The locomotion of the lorises, with special reference to the potto. *E. Afr. Wildl. J.* **7**, 1–5.

—— (1974). Locomotor adaptations in past and present prosimian primates. In F.A. Jenkins, Jr. (ed.) *Primate Locomotion* (New York, Academic Press), pp. 359–81.

Waters, J.A. (1977). Quantification of shape by use of Fourier analysis: the Mississippian blastoid genus *Pentremites. Paleobiology* **3**, 288–99.

Webber, R.L. and Blum, H. (1979). Angular invariants in developing human mandibles. *Science* **206**, 689–91.

Wechsler, H. (1976). A fuzzy approach to medical diagnosis. *Int. J. Bio-Med. Comput.* **7**, 191–203.

Weinberger, H. and Almi, U. (1971). Interference method for pattern comparison. *Appl. Opt.* **10**, 2482–87.

Wells, J.P., Wood, G.A. and Tebbetts, G. (1977). The vertical leap of *Cercopithecus aethiops sabaeus:* biomechanics and anatomy. *Primates* **18**, 417–34.

Welsch, R.E. (1976). Graphics for data analysis. *Comput. Graphics* **2**, 31–37.

White, H.S. (1970). Finding events in a sea of bubbles. *Symp. on Feature Extraction and Selection in Pattern Recognition* (Illinois, Argonne National Laboratories) Univ. Calif. Radiation Lab. 20216 preprint, pp. 1–12.

Whitehouse, W.J. and Dyson, E.D. (1974). Scanning electron microscope studies of trabecular bone in the proximal end of the human femur. *J. Anat.* **118**, 417–44.

Wilson, E.O. (1978). *On Human Nature* (Cambridge, Mass., Harvard University Press).

Wislocki, G.B. and Straus, W.L. Jr. (1932). On the blood vascular bundles in the limbs of certain edentates and lemurs. *Bull. Mus. Comp. Zool. Harvard* **74**, 1–34.

Wood, B.A. (1973). A *Homo* talus from East Rudolf, Kenya. *J. Anat. Lond.* **117**, 203–204.

—— (1975) An analysis of sexual dimorphism in primates. (Ph.D. dissertation, University of London).

Woodger, J.H. (1945). On biological transformations. In W.E. Le Gros Clark and P. B. Medawar (eds.) *Essays on Growth and Form presented to D'Arcy Wentworth Thompson* (Oxford, Oxford University Press), pp. 95–120.

Wu, Rukang (1982). *Recent Advances in Chinese Palaeoanthropology* (Hong Kong, Hong Kong University Press).

Yang, H.C.L. (1978). The development of optical data analysis for the investigation of radiographic patterns: its application to the study of the lumbar vertebral column in man and the great apes (Ph.D. dissertation, University of Chicago).

—— and Oxnard C.E. (1979). The structure of cancellous networks in vertebrae as revealed by optical data analysis: implications for function. *Anat. Rec.* **193**, 726–27.

—— and Oxnard, C.E. (1983a). An approach to the study of trabecular pattern in radiographs: optical data analysis (Submitted)

—— and Oxnard, C.E. (1983b). The structure of cancellous bone as revealed by optical data analysis of radiographs: associations with stress-bearing (Submitted).

—— and Oxnard, C.E. (1983c). Optical data analysis of lumbar vertebrae in some hominoids: implications for the study of primate locomotion (Submitted).

Yerkes, R.M. (1925). *Almost Human* (New York, Century).

Zadeh, L.A. (1965). Fuzzy sets. *Inf. Control* **8**, 338–53.

Zelenka, J.S. and Varner, J.R. (1968). A new method for generating depth contours holographically. *Appl. Optics* **7**, 2107–10.

Zihlmann, A. and Brunker, L. (1979). Hominid bipedalism: then and now. *Yrbk. Phys Anthrop.* **22**, 132–62.

Zuckerman, S. (1932). *Social Life of Monkeys and Apes* (London, Kegan Paul).

—— (1933). *Functional Affinities of Man, Monkeys and Apes* (London, Kegan Paul).

—— (1954). Correlation of change in the evolution of higher primates. In J.S. Huxley, A.C. Hardy, and E. B. Ford (eds.) *Evolution as a Process* (London, Allen and Unwin), pp. 347–401.

—— (1970) *Beyond the Ivory Tower* (London, Weidenfeld and Nicolson).

——, Ashton E.H. Flinn, R.M. Oxnard, C.E., and T.F. Spence (1973). Some locomotor features of the pelvic girdle in primates. *Symp. Zool. Soc. London* **33**, 71–165.

——, Ashton E.H. and Pearson J.B. (1962). The styloid of the primate skull. *Biblio. Primatol.* **1**, 217–25.

General Index

Index of Animals

Author Index